Cyclic Separating Reactors

Cyclic Separating Reactors

Takashi Aida
Tokyo Institute of Technology
Tokyo, Japan

Peter L. Silveston
University of Waterloo
Waterloo, Ontario, Canada

Blackwell
Publishing

Blackwell Publishing Ltd
Editorial Offices:
Blackwell Publishing Ltd, 9600 Garsington Road, Oxford OX4 2DQ, UK
Tel: +44 (0)1865 776868
Blackwell Publishing Professional, 2121 State Avenue, Ames, Iowa 50014-8300, USA
Tel: +1 515 292 0140
Blackwell Publishing Asia Pty Ltd, 550 Swanston Street, Carlton, Victoria 3053, Australia
Tel: +61 (0)3 8359 1011

First published 2005 by Blackwell Publishing Ltd

Library of Congress Cataloging-in-Publication Data is available

ISBN-10: 1-4051-3156-X
ISBN-13: 978-14051-3156-8

A catalogue record for this title is available from the British Library

Set in 10/12pt Minion
by TechBooks Electronic Services, India
Printed and bound in India
by Replika Press Pvt. Ltd., Kundli

The publisher's policy is to use permanent paper from mills that operate a sustainable forestry policy, and which has been manufactured from pulp processed using acid-free and elementary chlorine-free practices. Furthermore, the publisher ensures that the text paper and cover board used have met acceptable environmental accreditation standards.

For further information on Blackwell Publishing, visit our website:
www.blackwellpublishing.com

Contents

V OVERVIEW

About the Authors

Takashi Aida is currently an Associate Professor in the Department of Chemical Engineering at the Tokyo Institute of Technology in Japan. He is well represented in the reactor engineering literature with many contributions exploring the advantages of periodic operation of catalytic reactors and the effect of the presence of an adsorbent on performance under cyclic operation. In 1999, Dr. Aida served as a Visiting Professor at the University of Waterloo. His work there initiated the collaboration of which this book is a product. Dr. Aida earned a PhD degree from Tokyo Institute of Technology.

Peter L. Silveston is a Distinguished Professor Emeritus at the University of Waterloo in Canada. He is a pioneer of the periodic operation of catalytic reactors. His book on the topic, *Composition Modulation of Catalytic Reactors*, appeared in 1998 and he has authored or co-authored nearly 200 research papers on periodic operation. During the 1980s he was a fellow of the Japan Society for the Promotion of Science. He is a graduate of M.I.T. and holds a doctorate from the Technical University of Munich (Germany). He was recently made a Fellow of the AIChE.

Preface

Over 200 research papers on periodically operated separating reactors, employing mixtures of catalyst and adsorbent, have appeared over the last few decades in the engineering and scientific literature. This important literature is described in this book. Although periodically operated separating reactors using catalysts and adsorbent are just one of several ways of combining chemical reaction with the simultaneous separation of reactants and products, it is important because this combination is an effective means of process intensification for many chemical reaction systems. The purpose of combining catalyst and adsorbent is to alter selectivity for multiple reactions or increase conversion when a single reaction is equilibrium limited. In some systems, separating a product within a reactor may be more energy efficient than carrying out the separation in a separate process step. A periodically operated separating reactor is an example of a multifunctional reactor, a reactor type that is now being widely considered for process intensification as a strategy for reducing the size and the energy requirements of a process plant while meeting production objectives. Higher yields made possible by process intensification lower materials cost.

Separating reactors operating with discontinuous or periodic contacting fall into two groups: chromatographic reactors and reactors based on pressure swing adsorption. These differ by the requirement for continuous carrier gas flow in the former. Nevertheless, both groups of reactors exploit local nonequilibrium. They are dynamic (that is they are described by time varying variables), but may exhibit stationary cyclic states. This logical division of periodic separating reactors into groups provides the structure for this monograph. Following an introduction to multifunctional reactors and to periodic operation, we first consider chromatographic reactors and then those based on pressure swing adsorption. Periodically pulsed reactors fall between the two groups. We examine this reactor type in the pressure swing group. We also discuss trapping reactors and periodically operated extractive reactors there.

Our objective in this book is to establish, when possible, the magnitude of performance improvement with respect to a continuous reactor. We will compare simulation and experiment to consider the adequacy of present models and we will attempt to identify gaps in our understanding or in experimental verification of model predictions. Finally, we will suggest several research tasks and opportunities for innovation. To accomplish these objectives we will review many of the research papers by examining the models and calculation techniques used and the experiments conducted. Our primary interest is in the computational or experimental results obtained and the insights these provide into reactor operation and control of performance. More attention will be paid to recent papers, particularly with respect to chromatographic reactors. Many early chromatographic reactor papers employed simplified models. More recent papers have scrutinized the knowledge gleaned from these early studies.

Our separation of periodically operated separating reactors into chromatographic and pressure swing groups is then further divided into the different designs or types of operation in each group. These form our chapters. For example, simulated countercurrent moving-bed chromatographic reactors are discussed in one chapter. We assess what has been discovered, prospects for commercialization, research needs and opportunities in a final chapter in each division. A final division examines research on separating reactors and development strategy. We predict progress and research directions in our last chapter, entitled "Quo Vadis".

Acknowledgments

The authors are grateful for the support of Canadian Natural Sciences and Engineering Research Council for their financial support of one of us (T.A.) during the preparation of this review.

It is a pleasure to acknowledge the work of Lorraine MacKenzie who re-lettered and often redrew the figures used in this book. We also acknowledge the support of Prof. Hiroo Niiyama at the Global Scientific Information and Computing Center, Tokyo Institute of Technology, and the assistance by his secretary, Mrs. Kazuko Ito, in our literature survey. We acknowledge as well the gracious consent of the copyright holders of the artwork, who have allowed us to modify or reproduce without change their property. The copyright holder is identified in the caption under each figure.

Finally, we are grateful for the ever-present encouragement of our spouses and companions: Kimmy Aida and Renate Anderson.

Part I
Introduction

Chapter 1
Separating Reactors

1.1 What are they?

Conventional processes for converting material consist of a preparation section to prepare the material for conversion through, perhaps, purification or heating, a reactor in which conversion takes place, a separation unit and a finishing section. The penultimate section separates products from the original material and other reactants. It may also separate a desired product from by-products. Unconverted material could be and often is recycled to the reactor to increase conversion and yield. Figure 1.1 shows schematically a sequence of single unit operations that might be used to go from reactants to a desired product.

In some reaction systems it is possible to integrate reaction and separation operations in a single vessel or in a cascade of vessels making up the reactor referred to as a separating reactor. However, many alternatives are possible.

In this book, we examine the rapidly burgeoning literature on chemical reactors that use either discontinuous exposure of reactants to a mixed bed of catalyst and adsorbent or moving-bed reactors in which catalyst and adsorbent are exposed to a continually changing fluid environment. These adsorbent and catalyst combinations operate as separating reactors.

Of course, reactors employing discontinuous feed or product withdrawal, or a moving bed are just several of the many possible systems that combine reaction and separation in a single unit. Many systems are continuous rather than discontinuous in their operation. Some are quite familiar, for example, absorbers employing absorbents that chemically bind the absorbate use reaction to enhance absorption. Others, such as membrane reactors or reactive (catalytic) distillation, have been widely discussed in the literature. In the former, separation is provided in a reactor by introducing a membrane that selectively passes one or more products. Membrane reactors have received much attention, particularly with respect to dehydrogenation reactions. See, for example, papers by Shinji et al. (1982), Itoh (1987), Sun and Khang (1988), Champagnie et al. (1990), Matsuda et al. (1993), Ziaka et al. (1993), Dalmon (1997), Saracco et al. (1999), Dittmeyer et al. (1999) and Clausen and Nielsen (2000).

Alternatively, if the reaction proceeds in the liquid phase and one or more of the products are volatile, reaction and separation can be achieved through volatilization of those products. If there is a large difference in the volatility of reactant and product, then stripping of the liquid phase with a noncondensible gas can be used to remove the product and force the reaction to completion if it is equilibrium limited. On the other hand if volatility differences are small, multiple stages are needed. If heat is added or removed to affect evaporation, the process is commonly referred to as reactive or catalytic distillation. It has also been much studied (Geelen and Wijffels, 1964; Obenaus and Droste, 1980; Sheldon and Stringaro,

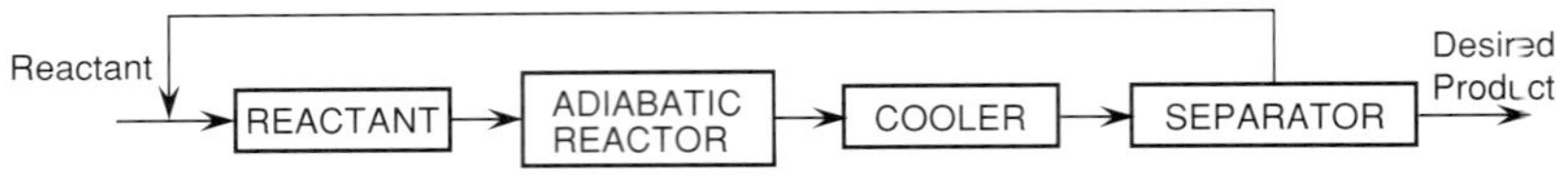

Figure 1.1 Schematic of a conventional process.

1992; de Garmo et al., 1992; Sanfillippo et al., 1992; Sundmacher and Hoffmann, 1996; Sneesby et al., 1999; Bessling et al., 1997; Higler et al., 1999; Sundmacher et al., 1999; Hanika et al., 1999; Tuchlenski et al., 2001). Like membrane reactors, catalytic distillations operate continuously. Unlike the former, catalytic distillation utilizes two fluid phases.

Extraction can also be used if an immiscible solvent for one of the products is available. Gorissen (2003) illustrates a reactive extraction process taking the alkylation of benzene by chloroform using an $AlCl_3$ catalyst as an example. In his interesting paper, he proposes the use of an extraction factor to guide system design, that is, the point in the countercurrent cascade of contactors where the extractant for the desired product should be introduced and the extractant/feed ratio that is optimal. Liquid phase separating reactors are an efficient technique for controlling selectivity when multiple reactions occur. Applications to biochemical systems appear promising and are being pursued by several research groups. Butyl acetate has been used by den Hollander et al. (2002) as an extractant to increase conversion in the hydrolysis of penicillin G to yield the industrially important intermediate 6-aminopenicillanic acid. Wasewar et al. (2003) describe the use of a liquid ion exchange material to extract lactic acid and thereby increase the enzymatic conversion of glucose.

Other separation operations can be used. Harmsen and Chewter (1999) describe controlling reactor operating conditions so that a product precipitates out of the reaction mixture. This is then a reactive precipitation process.

Table 1.1 summarizes many of the separating reactor alternatives possible. References to typical examples are given.

Separating reactors are discussed in a review that recently appeared (Kulprathipanja, 2001).

1.2 Process intensification and multifunctionality

Separating reactors are one type of multifunctional reactors. This term was introduced more then a decade ago by Agar and Ruppel (1988a,b). Initially, these authors considered the type and direction of heat and mass flows and showed that depending on these characteristics of the process different operations could be combined in a single vessel. More recently, Hoffmann and Sundmacher (1997), Agar (1999) and Krishna (2002) consider multifunctional reactors to be a combination of a reactor with another unit operation such as absorption or distillation. Westerterp (1992) has published a review of such reactor types. A review given by Krishna (2002) to open a conference on multifunctional reactors provides an excellent summary of operations that can be employed. Toward a methodology for developing effective multifunctional reactions, Agar (1999) suggests considering phases along with the type and direction of mass and heat flows. Neglecting heat flows, Figure 1.2 shows how several multifunctional reaction processes arise when more than a single

Table 1.1 Types of separating reactors proposed

Type	Description	References
Continuous		
Membrane reactor	Catalyst loaded inside membrane tube, product selectively withdrawn through shell	Clausen and Nielsen (2000)
	Hollow membranes tubes embedded in catalyst bed	Dittmeyer et al. (1999)
Stripping reactor	Noncondensible gas strips volatile product from a liquid phase, multistage reaction	Yu et al. (1997)
Catalytic distillation	Catalyst contained in structured packing in a packed-bed distillation column	Sheldon and Stringaro (1992); Tuchlenski et al. (2001)
	Catalyst as a slurry on distillation plate in a multiplate distillation column	Sanfillippo et al. (1992)
Catalytic extraction	Inert extractant circulating through catalyst bed along with reactants extracts or absorbs reaction product	Struijk (1992); Wasewar et al. (2003)
Reactive precipitation	This is usually noncatalytic. Reaction results in a product that precipitates from a liquid phase reaction in a stirred tank reactor	Harmsen and Chewter (1999)
Partially condensing reactor	Unusual combination in which condensation removes a product from the gas phase	Krishna (2002)
Periodic		
Countercurrent chromatographic reactor	Reactants' pulses injected into an inert carrier flowing through a packed bed of catalyst and adsorbent	Wetherold et al. (1974)
Countercurrent moving bed chromatographic reactor	Continuous reactant and carrier feed into a downward or rotating moving bed of catalyst and adsorbent	Cho et al. (1980)
Simulated countercurrent moving-bed chromatographic reactor	Fixed-bed of catalyst and adsorbent with many ports for injection of carrier gas, reactant and withdrawal of products. Ports are opened and closed in a timed cycle	Hashimoto et al. (1983)
Pressure swing reactor	Reactant introduced into a bed of catalyst under pressure and product is withdrawn at lower pressure from the end of the bed. When bed is pressurized, exit port at the front of the bed is opened and the pressure is reduced to regenerate the adsorbent (various cycles using purge are possible)	Vaporciyan and Kadlec (1989); Ding and Alpay (2000)
Temperature swing reactor	Reactant introduced into a bed of catalyst under pressure and product is withdrawn at lower pressure from the end of the bed. When bed is pressurized, flow direction is reversed and hot carrier gas is used to regenerate the adsorbent	Elsner et al. (2002)
Combined pressure and temperature swing reactor	This is a pressure swing system with a multistep cycle using a hot carrier gas to complete the stripping of the adsorbent	Hufton et al. (1999); Xiu et al. (2002a)
Pulse reactor	Catalyst/adsorbent bed is alternately exposed to reactant and desorbent	Mensah and Carta (1999)

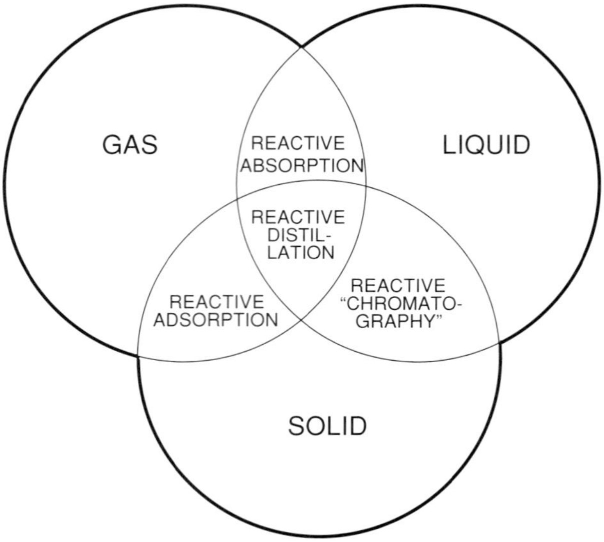

Figure 1.2 Different separating reactor systems arising from phase considerations. (Figure adapted from Agar (1999) with permission. ©1999 Elsevier Science Ltd.)

phase is present. Reactive adsorption and chromatography are the topics of this book. Although the figure assigns the processes to gas and liquid phases, this is misleading. Both processes employ either liquids or gases. Indeed, the liquid phase in the figure can be replaced by a supercritical phase. When a catalyst and an adsorbent are used there are two solid phases in the reaction–separation system. This is the usual arrangement treated in this book.

Multifunctional reactors or more generally, multifunctional process units, have emerged in the last few years as a process design paradigm. It replaces separate or unit operations design in which a process consisted of a cascade of vessels. Each vessel in that cascade had a single function, e.g. a reactor, a heat exchanger or an absorber as illustrated in Figure 1.1. The change in paradigm has been driven by commercial pressure to reduce capital and operating costs and has been enabled by the availability of high level models and sophisticated process analysis (CAD) software. Commercial pressure is behind the current development activity in microreactors. An attractive feature, indeed, of such reactors is that they make integration of reaction and other unit operations efficient (Jensen, 2001). Rapid heat addition or removal from a catalyst surface for a gas–solid reaction becomes possible in these microunits. Both multifunctional reactors and microreactors have come about through the need for process intensification.

Dautzenberg and Mukherjee (2001) describe process intensification as a strategy for reducing the size and the energy requirements of a process plant while meeting production objectives. Higher yields made possible by process intensification lower materials' cost. Dautzenburg and Mukherjee (2001) maintain that multifunctionality is a central concept in process intensification. They point out that multifunctional reactors have been used for many years, although the benefits of multifunctionality have not been emphasized in the past. Catalytic reforming of heavy naphthas using an array of catalyst filled pipes mounted in a radiation furnace, for example, was introduced in the 1950s. Multifunctionality can also be built into catalysts. Some modern catalysts are designed to have both high activity and high attrition resistance. Recently, a catalyst coated with a semipermeable material has been developed to improve reaction selectivity (Agar, 1999).

1.3 Potential advantages of separating reactors

How does a separating reactor provide process intensification? Consider the simplest situation of combining catalyst and adsorbent in a reactor carrying out a single equilibrium-limited reaction. If the adsorbent sequesters a product reducing its concentration in the fluid phase, the rate of the reverse reaction will be reduced thereby increasing the net rate of the forward reaction. Consequently, the yield or conversion achieved for a specified weight or volume of catalyst increases. As a further result, the load on a downstream separating unit decreases. If unconverted reactant is recycled, then the amount of recycle is reduced. Higher conversion or yield in a reactor will generally decrease capital and operating costs. Increasing adsorbent capacity, its specificity or the amount of adsorbent enhances these effects. Of course, an adsorbent can be replaced by another means of separating one of the products. For example, a selective membrane could serve just as well. In the presence of multiple equilibrium-limited reactions, adsorption or removal of a product enhances selectivity, thus increasing the yield of a desired product. This will affect capital and/or operating costs.

Separation may be important even in the absence of an equilibrium limitation. If there are parasitic reactions in the system that consume product, removing that product from the fluid phase or reducing its concentration diminishes the adverse effect of these reactions on selectivity and yield.

In some systems, product removal within the reactor, be it by an adsorbent or a membrane, may reduce the cost or even eliminate the use of a downstream separation unit. Also, it could avoid the need for reactant recycle.

Do separating reactors always provide process intensification? In most systems, this will be the case in terms of reactor size at least. Separating reactors will usually decrease energy demand and raw materials utilization. They may not, however, reduce cost. If membranes are employed for separation, reactor costs can rise significantly because of membrane cost and the special fabrication required. Chromatographic reactors suffer from low volumetric efficiency, while pressure swing reactors may need additional compressors or blowers. Both these periodically operated systems require expensive timer operated valves. Because periodic systems are at best semi-continuous, surge vessels may be needed to smooth production.

Operating costs may not decrease. Separating reactors are more complex than simple reactors and as such they may have higher maintenance costs. Separation is not always complete within the reactor, particularly in pressure swing systems, so that further separation steps will be required. Frequently, product streams will be dilute or at low pressure. This is a troublesome problem when membranes are used for separation. Product compression is often necessary.

Development of processes based on separating reactors will probably be more expensive than that for a conventional process. Not only are reaction and separation data needed, the interaction of reaction and separation must be studied. System design is challenging.

Even though modeling of individual separation and reaction systems has become quite sophisticated, models for separating reactors have not always been verified by experiment, as will be seen later in this book. Consequently, careful scale-up will be needed before a commercial reactor can be built. A further difficulty is that processes utilizing separating reactors are not universally better than conventional ones so that in many situations a conventional process may have to be developed alongside the process employing a separating reactor.

1.4 The trapping reactor

Although our main consideration in this book is periodically operated separating reactors, a diametrically different periodically operated reactor, the trapping reactor, has been investigated during the last decade. In the simplest situation, an adsorbent is combined with a catalyst. The adsorbent traps a dilute reactant as it flows through the reactor. Periodically the adsorbate is desorbed and reacts as it passes over the catalyst, also contained in the reactor. The trapping concept was suggested first by Agar and Ruppel (1988b) who combined the adsorbent with periodic flow reversal and applied it to catalytic deNO_x. Current interest in trapping is to reduce vehicle emissions on cold starting lean burn operation. Hydrocarbons and some CO and NO_x are trapped by the adsorbent and periodically burned off.

Trapping reactors do not fit into the structure of this book. They are neither chromatographic or pressure swing reactors. Nevertheless, they are potentially important cyclic reactors. We will consider these reactors in Chapter 12, albeit rather briefly.

1.5 Some examples of separating reactors

Reactive or catalytic distillation is probably the best-known example of a separating reactor process. It has been applied successfully to the commercially important production of high purity methyl acetate (Agreda et al., 1990; Siirola, 1996). This application is shown schematically in Figure 1.3. It can be seen from this figure that the upper part of the column distills acetic acid from the reaction mixture. The section just below strips water from this mixture. The main region in the center of the column combines acid catalyzed esterification with a small amount of separation. The bottom of the column strips methanol from the mixture and carries it back into the reaction zone. These distinct functions are illustrated by the concentration profiles shown in Figure 1.3b.

The reason for the change in diameter seen in Figure 1.3c is that the liquid phase esterification reaction is relatively slow so that a high liquid holdup on the plates of the distillation column is needed. This can be accomplished by lowering the gas superficial velocity and increasing the length of liquid passage across a plate. Increasing plate diameter achieves both these conditions.

Esterification can also be conducted in a pervaporation membrane reactor as has been demonstrated by Zhu et al. (1996) and Tanaka et al. (2002). The system employed is illustrated in Figure 1.4. In this system, water passes through a polyetherimide membrane into a vapor sweep. The remaining species in the reaction system are retained inside the reactor bed. In addition to this method of esterification and reactive distillation, the reaction can

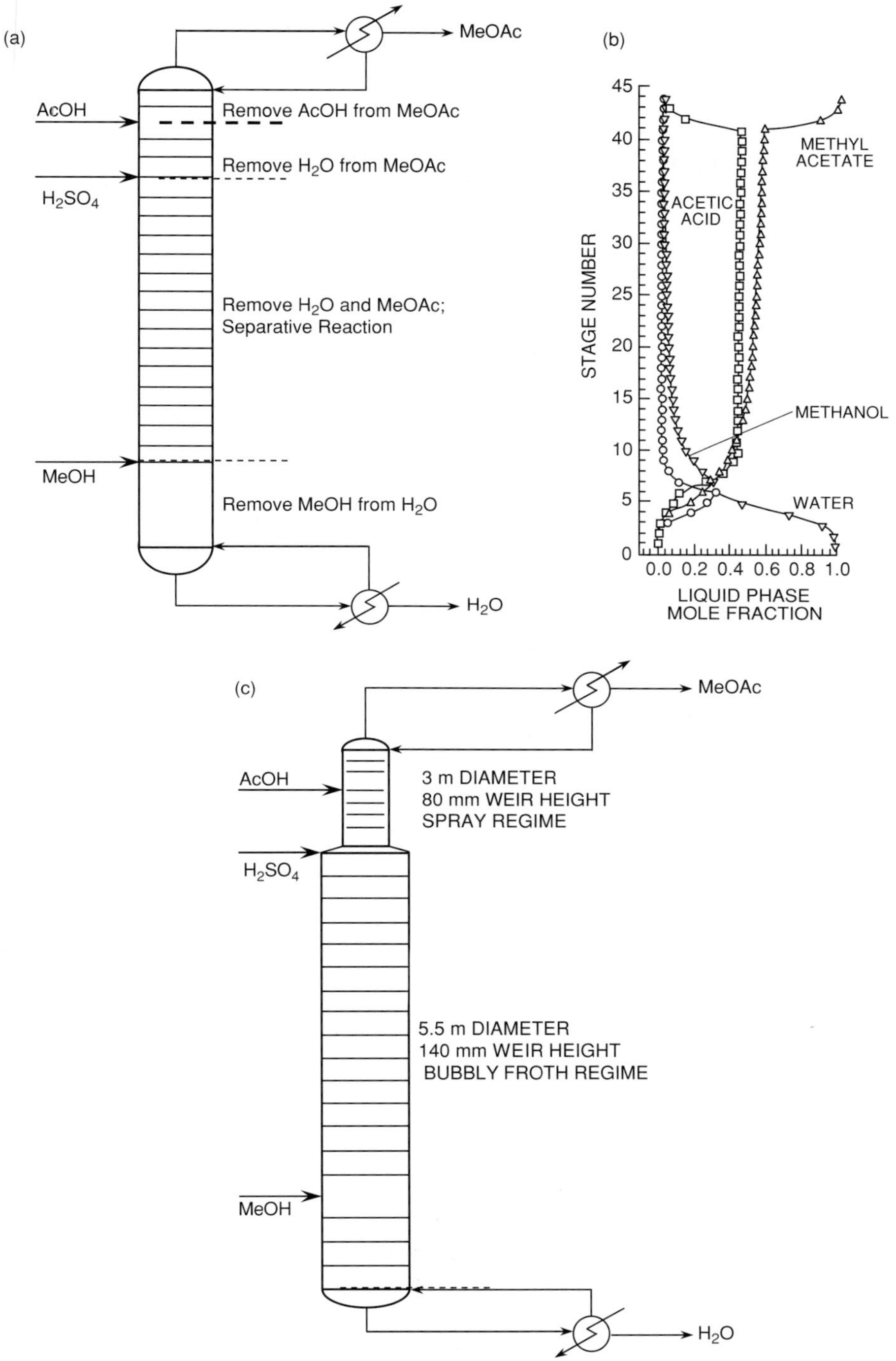

Figure 1.3 Schematic of the catalytic esterification of acetic acid with methanol in a reactive distillation column, showing (a) functions of different parts of a column, (b) concentration profiles within a column and (c) design of the column to integrate reaction and separation. (Figure adapted from Krishna (2002) with permission. ©2002 Elsevier Science Ltd.)

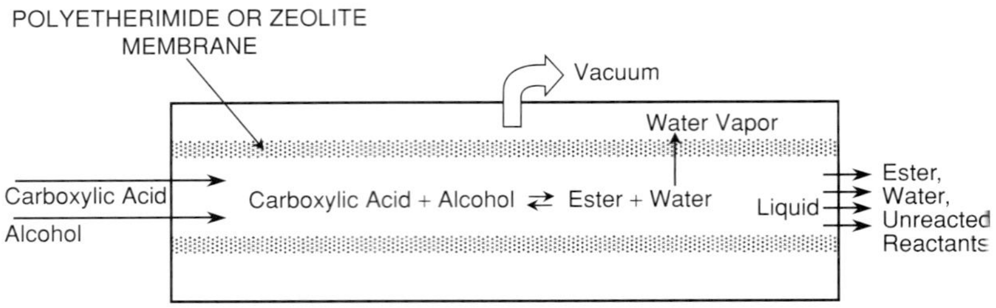

Figure 1.4 Schematic of a pervaporation membrane reactor for the esterification of acetic acid and methanol using polyetherimide impregnated onto a ceramic support material. (Figure adapted from Krishna (2002) with permission. ©2002 Elsevier Science Ltd.)

also be undertaken in a simulated countercurrent moving-bed chromatographic reactor, as we will discuss in Chapter 7.

Perhaps the greatest research interest in membrane reactors has been for equilibrium-limited endothermic dehydrogenation reactions. In these reactions, palladium membranes that selectively remove hydrogen are used. A thin coating of the metal is applied to a ceramic supporting material. The moderately high temperatures needed for dehydrogenation require membranes made of metals or perhaps fine pore zeolites (Saracco et al., 1999; Clausen and Nielsen, 2000; Julbe et al., 2001). Figure 1.5 illustrates the operation of this type of a membrane reactor.

Unfortunately, the high cost of palladium and difficulty in producing reliable very thin, dense films has held back commercialization. Dehydrogenation can also be undertaken in an adsorptive reactor. We will discuss research on this approach in Chapters 12 and 14.

Removal of a product to force a reaction to completion can be undertaken at low temperatures through the use of extraction. This results in an extractive reactor. A large literature is being developed on the use of such reactors. Figure 1.6 shows the application to hydrolysis of penicillin G (Pen G), an important step in production of synthetic antibiotics. The hydrolysis reaction forms two useful products, 6-aminopenicillanic acid (6-APA) and phenyl acetic acid (PAA). This reaction is equilibrium limited but the phenyl acetic acid is soluble in butyl acetate so that the product can be separated thus allowing higher conversion of penicillin G.

Hydrolysis of penicillin G can be carried out as well in a chromatographic reactor so we return to this reaction in Chapters 4 and 6.

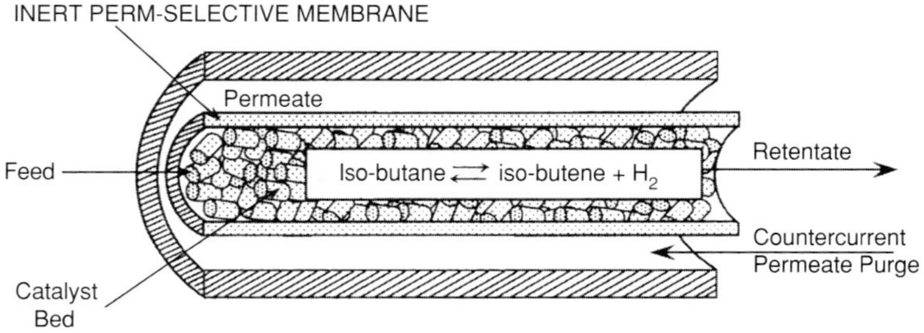

Figure 1.5 Schematic of the operation of a membrane separating reactor applied to catalytic dehydrogenation. (Figure adapted from Krishna (2002) with permission. ©2002 Elsevier Science Ltd.)

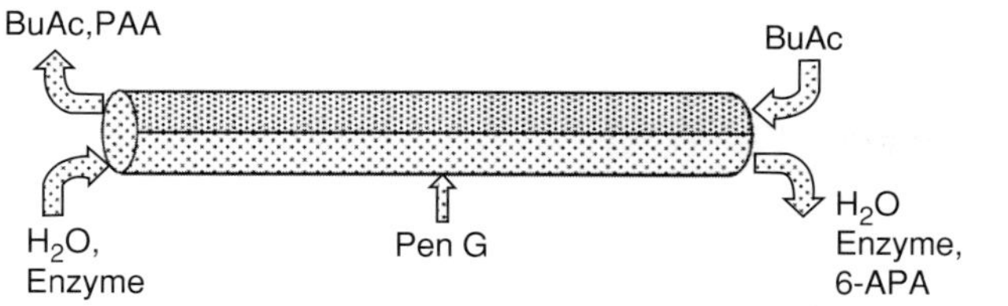

Figure 1.6 Schematic of a continuous extractive reactor for hydrolysis of penicillin G. (Figure adapted from Krishna (2002) with permission. ©2002 Elsevier Science Ltd.)

Extractive reactors are also applicable to several chemical reactions that occur at low temperatures. One of these processes, hydroformylation of olefins to form aldehydes, has been commercialized by Ruhrchemie-Rhône-Poulenc.

Finally, Figure 1.7 illustrates a reactor-crystallizer that has been studied for producing *p*-acetamidophenol (Seidlitz et al., 2001). This interesting reactor circulates fluid between a heated temperature region and a chilled one. In the former, acetic acid reacts with *p*-aminophenol and in the latter the product *p*-acetamidophenol precipitates. The heated portion is held at 100°C while crystallization is carried out at 30°C. The reaction is equilibrium limited so that removing the product forces the reaction toward completion.

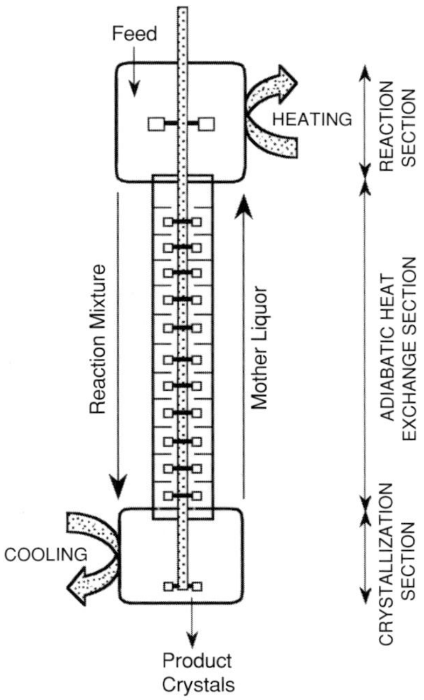

Figure 1.7 Schematic showing the operation of a crystallizing reactor. (Figure adapted from Krishna (2002) with permission. ©Elsevier Science Ltd.)

Chapter 2
Periodic Operation

Periodic operation of a process unit is often described using a different terminology, indeed a different viewpoint, therefore, it is a less familiar concept than steady state or continuous operation. Consider a batch process, such as a batch distillation or a batch reactor. It is essentially a multistep periodic operation. The vessel, say a reactor, is filled in the first step and heated to a specified temperature in a second step. The reaction proceeds for a set duration in a third step and vessel contents are cooled to ambient temperature in a fourth step. The vessel is emptied in a fifth step and finally cleaned in a sixth step. In a production setting, the sequence would be repeated periodically to achieve a production target. In this case, the vessel is "fixed" in position.

There is another equally familiar periodic operation, the circulating fluidized bed or more correctly a version of this reactor type, in which the solid phase, in this case a bed of catalyst, is not "fixed" in a location, but moves with time. Consider a fluidized catalytic cracking unit: catalyst passes upward through a riser cracking section, then through a disengager. From the disengager it drops into a fluidized bed regenerator where it is heated. The catalyst then passes through a steam stripper before it once again enters the riser. From a Lagrangian viewpoint, the moving element, the catalyst particle, is periodically exposed to a sequence of environments.

As we treat just periodically operated separating reactors in this book, it is proper to describe what we mean by periodic operation beyond the two examples just given.

2.1 Operation options for periodic separating reactors

2.1.1 *Constraints on options*

A periodic operation is an unsteady state operation. Clearly, there are an infinite number of unsteady state strategies available to run a process unit such as a reactor. Pulses of different magnitude can be imposed on an input; there can be steady gradual changes in that input or the input could vary corresponding to an analytical function. Of course, there is only one continuous or steady-state operation. The range of steady input values is infinite, however.

In a commercial situation, not all the unsteady state strategies are feasible, just as in the continuous case not all possible values of input variables can be used. A commercial processing situation imposes constraints. These arise from:

1. product property specifications
2. production quantity requirements
3. safety considerations

4. properties or cost of raw materials
5. catalyst and/or utilities availability and cost
6. equipment cost

Of these considerations, unsteady state operation is uniquely constrained by the need to meet production quantities. This constraint eliminates practically all but periodic unsteady state operations. If an input variable, such as a reactant feed rate, varies randomly with time, production quantities will not be met. The production constraint is most conveniently met by cycling an input variable about some mean, a mean chosen to satisfy the production requirement.

The frequency used in a periodic operation may be limited by the product properties constraint. Thus, low frequencies, relative to the natural time constant of a reactor, may lead to unacceptable variations in the product, such as the molecular weight distribution of a polymer. On the other hand, too high a frequency may force the reactor into a relaxed steady state that does not satisfy product specifications. Bailey (1977) offers an illuminating discussion of this high frequency state.

Amplitudes of the periodic variation could be constrained by safety considerations. Other constraints impact steady or continuous and unsteady state operations in about the same way, and need not be considered.

2.1.2 *Establishing periodic operation*

A reactor or a separating reactor may be brought into periodic operation by cyclically switching an input. A schematic diagram representing this operation and the nomenclature to be used is shown in Figure 2.1. An input that affects system performance is selected for manipulation. This input is referred to as the manipulated input or variable. With separating reactors there is often more than a single manipulated variable. As shown in the figure, this input is modulated by a square wave function. Others functions can be used, but the square wave is most effective (Silveston, 1998). A reactor with a periodic square wave input can be termed a periodically pulsed reactor. A chromatographic reactor is a subclass of the periodically pulsed reactors. A variety of inputs are suited to manipulation as will be evident in the chapters that follow. Some of these inputs are flow rate, composition, temperature, pressure and the flow direction. Heat flow through the reactor walls can also be manipulated. Apart from the manipulated inputs, other inputs are held at constant values. Modulation of an input causes a periodic variation in the system output. This variation has the same frequency as the modulation, but it is often out of phase and sinusoidal. Amplitude changes as well. Several output variables may be modulated. Phase shifts and amplitudes may differ.

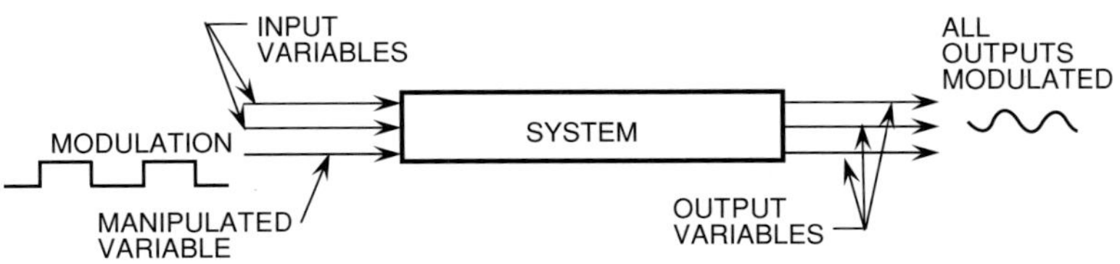

Figure 2.1 Schematic of a square-wave modulated system.

2.1.3 Reactor type and effect

Separating reactors or separating systems functioning alone can be of different type, that is they can exhibit different contact patterns or degrees of internal mixing. Those exhibiting complete mixing are isotropic, meaning that the composition, temperature and distribution of catalyst are uniform through the reactor space. They are modeled as fully back-mixed and are termed CSTRs. On the other extreme, there is no internal mixing so that composition and temperature may vary within the reactor space in the flow direction. This is the plug flow situation. The internal variables are said to be distributed in this case, while in the former, they are deemed to be lumped. Most real reactors lie between these extremes. They are frequently represented by a cascade where elements of the cascade are well mixed, that is, a tank-in-series model. These types are shown schematically in Figure 2.2. Only a serially connected cascade is shown. Parallel and serial parallel connections are also possible.

Any system of these reactor types can be operated continuously or periodically as is suggested on the left-hand side of Figure 2.2. As might be expected, the reactor types behave differently under periodic operation. Although input frequency is always preserved in the output, phase lag and amplitude depend on reactor type.

Different classes of periodically operated separating reactors correspond to the different reactor types as suggested by the text on the right-hand side of the figure. Lumped reactors are used for screening candidate systems for combined separation and reaction. Screening systems are discussed in Chapter 14. Cascades of packed beds, that is distributed systems, are probably the most promising type of chromatographic reactor and they have been also examined for pressure swing as well as combined pressure and temperature swing reactors. The chromatographic application is discussed in Part II, while pressure swing and other cascades are examined in Part III. The periodically pulsed reactor for which the chromatographic reactor is a subclass is also described in that section. The single packed bed, a distributed system, is the quintessential chromatographic and moving-bed chromatographic reactor (Chapters 4–6) and it is also the most heavily studied pressure swing reactor system (Chapter 9).

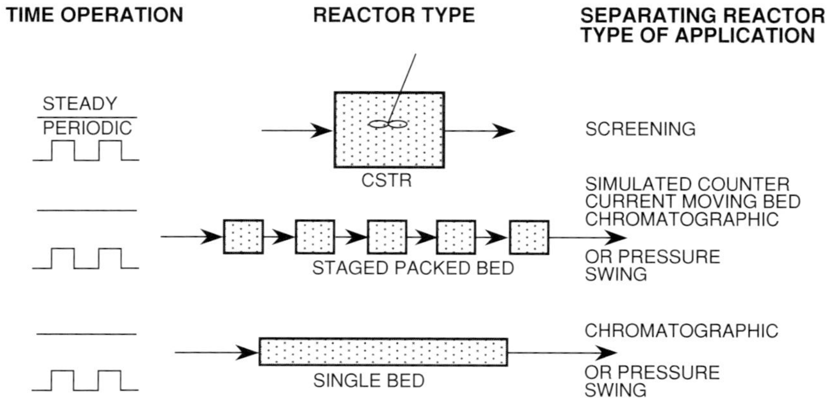

Figure 2.2 Schematic of options for separating reactors.

There are also continuously operated separating reactors that can be classified by reactor type as set out in Figure 2.2. The single stage, well-mixed reactor corresponds to a single stage stripping or extraction reactor, the cascade to reactive or catalytic distillation, while the packed bed distributed parameter system represents either reactive distillation or a membrane reactor.

2.1.4 Manipulated inputs

We have already mentioned that various systems inputs can be modulated to obtain periodic operation. Table 2.1 lists examples of these inputs in the second column and, where

Table 2.1 Examples of modulated reactor and separating reactor systems

System	Manipulated variables	Performance enhancement	Reference
Cyclohexane dehydrogenation in a chromatographic reactor	Flow at inlet and outlet of the chromatographic bed	Nearly complete conversion of cyclohexane	Matsen et al. (1965)
Hydrogenation of Mesitylene in a simulated counter-current moving bed chromatographic reactor	Flow and composition at 10 inlet and outlet ports in the five-stage reactor	Mesitylene conversion doubled	Ray et al. (1994)
Water gas shift reaction in a pressure swing reactor	Pressure, composition and flow at three inlet/outlet ports	Exceeded equilibrium conversion	Carvill et al. (1996)
Steam reforming of methane in a fluidized bed	Composition	Increased conversion	Brun-Tsekhovoi et al. (1986)
Synthesis of ammonia in a packed bed, catalytic reactor	Composition	100-fold increase in nitrogen conversion	Rambeau and Amariglio (1981)
VOC combustion in a packed bed, catalytic reactor	Flow direction	Permits combustion of dilute VOC	Boreskov et al. (1984)
CO oxidation in a packed bed, catalytic reactor	Temperature	None	Abdul-Kareem et al. (1980)
SO_2 absorption and oxidation to sulfuric acid in a trickle bed reactor	Flow	40% increase in SO_2 removal	Haure et al. (1989)
NO_x reduction by NH_3 in a packed bed, catalytic reactor	Flow and flow direction	Reduced ammonia consumption	Agar and Ruppel (1988b)

appropriate, the range of performance enhancement that has been observed in the third column. Only stand-alone reactors and separating reactors are included. All examples are taken from experimental studies.

As may be seen in the table, multiple inputs are manipulated generally in studies of separating reactors, whereas just single inputs are modulated in stand-alone reactor investigations. There is a large literature on the periodic operation of stand-alone reactors. A book is devoted to the subject (Silveston, 1998).

There is no reason why manipulation of multiple inputs cannot be used on reactors. Apart from the Agar and Ruppel (1988b) exploration given in the table and studies by Lynch and coworkers (Graham and Lynch, 1990; Sadhankar and Lynch, 1996), this mode of operation remains unexplored.

2.2 Characteristics of periodic process

2.2.1 *Cycle structure*

A periodic operation utilizes repetitive cycles imposed on one or more input variables of the reactor or separating reactor system. Inputs that function through a flow port may be rather complicated for separating reactors, such as a pressure swing reactor, so we will describe these cycles for stand-alone reactors only. Cycles can be much simpler for reactors, usually consisting of just steps between two values of the input variable.

The simplest input cycle is illustrated in Figure 2.3. The cycle is described by a period, τ, a cycle split, s, an amplitude, A, and a mean value of the variable, $\overline{x}$. Each of these cycle parameters is illustrated in the figure. Period is the reciprocal of frequency. Split establishes the symmetry of a cycle, but must be defined with respect to a direction of the step change. If it is the step up, s is the duration of the high value of the variable relative to the cycle period. Thus, $s = 0.2$ is a cycle with a short pulse at the high value of the variable. A symmetrical pulse is represented by $s = 0.5$. A pulse definition of amplitude is used. The amplitude, thus, is one half of the pulse height. The time average value of x, $\overline{x}$, becomes a function of the cycle split, amplitude and either the higher or lower value of the input variable. The time average value is useful when a periodic operation is compared to a continuous one.

When more than one input is manipulated, phase lag or shift, ϕ, becomes a consideration and another factor in the structure of the input cycle. Lag or shift is represented by the

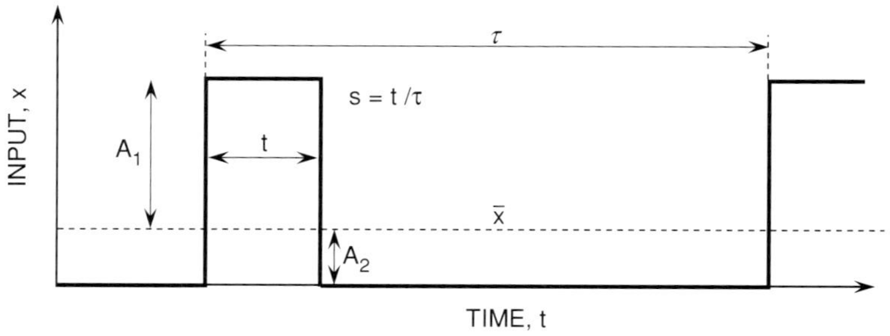

Figure 2.3 Input cycle structures.

duration between the initiation of the changes in the first and second manipulated input in a cycle relative to half of the cycle period, τ. It is often expressed in degrees or radians where 180° or π radians is a half cycle. If manipulation of the second input occurs at a time $1/4\tau$, the phase lag or shift, ϕ, is 90°.

2.2.2 Transients and the cyclic stationary state

Startup of a periodically operated process often occurs from a steady state. There is a transient of several cycles before a stationary cyclic operation is established. This is illustrated in Figure 2.4 assuming a periodically forced reactor. Transient behavior of periodically operated separating reactors has not been well investigated in the literature. On the other hand, there have been several studies, experimental and simulations, of startup or transient behavior in periodically operated reactors (Silveston, 1998). Most investigations find that a stationary cyclic operation appears after just two or three cycles. Although there are both simulations and experiments that report hundreds of cycles are needed (Hugo et al., 1986).

A stationary cyclic operation is not a steady state one. Conditions within the reactor as well as reactor output are varying with time. However, the states that are found at a point within the reactor at a specific time after commencement of each cycle are identical so that variation of a reactor output exhibits the same pattern with time from one cycle to the succeeding one.

The behavior shown in Figure 2.4 illustrates a slow approach to a stationary state. Under- and overshoot can also occur. With a packed-bed reactor, the output signal will lag behind the input, but frequency is preserved.

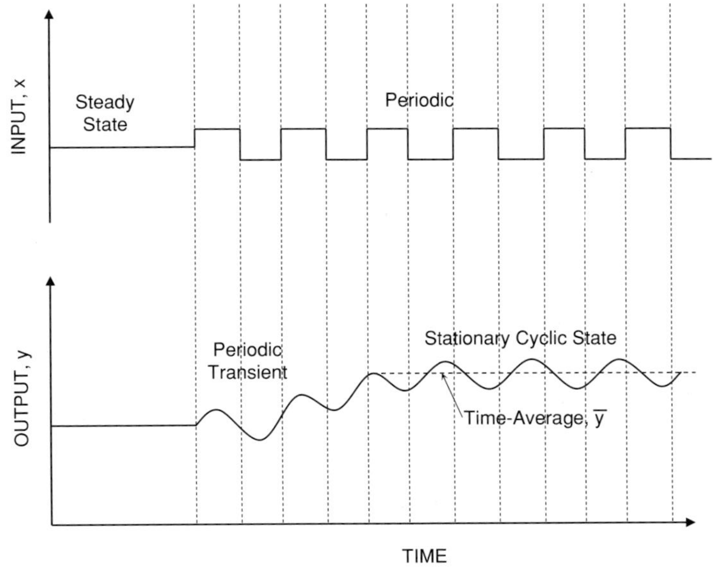

Figure 2.4 Transient output response to a periodically switched input.

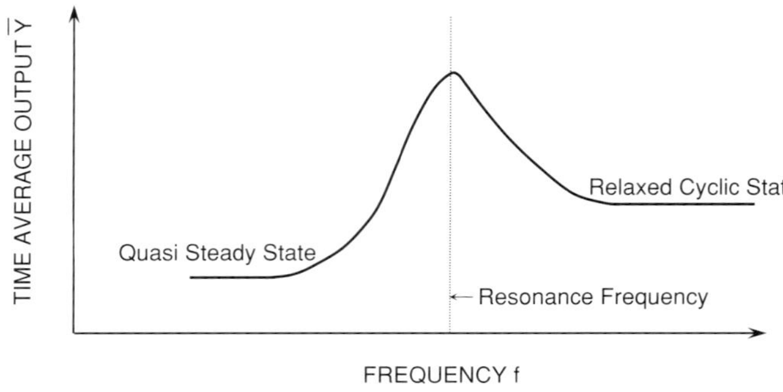

Figure 2.5 Resonance and asymptotic behavior.

2.2.3 Frequency behavior

Frequency effect on the performance of separating reactors has been thoroughly explored for chromatographic reactors. Less has been done with pressure and temperature swing alternatives. Frequency has been a central concern in the periodic operation of stand-alone reactors. Figure 2.5 shows the behavior that has attracted the attention of researchers. At low frequencies, the system illustrated in Figure 2.1 shifts between two steady states once a stationary cyclic has been reached as in Figure 2.4. We have referred to this as a quasi steady state earlier. Another asymptotic state is observed at high frequencies. This is the relaxed cyclic state. Unlike the quasi steady state asymptote, output variables do not change with time even though an input is modulated. Both the time average output at quasi steady state and the output at the relaxed cyclic state are independent of frequency. Frequently, there is no monotonic change between the two asymptotes. Instead, a maximum appears in a narrow frequency range. The maximum is referred to as resonance. It is shown in Figure 2.5.

2.2.4 Amplitude behavior

Amplitude effects have not been heavily explored in periodically operated separating reactors. A few experimental studies, however, have been undertaken on periodically operated reactors. Two phenomena have been observed: a threshold and a saturation effect.

The first of these is that some minimum amplitude must be imposed on the input modulation to achieve a time varying output. The effect certainly arises through mixing processes as a reacting fluid moves through a catalyst bed. Mixing processes occurring on catalyst surface may also contribute. The amplitude needed to initiate an output variation, the threshold amplitude, depends upon the extent of mixing and the modulation frequency. Mixing always smoothes the input variation as it penetrates the catalyst bed, but the smoothing and blending of successive input changes extends to the output only if $\tau < 2\bar{t}$. If the cycle period is brief enough, the threshold amplitude can be very large for a CSTR where mixing is instantaneous. On the other hand, if τ is large enough relative to $\bar{t}$, even a CSTR is forced into a quasi steady state mode and the output varies with time. Silveston (1998) describes limited observations of threshold amplitudes for the feed composition modulation of the oxidation of SO_2 over a promoted vanadia catalyst.

The saturation effect describes a limit on the time variation of an output variable with the amplitude of the modulated input. Triangular wave modulation of the temperature of a dilute CO–O_2 mixture in a carrier gas results in large time-dependent temperature variations within a packed bed of Pt/Al_2O_3 catalyst. Studying the influence of the input amplitude on this variation, Jaree et al. (2001) observed that the variations increased in magnitude with amplitude, but approached a limit asymptotically. The ratio of the output amplitude to the input amplitude, which is expressed as gain, decreased inversely with amplitude. The Jaree study demonstrating saturation appears to be the only investigation of amplitude effects on periodically operated reactors in the literature. Nevertheless, saturation is a well-known phenomenon in physics. It is a characteristic of nonlinear systems, particularly those of the Turing type where there is an interplay of accelerating and inhibiting forces.

2.2.5 *Phase lag*

Mentioned in Section 2.2.1, this cycle parameter is certainly important in separating reactors where multiple inputs are manipulated, such as simulated countercurrent moving-bed chromatographic reactors or pressure swing reactors. Phase lag has not been systematically studied. Like amplitude, it has been given some study in periodically operated reactors. Lynch and coworkers (Graham and Lynch, 1990; Sadhankar and Lynch, 1996) have shown that optimal values of ϕ are observed, however no general variation of performance with this parameter has been observed.

2.2.6 *Complications*

Several types of complications arise when attempting to describe or characterize a periodic operation. The first of these is seen when a single input is manipulated. Because output in a reactor system depends primarily on the mean value of the manipulated input, this mean value is normally fixed or specified. The cycle parameters discussed in Section 2.2.1, other than frequency, are not independent. Cycle split will be fixed by specifying amplitude, for example.

Cycle characterization is complicated when more than a single input is manipulated. A three-dimensional graph shows how system performance depends on the cycle when there is a single manipulated input. With two or more manipulated inputs, phase plots must be used. An overview of the influence of cycle parameters on performance becomes progressively more difficult as the number of manipulated inputs increases. With the exception of flow direction manipulation, each input introduces amplitude besides phase lag. In principle, each input could also have a frequency and a mean or cycle split parameter associated with it. Such systems have not yet been studied and thus the problem of an efficient way of representing them has not been tackled.

The importance of mixing goes beyond reactor characterization and the presence of threshold amplitudes already discussed. Mixing is important for the asymptotic operating states at high and low frequencies. At low levels of mixing as plug flow is approached, the quasi steady state can be reached using small cycle periods. Relaxation times associated with the various rate processes in the overall reaction become important. The cycle period must be greater than the largest time constant or relaxation time of one of the steps. On the other hand, at a high level of mixing in which the reactor becomes almost fully back-mixed, relaxation times lose significance and it is the mean residence time of reactant in the reactor that determines the cycle period or frequency for the quasi steady state. At the relaxed cyclic state asymptote, it is again the time constant of the individual rate processes that determine the cycle period or modulation frequency for this state. With a high level of mixing, the relaxed cyclic state is attained for cycle periods of the order of 1/10 of the mean residence time $\bar{t}$. In ordinary situations, however, this relaxed cyclic state is identical to steady-state operation at the time average feed condition. Only when the relaxation time of the slowest rate process in the reaction sequence is larger than the average residence time, there can be a difference between the performances at these two states.

2.3 Advantages of periodic processes and basis for choice

In Chapter 1 we mentioned that there are continuously operating separating reactor alternatives for each of the periodically operated separating reactors that we consider in the chapters that follow. Furthermore, perhaps all of the process enhancement derived from separating reactors can be obtained using a conventional process design with a separator following a reactor to recover and recycle unconverted reactants or less desirable products. What then are the reasons for choosing a periodically operated system? Industry attempts to make choices of this type on the basis of total cost per unit of product. When processes are under development, total cost may be uncertain so choice must be made on other criteria.

2.3.1 *Process enhancement*

There are industrial processes in which a major portion of the total cost per unit of product is the cost of raw materials or, occasionally, the cost of disposing of worthless reaction by-products. When this is clearly the case, process choice is governed by yield considerations. Or when there is just a single reaction, choice will be governed by conversion or rate results. As we will show in the chapters that follow, periodically operating separating reactors can provide large improvements in performance. Some of the enhancements demonstrated experimentally are given in Table 2.1. Regrettably most of these enhancements are measured against a simple reactor operating at steady state but without a sequential separator and recycle of unconverted reactant or waste products.

To properly assess the attractiveness of periodically operated separating reactors, performance results should be compared to performance results for continuous, conventional reactor–separator sequence alternatives. Unfortunately, R&D teams rarely have such comparisons for the alternatives using identical feeds and suitable operating conditions.

2.3.2 *Process stability*

Steady-state reactor systems involving feedback of heat through axial mixing or by using an outlet stream to heat the reactor feed can be sensitive to input disturbances and through these may be forced into a less desirable, but nonetheless steady state operation. Generally, the existence of multiple steady states is undesirable. In addition to multiplicity, an operating state may be unstable with respect to a process perturbation. Such a state can be considered as unstable. Stability is desirable in a reactor as well as in a separating reactor. Stability is thus a consideration in choice among alternatives, but it is rarely the most important one.

There are studies that suggest periodic operation increases process stability. Control of periodically operated separating reactors is not discussed in the reactor or process literature. However, some early studies of control for a periodically operated reactor have appeared (Liaw and Bailey, 1974). Most deal with periodic flow reversal (Nieken et al., 1994; Řeháček et al., 1998). Despite a paucity of experimental research, much has been written about vibrational or "chattering" control. Imposing a low amplitude modulation on a process input has been shown by both simulation and experiment to lessen the response to an input disturbance (Çinar et al., 1987a,b; Özgülşen et al., 1993).

2.3.3 On-line optimization

A further consideration favoring periodically operated separating reactors among other alternatives is that periodic operation often introduces several control parameters, such as frequency or phase shift, that can be tuned while the separating reactor is on-line. These are additional control parameters besides those always present, like feed temperature or total pressure, that can be varied to optimize performance. Indeed operating conditions such as feed or coolant temperatures and total pressure are common to both periodically and continuously operated separating reactors or even conventional reactor–separator sequences with recycle. However, unlike the periodic control parameters, tuning of these conditions or parameters affects other parts of the process severely, limiting the application of on-stream optimization.

The possibility of on-line tuning of the periodic control parameters suggests that a periodically operated separating reactor may be better able to respond to changing feed conditions or throughput requirements than continuously operated alternatives for which tuning opportunities may be limited. Changes are probably inevitable in an industrial operation.

2.3.4 Problems with periodic operation

The discussion in the previous section has been essentially an enumeration of the advantages of periodic operation of separating reactors vis-à-vis continuously operated alternatives. There are, of course, problems or disadvantages associated with a periodic operation.

Periodic operation of any system can involve time fluctuating flows and composition. For downstream processing, these fluctuations must be smoothed or eliminated. This is most readily accomplished by adding holdup vessels with provision for mixing after the periodically operated unit. The volume required depends on the frequency employed in the modulated unit. Holdup can be quite small as a fraction of the flow rate if the modulation frequency is high. Separating reactors examined in the following chapters often operate at low frequencies, about 0.001–0.01 Hz, so that the holdup requirements might be large and thereby costly. Another way of meeting the problem of fluctuation is to use double or even multiple reactor–separators and operate the array by phasing the cycles in each to minimize the overall flow and composition variation with time. The choice of option depends upon cost and compromises employing parallel reactor separators together with holdup are possible.

Periodic operation requires close flow control as well as timing so that computer-operated controls will be necessary. Quite likely controls for a periodically operated separating reactor will be more extensive and costly than controls needed for continuous operation On the other hand, more control offers an opportunity for on-line optimization as discussed in the previous section.

Periodic temperature variation imposed on the separating reactor in a temperature-swing process or resulting from concentration variation through the coupling of reaction and heat-generation or consumption in the reactor or from heat effects in adsorption can lead to a loss of crushing strength. Particles are under stress in packed beds so that a reduction of crushing resistance may increase breakage and solids loss. The importance of this problem will depend on both frequency and amplitude of periodic changes and are difficult to assess without data.

Finally, periodically operated systems are more difficult and thus more costly to develop than ones operating continuously. There are several reasons for this: periodic operation is more complex than continuous operation, models require more effort to validate, and measurements of transient processes will be required. Such measurements are more time-consuming and more costly than those made at steady state.

2.4 Moving-bed systems

Problems of holdup vessels and/or parallel production lines to smooth composition and property variation with time as well as the cost of control and higher valve performance standards that arise can be avoided by designing separating reactors with countercurrent moving fluid and solids. Feed and product withdrawal then becomes continuous.

Moving-bed systems have never occupied a major role in reactor design because problems that this design alleviates are replaced by other vexing ones. Some separating reactors, such as countercurrent moving-bed chromatographic reactors, involve moving solids. Moving solids often requires expensive machinery and storage in the cycle or out of the cycle for solids replacement. Attrition in a moving bed means reduced catalyst and adsorbent life, and thus higher replacement costs or it might entail use of more costly attrition resistant materials. Attrition implies the generation of fines or dust that must be removed from product streams by filters or other collection devices, adding cost to the process.

2.5 Neglect of periodic processes

Problems with periodic operations enumerated above and the uncertain advantages such systems offer may explain why less attention has been given to multifunctional separating reactors employing cyclic operation or moving beds than to other alternatives. At play is also an engineering paradigm that steady state is always desirable. Certainly much more attention has been paid to separating reactors based on membranes than to pressure swing systems. We argue that a periodically operated reactor can achieve the level of performance of a membrane reactor and offer operational advantage, such as no membrane integrity problems when high permeation rates are needed or permeation rates problems when membrane integrity is essential. Some membrane systems require high purge rates, possibly creating another separation problem. Membranes often involve a limited choice of materials. On the other hand, there are normally a variety of adsorbents that could be employed in a pressure swing process. Furthermore, membrane reactors may be difficult to integrate with heat transfer needs when a reaction is highly endothermic or exothermic.

We believe more attention should be focused on separating reactors using discontinuous or periodic contacting. It is this, ultimately, that is the justification for this book. A useful undertaking would be to compare energy consumption for the different types of separating reactors or their exergy implications per mole of product recovered. We return to this suggestion in our summary and assessment of research accomplished so far in the final chapter of this book.

Part II

Chromatographic Reactors

Chapter 3

Introduction to Chromatographic Reactors

The chromatographic reactor is the first separating reactor to be described in the literature. Shortly after the introduction of chromatography as a separation technique in the 1960s, Roginskii et al. (1962) suggested that an equilibrium-limited reaction could be forced toward completion by carrying out the reaction in a chromatographic column. Magee (1963) contributed further to this idea by proposing a simplified model to explain the operation. Of course, the use of separation to promote a reaction is an even older concept. Glueckauf and Kitt (1957) and Thomas and Smith (1959) discussed a pulse reactor in which adsorption on the catalyst shifted the reaction toward completion. The concept was applied to the equilibrium reaction, $2HD \leftrightarrow H_2 + D_2$. The pulse reactor was developed ultimately and demonstrated experimentally by Hattori and Murakami (1968) using the dehydrogenation of cyclohexane to benzene. They studied the movement of single pulses of cyclohexane through a bed of 0.6 wt.% Pt/Al_2O_3 in which the alumina behaved as an adsorbent.

Reviews of work prior to 1990 have been published by Barker et al. (1987a,b, 1992b), Barker and Ganetsos (1988) and by Carta (1991). Simple chromatographic reactors (CR), the countercurrent moving-bed version (CMCR) and the simulated countercurrent moving-bed designs (SCMCR) are described and models given, but many studies have been omitted. Emphasis in these reviews is on biotechnology applications.

Table 3.1 summarizes the chromatographic reactor literature to date. There are a large number of papers in this literature so the table is not exhaustive. Bibliographic details are given in the Reference section.

3.1 Concept and types

The principle of the chromatographic reactor (CR) is illustrated in Figure 3.1. We use as an example the catalytic dehydrogenation of cyclohexane. A pulse of cyclohexane (A) in a nitrogen carrier gas is introduced into a fixed bed of mixed catalyst and adsorbent in Figure 3.1a. Above 200°C, cyclohexane chemisorbs on the catalyst and breaks down to benzene (B) and atomic hydrogen. The latter rapidly combines to form molecular hydrogen and desorbs because molecular hydrogen (C) is just weakly adsorbed on the catalyst. However, benzene (B) is strongly adsorbed. This difference inaugurates separation between cyclohexane (A), benzene (B) and hydrogen (C). Because it is just weakly adsorbed, hydrogen moves through the reactor at almost the space velocity of the carrier gas. Cyclohexane (A), more strongly adsorbed, moves more slowly through the reactor. Thus, reactant and products are separated. Cyclohexane continues to breakdown because the back reaction decreases

Table 3.1 Chromatographic Reactor literature

Authors	Type of study	Reactions	Remarks
Chromatographic reactors (CR)			
Klinkenberg (1961)	Calculated	General	Reaction effect on chromatographic band width
Gaziev et al. (1963)	Experimental	Dehydrogenation of C_6H_{12}	Demonstrated C_6H_6 yield in excess of equilibrium
Roginskii et al. (1961)	Calculated	General	Interpretation of peak shape
Roginskii et al. (1962)	Experimental	As above	As above
Roginskii and Rozental (1962)	Calculated	General	Model development
Magee (1963)	Calculated	General reversible reaction	Proposed simplified model
Semenenko et al. (1964)	Calculated	Dehydrogenation of n-C_4H_{10}	Yields in excess of equilibrium
Roginskii and Rozental (1964)	Calculated	General	Derivation of kinetic models
Matsen et al. (1965)	Experimental	C_6H_{12} dehydrogenation	Conversion in excess of equilibrium
Saito et al. (1965)	Calculated	General	Model development
Gore (1967)	Calculated	General	Simulation of CR
Kocirik (1967)	Calculated	General	As above
Hattori and Murakami (1968)	Calculated	General	Analysis of pulse reactors and CRs
Chu and Tsang (1971)	Calculated	General	Effect of operating parameters on CR performance
Wetherold et al. (1974)	Experimental and calculated	Hydrolysis of methyl formate	Conversion in excess of equilibrium; model validation
Unger and Rinker (1976)	Experimental	Ammonia synthesis	Demonstrated increased conversion
Schweich and Villermaux (1978)	Experimental and calculated	C_6H_{12} dehydrogenation	Model development and validation
Sardin and Villermaux (1979)	Experimental	Esterification	Increased conversion with no separation of reactants
Schweich et al. (1980)	Calculated	General	Calculation method
Schweich and Villermaux (1982a,b,c)	Calculated	C_6H_{12} dehydrogenation	Assumption of local reaction equilibrium questionable
Antonucci et al. (1978)	Experimental	C_2H_6 dehydrogenation	Conversion in excess of equilibrium
Cho and West (1986)	Experimental	CO oxidation	Chromatographic effects within a catalyst pellet
Zafar and Barker (1988)	Experimental and calculated	Enzymatic sucrose polymerization	Validation of a plug flow model
Liden and Vamling (1989)	Calculated	General consecutive reactions	Improved selectivity over FFR
Sad et al. (1996)	Experimental	Dehydroisomerization of n-C_4	Use of composite catalysts and different adsorbents
Mazzotti et al. (1997a)	Experimental and calculated	Esterification of alcohol with acetic acid	Back-mixing due to density differences in displacement front

(Continued)

Table 3.1 Chromatographic Reactor literature *(Continued)*

Authors	Type of study	Reactions	Remarks
Wu (1998)	Experimental	Enantioselective esterification	Analytical method of measuring enantioselectivity
Falk and Seidel-Morgenstern (1999)	Calculated	Hydrolysis of methyl formate	Conversion improvement due to a dilution effect
Wu and Liu (1999)	Experimental	Esterification of racemic naproxen	Analytical method of measuring enantioselectivity
Migliorini et al. (2000)	Experimental and calculated	Enzymatic diol esterification	Importance of water content; model validation
Silva and Rodrigues (2002)	Experimental and calculated	Diethyl acetal synthesis	Model validation and parameter measurement
Falk and Seidel-Morgenstern (2002)	Experimental and calculated	Hydrolysis of methyl formate	Product separation and recovery more important than conversion
Gelosa et al. (2003)	Experimental and calculated	Triacetine synthesis	Conversion greatly in excess of equilibrium
Countercurrent moving-bed chromatographic reactors (CMCR)			
Viswanathan and Aris (1974)	Calculated	General irreversible reaction	No improvement in conversion, existence of shock fronts
Takeuchi and Uraguchi (1976a,b)	Calculated	General	Calculation of separation based on moments
Takeuchi and Uraguchi (1977a,b)	Experimental and calculated	General, CO oxidation	No improvement in conversion. Model validation
Takeuchi et al. (1978)	Calculated	General consecutive reactions	Selectivity improvement over PFR
Cho et al. (1982)	Calculated	General reversible reaction	Discontinuity identified and discussed
Altshuller (1983)	Calculated	General	Treatment of a complex isotherm
Petroulas et al. (1985a,b)	Experimental and calculated	Hydrogenation of mesitylene	Improved product purity and conversion over equivalent PFR
Fish et al. (1986)	Experimental and calculated	As above	Description of reactor design and trouble shooting
Fish and Carr (1989)	Experimental and calculated	As above	Model validation
Lode et al. (2003a)	Calculated	General	Comparison of CMCR and SCMCR design
Rotating annular chromatographic reactors (CRAC)			
Cho et al. (1980a,b)	Experimental and calculated	Hydrolysis of methyl formate	Conversion in excess of equilibrium
Wardwell et al. (1982)	Experimental	C_6H_{12} dehydrogenation	As above
Sarmidi and Barker (1993a)	Experimental	Saccharification of starch	High maltose yield with good dextrin separation
Sarmidi and Barker (1993b)	Experimental	Inversion of sucrose	Complete conversion, model validation
Herbsthofer and Bart (2003)	Calculated and experimental	Redox reaction of iridium and iron	Performance comparison of a CRAC and a PFR followed by a chromatographic column

(Continued)

Table 3.1 Chromatographic Reactor literature *(Continued)*

Authors	Type of study	Reactions	Remarks
Variations on countercurrent moving-bed chromatographic reactors			
Kuczynski et al. (1987b)	Experimental	Methanol synthesis	Complete CO conversion with high adsorbent circulat on
Westerterp and Kuczynski (1987a)	Calculated	Methanol synthesis	Complete CO conversion
van der Wielen et al. (1990)	Experimental and calculated	Glucose isomerization	Multistage fluidized beds with moving adsorbent
van der Wielen et al. (1996)	Experimental and calculated	Penicillin G deacylation	Identification of key operating parameters
van der Wielen et al. (1998)	Experimental and calculated	None	Hydrodynamics and modeling
Simulated countercurrent moving-bed chromatographic reactors (SCMCR)			
Zabransky and Anderson (1977)	Patent	Zeolite-catalyzed alkylation	Increased conversion
Hashimoto et al. (1983)	Experimental and calculated	Glucose isomerization	Increased conversion
Fish et al. (1988)	Calculated	General	Comparison with an equivalent CMCR
Akintoye et al. (1991)	Experimental	Inversion of sucrose	Improved yields
Barker et al. (1992c)	Experimental	Production of dextran	Improved yields
Tonkovich et al. (1993)	Experimental	Oxidative coupling of methane	Improved C_2 yield
Ray et al. (1994)	Calculated	General	Equilibrium stage model proposed
Kruglov (1994)	Calculated	Methanol synthesis	Treated nonisothermal segregated reactor–adsorber beds
Tonkovich and Carr (1994a)	Experimental	Oxidative coupling of methane	As above
Tonkovich and Carr (1994b)	Calculated	As above	Conversion in excess of equilibrium
Ray and Carr (1995a)	Experimental	Hydrogenation of mesitylene	As above
Ray and Carr (1995b)	Calculated	Hydrogenation of mesitylene	As above
Bjorklund and Carr (1995)	Experimental	As above	Increased conversion
Shieh and Barker (1995)	Experimental	Saccharification of starch to maltose	Effect of operating variables and comparison with an equivalent CRAC
Kawase et al. (1996)	Experimental	Synthesis of β-phenetyl acetate	Improved purity of product, conversion in excess of equilibrium
Shieh and Barker (1996)	Experimental	Hydrolysis of lactose	Batch and continuous operation compared
Kruglov et al. (1996)	Calculated	Oxidative coupling of methane	Optimization of a reactor–adsorber system operating at different temperatures

(Continued)

Table 3.1 Chromatographic Reactor literature *(Continued)*

Authors	Type of study	Reactions	Remarks
Mazzotti et al. (1996)	Experimental	Esterification of acetic acid	Demonstration of complete conversion
Ching and Lu (1997)	Experimental and calculated	Inversion of sucrose	Effect of operating variables on performance
Kawase et al. (1999)	Calculated	Production of bisphenol A	Good product separation and conversion in excess of equilibrium
Mensah and Carta (1999)	Experimental and calculated	Esterification with immobilized enzymes	Two-zone version overcomes water deactivation of enzymes
Fricke et al. (1999)	Calculated	General	Effect of process parameters
Migliorini et al. (1999a)	Calculated	General	Parametric analysis
Dünnebier et al. (2000)	Calculated	Inversion of sucrose and esterification	SCMCR optimization strategy
Kawase et al. (2001)	Experimental	Lactosucrose production	High yields limited by side reactions
Lode et al. (2001)	Experimental and calculated	Methyl acetate synthesis	Model testing and process optimization
Zhang et al. (2001)	Calculated	Synthesis of MTBE	Effect of operating variables on reactor performance
Falk and Seidel-Morgenstern (2002)	Experimental and calculated	Hydrolysis of methyl formate	Performance evaluation and model testing
Zhang et al. (2002)	Calculated	Synthesis of MTBE	Development of an optimization approach
Bjorklund and Carr (2002)	Experimental	Partial oxidation of CH_4 to CH_3OH	High conversion and selectivity observed
Lode et al. (2003a)	Calculated	General	Optimization of a linear model fails to represent SCMCR performance
Lode et al. (2003b)	Experimental and calculated	Methyl acetate synthesis	Performance evaluation and model testing
Fricke and Schmidt-Traub (2003)	Calculated	General	Use of triangular theory for design

as the reaction products become separated. This is the situation shown in Figure 3.1b. The sequence continues as the cyclohexane pulse moves through the reactor. Products continue to form at each point in the reactor from the moving pulses of cyclohexane. This causes spreading of the benzene and hydrogen pulses. The spreading can be seen in Figure 3.1c.

Once benzene reaches the exit of the catalyst and adsorbent fixed bed, a new pulse of cyclohexane can be introduced at the entrance (Fig. 3.1d). It is evident that nearly complete conversion may be obtained in pulse operation provided the bed is long enough. However, there is a throughput penalty because reactant cannot be fed continuously to the reactor. Much of the initial development of chromatographic reactors dealt with overcoming the throughput limitation. The early literature on chromatographic reactors, particularly the

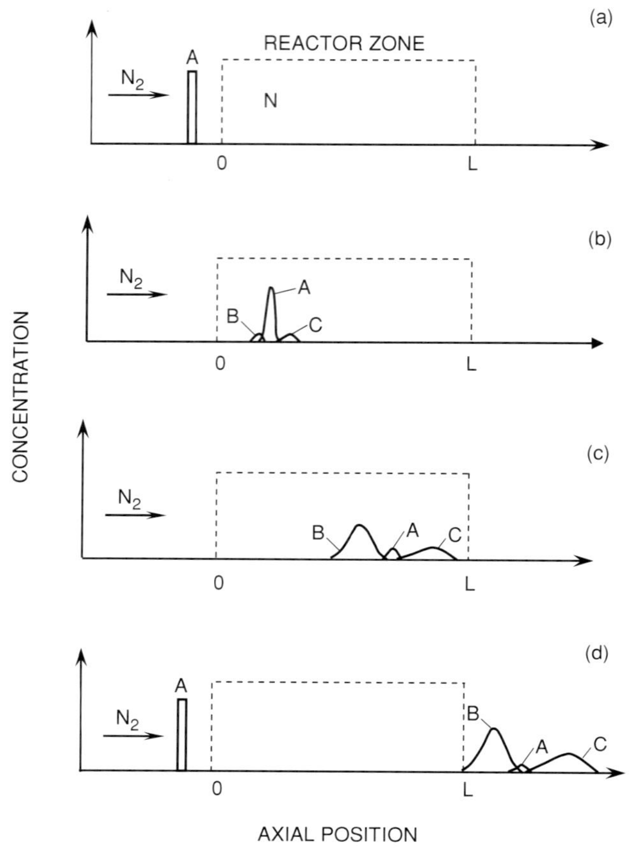

Figure 3.1 Principle of the chromatographic reactor: Sketches (a)–(d) represent pulse position and shape of reactant and product species at successive times after injection of A for the reaction A → B + C where B is most strongly adsorbed and C is just weakly adsorbed.

work of Schweich and Villermaux (1978, 1982a,b,c), deals with the single pulse version of the chromatographic reactor. Chapter 4 is devoted to chromatographic reactors (CR); both single pulse and repetitive pulsed versions are considered.

The chromatographic reactor is a special case of a periodically pulsed reactor in which the relative or fractional duration of reactant feed is small. Defining a cycle split, s, as the ratio of the duration of reactant feed to the cycle period, s is usually ≤ 0.1 for chromatographic reactors. The periodically pulsed reactor is discussed in Chapter 12.

One technique for increasing throughput has been to maintain a constant feed flow rate and move the catalyst bed pass the feed point. Use of a moving bed, extensively explored by Carr, Aris and coworkers, is called a countercurrent moving-bed chromatographic reactor (CMCR). It is shown in Figure 3.2. Gaseous reactant with carrier gas moves upward, while the solid catalyst moves downward. Here, we use the dehydrogenation of cyclohexane again. In Figure 3.2a, cyclohexane is introduced at the bottom of the reactor with carrier gas. Reaction products, hydrogen and benzene, are separated by the chromatographic effect. H_2

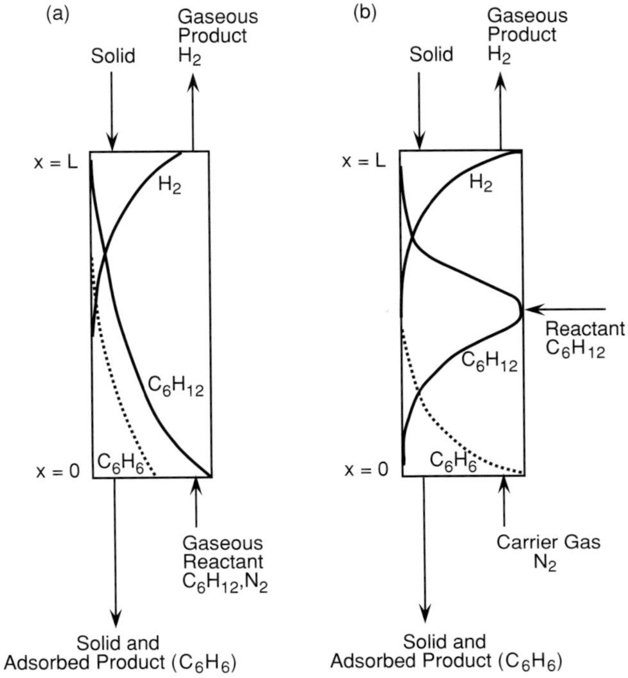

Figure 3.2 Principle of the countercurrent moving-bed chromatographic reactor: Steady state concentration profiles in the moving bed for the dehygrogenation of cyclohexane to benzene over a Pt/Al_2O_3 catalyst: (a) feed and carrier gas entering at the bottom of the bed, (b) feed entering midway and carrier gas entering at the bottom of the bed.

is driven upward through the bed by the carrier gas and emerges in the effluent at the top of the reactor. Catalyst particles are fed at the top, move downward and are withdrawn from the bottom. Benzene strongly adsorbs on the catalyst and thus moves downward through the bed, exiting at the bottom. The two products can be separated, however, cyclohexane, which adsorbs on the catalyst too, also emerges at the bottom as adsorbate. Cyclohexane and benzene cannot be separated in the countercurrent moving bed shown in Figure 3.2a. In Figure 3.2b, another configuration is illustrated. Cyclohexane is fed at the middle of the reactor, while the carrier gas enters from the bottom. High purity of the two products (exclusive of the carrier gas) and high conversion of the reactant are expected from this configuration. A variation of this contacting technique is the continuous rotating annular chromatographic reactor or rotating disc chromatographic reactor. The principle is the same of course, but in the rotating bed the individual catalyst/adsorbent particle is motionless relative to the bed. The bed itself is carried past the feed inlet and the product outlet.

In these moving-bed systems, all flows are continuous. Reactant is fed to the reactor continuously and products are withdrawn continuously. Use of a countercurrent moving bed of catalyst and adsorbent converts the reactor system from unsteady state to steady state. As discussed already, elements of catalyst and adsorbent circulating through the reactor encounter varying, indeed, periodically varying fluid environments. We treat countercurrent moving-bed chromatographic reactors (CMCR) in Chapter 5 and variants of the design in Chapter 6.

The third version is the simulated countercurrent moving-bed chromatographic reactor (SCMCR) in which the bed is stationary but the feed entrance and product exits shift with time. This operation can be carried out on a single fixed bed or on a rotating annular bed in which the feed and exit connections move with time across the axial faces of the annular structure. The operation in the single bed reactor is illustrated in Figure 3.3, adapted from Ray et al. (1990). The reactor bed is divided into N_s segments. Each segment has a feed entrance or a product exit. At the mth stage of operation, the reactant is fed into the jth segment and then the product is taken out from the kth segment. The entrances or exits of other segments are closed. After a certain period, the feed and product segments are shifted from jth to $(j-1)$th and from kth to $(k-1)$th position, respectively. Shifting of the exit and feed points is periodic. A SCMCR still retains the characteristics of a countercurrent moving-bed chromatographic reactor without actual movement of the solid. This operation solves the problem of catalyst attrition due to the movement of the solid particles. A further advantage is that several configurations are possible. Multistage beds can be employed as alternatives to the single bed with several segments. In the multistage configuration, different arrangements are possible for each stage. Thus, the types of solid, the volume, the temperature, etc. of

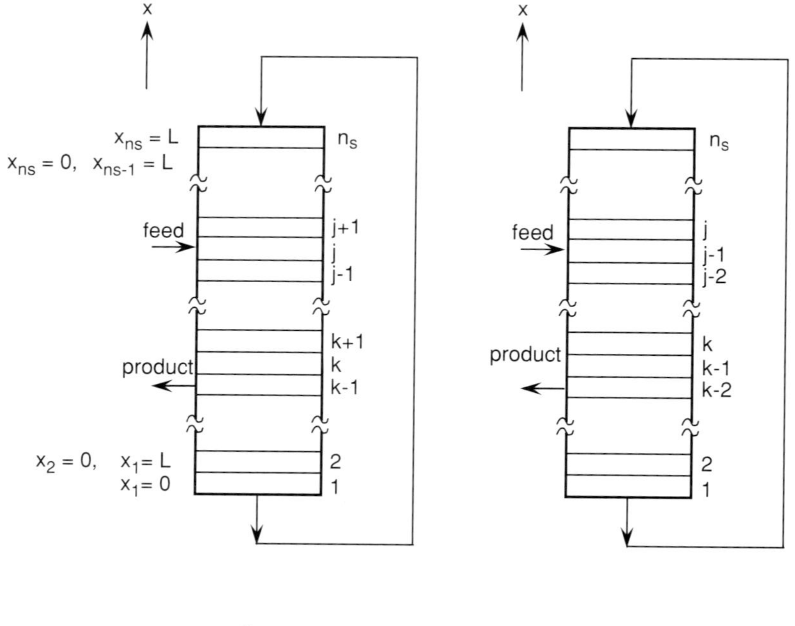

(a) m[th] STEP (b) (m+1)[th] STEP

Figure 3.3 Principle of the simulated countercurrent moving-bed chromatographic reactor in which the feed and product removal ports in a single-bed reactor change periodically: (a) location of the ports in the mth time step, (b) location of the ports in the $(m-1)$th time step. (Figure adapted from Ray et al. (1990) with permission. ©1990 Pergamon Press Plc.)

each stage can be varied. The SCMCR provides the widest possible range of operation for a chromatographic reactor. Chapter 7 deals with the state of the art for SCMCRs.

The examples discussed so far in this chapter employed a solid phase that functioned both as catalyst and adsorbent. Often, it is simpler to pack the reactor with two solids, one functioning as a catalyst and the other as an adsorbent. The latter should exhibit a larger affinity for the product than for the reactant in a simple isomerization reaction. When there are two products, the adsorbent must have a greater affinity for one of the products if separation as well as yield improvement is to occur. Separate catalyst and adsorbent particles are not mandatory, of course. Many catalysts adsorb products and often display large differences in adsorption strength as the earlier examples show. Different solids add flexibility, however.

There are constraints on the choice of adsorbent. Weak adsorption of the key product increases bed length and thus capital expense, while strong adsorption raises the carrier gas or solvent requirement per unit of reactant processed and results, also, in a more dilute product. This increases the cost of product concentration and recovery. A further constraint arises for single reactions with multiple reactants. Significantly different affinities of the adsorbent for the reactants can cause their separation and a retardation of the reaction rate. This is not a problem when one reactant serves as the carrier or solvent. Screening of candidate adsorbents is an important step in development of chromatographic reactor systems. We discuss screening reactors in Chapter 14.

3.2 General models

3.2.1 *Distributed systems*

Figure 3.4 shows schematically the packed bed of catalyst and adsorbent that compromises a separating reactor and defines the variables used in the model. The separating reactor can either be a CR with a fixed bed or a section between ports of a CMCR or a SCMCR. In the latter two cases the solid phase moves or is in virtual motion. This is the case shown in the figure. The schematic is broken into parts to indicate internal variables in the bed. Symbols at the ends indicate input variables, some of which will be functions of time. Reactor parameters, treated as constant in the model, are the symbols shown in the boxes. Assumed is that a single solid phase serves as catalyst and adsorbent. Dispersive fluxes resulting from concentration or temperature gradients are not given in the schematic.

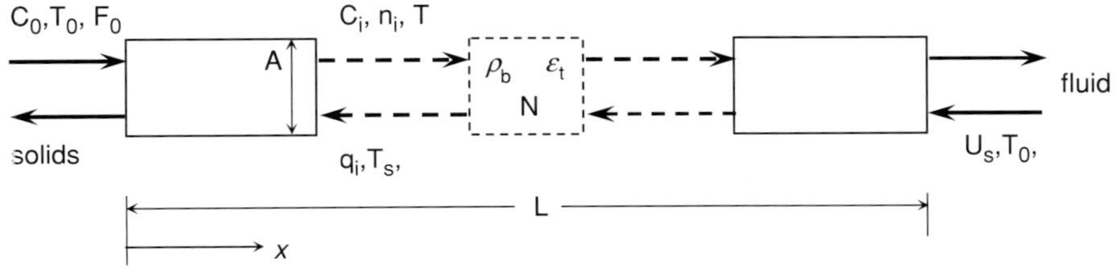

Figure 3.4 Schematic of the catalyst–adsorbent bed of cross section A and length L for a separating reactor showing bed parameters, inlet and bed variables.

Because isotropic beds of catalyst and adsorbent are generally used, all chromatographic reactors are described by the same set of heat and mass balances. A momentum balance should be part of the set, but in practice it is often replaced by a phenomenological equation. Table 3.2 summarizes a relatively simple description of a separating reactor employing a packed bed of adsorbent and catalyst. Equations in the table have several important assumptions: namely a single solid material serves simultaneously as catalyst and adsorbent, the bed is isotropic and intraparticle diffusion proceeds rapidly. Only a single reaction occurs. Particles are isothermal and uniformly bathed by the fluid phase, and the bed has no radial gradients. Furthermore, mass and heat transfer between fluid and solid is fast enough so that concentration and temperature differences between the phases are negligible. These assumptions permit the use of a pseudo homogeneous, one-dimensional model. It is further assumed that fluid and solids move: when the solid moves, it moves in plug flow

Table 3.2 Partial differential equations of pseudo-homogeneous models for chromatographic and swing reactors

Mass balance for component *i*	$\varepsilon_t \dfrac{\partial C_i}{\partial t} + \rho_b \dfrac{\partial q_i}{\partial t} = D_x \dfrac{\partial^2 C_i}{\partial x^2} - \dfrac{1}{A}\dfrac{\partial n_i}{\partial x} + U_s \rho_b \dfrac{\partial q_i}{\partial x} + \rho_b \nu_i r$	(3.1)
For CR, SCMCR, PSR, TSR, PPR	$U_s = 0$	
For CMCR at steady state	$\varepsilon_t \dfrac{\partial C_i}{\partial t} + \rho_b \dfrac{\partial q_i}{\partial t} = 0$	
Overall mass balance for components	$\varepsilon_t \dfrac{\partial C}{\partial t} + \rho_b \sum_{i=1}^{N_C} \dfrac{\partial q_i}{\partial t} = D_x \dfrac{\partial^2 C}{\partial x^2} - \dfrac{1}{A}\dfrac{\partial n}{\partial x} + U_s \rho_b \sum_{i=1}^{N_C} \dfrac{\partial q_i}{\partial x} + \rho_b \sum_{i=1}^{N_C} \nu_i r$	(3.2)
For CR, SCMCR, PSR, TSR, PPR	$U_s = 0$	
For CMCR at steady state	$\varepsilon_t \dfrac{\partial C}{\partial t} + \rho_b \sum_{i=1}^{N_C} \dfrac{\partial q_i}{\partial t} = 0$	
Energy balance	$\left(\varepsilon_t C_{p_g} C + \rho_b C_{p_s}\right) \dfrac{\partial T}{\partial t} = k_x \dfrac{\partial^2 T}{\partial x^2} - C_{p_g} \dfrac{n}{A}\dfrac{\partial T}{\partial x} + U_s \rho_b C_{p_s} \dfrac{\partial T}{\partial x} - \dfrac{P}{A}\dfrac{\partial}{\partial x}\left(\dfrac{n}{C}\right)$	
	$-\rho_b \sum_{i=1}^{N_C} H_{a_i} \dfrac{\partial q_i}{\partial t} + \rho_b \Delta H_R r - \dfrac{4h_0}{d_c}(T - T_a)$	(3.3)
For CR, SCMCR, PSR, TSR, PPR	$U_s = 0$	
For CR, CMCR, SCMCR	$\dfrac{\partial}{\partial x}\left(\dfrac{n}{C}\right) = 0$	
For CMCR at steady state	$\left(\varepsilon_t C_{p_g} C + \rho_b C_{p_s}\right) \dfrac{\partial T}{\partial t} = 0$	
Flow model: Ergun correlation (instead of momentum balance)	$\dfrac{dP}{dx} = -J_v u - J_k u^2$	(3.4)
	$J_v = \alpha \dfrac{\mu_g \left[\lambda_s (1 - \varepsilon_b)\right]^2}{d_p^2 \varepsilon_b^3}$	(3.5)
	$J_k = \dfrac{\lambda_s (1 - \varepsilon_b) \rho_g}{d_p \varepsilon_b^3}$	(3.6)

but axial dispersion can occur for the fluid phase. As usual, transport and thermodynamic properties are independent of temperature, pressure and composition. Fluids and mixtures exhibit ideal behavior. Furthermore, there are no entrance and exit effects. There is of course a dilemma with a pseudo homogeneous model. The rate constant, k, in r cannot be taken from independent rate measurements because the term is formulated in terms of the bulk density of a combined catalyst and adsorbent and because the presence of an adsorbent alters the concentration in the neighborhood of the catalyst surface. The rate constant must be evaluated for the catalyst-adsorbent mixture employing the pseudo homogeneous model.

The mass balances in Table 3.2 do not assume rapid adsorption (so that equilibrium is attained), however, that assumption is often made when the equations are applied. Some researchers separate Equations 3.1 and 3.2 into separate balances for the fluid and solid phases. This results in a heterogeneous model. Of course, when this is done, the balances are coupled by a rate of transport and/or an adsorption term. Separate balances for each phase are necessary when transport between the phases and/or adsorption is rate controlling and when measurements of the transport or adsorption steps are to be made. Separate balances are often used when the solid phase is stationary so that adsorbate measurements are not possible. In this case, the adsorbate flux terms disappear from Equations 3.1 and 3.2. This is the situation found for CR and SCMCR systems. The heterogeneous model is presented in Table 3.3. It is assumed that fluid-particle transport is slow when compared to adsorption rates.

Models in the following tables also represent the swing reactors, pressure swing (PSR) and temperature swing (TSR), as well as periodically pulsed reactors (PPR). These separating reactors are discussed in Part III. In the table, the total void fraction $\varepsilon_t = \varepsilon_b + (1 - \varepsilon_b)\varepsilon_p$. Usually, an intimate mixture of catalyst and adsorbent, two solid phases, will be used. When this is the case, ρ_b modifying the adsorbate density q_i terms must be replaced by $(\rho_b)_{\text{ads}}$ and the ρ_b modifying the reaction rate term must be replaced by $(\rho_b)_{\text{cat}}$. These "bulk" densities are the product of a volume fraction and the particle density. Thus, $(\rho_b)_{\text{ads}} = \varepsilon_{\text{ads}}\ (\rho_p)_{\text{ads}}$. The total void fraction term also changes $\varepsilon_t = \varepsilon_b + \varepsilon_{\text{ads}}\varepsilon_{p_{\text{ads}}} + \varepsilon_{\text{cat}}\varepsilon_{p_{\text{cat}}}$. The dispersion term above is based on the fluid volume. It may also be based on the reactor volume. When this definition is used D_x should be replaced by $\varepsilon_t D_x$. In Equation 3.3, both specific heat and density may not be the same for catalyst and adsorbent. In this case, $\rho_b(C_p)_s$ should be replaced by $\rho_{\text{ads}}(C_p)_{\text{ads}} + \rho_{\text{cat}}(C_p)_{\text{cat}}$. The two other ρ_b terms also change because they become specific to either catalyst or adsorbent.

The heat and material balances in Table 3.3 assume a single solid phase serving both as catalyst and adsorbent. As discussed above in reference to Table 3.2, different materials are normally used. We have presented the model version above because it generates a simple set of dimensionless groups that are important for predicting reactor behavior and performance. It is the usual practice to assume that mass transfer is limiting only for adsorption. It is generally observed that heat transfer between the phases is rapid so that a pseudo-homogeneous assumption can be made for the energy balance. Lumping catalyst and adsorbent causes problems in defining a rate of reaction term, as mentioned above, because the surface concentration, q_i, represents an adsorbate concentration rather than the concentration on the surface of the catalyst. A similar problem arises with the flux terms. Although the mass and heat transfer coefficients for catalyst and adsorbent will be similar, the driving forces will not be. An advantage of distinguishing between catalyst and adsorbent is that the rate expression and constants for r can be taken from independent rate measurements on the

Table 3.3 Partial differential equations of heterogeneous models for chromatographic and swing reactors

Mass balance for component i in the fluid phase	$\varepsilon_t \dfrac{\partial C_i}{\partial t} = D_x \dfrac{\partial^2 C_i}{\partial x^2} - \dfrac{1}{A}\dfrac{\partial n_i}{\partial x} - k_m a_m \rho_b \left(C_i - \dfrac{q_i}{K_i} \right)$ (3.7)
For CMCR at steady state	$\dfrac{\partial C_i}{\partial t} = 0$
Mass balance for component i in the solid phase	$\rho_b \dfrac{\partial q_i}{\partial t} = U_s \rho_b \dfrac{\partial q_i}{\partial x} + k_m a_m \rho_b \left(C_i - \dfrac{q_i}{K_i} \right) + \rho_b \nu_i r$ (3.8)
For CR, SCMCR, PSR, TSR, PPR	$U_s = 0$
For CMCR at steady state	$\dfrac{\partial q_i}{\partial t} = 0$
Overall mass balance for components in the fluid phase	$\varepsilon_t \dfrac{\partial C}{\partial t} = D_x \dfrac{\partial^2 C}{\partial x^2} - \dfrac{1}{A}\dfrac{\partial n}{\partial x} - a_m \rho_b \sum_{i=1}^{N_C} k_{m_i} \left(C_i - \dfrac{q_i}{K_i} \right)$ (3.9)
For CMCR at steady state	$\dfrac{\partial C}{\partial t} = 0$
Overall mass balance equation for components in the solid phase	$\rho_b \sum_{i=1}^{N_C} \dfrac{\partial q_i}{\partial t} = U_s \rho_b \sum_{i=1}^{N_C} \dfrac{\partial q_i}{\partial x} + a_m \rho_b \sum_{i=1}^{N_C} k_{m_i} \left(C_i - \dfrac{q_i}{K_i} \right) - \rho_b \sum_{i=1}^{N_C} \nu_i r$ (3.10)
For CR, SCMCR, PSR, TSR, PPR	$U_s = 0$
For CMCR at steady state	$\sum_{i=1}^{N_C} \dfrac{\partial q_i}{\partial t} = 0$
Energy balance for the fluid phase	$(\varepsilon_t C_{p_g} C) \dfrac{\partial T}{\partial t} = k_x \dfrac{\partial^2 T}{\partial x^2} - C_{p_g} \dfrac{n}{A}\dfrac{\partial T}{\partial x} - \dfrac{P}{A}\dfrac{\partial}{\partial x}\left(\dfrac{n}{C} \right) + a_m h \rho_b (T_s - T)$ (3.11)
For CR, CMCR and SCMCR	$\dfrac{\partial}{\partial x}\left(\dfrac{n}{C} \right) = 0$
For CMCR at steady state	$\dfrac{\partial T}{\partial t} = 0$
Energy balance for the solid phase	$\rho_b C_{p_s} \dfrac{\partial T_s}{\partial t} = k_s \dfrac{\partial^2 T_s}{\partial x^2} + U_s \rho_b C_{p_s} \dfrac{\partial T_s}{\partial x} - a_m h \rho_b (T_s - T) - \rho_b \sum_{i=1}^{N_C} H_{a_i} \dfrac{\partial q_i}{\partial t} - \rho_b \Delta H_R r$ (3.12)
For CR, SCMCR, PSR, TSR, PPR	$U_s = 0$
For CMCR at steady state	$\dfrac{\partial T_s}{\partial t} = 0$
Flow model: Ergun correlation (instead of momentum balance)	$\dfrac{dP}{dx} = -J_v u - J_k u^2$ (3.4)
	$J_v = \alpha \dfrac{\mu_g [\lambda_s (1 - \varepsilon_b)]^2}{d_p^2 \varepsilon_b^3}$ (3.5)
	$J_k = \dfrac{\lambda_s (1 - \varepsilon_b) \rho_g}{d_p \varepsilon_b^3}$ (3.6)

catalyst. Lumping the materials into a single solid phase means that the rate constants must be extracted from measurements on the adsorbent–catalyst mixture using the heterogeneous model.

Rather than create another table, we present the equations in Table 3.3 in an amended form now assuming catalyst and adsorbent are different materials. Volume fractions and particle densities are used in place of "bulk" densities.

Fluid phase material on component i:

$$\varepsilon_t \frac{\partial C_i}{\partial t} = D_x \frac{\partial^2 C_i}{\partial x^2} - \frac{1}{A}\frac{\partial n_i}{\partial x} - k_m a_m (\varepsilon_{\text{ads}} \rho_{p_{\text{ads}}}) \left(C_i - \frac{q_i}{K_i} \right) + \varepsilon_{\text{cat}} \rho_{p_{\text{cat}}} \nu_i r \tag{3.7a}$$

Solid phase material balance on component i:

$$\varepsilon_{\text{ads}} \rho_{p_{\text{ads}}} \frac{\partial q_i}{\partial t} = U_s \varepsilon_{\text{ads}} \rho_{p_{\text{ads}}} \frac{\partial q_i}{\partial x} + k_m a_m \varepsilon_{\text{ads}} \rho_{p_{\text{ads}}} \left(C_i - \frac{q_i}{K_i} \right) \tag{3.8a}$$

Changes to the total mass balances, Equations 3.9 and 3.10, must also be made. What these must be are readily seen from the amended component material balances. The pseudo-homogeneous assumption for the energy balance means that it is given by simple changes in Equation 3.3. Thus,

$$\begin{aligned}(\varepsilon_t C_{p_g} C + \varepsilon_{\text{cat}} \rho_{p_{\text{cat}}} C_{p_{\text{cat}}} + \varepsilon_{\text{ads}} \rho_{p_{\text{ads}}} C_{p_{\text{ads}}}) \frac{\partial T}{\partial t} &= k_x \frac{\partial^2 T}{\partial x^2} - C_{p_g} \frac{n}{A}\frac{\partial T}{\partial x} + U_s (\varepsilon_{\text{ads}} \rho_{p_{\text{ads}}} C_{p_{\text{ads}}} \\ &+ \varepsilon_{\text{cat}} \rho_{p_{\text{cat}}} C_{p_{\text{cat}}}) \frac{\partial T}{\partial x} - \frac{P}{A}\frac{\partial}{\partial x}\left(\frac{n}{C}\right) - \varepsilon_{\text{ads}} \rho_{p_{\text{ads}}} \sum_{i=1}^{N_C} H_{a_i} \frac{\partial q_i}{\partial t} + \varepsilon_{\text{cat}} \rho_{p_{\text{cat}}} \Delta H_R r - \frac{4h_0}{d_c}(T - T_a)\end{aligned} \tag{3.11a}$$

Of course, an energy balance is no longer needed for the solid phase.

Many of the early modeling studies employed only the mass balances because isothermal and isobaric conditions were assumed. Equations 3.1 to 3.3, 3.7, 3.9 and 3.11 represent the fluid velocity by the total molar flow rate for the fluid n. Velocity is affected by temperature and pressure. If isothermal and isobaric conditions are assumed and the fluid phase is dilute, fluid velocity can replace the molar flow rate.

All the models presented in the chapters which follow are versions of those in Tables 3.2 and 3.3. These versions reflect different assumptions and, sometimes, different definitions of model parameters.

Model variables are usually rendered dimensionless by characteristic length and time resulting in the partial differential equations shown in Tables 3.4 and 3.5. These tables assume a single solid acting as both catalyst and adsorbent. Simplifications for the differing types of separating reactors have been left out of Tables 3.4 and 3.5, but they are readily developed. All rates of reaction are assumed to be first order in species i and irreversible. Normalization variables employed in the tables are the length of the bed (L), the space time ($\varepsilon_t AL/F_0$), the average velocity ($F_0/\varepsilon_t A$), and the capacity of the adsorbent for species i ($\rho_t N$ or $\varepsilon_t N$). The temperature in Tables 3.2 and 3.3 is replaced by a temperature departure ($T - T_0$) and made dimensionless by the feed or inlet temperature (T_0) in Tables 3.4 and 3.5.

Table 3.6 defines the dimensionless variables in the previous tables. Other conventions for rendering the differential equations dimensionless have been used so that other

Table 3.4 Dimensionless form of a pseudo-homogeneous separating reactor model

Mass balance for component i	$\frac{\partial \gamma_i}{\partial \theta} + \frac{\partial \varphi_i}{\partial \theta} = N_d \frac{\partial^2 \gamma_i}{\partial \lambda^2} - \frac{\partial \nu_i}{\partial \lambda} + \sigma \frac{\partial \varphi_i}{\partial \lambda} - N_{\mathrm{Da}_i} \varphi_i$	(3.13)
Overall mass balance for components	$\frac{\partial \gamma}{\partial \theta} + \sum_{i=1}^{N_C} \frac{\partial \varphi_i}{\partial \theta} = N_d \frac{\partial^2 \gamma}{\partial \lambda^2} - \frac{\partial \nu}{\partial \lambda} + \sigma \sum_{i=1}^{N_C} \frac{\partial \varphi_i}{\partial \lambda} - \sum_{i=1}^{N_C} N_{\mathrm{Da}_i} \varphi_i$	(3.14)
Energy balance	$(\gamma + \omega) \frac{\partial \phi}{\partial \theta} = N_h \frac{\partial^2 \phi}{\partial \lambda^2} - (\nu + \sigma \omega) \frac{\partial \phi}{\partial \lambda} - \psi Z \frac{\partial (\nu/\gamma)}{\partial \lambda} - \sum_{i=1}^{N_C} \left(\kappa_{a_i} \frac{\partial \varphi_i}{\partial \theta} \right)$	
	$- \sum_{i=1}^{N_C} \left(\kappa_{a_i} N_{\mathrm{Da}_i} \varphi_i \right) - \xi (\phi - \phi_a)$	(3.15)
Ergun correlation	$\frac{dZ}{d\lambda} = -\zeta_v \eta - \zeta_k \eta^2$	(3.16)

dimensionless groups arise. Frequently, fluid concentration is normalized by the inlet concentration of a key reactant.

We suggest naming the solid–fluid velocity ratio, so important in characterizing moving-bed chromatographic reactors, the Aris Number (N_{Ar}) to honor Aris's important contributions to the study of chromatographic reactors. We suggest names for other dimensionless groups important for pressure swing reactors later on. Use of the Damköhler number is well established. However the first form shown in Table 3.6 is strongly temperature dependent. Frequently the rate terms in Tables 3.4 and 3.5 are represented by $N'_{\mathrm{Da}} e^{-\beta/\phi+1}$ where the Damköhler number is now independent of temperature. β is the Prater number.

Table 3.5 Dimensionless form of a heterogeneous separating reactor model

Mass balance for component i in fluid phase	$\frac{\partial \gamma_i}{\partial \theta} = N_d \frac{\partial^2 \gamma_i}{\partial \lambda^2} - \frac{\partial \nu_i}{\partial \lambda} - N_{m_i} \left(\gamma_i - \frac{\varphi_i}{K_i} \right)$	(3.17)
Mass balance for component i in solid phase	$\frac{\partial \varphi_i}{\partial \theta} = \sigma \frac{\partial \varphi_i}{\partial \lambda} + N_{m_i} \left(\gamma_i - \frac{\varphi_i}{K_i} \right) - N_{\mathrm{Da}_i} \varphi_i$	(3.18)
Overall mass balance for components in fluid phase	$\frac{\partial \gamma}{\partial \theta} = N_d \frac{\partial^2 \gamma}{\partial \lambda^2} - \frac{\partial \nu}{\partial \lambda} - \sum_{i=1}^{N_C} N_{m_i} \left(\gamma_i - \frac{\varphi_i}{K_i} \right)$	(3.19)
Overall mass balance for components in solid phase	$\sum_{i=1}^{N_C} \frac{\partial \varphi_i}{\partial \theta} = \sigma \sum_{i=1}^{N_C} \frac{\partial \varphi_i}{\partial \lambda} + \sum_{i=1}^{N_C} N_{m_i} \left(\gamma_i - \frac{\varphi_i}{K_i} \right) - \sum_{i=1}^{N_C} N_{\mathrm{Da}_i} \varphi_i$	(3.20)
Energy balance for the fluid phase	$\gamma \frac{\partial \phi}{\partial \theta} = (N_h)_g \frac{\partial^2 \phi}{\partial \lambda^2} - \nu \frac{\partial \phi}{\partial \lambda} - \psi Z \frac{\partial (\nu/\gamma)}{\partial \lambda} + N_h (\phi_s - \phi)$	(3.21)
Energy balance for the solid phase	$\omega \frac{\partial \phi_s}{\partial \theta} = (N_h)_s \frac{\partial^2 \phi_s}{\partial \lambda^2} - \sigma\omega \frac{\partial \phi_s}{\partial \lambda} - N_h (\phi_s - \phi) - \sum_{i=1}^{N_C} \left(\kappa_{a_i} \frac{\partial \varphi_i}{\partial \theta} \right)$	
	$- \sum_{i=1}^{N_C} \left(\kappa_{a_i} N_{\mathrm{Da}_i} \varphi_i \right)$	(3.22)
Ergun correlation	$\frac{dZ}{d\lambda} = -\zeta_v \eta - \zeta_k \eta^2$	(3.16)

Table 3.6 Definition of dimensionless groups and parameters in Tables 3.4 and 3.5

Nondimensional pressure	$Z = \dfrac{P}{P_f}$	(3.23)
Nondimensional concentration for component i in fluid phase	$y_i = \dfrac{C_i}{\rho_b N}$	(3.24)
Nondimensional concentration for all components in fluid phase	$y = \dfrac{C}{\rho_b N}$	(3.25)
Nondimensional concentration for component i in solid phase	$\varphi_i = \dfrac{q_i}{\varepsilon_t N}$	(3.26)
Nondimensional temperature departure	$\phi = \dfrac{T - T_0}{T_0}$	(3.27)
Nondimensional length	$\lambda = \dfrac{x}{L}$	(3.28)
Nondimensional flow rate (molar basis) for component i	$\nu_i = \dfrac{n_i}{\rho_b N F_0}$	(3.29)
Nondimensional flow rate (molar basis) for all components	$\nu = \dfrac{n}{\rho_b N F_0}$	(3.30)
Nondimensional velocity of fluid	$\eta = \dfrac{\varepsilon_t A u}{F_0}$	(3.31)
Nondimensional time	$\theta = \dfrac{F_0 t}{\varepsilon_t A L}$	(3.32)
Bodenstein number for mass	$N_d = \dfrac{D_x A}{L F_0}$	(3.33)
Bodenstein number for heat (gas)	$(N_h)_g = \dfrac{k_x}{C_{p_g} \rho_b N} \dfrac{A}{L F_0}$	(3.34)
Bodenstein number for heat (solid)	$(N_h)_s = \dfrac{k_s}{C_{p_s} \rho_b N} \dfrac{A}{L F_0}$ *or* $= \dfrac{k_s}{C_{p_{cat}} \varepsilon_{cat} \rho_{p_{cat}} + C_{p_{ads}} \varepsilon_{ads} \rho_{p_{ads}}} \dfrac{A}{N L F_0}$	(3.35)
Damköhler number for the reaction of component i assuming an irreversible, first order rate model	$N_{Da_i} = \dfrac{\varepsilon_t A L k_{r_i}}{F_0}$ or $k_{y_i} \bar{t}$	(3.36)
Temperature independent Damköhler number	$N'_{Da_i} = \dfrac{\varepsilon_t A L k'_i}{F_0}$	(3.38)
Prater number	$\beta = \dfrac{E}{R T_0}$	(3.39)
Velocity ratio of solid and fluid or the Aris number (N_{Ar})	$\sigma = \dfrac{\varepsilon_t A U_s}{F_0}$	(3.40)
Heat capacity ratio of solid and fluid	$\omega = \dfrac{C_{p_s}}{\varepsilon_t C_{p_g} N}$ *or* $= \dfrac{(\varepsilon_{cat} \rho_{p_{cat}} C_{p_{cat}} + \varepsilon_{ads} \rho_{p_{ads}} C_{p_{ads}})}{\varepsilon_t C_{p_g} N \rho_{b_{ads}}}$	(3.41)
Ratio of energies in the form of pressure and heat	$\psi = \dfrac{P_f}{C_{p_g} \rho_b N T_0}$	(3.42)
Ratio of heats escaping from the wall and passing along the reactor	$\xi = \dfrac{\pi h_0 d_c L}{C_{p_g} \rho_b N F_0}$	(3.43)
Ratio of fluid–solid heat transfer to the axial convective heat flux	$N_h = \dfrac{a_m h A L}{C_{p_g} F_0 N}$	(3.44)

(Continued)

Table 3.6 Definition of dimensionless groups and parameters in Tables 3.4 and 3.5 *(Continued)*

Ratio of fluid–solid volumetric mass transfer coefficient to space velocity	$N_m = \dfrac{k_m a_m \rho_b L\,A}{F_0}$	(3.45)
Normalized heat of adsorption of component *i*	$\kappa_{a_i} = \dfrac{\Delta H_{a_i}}{C_{p_g} T_0}$	(3.46)
Normalized heat of reaction of component *i*	$\kappa_{r_i} = \dfrac{\Delta H_{r_i}}{C_{p_g} T_0}$	(3.47)
Normalized viscous pressure loss term for fluid flow in a packed bed	$\zeta_v = \dfrac{J_v F_0 L}{\varepsilon_t P_f A}$	(3.48)
Normalized kinetic pressure loss term for fluid flow in a packed bed	$\zeta_k = \dfrac{J_k F_0^2 L}{\varepsilon_t^2 A^2 P_f}$	(3.49)

The different types of operations are distinguished from one another through the initial and boundary conditions that we summarize by operation in Table 3.7. Just those for chromatographic reactors and periodically pulsed reactors are given in this table. Boundary conditions for pressure and temperature swing reactors are given later on. The chromatographic reactor (CR) is operated under unsteady boundary condition: namely the reactant fraction in the feed, $C_i|_{0-}$ and/or $n_i|_{0-}$, is a function of the elapsed time. Usually, $C_i|_{0-}$ and/or $n_i|_{0-}$ is modulated as a highly asymmetric square wave. In the table we assume that the cycle split, s, is the cycle fraction of the duration of feed to the reactor containing reactant. Thus, during times $n\tau \leq t \leq s(n+1)\tau$ reactant flows to the reactor, while during $s(n+1)\,\tau \leq t \leq (n+1)\tau$ only carrier gas feeds the reactor. In these relations, n is an integer with an initial value of zero that is augmented on each cycle. We further assume that the volumetric flow to the reactor, Q_0, remains constant so that the flow of carrier gas increases whenever reactant is withdrawn from the feed.

Only the boundary conditions for the case (a) in Figure 3.2 are described in Table 3.7 for the CMCR. For SCMCR, we assume that each segment has the same volume, solid composition and operating conditions. Each segment, including the feed and product ones, periodically shifts to a segment of lower number. The papers published to date on SCMCR have assumed isobaric condition so that no pumping system for the recycle is shown in Figure 3.3. Thus, we reserve description of the boundary conditions for momentum equations because the pumping system is undefined. Boundary conditions given in the table below are for the pseudo-homogeneous model presented in Table 3.2. These are essentially the same for the heterogeneous model given in Table 3.3, just the fluid and solid terms must be separated when applied to a CMCR.

3.2.2 Lumped models

An alternative to the representation of a packed bed of mixed adsorbent and catalyst by a distributed model is a cell model in which the bed is represented by a cascade of cells, each of which has a uniform temperature and composition. Thus, each cell can be described by a lumped model. Cascade models are best suited to fixed-bed chromatographic reactors and to simulated countercurrent moving-bed chromatographic reactors. Isothermal operation

Table 3.7 Initial and boundary conditions for chromatographic reactors

Chromatographic reactor (CR)

At $x = 0$

Pressure

$$P = P_0 \qquad (3.50)$$

Concentration of components

$$D_x \frac{d}{dx}(C_i)|_{0+} = \frac{1}{A}(n_i|_{0+} - n_i|_{0-})$$
$$n_i|_{0-} = (n_i)_0 \quad \text{for } n\tau \le t \le s(n+1)\tau$$
$$n_i|_{0-} = 0 \quad \text{for } s(n+1)\tau \le t \le (n+1)\tau, \text{ where} \qquad (3.51)$$
$$n = 0, 1, 2, 3, \ldots$$

Temperature

$$k_x \frac{dT}{dx}\Big|_{0+} = \frac{nC_{pg}}{A}(T|_{0+} - T_0|_{0-}) \qquad (3.52)$$

At $x = L$

$$\frac{dT}{dx} = \frac{dP}{dx} = \frac{dC_i}{dx} = 0 \qquad (3.53)$$

Countercurrent moving-bed chromatographic reactor (CMCR)

At $x = 0$ (at bottom, inlet for fluid and outlet for solid)

Pressure

$$P = P_0 \qquad (3.54)$$

Concentration of components

$$\varepsilon_t D_x \frac{d}{dx}(C_i)\Big|_{0+} = \frac{\varepsilon_t}{A}(n_i|_{0+} - n_i|_{0-}) - \frac{\rho_b U_s}{A}(q_i|_{0+} - qi|_{0-}) \qquad (3.55)$$

Temperature

$$\varepsilon_t k_x \frac{dT}{dx}\Big|_{0+} = \frac{\varepsilon_t n C_{pg}}{A}\left(T|_{0+} - T_g|_{0-}\right) - \frac{\rho_b U_s C_s}{A}(T|_{0+} - T_s|_{0-}) \qquad (3.56)$$

At $x = L$ (at top, outlet for fluid and inlet for solid)

Pressure

$$\frac{dP}{dx} = 0 \qquad (3.57)$$

Concentration of components

$$\varepsilon_t D_x \frac{d}{dx}(C_i)\Big|_{L-} = \frac{\varepsilon_t}{A}(n_i|_{L-} - n_i|_{L+}) - \frac{\rho_b U_s}{A}(q_i|_{L-} - qi|_{L+}) \qquad (3.58)$$

Temperature

$$\varepsilon_t k_x \frac{dT}{dx}\Big|_{L-} = \frac{\varepsilon_t n C_{pg}}{A}\left(T|_{L-} - T_g|_{L+}\right) - \frac{\rho_b U_s C_s}{A}(T|_{L-} - T_s|_{L+}) \qquad (3.59)$$

Simulated countercurrent moving-bed chromatographic reactor (SCMCR)

At $x = 0$ for each segment

For the inlet segment

Concentration

$$D_x \frac{d}{dx}(C_i)_1\Big|_{x_1=0+} = \frac{1}{A}\left((n_i)_1|_{x_1=0+} - (n_i)_{n_s}|_{x_{n_s}=L+} - (n_i)_f\right)$$
$$j = 1 \qquad (3.60)$$
$$D_x \frac{d}{dx}(C_i)_j\Big|_{x_j=0+} = \frac{1}{A}\left((n_i)_j|_{x_j=0+} - (n_i)_{j-1}|_{x_{j-1}=L+} - (n_i)_f\right)$$
$$j = 2, 3, \ldots, n_s \qquad (3.61)$$

Temperature

$$k_x \frac{dT_1}{dx}\Big|_{x_1=0+} = \frac{C_{pg}}{A}\left((nT)_1|_{x_1=0+} - (nT)_{n_s}|_{x_{n_s}=L+} - (nT)_f\right)$$
$$j = 1 \qquad (3.62)$$
$$k_x \frac{dT_j}{dx}\Big|_{x_j=0+} = \frac{C_{pg}}{A}\left((nT)_j|_{x_j=0+} - (nT)_{j-1}|_{x_{j-1}=L+} - (nT)_f\right)$$
$$j = 2, 3, \ldots, n_s \qquad (3.63)$$

For the segment following the outlet segment

Concentration

$$D_x \frac{d}{dx}(C_i)_1\Big|_{x_1=0+} = \frac{1}{A}\left((n_i)_1|_{x_1=0+} - \kappa(n_i)_{n_s}|_{x_{n_s}=L+}\right)$$
$$k = n_s \qquad (3.64)$$
$$D_x \frac{d}{dx}(C_i)_{k+1}\Big|_{x_{k+1}=0+} = \frac{1}{A}\left((n_i)_{k+1}|_{x_{k+1}=0+} - \kappa(n_i)_k|_{x_k=L+}\right)$$
$$k = 1, 2, \ldots, n_s - 1 \qquad (3.65)$$

Temperature

$$k_x \frac{dT_1}{dx}\Big|_{x_1=0+} = \frac{C_{pg}}{A}\left((nT)_1|_{x_1=0+} - \kappa(nT)_{n_s}|_{x_{n_s}=L+}\right)$$
$$k = n_s \qquad (3.66)$$
$$k_x \frac{dT_{k+1}}{dx}\Big|_{x_{k+1}=0+} = \frac{C_{pg}}{A}\left((nT)_{k+1}|_{x_{k+1}=0+} - \kappa(nT)_k|_{x_k=L+}\right)$$
$$k = 1, 2, \ldots, n_s - 1 \qquad (3.67)$$

(Continued)

Table 3.7 Initial and boundary conditions for chromatographic reactors *(Continued)*

For the other segments		
Concentration	$D_x \frac{d}{dx}(C_i)_1\big\rvert_{x_1=0+} = \frac{1}{A}\left((n_i)_1\rvert_{x_1=0+} - (n_i)_{n_s}\rvert_{x_{n_s}=L+}\right)$	
	$i = 1$	(3.68)
	$D_x \frac{d}{dx}(C_i)_l\big\rvert_{x_l=0+} = \frac{1}{A}\left((n_i)_l\rvert_{x_l=0+} - (n_i)_{l-1}\rvert_{x_{l-1}=L+}\right)$	
	$i = 2, 3, \ldots, n_s$	(3.69)
Temperature	$k_x \frac{dT_1}{dx}\Big\rvert_{x_1=0+} = \frac{C_{pg}}{A}\left((nT)_1\rvert_{x_1=0+} - (nT)_{n_s}\rvert_{x_{n_s}=L+}\right)$	
	$i = 1$	(3.70)
	$k_x \frac{dT_l}{dx}\Big\rvert_{x_l=0+} = \frac{C_{pg}}{A}\left((nT)_l\rvert_{x_l=0+} - (nT)_{l-1}\rvert_{x_{l-1}=L+}\right)$	
	$i = 2, 3, n_s$	(3.71)
At $x = L$ for each segment	$\frac{dT}{dx} = \frac{dP}{dx} = \frac{dC_i}{dx} = 0$	(3.72)

is usually assumed as representation of heat dispersion requires the use of pseudo heat transfer coefficients to represent conductive flows between cells. The cascade model is given by Figure 3.5. In this figure, the boxes are cells, $j-1$, j and $j+1$. There is a junction before each cell denoted by $j-1$ and j. The junctions allow the model to be applied to a SCMCR that will have addition/withdrawal ports located axially along the mixed catalyst and adsorbent bed.

Cell models are used mainly with dilute systems under isothermal and isobaric conditions. Assuming a pseudo-homogeneous system and thus the assumption listed for Table 3.2. A material balance for the ith component gives

$$V\left(\varepsilon_t \frac{dC_i^j}{dt} + \rho_{b_{\text{ads}}} \frac{dq_i^j}{dt}\right) = F\left(C_i^{j-1} - C_i^j\right) + \rho_{b_{\text{cat}}} \nu_i V_r \tag{3.73}$$

The equation assumes that adsorption units are per weight of adsorbent and that only a surface catalyzed reaction takes place. In this equation and others in this chapter we use bulk densities for the adsorbent and the catalysts. These are the product of the volume fraction and the particle density. Thus, $(\rho_b)_{\text{cat}} = \varepsilon_{\text{cat}}(\rho_p)_{\text{cat}}$. The superscripts identify the cell or junction. The overall balance is

$$V\left(\varepsilon_t \frac{dC_i^j}{dt} + \rho_{b_{\text{ads}}} \sum_{i=1}^{N_c} \frac{dq_i}{dt}\right) = F\left(C_i^{j-1} - C_i^j\right) - \rho_{b_{\text{cat}}} V_r \sum_{i=1}^{N_c} \nu_i \tag{3.74}$$

Mass balances at junctions between cells, $j-1$, j, $j+1$, in Figure 3.5 are algebraic relations.

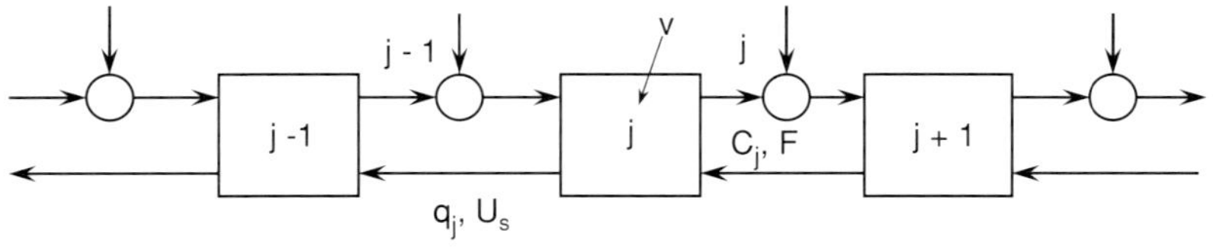

Figure 3.5 Cascade representation of a separating reactor. Each cell, $(j-1)$, j, $(j+1)$, is assumed to have a uniform concentration and temperature. Junctions on either side of all but the first and final cells permit fluid addition or withdrawal.

3.3 Cyclic steady state

With few exceptions, periodically operated reactors are run continuously and so eventually reach what can be called a cyclic steady state. In this state, concentration of any component, C_i, temperature, T, velocity, u, quantity of an adsorbate on a surface, q_i, are constant at identical times measured from the start of the cycle. The conditions can be formulated for a two-part cycle of period, τ, and cycle split, s, assuming that s measures the fractional duration of the first part of the cycle:

$$\text{For } n\tau \leq t \leq sn\tau. \quad C_i(x, t) = C_i(x, t + \tau), \quad q_i(x, t) = q_i(x, t + \tau), \quad u(x, t) = u(x, t + \tau), \quad T(x, t) = T(x, t + \tau) \tag{3.75}$$

$$\text{For } sn\tau \leq t \leq (n + 1)\tau. \quad C_i(x, t) = C_i(x, t + \tau), \quad q_i(x, t) = q_i(x, t + \tau), \quad u(x, t) = u(x, t + \tau), \quad T(x, t) = T(x, t + \tau) \tag{3.76}$$

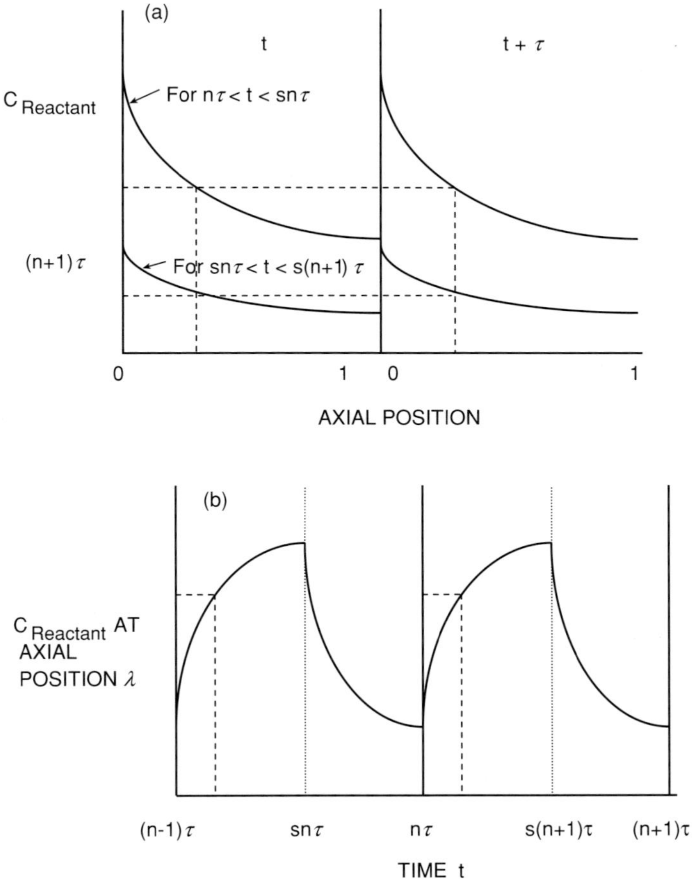

Figure 3.6 Schematic representation of the cyclic steady state for a reactant: (a) at distinct times, (b) at an axial point λ within the reactor. Two cycles of period τ and cycle split s are shown.

In these expressions, n is a large integer. Of course, the conditions apply as well to reaction rates, adsorption rates and transport properties and thus imply the absence of fouling and deactivation processes.

Figure 3.6 illustrates the cyclic steady state with respect to a reactant.

The conditions of a cyclic steady state can be exploited during integration of the model equations to accelerate conversion, but at the expense of losing description of system performance during the time after startup until the cyclic steady state is attained.

Chapter 4
Chromatographic Reactors (CR)

In the preceding chapter we reviewed types of chromatographic reactors and how each operates. Representation of these reactors was also examined. In this and the succeeding chapters we examine the state of the art for each type with respect to modeling and experimental performance. The chromatographic reactor is the simplest of the genre and was also the first to be studied.

4.1 Modeling studies

Modeling of a chromatographic reactor (CR) was attempted first by Magee (1963) who employed a simplified model for an equilibrium limited reaction, $A \leftrightarrow B + C$. Isothermal and isobaric conditions were assumed and axial dispersion was neglected. Rates of movement of each component through the reactor were the variables instead of concentration or partial pressure, so Equations 3.1 and 3.2 in Table 3.2 were modified. Component fluxes can be used as variables if adsorption is assumed to be at equilibrium. A solution was obtained for instantaneous establishment of reaction equilibrium and the same rate of movement for A and B. Calculations showed that a CR can force a fast equilibrium limited reaction to completion. The reactor is more effective when the difference in the rate of movement between A and C is large. For example, 70–95% conversion can be expected even with an equilibrium constant, $K_R = 2 \times 10^{-7}$ atm, when a 1-m-long reactor was employed and the rate of movement for A and B was 3 cm/s while for C it was 9 cm/s. Magee's simulation thus confirmed Roginskii's proposal and various patent claims (Dinwiddie and Morgan, 1961; Magee, 1961; Broughton, 1961; Gaziev et al., 1963).

Several irreversible and reversible reactions were considered by Saito et al. (1965) and Hattori and Murakami (1968) for a single pulse and for consecutive pulses, that is, a chromatographic reactor. For first order reactions, the chromatographic reactor (CR) and a plug flow reactor (PFR) give identical conversions for irreversible and reversible reactions. The selectivity to the intermediate products was also the same for consecutive reactions. However, if the order was different from one or zero, conversions in a CR were higher than that for a PFR. This was also true for reversible reactions. For irreversible consecutive reactions, $A + B \rightarrow R$, $R + B \rightarrow S$: $A \rightarrow R$, $2R \rightarrow S$, Hattori and Murakami found that pulse operation increases the yield of the intermediate product. They explained their observations on the basis of the concentration changes due to adsorption and to longer contact times for the adsorbed components.

Chu and Tsang (1971) considered the reaction, $A \leftrightarrow B + C$ and developed a model assuming a Langmuir–Hinshelwood reaction mechanism where A and B adsorb on the same sites of the catalyst, while C does not adsorb. Their model, based on material balances for each reacting species, is the same as given by Equation 3.1 in Table 3.2, neglecting the term for moving solids, that is,

$$\varepsilon_t \frac{\partial C_i}{\partial t} + \rho_b \frac{\partial q_i}{\partial t} = D_x \frac{\partial^2 C_i}{\partial x^2} - \frac{1}{A}\frac{\partial n_i}{\partial x} - \rho_b r_i \tag{4.1}$$

They assumed an isothermal system so the energy balance, Equation 3.3, was not used. The authors reduced the model to a dimensionless form but defined their dimensionless group somewhat differently than those given in Table 3.4. Chu and Tsang investigated the effects on CR performance of input waveform, reverse reaction rate constant, average reactant concentration in the feed, adsorption equilibrium constants and active center concentration. They also made a limited study on the effect of axial dispersion. Chu and Tsang showed, using a sine wave to represent variation of reactant concentration in a pulse, that the chromatographic separation obtained for the first pulse was destroyed by a subsequent pulse. A "delayed" sine curve, which has one impulse every several cycles, improved the time-average production of the desired product over that of a plug flow reactor operating at the average concentration of the sinusoidal varying reactant. A feed pulse in every four cycles gave the maximum improvement in the time-average production under their assumptions. When the number of missed pulses exceeded four, the time-average production in the CR was lower because during most of the cycle period the reactor operated at a lower reactant concentration and therefore at a lower production rate. In other words, the reactor was inadequately utilized. Chu and Tsang identified reduced throughput as a major drawback of chromatographic reactors.

Chu and Tsang found that the average conversion was improved in a CR as the reverse reaction rate constant increased. Conversion also improved as the average concentration of the reactant in the feed increased, particularly when the reverse reaction rate constant was comparable to the forward one. Magnitude of the adsorption equilibrium constants showed a significant effect on the improvement because they affect the separation of B and C. Largest improvement was achieved when B adsorbed on the catalyst more strongly than A. Furthermore, the average conversion was improved as the number of active sites on the catalyst increased up to a certain value. At higher values, conversion dropped. Chu and Tsang observed that axial dispersion had a negligible effect on throughput and conversion for reasonable values of the dispersion coefficient.

Schweich, Villermaux and coworkers (Schweich and Villermaux, 1978) in a series of papers extending over several years were concerned with calculational techniques to ease the burden of integrating the CR reactor model represented by Equations 3.1 and 3.2 in Table 3.2 or by Equation 4.1 above, and the initial and boundary conditions in Table 3.7. Schweich and Villermaux (1978) formulated a CR model in terms of fully back mixed cells. This replaces each of the partial differential equations in Table 3.2 by a set of ordinary differential equations, one for each cell, giving Equations 3.73 and 3.74. Restating the single component balance,

$$QC_{i,k-1} \pm r\frac{V}{J} = QC_{i,k} + (1+\alpha_i)\frac{V}{J}\frac{dC_{i,k}}{dt} \tag{4.2}$$

In this equation, V is the reactor volume, J is the numbers of cells and α_i is an equilibrium constant in a linear adsorption isotherm. This formulation is convenient for numerical integration. At steady state, the model is algebraic. Boundary conditions are no longer required. Schweich and Villermaux (1978) assumed square waves, both symmetrical and asymmetrical as inputs to the model.

Because of the time and effort still needed to solve the cell model, Schweich and Villermaux (1978) introduced a semi-batch reactor to model the physical changes occurring in each CR cell. This approach was based on an observation of Hattori and Murakami (1968) that the shape of the chromatogram was not important to the total conversion of the reactant. The observation is indeed true if the reaction is first order with respect to the reactant, namely if the physical system is linear. In the Schweich and Villermaux model, the shape of the chromatogram is neglected. This provided a significant simplification: only one parameter is unknown in this model. Figure 4.1 illustrates the simplified model. Schweich and Villermaux (1978) compared results calculated by the "rigorous", back-mixed cell model and the simplified model for a dehydrogenation reaction. They showed good agreement when the single parameter was optimized.

In a further exploration of calculation methods, Schweich et al. (1980) also used the set of partial differential equations in Table 3.2 or given by Equation 4.1. For linear isotherms of the adsorbed species and for linear reaction rate expressions, the PDEs can be reduced to a system of ordinary differential equations by the method of characteristics. Schweich et al. introduced Lax's theorem (Lax, 1957), which allows conversion of the ODEs into algebraic

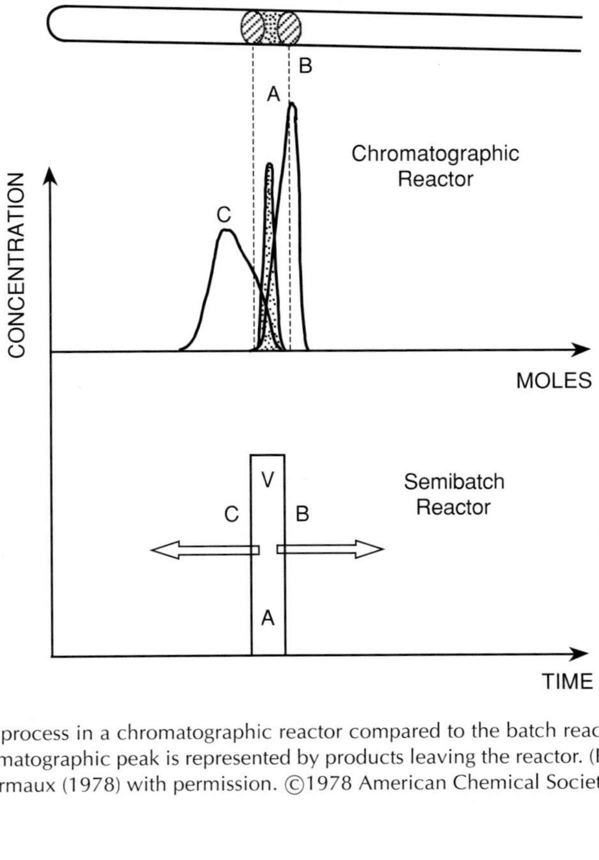

Figure 4.1 Actual process in a chromatographic reactor compared to the batch reactor model in which broadening of chromatographic peak is represented by products leaving the reactor. (Figure adapted from Schweich and Villermaux (1978) with permission. ©1978 American Chemical Society.)

equations known as the characteristic equations. Solutions of the algebraic equations give time–position behavior thereby predicting the shape of the concentration waves moving through the reactor. The method predicts the existence of step changes or shock waves moving through the CR. Vanishing species are also predicted just as with the method of characteristics. The authors applied Lax's theorem to adsorption governed by Langmuir isotherms but concluded that it is not possible to determine the shape of chromatographic peaks for this case.

A first-order reaction inhibited by the product was examined by Schweich and Villermaux (1982a). They observed that performing this reaction in a chromatographic reactor did not increase conversion. This result contradicts the observation of Unger and Rinker (1976) who found experimentally a conversion improvement for ammonia synthesis at low conversions. Ammonia synthesis at low conversion is an irreversible reaction inhibited by the ammonia product. Schweich and Villermaux concluded that the experimental observation could not be explained by a chromatographic effect.

Liden and Vamling (1989) examined an irreversible consecutive-parallel reaction, $A + B \rightarrow R$, $R + B \rightarrow S$, considered earlier by Saito et al. (1965) and Hattori and Murakami (1968). Homogeneous reactions in which rate is directly proportional to gas phase concentrations and heterogeneous reactions in which rates are proportional to fractional coverage of the adsorbed reactants were considered. Liden and Vamling only carried out simulations of dilute systems for which the Langmuir isotherm reduces to a linear isotherm. In this case, homogeneous and heterogeneous mechanisms become identical. Simulations showed that the chromatographic reactor achieved a higher selectivity to the intermediate than the plug flow reactor although the amount of improvement depended on how comparison between the reactors is made. They found that improvement depended on the adsorption capacity of the catalyst for the intermediate (it was assumed that reactants did not adsorb). Improvement in selectivity also depended on the cycle split. This agreed with the results of Hattori and Murakami who found that narrow pulses gave the largest selectivity improvement. Liden and Vamling also observed that the reactants must be pulsed together. If only the B component is pulsed, there is no improvement in yield. This confirmed earlier results by Schweich and Villermaux (1982a). Liden and Vamling explained their results by noting that the product, R, was separated from the reactants, A and B, so that the subsequent reaction of R and B was inhibited. Namely, the first reaction step was not affected by pulsed operation but the second step was.

Only one investigation of a biochemical application was found in the literature. Zafar and Barker (1988) applied Equation 4.1 to the enzyme-catalyzed formation of a glucose polymer, dextran, from sucrose. The authors neglected the dispersion term in the equation and assumed a linear adsorption isotherm for the by-product fructose on the ion-exchange resin used as an adsorbent. The enzymatic reaction occurs homogeneously so the bed density term in the model disappears. The rate expression utilized the Michaelis–Menten relationship, thus the reaction is reactant inhibited. Model predictions were tested by experiments and are discussed at the end of the next section.

Research on the simple chromatographic reactor has lapsed until rather recently. Mazzotti et al. (1997) examined a chromatographic reactor employing a swellable polymer, a cationic exchange resin, as a catalyst for the liquid phase reaction: esterification of ethanol and acetic acid to form ethyl acetate and water. The authors assumed an isothermal system but they allowed for axial dispersion. The swelling resin holds both reactant and product and thus resembles an adsorbent. The authors treat the reaction as occurring in the polymer phase. The back reaction depends on the equilibrium constants as well as on the activities of products and reactants. The latter were predicted using UNIFAC in the framework of the extended Flory–Huggins model of the polymer phase. The polymer has elasticity and thus expands or contracts as the liquid phase composition changes. Consequently, the void fraction in the packed bed becomes a function of time and position. The Mazzotti et al. model was based on Equation 4.1 or rather Equation 3.1 in Table 3.2. Mathematical details are given by the authors. The significance of this contribution is that it represents a remarkably detailed description of the reactor system. The method of lines was used for the solution and space was discretized using finite central differences. The resulting ordinary differential equations are stiff and were integrated using an appropriate algorithm. The model was applied to breakthrough behavior for step changes.

The simulation showed a sequence of concentration fronts moving through the polymer filled column. These are illustrated in Figure 4.2a shows the movement of acetic acid through the reactor. The rapid drop in the front of the reactor is due to esterification. This

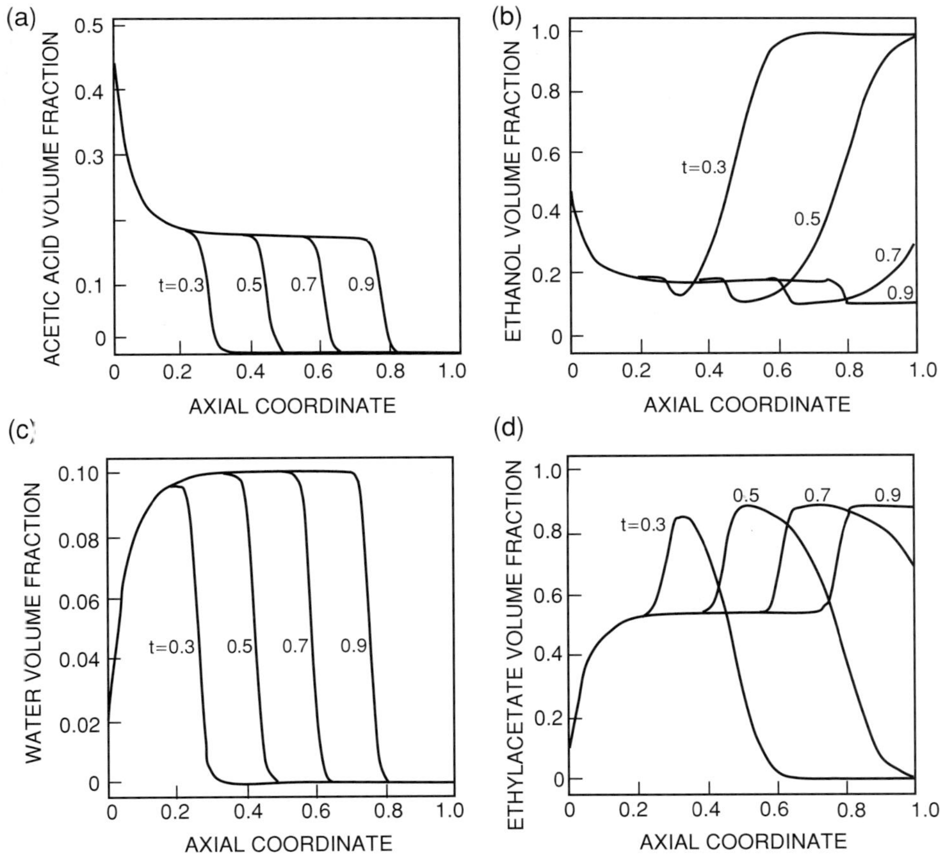

Figure 4.2 Time and position variations of reactant and product compositions for the esterification of ethanol by acetic acid over a cationic exchange resin catalyst: (a) acetic acid, (b) ethanol, (c) water and (d) ethyl acetate. (Figure adapted from Mazzotti et al. (1997a) with permission. © 1997 American Chemical Society.)

proceeds until equilibrium is reached. Continued feeding of the reactant forces acetic acid through the column as a shock front that propagates at a constant velocity and becomes steeper as it moves. Reaction occurs at a low rate during front movement because the composition in the column is changing axially.

Figure 4.2c plots the concentration profiles of water with time in the reactor. The initial rapid rise results from esterification that proceeds to equilibrium. Water concentration then drops rapidly to form a moving front due to sorption by the resin. Water is the most strongly adsorbed component. In Figure 4.2d, the concentration profiles of ethyl acetate appear. The initial sharp rise corresponds to a reaction zone. This is a stationary front. The concentration of ethyl acetate continues to build up to form a moving front due to further reaction as acetic acid moves into polymer filled with ethanol. Ethyl acetate is just weakly adsorbed and so enters the flowing fluid phase. The front formed is purely separative. No back reaction takes place because water is held back by the resin and moves more slowly than ethyl acetate. The propagation velocity of the ethyl acetate front is greater than that of other species due to desorption and the front expands rather than sharpens. With respect to ethanol, Figure 4.2b exhibits a steady state reaction zone as mentioned for the other reactants. There is a further small drop following this zone that propagates downstream and then the concentration front appears which moves through the reactor. The ethanol front moves faster than water. The authors explain the small drop as the result of the equilibrium composition moving through the reactor contacting a phase rich in ethyl acetate while at the same point the polymer phase contains mainly ethanol that originally saturated the column. This results in the adsorption of water and some acetic acid as well as desorption of ethyl acetate whose second peak was mentioned earlier. The condition leads to further esterification causing a decrease of ethanol. The authors point out that the situation leads to a number of different types of fronts in a chromatographic reactor: a stationary front in which reaction proceeds, moving reactive fronts where sharpening of the front takes place and moving separative fronts The behavior of these fronts depends upon the composition of the feed to the chromatographic column and also on whether or not the column is initially saturated with ethanol or acetic acid.

Esterification of glycerol with acetic acid proceeds through the monoester, monoacetine, the diester, diacetine, until the triester, triacetine is formed. Each step is equilibrium limited so that in a batch reactor very little triacetine is produced. Gelosa et al. (2003) examined a CR for this reaction system in which an ion-exchange resin serves as a combined catalyst and adsorbent. Batch experiments carried out by these researchers indicated that mass transfer from the liquid to the resin phase was significant. Consequently, for their simulation, Gelosa et al. used a heterogeneous model based on Equations 3.7 and 3.8 in Table 3.4. However, given a swellable resin and considering component concentration in the pores introduce such changes that the model is worth restating:

$$\varepsilon_b \frac{\partial C_i}{\partial t} + u\frac{\partial C_i}{\partial z} - D_{ax_i}\frac{\partial^2 C_i}{\partial z^2} = -k_{m_i} a_V (1-\varepsilon_b)(C_i - C_i^*) \tag{4.3}$$

In this relation, C_i^* is the concentration of species i in the pores of the resin. The model for the resin phase is

$$\varepsilon_p q \frac{\partial C_i^*}{\partial t} + (1-\varepsilon_p)\frac{\partial C_i^P}{\partial t} = k_{m_i} a_V q (C_i - C_i^*) + (1-\varepsilon_p)\rho_{\text{resin}} \sum_{k=1}^{K} \nu_{i_k} r_k \tag{4.4}$$

Now C_i^P is the concentration in the resin phase, ε_p is the pore or internal porosity of the resin and q is the swelling ratio. Gelosa assumes equilibrium exists between the resin phase and the liquid within the pore structure. This equilibrium is expressed through a multicomponent Langmuir isotherm,

$$C_i^P = \frac{(K_{\text{ads}})_i (C_\infty^P)_i C_i^*}{1 + \sum_{j=1}^{N} K_{\text{ads}_j} C_j^*} \tag{4.5}$$

where $(C_\infty^P)_i$ is the resin capacity for species i, $(K_{\text{ads}})_i$ is the adsorption equilibrium constant. The mass transfer resistance arises according to Gelosa through diffusion in the pore structure. The complexity this introduces is avoided by using a linear driving force model following the classical treatment (Glueckauf, 1955). Kim (1989) provides justification for this simplification.

Gelosa et al. (2003) have tested the model just discussed. We examine their results in Section 4.2.1.

Falk and Seidel-Morgenstern (1999, 2002) reexamined the performance of a simple chromatographic reactor with respect to a comparable fixed-bed reactor. Hydrolysis of methyl formate was used as the test reaction for both simulation and experiment; this is an interesting system because the carrier, water, is a reactant, thus it is unlike pulse chromatographic systems examined earlier.

The authors assumed the reactor to be a cascade of back-mixed cells so Equation 3.73 was employed, but the adsorption constant was based on the solid phase rather then the liquid phase. Falk and Seidel-Morgenstern allowed for a homogeneous reaction as well as for a surface catalyzed one. This results in

$$QC_{i,k-1} \pm \left(\varepsilon r_i^{\text{hom}} + (1-\varepsilon) r_i^{\text{het}}\right)\frac{V}{J} = QC_{i,k} + (\varepsilon + (1-\varepsilon)\alpha_i')\frac{V}{J}\frac{dC_{i,k}}{dt} \tag{4.6}$$

where α'_i is the slope of the adsorption isotherm at a concentration, C_i, or the equilibrium adsorption constant if the isotherm is linear. ε is the void fraction in the bed. The formulation assumes the adsorbent is also the catalyst. If separate catalysts and adsorbents are used, the $(1-\varepsilon)$ term on the LHS of Equation 4.6 must be replaced by ε_{cat}, while the term on the RHS by ε_{ad}. Of course, $1 - \varepsilon = \varepsilon_{\text{cat}} + \varepsilon_{\text{ad}}$.

Falk and Seidel-Morgenstern (1999) commented that when a carrier is also a reactant the improvement of conversion in a CR results mainly from dilution of the other reactant, in their case, methyl formate. They demonstrated this by comparing their CR model with a PFR model assuming that the same amount of feed was introduced to both reactors, even though the feed for the chromatographic reactor was injected as pulses. It was further assumed that both reactors had the same mean residence time as measured by their input volumetric flow rates. The properties of the solid as either a catalyst in the PFR or catalyst and adsorbent together in the CR were the same. Of course, reactor dimensions and operating temperatures were identical. In order to meet these assumptions the concentration of methyl formate in the feed to the PFR was related to the concentration of the feed to the CR by the ratio of the injection time to the cycle period. The cycle period in turn was set by the time between the beginning of the methyl formate elution in the CR and the end of elution. Shorter periods lead to pulse overlapping and a loss of the CR effectiveness. The authors demonstrate that for high injection durations relative to the cycle length, the PFR provides higher conversion than a CR. The greater the injection time, the more concentrated is the methyl formate in the reactor relative to the carrier, water, and this depresses methyl formate conversion. Indeed, it is only at short injection pulses that a chromatographic reactor outperforms a fixed-bed reactor.

A more recent paper (Falk and Seidel-Morgenstern, 2002) develops the dilution effect for methyl formate hydrolysis further. They define a dilution ratio, $DR = 1 - \Delta t_{\text{injection}}/\tau$, where $\Delta t_{\text{injection}}$ is the duration of the input pulse assuming a square wave injection. Then they examine the effect of DR on the production rate (formic acid), conversion of the second reactant (methyl formate) and the recovery of product at a specified purity, for example, formic acid at a mole fraction = 0.8. From simulations using their CSTR cascade model and experimentally determined kinetics and adsorption constants, Figure 4.3 was obtained. This figure demonstrates the effect of the dilution ratio, DR, on methyl formate conversion at four different liquid flow rates, that is, decreasing residence times. The CR and PFR are identical at a $DR = 0$. The "diluted" PFR assumes a continuous feed at the time average feed concentration for the pulsing used in a CR. It is evident that a chromatographic reactor displays higher conversion than a continuously operated PFR only at high dilution rates or short feed pulses relative to cycle time. Note that at short residence times (Fig. 4.3d) there is little difference between a CR and PFR except at zero injection time. The number of equilibrium stages for separation in the packed bed decrease with increasing flow rate. There are 1000 stages at the lowest flow rate. This drops to 200 at the highest rate.

The dilution ratio also affects the production rate of formic acid and product recovery where the latter is the moles of product (formic acid) above a specified purity obtained per

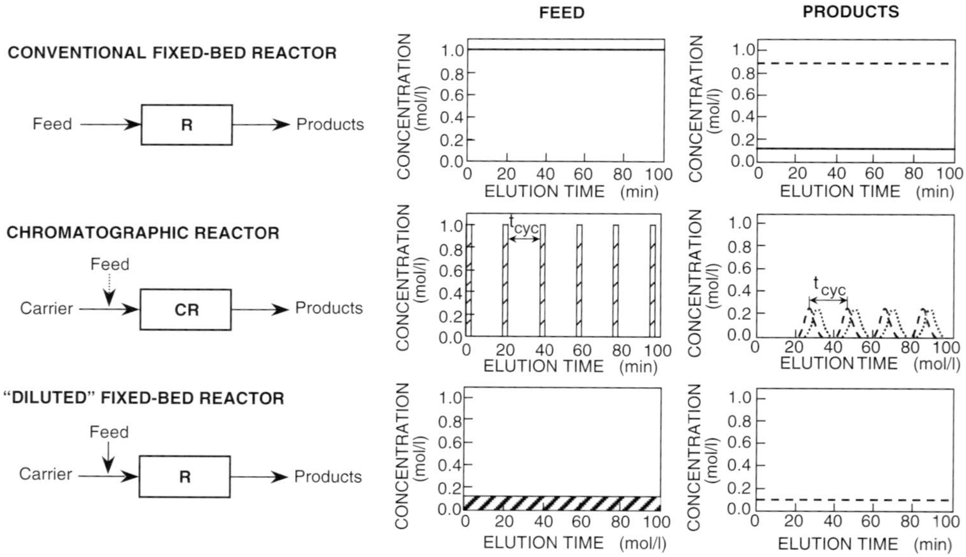

Figure 4.3 Effect of dilution rate on methyl formate conversion for plug flow (PFR) and chromatographically operated (CR) packed-bed reactors as a function of liquid flow rate. In the diluted PFR, the continuous feed concentration is at the time average feed concentration of the periodically pulsed chromatographic reactor. The solid line shows performance of the CR. (Figure adapted from Falk and Seidel-Morgenstern (2002) with permission. © 2002 Elsevier Science Ltd.)

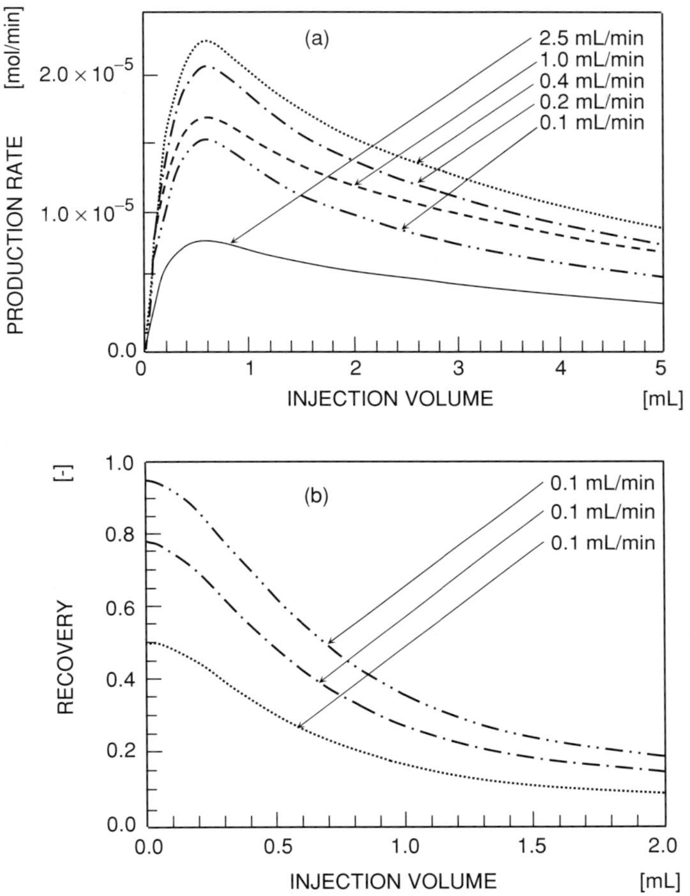

Figure 4.4 Effect of injection time ($\Delta t_{injection}$) or injection volume on production rate in a cycle (a), and on product recovery (b), in a periodically pulsed packed-bed reactor. The influence of liquid flow rate or reciprocal residence time is shown. Simulation results are for hydrolysis of aqueous methyl formate over Dowex 50W-X8 ion-exchange resin for a methyl formate feed concentration of 1 mol/L. In (b) formic acid mole fraction in the product is >0.99. (Figure adapted from Falk and Seidel-Morgenstern (2002) with permission. © 2002 by Elsevier Science Ltd.)

mole of feed. The injection volume of methyl formate in a pulse fed to a CR is roughly reciprocally related to the dilution ratio. Thus Figure 4.4a demonstrates that high dilution ratios or short pulses provide the highest production rates. However, as injection times become very short, production rates drop rapidly because of low reactant throughput. Similarly, high dilution rates result in the highest product recovery. Maximum recovery occurs as the injection times approaches zero. Liquid flow rates influence both production rate and product recovery. The number of stages increases with decreasing flow rate.

It must be recognized that the diluent ratio effects discussed by Falk and Seidel-Morgenstern apply strictly to systems in which the fluid carrier in the reaction is also a reactant. Nevertheless, dilution ratio adversely affects production rates at any residence time as indicated by the region around an injection volume of 0 in Figure 4.4a.

4.2 Experimental studies

There is a fairly large literature that we will examine by the reaction considered rather than in a historical sequence. It is convenient for this purpose to separate the literature by class of reaction.

4.2.1 *Catalyzed chemical reactions*

There are several early studies of these simple chromatographic reactors. We will discuss them just briefly because Langer and Patton (1974) reviewed these studies years ago. Bassett and Habgood (1960) investigated the irreversible isomerization of cyclopropane to propylene. From their measurements, they were able to deduce a rate constant for the reaction. Roginskii and coworkers (Gaziev et al., 1961; Roginskii et al., 1962) studied the dehydrogenation of cyclohexane experimentally. However, the first reasonably comprehensive investigation in English was published by Matsen et al. (1965). It also dealt with cyclohexane dehydrogenation. Matsen et al. employed a 0.6 wt% Pt/Al_2O_3 catalyst. Cyclohexane and benzene adsorbed on the alumina support so that the catalyst functioned as an adsorbent as well. Carrier gas was helium and the authors injected a pulse of cyclohexane at the top of the column. Volume of the pulse was a variable in their study along with the carrier gas flow rate and column temperature. The authors analyzed the eluted peaks to gauge column performance. Both single pulses and repeated pulsing were investigated. In the latter, the variable was pulse frequency. With a single pulse, they observed high conversion of cyclohexane if pulse size was kept small. Carrier gas flow rate was not an important variable. With repeated pulsing, an optimum frequency was identified. Above this frequency, hydrogen from the following pulse, catches up to the benzene peak moving through the CR and re-hydrogenates some of the benzene back to cyclohexane. Matsen et al. concluded that the simple chromatographic reactor provides a modest increase above equilibrium conversion for cyclohexane dehydrogenation. They observed, however, that a chromatographic reactor has the additional advantage that product and reactant are separated.

Murakami et al. (1968) followed up their analytical study (Hattori and Murakami, 1968) with an experimental investigation of cumene alkylation and disproportionation over a commercial silica–alumina cracking catalyst. Experimental variables for the pulsed chromatographic reactor were pulse width, temperature and weight of catalyst used. A careful injection system was developed by the investigators to ensure that a uniform square wave was fed to their reactor. At close to zero pulse width, dealkylation was irreversible and the authors were able to extract reaction rate constants from their data. Hattori and Murakami pointed out in their theoretical study that a chromatographic reactor may be useful for measuring reaction kinetics.

Murakami et al. also observed that equilibrium conversion could be exceeded in single pulse operation. Temperature can be important because at low temperature (300°C) rate and conversion are low so that dealkylation appears to be irreversible. Equilibrium conversion is not achieved. However, as temperature increases (350°C), pulsing results in conversion of about 80%, well above the equilibrium conversion of 60%, provided the pulse width is kept less than 90 s. Diisopropylbenzene was also observed as a reaction product in their pulsing experiments. Murakami et al. used this result to suggest that diisopropylbenzene arises from disproportionation of cumene.

Wetherold et al. (1974) investigated the liquid phase, HCl catalyzed hydrolysis of methyl formate in a simple chromatographic reactor using activated charcoal as the adsorbent. The reaction is known to be reversible and equilibrium limited even when carried out in excess water. In this case, hydrolysis kinetics is first order in methyl formate, while the reverse reaction is second order overall. Adsorption follows the Freundlich isotherm with a fractional concentration exponent. Experiments were performed in a 107-mm length of PVC pipe with carrier flow rate (water) and pulse size as experimental variables. Just single pulses were used. The objective was to demonstrate conversion in excess of the equilibrium limit and to verify the authors' model, discussed in the previous section. Their model assumed instantaneous adsorption, negligible axial dispersion and noncompetitive adsorption.

Wetherold et al. found that their model described the experimental effluent chromatogram closely if a quite narrow input pulse was used but that the description was noticeably poorer for longer, but more practical input pulses. To correct for pulse size, the authors introduced an empirical factor to allow for adsorption interference. This factor was empirically determined. When used for the medium pulse size, a good fit of the effluent chromatogram was achieved as shown in Figure 4.5. The authors attempted to assess the chromatographic reactor performance through an efficiency defined as the product of the input pulse duration and the conversion divided by the product of the elution duration and the equilibrium conversion. Elution duration fixes the pulse frequency because higher frequencies would result in an overlap of successive pulses. Essentially, Wetherold's efficiency compares production achieved in a chromatographic reactor with that in a fixed-bed reactor in which conversion equilibrium is attained. Wetherold et al. found that their experiments gave efficiencies ranging from 14 to 42%. Thus, experimentally, production in a CR failed to match the production achieved in an equal sized PFR. In Figure 4.5, the efficiency is 35%.

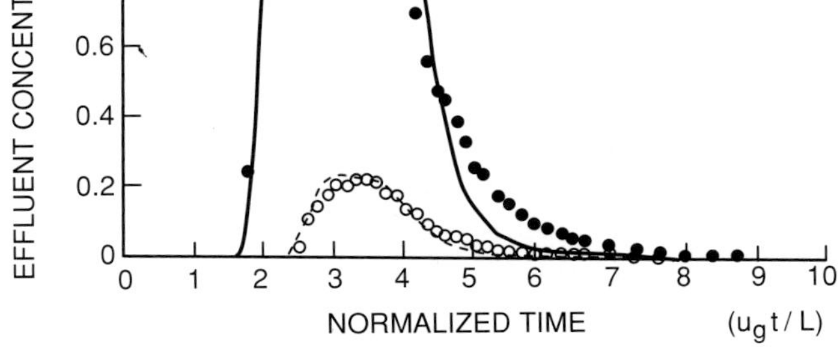

Figure 4.5 Comparison of predicted and experimental chromatograms for the HCl catalyzed hydrolysis of methyl formate with an activated carbon adsorbent. (Figure adapted from Wetherold et al. (1974) with permission. © 1974 American Chemical Society.)

Hydrolysis of methyl formate has also been used by Cho et al. (1980) in their study of a rotating annular bed chromatographic reactor. The Cho work is discussed in Chapter 6. This reaction system has since been employed by Falk and Seidel-Morgenstern (1999, 2002) who substituted an ion-exchange resin for the activated carbon employed by Wetherold et al. (1974).

The primary objective of Falk and Seidel-Morgenstern was to obtain adsorption and kinetic data for model validation and for their parametric study of chromatographic reactor performance through simulation, although at least one CR experiment was undertaken. These investigators carried out the hydrolysis on Dowex 50W-X8, a strongly acidic cation exchange resin employing both a batch reactor and a chromatographic column. The batch reactor was used to measure reaction rates while the chromatographic column, 4.6 mm i.d. and 250 mm in length, with a pulse input of methyl formate into the carrier liquid (water), yielded equilibrium constants for the various reaction species. Introducing a single pulse into the column gave the results shown in Figure 4.6 for a 20-μL pulse with methyl formate concentrations varying between 0.8 and 1.0 mol/L. It can be seen from Figure 4.6a that quantitative conversion of methyl formate is possible for a sufficiently low flow rate or a large residence time. Separation of the formic acid and methanol products is possible up to a flow rate of about 1 mL/min, but conversion is low and separation becomes poor when

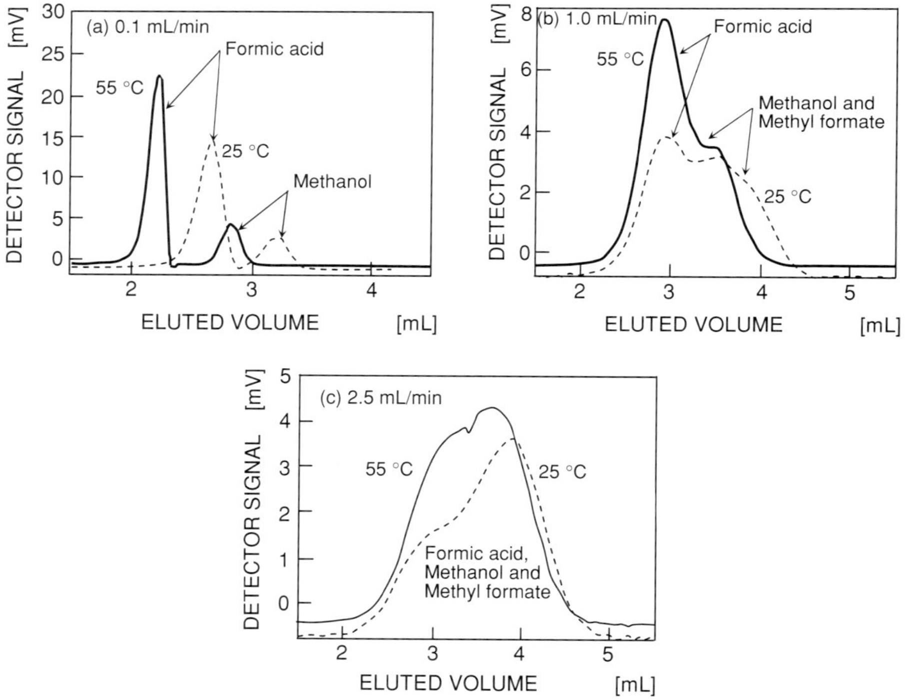

Figure 4.6 Elution patterns from single pulse experiments on a 250-mm-deep bed of Dowex 50W-X8 ion-exchange resin using a 20 μL sample at 0.8–1 mol/L of methyl formate in water. (Figure adapted from Falk and Seidel-Morgenstern (2002) with permission. © 2002 Elsevier Science Ltd.)

the residence time is further decreased (Fig. 4.6c). Temperature, that is, the hydrolysis rate is clearly important as well.

The elution patterns just shown were used by Falk and Seidel-Morgenstern (1999) to demonstrate that a linear adsorption isotherm applied in the range of their experiments. Adsorption equilibrium constants, α', were obtained. Batch experiments at different solution/resin ratios and different methyl formate concentrations were used to evaluate rate and chemical equilibrium constants for the homogeneous, autocatalytic and Dowex catalyzed reactions identified by other investigators (Falk and Seidel-Morgenstern, 2002). It was assumed that the forward and reverse reactions were first order in each reactant and that inhibition by reactant or product was absent.

The authors' CSTR cascade model, discussed in Section 4.1, was validated by single pulse experiments as demonstrated in Figure 4.7. Agreement of model and experiment is satisfactory. Only the total response could be measured at a flow rate of 2.47 mL/min in Figure 4.7a. Calculations indicated that the contributions of the homogeneous and autocatalyzed hydrolysis are small and can be neglected. As shown in Figure 4.6b, methyl formate conversion is complete at 0.1 mL/min and product concentrations become measurable. Experimental results are closely predicted by the CSTR model. Once again only the catalytic reaction is important. Periodically pulsed, experimental product exit concentrations are given in

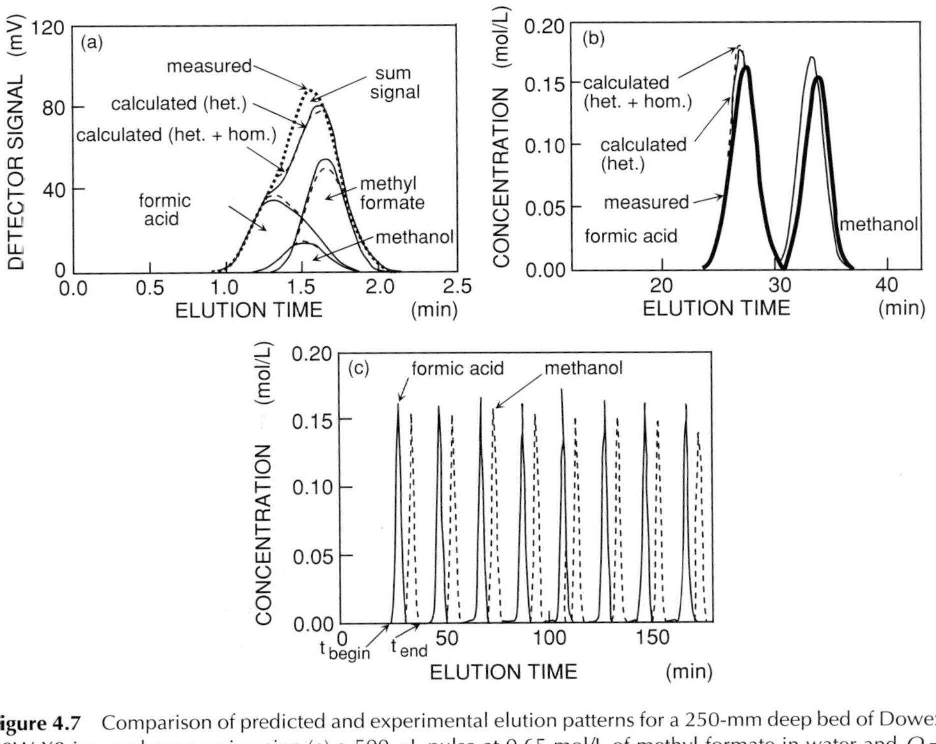

Figure 4.7 Comparison of predicted and experimental elution patterns for a 250-mm deep bed of Dowex 50W-X8 ion-exchange resin using (a) a 500 μL pulse at 0.65 mol/L of methyl formate in water and $Q =$ 2.47 mL/min, and (b) a 50-μL pulse at 0.94 mol/L of methyl formate in water and $Q = 0.1$ mL/min. In (c), experimental data for continuous pulsing under conditions used in (b). (Figure adapted from Falk and Seidel-Morgenstern (2002) with permission. © 2002 Elsevier Science Ltd.)

Figure 4.7c for operation at the lowest liquid flow rate. It can be seen that the simple chromatographic reactor provides complete conversion of methyl formate and good separation of the formic acid and methanol products.

Sardin and Villermaux (1979) undertook an experimental study of the liquid phase, reversible esterification of acetic acid by ethanol catalyzed by a cationic exchange resin in a layered bed of catalyst and an activated alumina adsorbent. Experiments were carried out at 70°C with a carrier mixture of dioxane and heptane where the ratio of these compounds affects separation of reactants and products. Conversion of acetic acid to the acetate and purity of the acetate were measured. Experimental variables were the volume of reactants injected, ratio of dioxane and heptane in the carrier mixture and water concentration in the feed. The authors demonstrated that acetic acid conversions well in excess of the equilibrium limit of about 67% could be obtained. Indeed, conversions of 90% were achieved. Purity of ethyl acetate, the reaction product, exceeded 90% at 95% recovery of ester. If just 90% is recovered, ester purity approaches 100%, however, the ester must be separated from the carrier solvents.

The experimental data obtained by Sardin and Villermaux (1979) were used in a modeling investigation considered in the previous section (Schweich et al., 1980). The model proposed by these authors was ideal in the sense that adsorption equilibrium was established instantaneously and axial dispersion was neglected. As discussed in the section on simulation, this can lead to discontinuities or shock fronts inside the chromatographic reactor. The experiments of Sardin and Villermaux did not show shock fronts but they did demonstrate rapid changes in concentration of ethyl acetate at two locations where discontinuities were anticipated. Neither reactant, ethanol or acetic acid, showed rapid changes. The authors attributed this to nonlinear isotherms for these components.

Esterification of acetic acid with ethanol in a CR was reexamined by Mazzotti et al. (1996, 1997a,c) employing downflow through a 38 cm i.d. bed of a cationic exchange resin based on polystyrene-divinylbenzene (Amberlyst 15). The ion-exchange resin, amounting to 130 cm^3, with $d_p = 6.6$ mm, $\varepsilon_b = 0.42$ and $\varepsilon_p = 0.36$, served as both catalyst and adsorbent for water. The Amberlyst resin has elastic properties so it expands or contracts, as the composition of the liquid phase changes, thus the bed length was not constant. Several types of experiments were performed. Reaction and adsorption experiments introduced an equimolar mixture of reactants to the bed and measured the effluent concentrations leaving as a function of time: a step change experiment rather than a pulse one. Regeneration experiments were performed by introducing the eluent, ethanol, into the bed after breakthrough of acetic acid occurred in the reaction and adsorption experiment. Regeneration saturated the bed with ethanol, causing it to expand, and then introducing a mixture of ethanol and acetic acid. Reaction begins immediately at the top of the bed. Resin traps the water product but ethyl acetate is expelled and forms a downward moving front. Breakthrough of ethanol occurs first because it is already present at the beginning of the reaction. However, its concentration in the effluent drops rapidly due to reaction. The second peak emerging is ethyl acetate and it is followed by acetic acid and eventually water. By the time acetic acid appears in the effluent, the reactor approaches fixed-bed behavior. The progress and shape of these moving fronts were well predicted by the authors' model (Mazzotti et al., 1997a) as discussed below.

The presence of moving fronts can result in flow instability caused by a buoyancy effect. If the chromatographic reactor operates in downflow with a water front following an ethanol front, the lower density of the leading ethanol front can lead to large-scale liquid mixing in the reactor. This was demonstrated by Mazzotti et al. (1997a) for a regeneration operation in which ethanol replaces the ethanol–acetic acid mixture stepwise in upflow through the reactor. In this case, the ethanol feed is less dense than the effluent containing water. The hydrodynamic instability caused mixing and the ethanol front emerged more rapidly from the reactor than predicted by theory. The experiment illustrates that density differences must be considered in the choice of flow direction in a CR. Stability is difficult to model and it is unlikely that it could be introduced into a useful simulation.

A CR model devised by Mazzotti et al. (1997a) and discussed in Section 4.1 was tested in the experiments. In the model, an equilibrium distribution of water between the resin, taken to be a polymer phase, and the solution is assumed. Activities for the solution are predicted by the UNIFAC group contribution method, while an extended Flory–Huggins model is used for the polymer phase. This model requires swelling data, which was acquired through experiments. Interaction parameters in the model were set equal to zero. CSTR measurements showed that the reaction rate was first order in each of the reactants. They also evaluated the rate constants. In the model, the reverse reaction was based on product–reactant activities. Since equilibrium between the solution and polymer phases was assumed, solution activities could be employed. Close prediction of the time varying concentrations verified both model and parameters.

Another challenging esterification reaction has been treated recently by the Politechnico di Milano–ETHZ team, the esterification of glycerol with acetic acid (Gelosa et al., 2003). This reaction proceeds through stepwise esterification until the triester, triacetine, is obtained. Formation of the triester is equilibrium limited and current technology produces this material via several reaction and separation steps. It appears to be an attractive candidate for reactive chromatography. Gelosa et al. used a 44 cm packed bed with a 1.5 cm i.d. Catalyst and adsorbent for the esterification was the same Amberlyst 15 ion-exchange resin used by Mazzotti et al. (1997a) in the studies discussed above. The bed was first loaded with acetic acid and then the feed stream of glycerol and acetic acid in a 1:4.5 mole ratio was introduced in upflow at a rate of 0.3 cm^3/min. Experiments were performed at 80°C and the acetic acid used to saturate the bed prior to start up contained a small amount of water.

Breakthrough behavior is shown in Figure 4.8. Curves are predictions from the simulations that used the model given in Equations 4.3–4.5. Model parameters were measured in batch experiments. The figure shows that glycerol is completely converted to ester. Triacetine, most weakly adsorbed, breaks through first. From 270 to 300 min, it can be obtained as a solution in acetic acid with just a small amount of contamination by the diacetine. After 370–380 min, the effluent is mainly the diacetine in acetic acid with some triacetine and monoacetine as minor components. Water breaks through after 380 min and by 420 min the effluent is approximately the equilibrium mixture for the residence time in the bed (about 4 mole% triacetine). The figure demonstrates the large improvement obtained and the good separation achieved among the esters. Of course, some of the conversion improvement comes from the dilution effect as mentioned above.

Schweich and Villermaux (1982b,c) investigated dehydrogenation of cyclohexane to benzene catalyzed by platinum on alumina. Their contribution is unusual because they were looking for a contribution of unsteady state operation to the rate of dehydrogenation. Change in rates of reaction due to unsteady state operation is well supported in the literature (Silveston, 1998). Their experimental reactor consisted of a 4.95 mm i.d. × 1-m tube filled

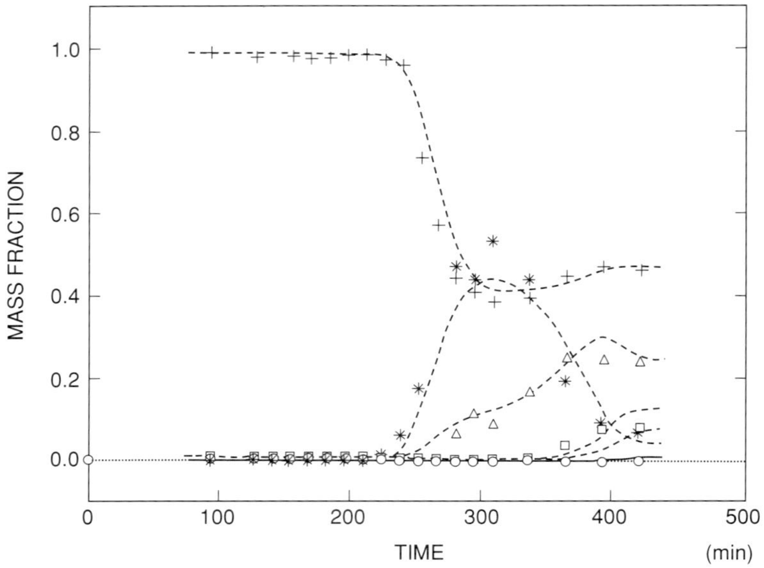

Figure 4.8 Breakthrough data for a step change experiment at 80°C using a packed bed of Amberlyst 15 ion-exchange resin prewetted with acetic acid. At $t = 0$, a 1:4.5 molar mixture of glycerol and acetic acid is introduced at the bottom of the bed and flows upward through the resin. Curves in the figure are predictions from a simulation. (Figure adapted from Gelosa et al. (2003) with permission. © 2003 American Chemical Society.)

with 0.1 wt% Pt/Al_2O_3. Catalyst was activated by H_2 at 350°C. Cyclohexane was periodically injected into a helium carrier stream so that the time average flow rate was 4 μL/min. Effluent from the reactor was trapped in liquid air, which meant that the time average yield of benzene was measured. Pulse shape was determined using continuous mass spectrometry.

The authors' observations are reproduced in Figure 4.9 for injection cycles of 1 and 2 min. Time average concentration is constant so the amplitude of the 2-min pulse is twice that of the 1-min injection cycle. For the latter, the cyclohexane peaks do not fall to zero so that interference of successive injections occurs. Both parts of Figure 4.9 show a lag between hydrogen and benzene indicating a chromatographic effect. The hydrogen and benzene chromatograms were used to calculate an equilibrium chromatogram for cyclohexane. This is compared in Figure 4.10 to the experimental chromatogram for both injection cycles. A large difference between predicted and experimental chromatograms in Figure 4.10a suggests that local equilibrium was not attained as the cyclohexane pulses moved through the reactor. However, in part (b) of the figure for a 1-min cycle, there is some agreement between the equilibrium assumption and measured composition in the valleys between the peaks.

Schweich and Villermaux (1982c) offer several explanations for this observation. The first is that the dynamic operation changes the state of the catalyst surface. Thus, the coverage of benzene may not correspond to the coverage predicted by a steady state model. They also suggest that the mechanism may be affected by dynamic operation. In the case of the chromatographic reactor, there are two competing reactions occurring at the same set of active sites: dehydrogenation of cyclohexane and hydrogenation of benzene. Thus, in a dynamic situation the balance between the competing reactions may be altered. This

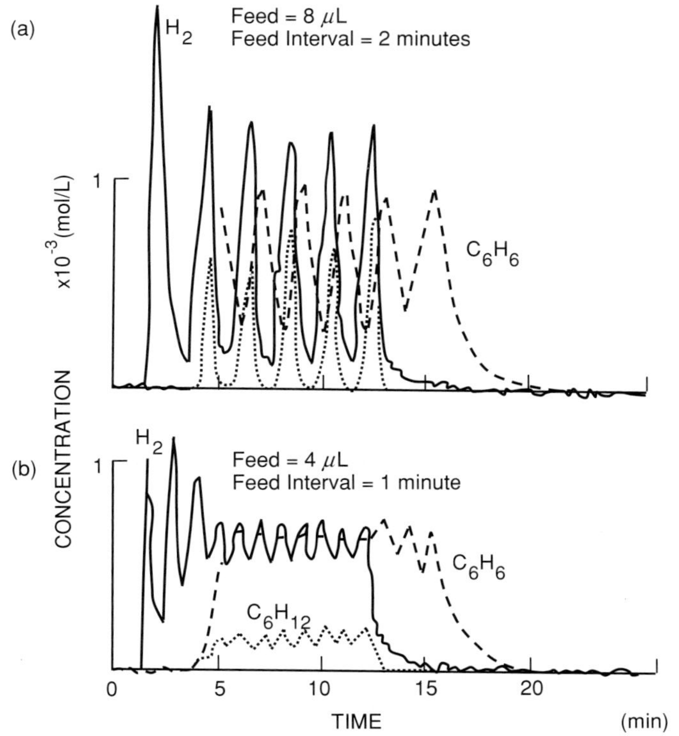

Figure 4.9 Benzene and hydrogen concentrations as a function of time in the effluent from a chromatographic reactor used for the catalytic dehydrogenation of cyclohexane: (a) pulse injection of cyclohexane every second minute, (b) pulse injection every minute. (Figure adapted from Schweich and Villermaux (1982c) with permission. © 1982 American Chemical Society.)

could give rise to a bifurcation situation. Also, the authors suggest that traces of oxygen in cyclohexane can affect rates during dynamic operation. Schweich and Villermaux rule out a temperature excursion explanation. The significance of their results is that it places doubt on the assumption of local reaction equilibrium made in several analytical studies of chromatographic reactors.

Dehydrogenation has also been investigated by Imai et al. (1985) and by Goto and his coworkers at Nagoya University (Goto et al., 1993, 1995, 1996). These investigators, unlike Matsen et al. (1965) and Schweich and Villermaux (1982b), added a hydrogen sorbent as a second solid phase to the supported Pt catalyst. Furthermore, their experimental studies employed consecutive pulses of reactant imposed on a constant flow of carrier gas. Because successive pulses overlap complete separation between reactant and products is not achieved. Sorption lowers the local hydrogen concentration and therefore shifts equilibrium in the forward direction. This type of periodic operation resembles what is undertaken in pressure swing reactors except that product partial pressure is periodically reduced by flushing with carrier gas rather than by reducing the total pressure. We consider this operation in Chapter 12.

Among the authors just mentioned, Imai et al. were interested in comparing the performance of various metallic alloys on cyclohexane conversion and correlating performance

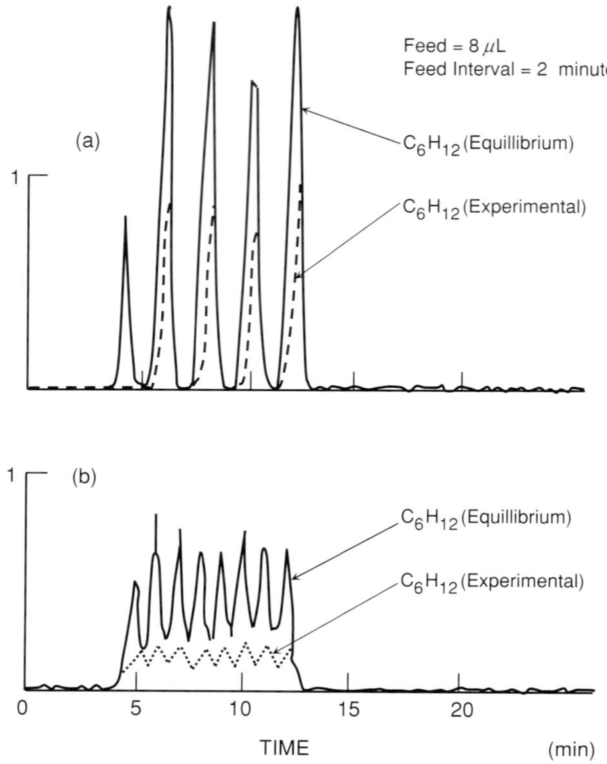

Figure 4.10 Comparison of predicted and measured cyclohexane concentration variations in the effluent from a chromatographic reactor: (a) pulse injection of cyclohexane every second minute, (b) pulse injection every minute. (Figure adapted from Schweich and Villermaux (1982c) with permission. © 1982 American Chemical Society.)

with the H_2 storage capacity and uptake rate of the alloys. Step change experiments were used and no attempt was made to observe how the alloys would perform in a CR. We will examine the Imai results in Chapter 14.

Goto et al. (1993) investigated cyclohexane dehydrogenation using a 5 wt% Pt on Al_2O_3 catalyst and the metallic alloy for hydrogen storage, $CaNi_5$ found to be the most effective alloy by Imai et al. (1985). They employed a packed column of catalyst and adsorbent that was continuously fed with helium and periodically pulsed with cyclohexane in helium. Pulsing was symmetrical so their system functioned as a periodically pulsed reactor (PRR) rather than as a CR. We examine their study in Chapter 12 along with the dehydrogenative aromatization experiments by the same team (Goto et al., 1995, 1996).

4.2.2 *Enzyme-catalyzed biochemical reactions*

In the early 1980s Barker and his coworkers in Birmingham (UK) extended their studies on preparative chromatography (Barker and Deeble, 1975; Barker and Ching, 1980; Barker et al.,

1983) to the simple chromatographic reactor. Their test reaction was the formation of a glucose polymer, dextran, from sucrose. The enzyme-catalyzed reaction can be represented as

$$n\ \text{sucrose} \xrightarrow{1,6\text{-}\alpha\text{-D-glucosyl transferase}} (\text{glucose})_n + n\ \text{fructose}$$

For their studies, Barker et al. (1987a,b) and Zafar and Barker (1988) employed jacketed columns packed with a calcium exchanged polystyrene ion-exchange resin. Column diameters ranged from 1 to 5.4 cm, while length was kept between 1.75 and 2.0 m. Experiments were conducted at 25°C. Startup was from a column saturated with the water carrier or eluent charged with the dextransucrase. Some of the enzyme adsorbed on the resin, but it was necessary, nonetheless, to continually feed the enzyme with the carrier at a concentration of 0.8 DSU/cm^3 (1 DSU/cm^3 is the amount of enzyme needed to convert 1 mg of sucrose in 1 h at 25°C and a pH of 5.2).

Commercial requirement is a high molecular weight dextran >150 000 Da. Fructose, however, interferes with the polymerization so that in a batch operation much low molecular product can form. In a CR, dextran is excluded from the resin by size, whereas fructose forms a complex with the calcium form of the resin. Separation of fructose allows glucose polymerization to proceed. At low sucrose concentration in the pulse (2 wt%), conversions to dextran greater than 150 000 Da both in a single pulse and in a batch reactor were over 90%, but when the concentration was raised to 20 wt% conversions declined to about 77% for the CR and to just 43% for the batch system. The investigators observed that increasing the pulse volume or the sucrose concentration in the pulse suppressed sucrose conversion to high molecular weight dextran. Decreasing the enzyme concentration in the eluent had a similar effect. Repetitive pulsing in place of a single pulse was also examined. As expected, pulse frequency becomes important. Low frequency pulsing gave identical results to those from the single pulse experiments. These effects were predicted by the Zafar–Barker model. Elutriated concentrations of products and the sucrose reactant were measured and good agreement between model predictions and measurements was found (Zafar and Barker, 1988). The model showed that carrying out the sucrose reaction in a once-through CR wasted enzyme.

Chapter 5

Countercurrent Moving-Bed Chromatographic Reactors (CMCR)

5.1 Introduction

The imagined problem of intermittent operation can be overcome by allowing the catalyst and adsorbent solids to flow through the reactor countercurrent to the fluid flow of reactants and products. When this is done, as discussed in Chapter 2, the feed to the reactor and withdrawal of products are continuous rather than periodic. The reactor operates at steady state and only ordinary differential equations describe reactor performance relative to fixed spatial coordinates. Catalyst and adsorbent, however, are not at steady state as they move through the reactor space and come into contact with different fluid environments. These solids experience cyclically changing environments.

Throughput through the reactor may be substantially increased by using a countercurrent moving bed. Product yield per weight of catalyst or adsorbent, however, may not change because in practice a much larger weight of solids will be needed when they circulate through the reactor. The moving-bed alternatives exchange the complexities and problems of periodic operation for those of moving solids. These problems are nonuniform flow and attrition of the solids. Attrition means solids replacement and the continuous removal of fines.

As discussed in Chapter 2, we consider continuous countercurrent moving-bed reactors in this monograph on periodic operation because of the periodic exposure of the solids as they cycle through the reactor. Indeed, this time-varying exposure is quite similar to the solids exposure when reactor inputs are manipulated periodically.

The countercurrent moving-bed chromatographic reactor (CMCR) is usually associated with the University of Minnesota research group directed initially by Aris and later by Carr. However, the concept was proposed at about the same time by a Japanese team (Takeuchi and Uraguchi, 1976a) who studied separations possible with moving beds and demonstrated throughput advantages. We will consider the Takeuchi and Uraguchi papers first and then return to the more detailed studies undertaken by Aris, Carr and their associates.

5.2 Modeling studies

Takeuchi and Uraguchi (1976a) investigated the design of CMCR as a simultaneous reactor–separator for an A $\rightarrow$ B reaction using an analysis based on the first absolute moment of a simplified form of Equation 3.1 in Table 3.2. They also allowed for diffusion in the solid phase. With an irreversible reaction, a CMCR cannot increase conversion, so the authors focused on separation. They assumed uniform gas velocity, isothermal operation, first order kinetics in terms of the adsorbed reactant and a linear adsorption isotherm. Takeuchi and Uraguchi showed that conditions for effective separation of reactant and product depends on a modified Thiele modulus, ϕ, the ratio of velocities of gas and solid phases, u_g/u_s, and, to a lesser extent, the ratio of reaction and adsorption constants, k_1/k_a. This is illustrated in Figure 5.1. The ratio of gas to solid velocity must lie between $(1-\varepsilon)(\varepsilon_p/\varepsilon_b)(1+K_A)$ and $(1-\varepsilon)(\varepsilon_p/\varepsilon_b)(1+K_B)$ for satisfactory separation of A and B. If the adsorption constant for product B is much greater than that for reactant A, separation requires low values of the modified Thiele modulus for the adsorption constants and overall porosity, ε_b, assumed. On the other hand, if reactant A is more strongly adsorbed, separation occurs in two zones

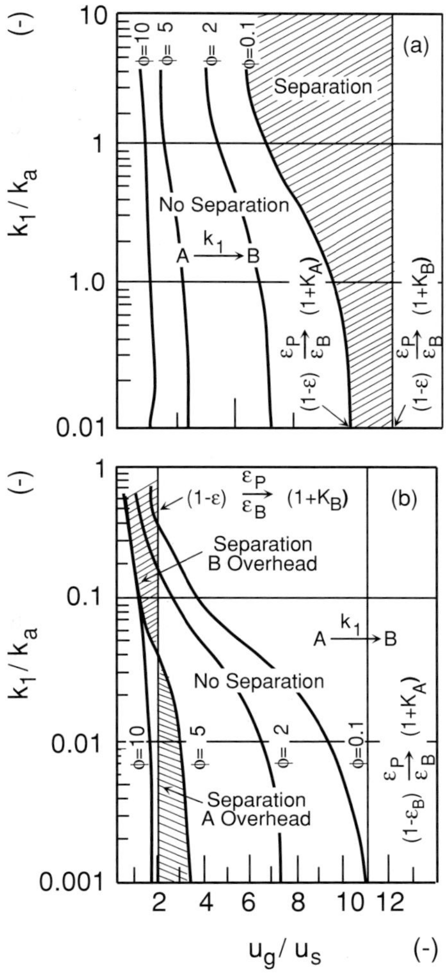

Figure 5.1 Regions of adequate reactant–product separation as a function of system parameters for an irreversible reaction A → B: (a) product B strongly adsorbed, (b) reactant A strongly adsorbed. (Figure adapted from Takeuchi and Uraguchi (1976a) with permission. © Society of Chemical Engineers of Japan.)

but now the modified Thiele modulus must be greater than 5 at low values of k_1/k_a. At higher ratios, separation requires a modified Thiele modulus below 2.

Design of a CMCR considered in a second paper by Takeuchi and Uraguchi (1976b) recognized discontinuities in reactant concentration at the top and in product concentration at the bottom of a downward moving bed of solids. A heterogeneous model was again used by Takeuchi and Uraguchi. They solved their model for steady state and from their solutions calculated conversion of reactant in the CMCR. Conversion is always less than in a fixed-bed reactor of the same length. Takeuchi and Uraguchi comment that lower conversion results from loss of reactant through adsorption on the solid leaving the bottom of the reactor. Thus, separation of product and reactant in the CMCR is at the cost of lower conversion.

The concept of a stripping section, which requires a side feed, was introduced by Takeuchi and Uraguchi (1977a). The advantage of a CMCR with a stripping section is that reactant adsorbed on the solid is not withdrawn from the reactor. Takeuchi and Uraguchi extended their analysis to the stripping section but neglected the mass transfer resistances. With this simplification, they were able to obtain analytical solutions for the stripping section. The researchers showed that neglecting mass transfer leads to a discontinuity at the feed location and one at the bottom outlet. With a stripping section, conversion in a CMCR can substantially exceed conversion in a PFR of equal size.

Takeuchi et al. (1978) extended their investigation to consecutive ($A \rightarrow B \rightarrow C$) and reversible ($A \leftrightarrow B$) reactions. They neglected all mass transfer resistances so that their model is pseudo-homogeneous. Because of discontinuities at both ends of the CMCR, selectivity in a consecutive reaction depends strongly on the relative velocity of the intermediate B in the column. Defining selectivity for the consecutive reaction as the amount of B produced for the amount of A consumed, they obtained a solution for selectivity as a function of conversion. Their results are shown in the two parts in Figure 5.2. In both, the selectivity for a PFR with the same catalyst amount is given for comparison purposes. Sign of the relative velocity of B is important: if it is negative, B is removed, adsorbed on the solid at the bottom of the reactor. Figure 5.2a shows that the selectivity of a CMCR in this case is always less than that of a PFR. Selectivity is a strong function of bed length. On the other hand, if u_B is positive, which means B moves upward in the column, CMCR selectivity can exceed that for a PFR. Conversion of reactant is limited by column height and the magnitude of u_B. When u_B is negative, conversion of A increases as u_B goes to zero. Without a stripping section, complete conversion cannot be reached for the range of L considered. For a CMCR with a stripping section, complete conversion of A is possible if $u_B > 0$. In this case, selectivity at high conversion can greatly exceed that attained in a fixed bed. This is shown in Figure 5.2b. The significance of zero velocity of the intermediate is that in this case it is completely consumed by the second reaction so that the selectivity to B is zero.

For a reversible reaction, Takeuchi et al. (1978) considered xylene isomerization as an example. They assumed that both forward and backward reactions are first order and adsorption follows a linear isotherm. This now linear model has an analytical solution that the authors present. Parameters based on experiments with xylene isomerization were used for their simulations. Takeuchi et al. demonstrate that conversion of xylene depends on bed length, fluid velocity and the adsorption capacity of the adsorbent. For example, the equilibrium conversion of *m*-xylene to *p*-xylene is 21% at 170°C. If bed length and fluid velocity are properly chosen, a PFR can reach this conversion. However, a CMCR with the

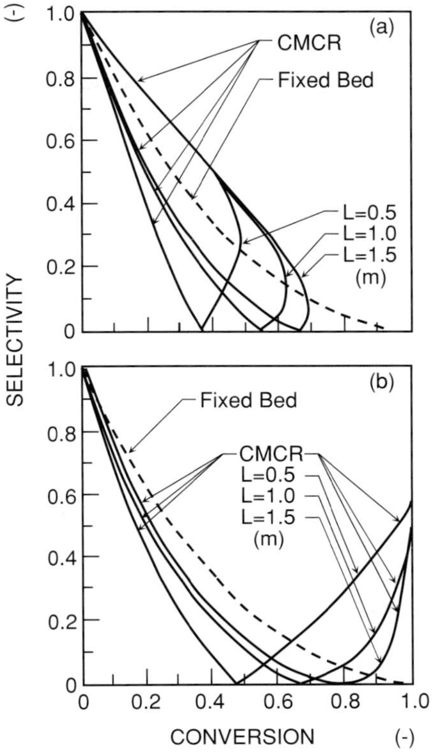

Figure 5.2 Comparison of selectivity to the intermediate B in the reaction A → B → C in a PFR and in a CMCR as a function of conversion and reactor length: (a) without a stripping section, (b) with a stripping section. Branches of the CMCR curves indicate positive and negative values of u_B. (Figure adapted from Takeuchi et al. (1978) with permission. © Society of Chemical Engineers of Japan.)

same bed length and carrier velocity provides a 76% conversion. Even for a shorter bed, conversion in a CMCR is 42% compared to a fixed-bed conversion of 12%.

The work of Takeuchi, Uraguchi and coworkers (Takauchi and Uraguchi, 1976a, b, 1977a; Takauchi et al., 1978) on the CMCR confirmed what was already known from studies on the simple chromatographic reactor, namely that irreversible reactions cannot be improved with respect to conversion by employing a CMCR. However, selectivity to an intermediate product in consecutive reactions and conversion for reversible reactions can be greatly improved. The Japanese workers also observed that discontinuities are present when improvements are found. It is not surprising then that the Minnesota researchers focused particularly on the existence of discontinuities or concentration jumps in the CMCR and their implications. Discontinuities are often referred to as shock fronts or simply as shocks. Of course, the existence of discontinuities in a countercurrent moving bed has been known for many years in the field of preparative chromatography. Their physical explanation is relatively simple and we use the presentation developed by Fish et al. (1986) below. These authors point out that if we assume adsorption equilibrium and the absence of axial dispersion, it is possible to write simple equations for the velocity at which a concentration discontinuity moves through a stationary bed of adsorbent. In the case of an A $\leftrightarrow$ B reaction, if both reactant and product adsorb on the same site, and if the Langmuir isotherm can be assumed, this velocity is

$$v_{s_A} = \frac{u_g}{1 + \left(\frac{1-\varepsilon}{\varepsilon}\right)\left(\frac{NK_A}{1+K_A C_A + K_B C_B}\right)} \tag{5.1}$$

If the adsorbent is moving downward through the column, the velocity of the front is affected and the above equation changes to

$$v_{s_A} = u_g \frac{\left(1 - \frac{u_s}{u_g}\left(\frac{1-\varepsilon}{\varepsilon}\right)\left(\frac{NK_A}{1+K_A C_A + K_B C_B}\right)\right)}{1 + \left(\frac{1-\varepsilon}{\varepsilon}\right)\left(\frac{NK_A}{1+K_A C_A + K_B C_B}\right)} \tag{5.2}$$

If we introduce a term, σ, to describe upward or downward velocity of the concentration front of a species, it takes the form for component A,

$$\sigma_A = \frac{u_s}{u_g}\left(\frac{1-\varepsilon}{\varepsilon}\right)\left(\frac{NK_A}{1 + K_A C_A + K_B C_B}\right) \tag{5.3}$$

The two relationships indicate that the velocity of the concentration front and the relative velocities of the species depend on concentrations. Assume that A is strongly adsorbed and B is just weakly adsorbed. In this case, the denominator of Equations 5.2 and 5.3 depends primarily on the concentration of A. If we now feed A at high concentration into the bottom of the CMCR, σ_A will be less than unity and A will move up the column. As this occurs, product B is formed. With unfavorable equilibrium B is always smaller than A and since K_B is smaller than K_A, σ_B will be smaller than σ_A. Consequently, the product B will move faster through the column than the reactant. Eventually, consumption of A by reaction increases σ_A. When σ_A reaches unity, the concentration front cannot move and a discontinuity appears. At the bottom of the column, some B will be present as soon as A enters the column but B cannot exit through the bottom because σ_B is less than 1. Thus, at the bottom, a discontinuity of B occurs. Of course, if the bed is not long enough, σ_A does not reach unity by the top of the column and there is no internal discontinuity.

Before continuing our analysis of CMCRs in terms of the capacity ratio of the phases, σ_i, we need to point out that another ratio, σ, the ratio of solid to fluid velocities or the Aris number, is also used to characterize the performance of countercurrent moving-bed systems. The latter ratio arises from the dimensionless form of the material balance. It is widely used for adsorptive separation or chromatographic systems. It is also used in the analysis of simulated countercurrent moving-bed chromatography and chromatographic reactors.

Viswanathan and Aris (1974) examined discontinuities or shock fronts within a moving-bed chromatographic reactor, addressing questions of whether or not they appear and their location in the reactor if they exist. They assumed an irreversible reaction, A $\rightarrow$ B, in which only A adsorbs. Adsorption–desorption equilibrium was instantaneous, the isotherm linear and axial dispersion negligible. The resulting PDE model is linear and the method of characteristics can be employed for integration. In this method, reactant concentration varies along a characteristic curve in the time–position plane. Viswanathan and Aris showed that a discontinuity could occur in a characteristic curve that physically is represented by an abrupt concentration change at a point within or at a boundary of a CMCR. This is the shock front. On startup, this front begins at a boundary. It usually moves through the reactor, but can become stabilized at an internal point. A critical parameter, σ_i, was identified by the authors. This is the ratio of the flow rate of a specific species, i, adsorbed on the downflowing solid to the flow rate of the same species in the upflowing gas phase. The parameter σ_i was introduced in Equation 5.3, but there it was defined for a Langmuir isotherm with competitive adsorption. This parameter must be constant at any point in the column at steady state. As discussed above, if σ_i for a reactant species i is less than 1 at all points within the CMCR, that is, the fluid velocity is fast enough compared to the velocity of solid in countercurrent operation, the discontinuity at the gas inlet moves through the reactor to the outlet so that a continuous profile within the reactor is established at steady state. For this to occur, the adsorbate concentration of i at the boundary of the solid outlet ($x = 0$ in Fig. 3.2a) must be greater than the gas phase concentration of i at that boundary. However, if the velocity of the solid is fast enough so that σ_i is greater than 1 at any point inside the column, a discontinuity will be observed in the bed at steady state. These situations are illustrated in Figure 5.3. $\sigma_i < 1$ in (a) and (b) of the figure, so the discontinuities are located at the reactor boundaries. Y and X are the dimensionless time and position variables, respectively. $X = 1$ is the top of the reactor where $z = L$ and $Y = \infty$ is steady state. In Figure 5.3c and d, σ_i exceeds 1 at a point within the reactor so an

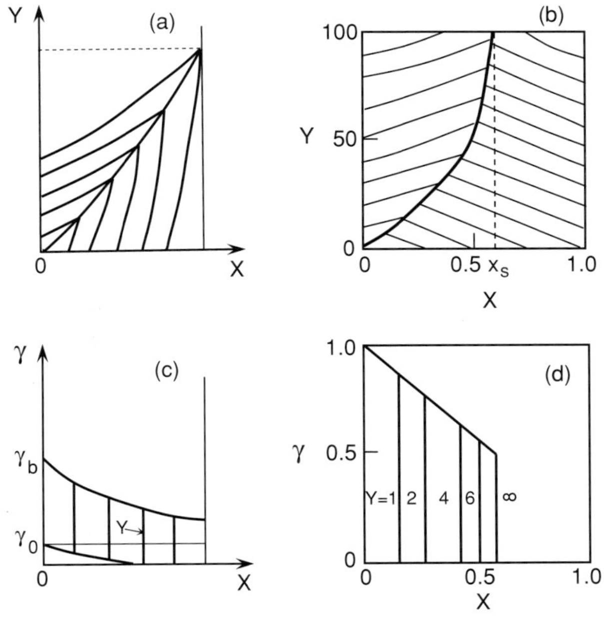

Figure 5.3 Phase plane representation (a, b) and dimensionless concentrations (c, d) versus dimensionless position for an ideal CMCR: (a, c) with $\sigma < 1$; discontinuities (shock waves) only at ends of the bed; (b, d) for $\sigma > 1$, path of a discontinuity in the phase plane is shown. It also appears in a dimensionless concentration versus position plot. (Figure adapted from Viswanathan and Aris (1974) with permission. © 1974 American Chemical Society.)

internal discontinuity forms. When adsorption–desorption rates and axial dispersion are important, concentration discontinuities are replaced by a more moderate concentration change. Viswanathan and Aris also examined situations where solids containing reactant i enter the chromatographic reactor. In these cases, the magnitude of the concentration discontinuity is affected.

The mathematical analysis was developed further by Cho et al. (1982) for the reversible reaction, A $\leftrightarrow$ B. They considered four cases and derived four sets of equations to describe the CMCR: (1) transient where adsorption equilibrium could not be assumed, (2) transient where adsorption equilibrium is achieved, (3) steady state where adsorption equilibrium was not reached and (4) steady state where adsorption equilibrium was attained. For the models describing the transient cases, they employed the method of characteristics and demonstrated that because concentrations must always be zero or positive the partial differential equations must be hyperbolic. Discontinuities are possible with hyperbolic PDEs. Cho et al. favored a hodographic transform and found this was convenient for describing movement and position of shock fronts. In the hodograph formulation characteristic curves are plotted in the concentration plane rather than in the time and position plane. The objective of the Cho study was to generalize shock front behavior and relate this behavior to the performance of a CMCR. They identified two critical parameters that govern both transient and steady state behavior: the capacity ratio, σ_i, defined earlier and κ, a relative adsorptivity, which is defined as the ratio of the adsorption constants for the reactant and product in the reversible reaction A $\leftrightarrow$ B.

Results obtained by Cho et al. (1982) are easiest to understand in terms of an example: the reaction A $\leftrightarrow$ B where product B is not favored. Catalyst and adsorbent enter at the top of the CMCR containing neither A nor B. A enters at the bottom with a dimensionless concentration of 7.5. The relative adsorptivity, $\kappa = 5$, the capacity ratio, $\sigma_A = 1.5$ and the ratio of the rates of adsorption = 1. The normalized capacity for A is also 1 and the reaction equilibrium constant is 0.5. Trajectories in the plane of γ_A and γ_B, the dimensionless concentrations of A and B in the gas phase, appear in Figure 5.4a. Curves marked as λ represent different ratios of the rates of adsorption and of reaction. $\lambda = 0$ means there is no adsorption on the catalyst surface so that reaction does not occur. $\lambda = \infty$ represents adsorption equilibrium. The figure shows a large change in concentration as the reaction is shifted toward B because of the chromatographic effect. The internal discontinuity exists because the reactor is long enough so that the change in concentration crosses the curve F_+. There is also an F_- curve close to the origin. The discontinuity extends from the characteristic, S_1, at $\lambda = \infty$ across the equilibrium line E to S_2 on the same characteristic. Discontinuities exist for both reactant A and product B. Figure 5.4c interprets the characteristic curves in (b) and shows the variation of A as a dimensionless concentration γ_A versus position in the CMCR. The discontinuity at $z = 5.3$ is evident. There is also a discontinuity at the top indicated by the F_- construct in the bottom corner of Figure 5.4a. This can be seen in the figure from the sharp drop in γ_3 at $\gamma_A = 0$. Of course, it does not appear in Figure 5.4c. The two discontinuities are illustrated in Figure 5.4d for the case of equilibrium adsorption. An internal discontinuity for B and a top discontinuity are shown in the figure. Figure 5.4d in this example demonstrates the importance of the internal discontinuity for the performance of the CMCR. If the height of the reactor is less than a critical value, there is no internal discontinuity but only one at the top. Reactant A cannot be taken to near zero concentration and thus the change in product B will be smaller. As a result, the ratio of product to reactant at the top of the column will

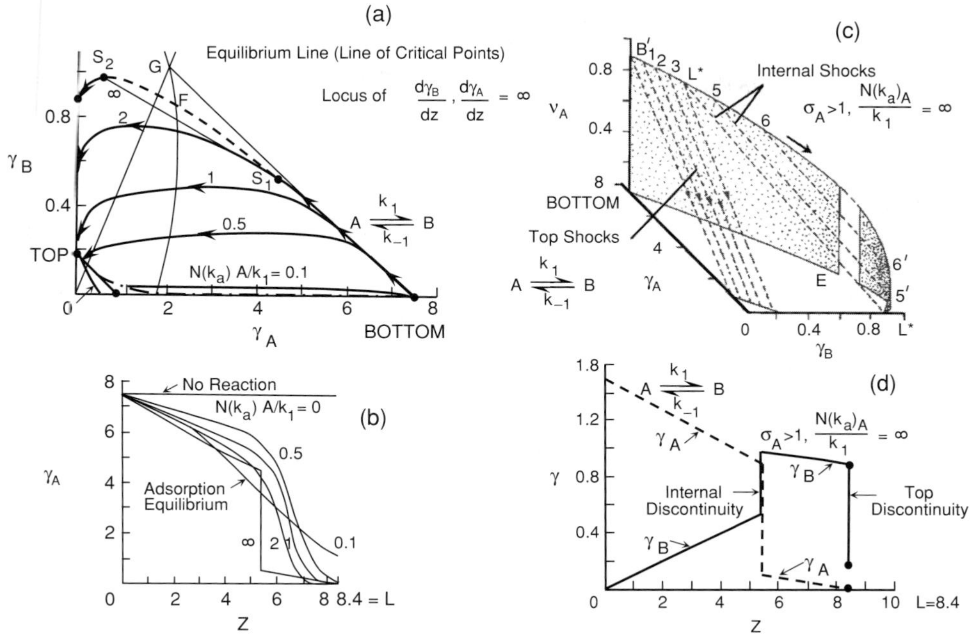

Figure 5.4 Trajectories in the concentration phase plane (a), and dimensionless concentration profiles (b) versus position in the reactor for the example. Note that B and T are bottom and top of the bed and S_1 and S_2 are discontinuity termini. (c) A three-dimensional representation for adsorbate A shown in (a) and (b). (d) A concentration versus position plot for the ideal CR with internal and boundary discontinuities. (Figure adapted from Cho et al. (1982) with permission. © 1974 American Chemical Society.)

be considerably less than the ratio if an internal discontinuity occurs; indeed that ratio for the example shown approaches infinity. In a three-dimensional plot, such as Figure 5.4c where the dimensionless adsorbate concentration of A is the vertical axis, both internal and top discontinuities can be shown. Lines on the diagram connect concentrations on either side of the discontinuity. Indeed, they show the transition at L^*, the critical height, from an internal to a top discontinuity.

Cho et al. (1982) concluded that an internal discontinuity representing a transition from reaction to adsorption control is essential in a CMCR for high product purity. They observed that the optimum reactor length is the minimum length needed for the existence of an internal discontinuity. Optimum in this case refers to a comparison with a fixed-bed reactor with respect to overall conversion and product purity. They demonstrate that under certain operating conditions as in the example described, pure product B can be obtained from the reactant A without the use of a subsequent separation operation. The significance of the Cho paper is that the authors provide methods for determining the presence of an internal discontinuity and mapping the concentration profiles in a CMCR. As mentioned and shown in Figure 5.4c, the authors also treated cases where adsorption rates must be considered and where axial dispersion occurs. These nonidealities modify the concentration profiles and discontinuities disappear to be replaced by rapid changes in concentrations. Axial dispersion enhances conversion in a CMCR but decreases product purity. In a similar way, finite adsorption rates can enhance conversion when rates are small relative to the reaction rate. Again, product purity decreases. It is possible to overcome negative effects of dispersion and a finite rate of adsorption by adjusting the critical parameters, σ and κ. The authors observed that σ has a critical value of about 1. Below this value the CMCR behaves like a PFR.

Petroulas et al. (1985a) considered the same system as Cho et al. (1982), a reversible reaction, A $\leftrightarrow$ B, with adsorption governed by the Langmuir isotherm and both reactant and product adsorbed at the same site. However, they employed stability analysis to investigate the existence of discontinuities and conditions under which these arise. For this, the model is linearized in the neighborhood of a critical point (chemical equilibrium), and the Jacobian matrix determined. In the usual way, eigenvalues for the characteristic equations are evaluated. These indicate the stability at the critical point. The authors pointed out that adsorption equilibrium cannot be maintained at an internal discontinuity so that only the starting and end points can lie on a Langmuir isotherm in the phase plane. In addition, the discontinuity at the bottom represents a transition ending on a Langmuir isotherm while a discontinuity at the top of the column must begin at an isotherm. The authors identified classes of discontinuity and from stability analysis determined which were feasible and which infeasible. They showed how to establish the relative position of the discontinuity from their analysis just as in the Viswanathan and Cho et al. papers. The method presented by Petroulas et al. is difficult to discuss succinctly in a review. The original paper should be consulted. Numerical examples are given. The first of these is similar to the one discussed above. Figure 5.5a shows locations of discontinuities for their example. Petroulas et al. observed that the internal discontinuity divides the reactor into two sections. The lower section is governed by equilibrium behavior and could be represented by a plug flow reactor. While in the region above the internal discontinuity, the chromatographic property of the system dominates and composition diverges from chemical equilibrium.

Existence of a critical length for optimal performance, as just discussed above, is illustrated in Figure 5.5b for this example. It can be seen that this optimum length is about 210 feet. Even at this length, conversion of A is less than that achieved in a PFR containing the same quantity of catalyst. The advantage offered by the CMCR is that product–reactant separation is achieved. Presumably, separation of product and reactant from the carrier gas is facile.

The question of performance criteria was raised by Cho et al. (1982) who point out that the conventional definition of conversion as reactant removed from the feed divided by reactant entering with the feed includes the contribution of reactants and products adsorbed on the catalyst leaving the bottom of the reactor. Petroulas et al. (1985a) suggest redefining conversion as product in the gas phase leaving the top of the bed divided by reactant entering at the bottom of the bed. These authors did not consider the problem of a measure for product purity. We suggest that this measure could be the separation factor used with pressure swing adsorbers. For a chromatographic reactor, this factor would be the ratio of reactant and product leaving at the top of the reactor divided by the ratio as adsorbates for the catalyst leaving at the bottom of the reactor. If pure A is obtained at the top of the reactor, or if only B appears at the bottom of the reactor, this separation factor would be infinite. On the other hand, if the compositions are reversed so that no reactant leaves at the top of the reactor, the separation factor would be zero.

The problem of adsorption on the catalyst and loss of product and reactant can be solved by adding a stripping section as mentioned earlier. This case was considered in an example

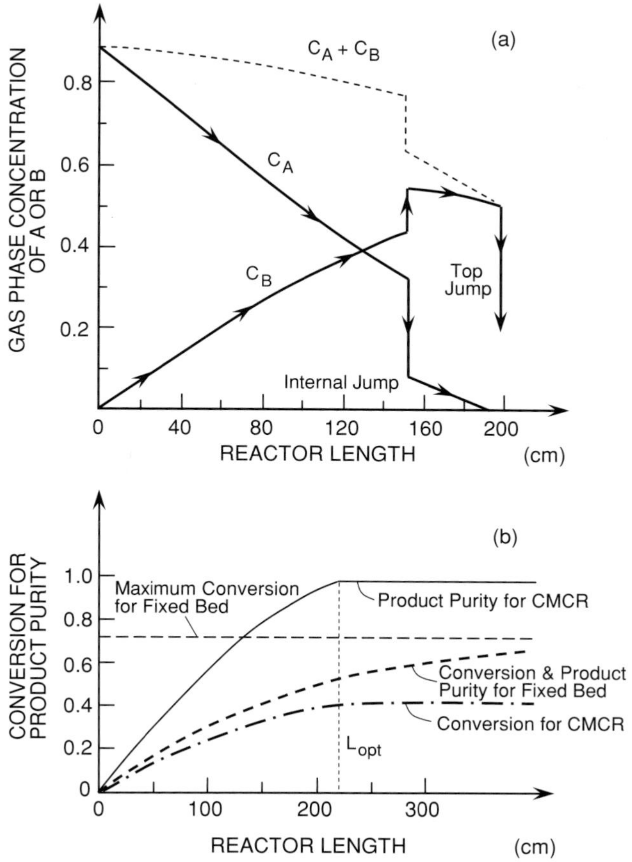

Figure 5.5 Variation of reactant and product concentration for A ↔ B in an ideal CMCR with a nonlinear adsorption isotherm showing internal and boundary discontinuities. (a) Comparison of conversion and product purity for a CMCR and a PFR as a function of reactor length (b). (Figure adapted from Petroulas et al. (1985a) with permission. © 1985 Pergamon Press Ltd.)

given by the authors who assumed that the flow rate of feed to the middle of the CMCR is small compared to the flow rate of carrier gas entering at the bottom. Indeed, pure reactant, A, is fed in the middle, while only a carrier gas is fed at the bottom. Catalyst enters at the top of the CMCR without any A or B. The authors do not give model parameters for their example but they appear to be the same as those considered by Cho et al. (1982). Figure 5.6 shows a schematic of the modified CMCR with midpoint feed and also the gas phase concentration profiles. There are four discontinuities for the column length chosen, a bottom jump to satisfy the inlet boundary conditions, a discontinuity at the reactant feed point, and an internal discontinuity toward the top of the column where the system switches from reaction equilibrium to chromatographic control. There is also a discontinuity jump at the top of the column. In Figure 5.7 these discontinuities are shown in a hodographic phase plane in terms of dimensionless concentrations. With middle feed and the internal discontinuity, the performance of the CMCR, in terms of conversion, is superior to that of

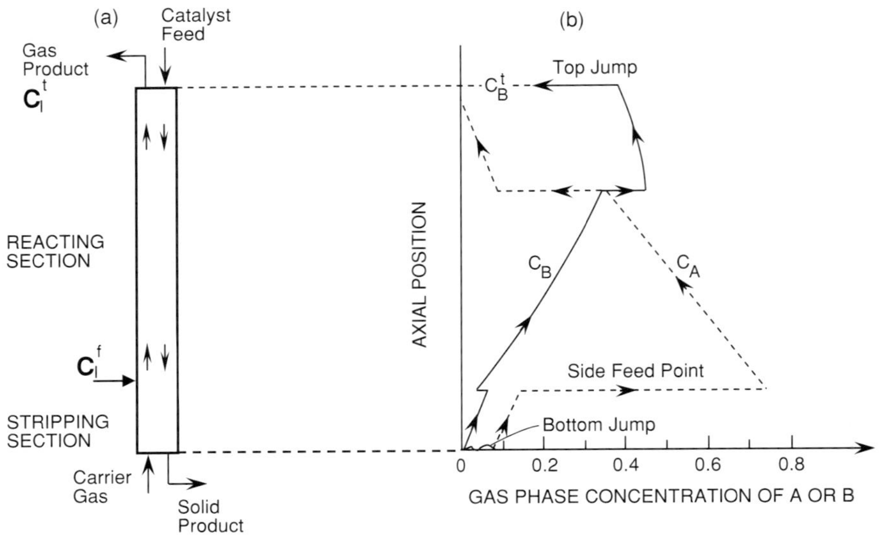

Figure 5.6 Schematic of a CMCR with a stripping section (a), and product–reactant trajectories in this CMCR with boundary and internal discontinuities (b). (Figure adapted from Petroulas et al. (1985a) with permission. © 1985 Pergamon Press Ltd.)

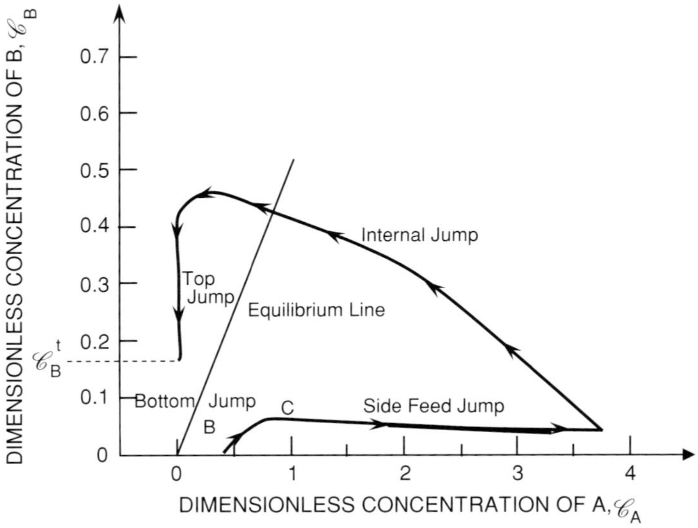

Figure 5.7 Representation of feed point and internal and boundary discontinuities in the phase plane for a CMCR with a stripping section. (Figure adapted from Petroulas et al. (1985a) with permission. © 1985 Pergamon Press Ltd.)

a comparable fixed-bed reactor. As the figure shows, loss of reactant is very small. The flow of stripping gas causes B to emerge from the top of the column. In this situation, with just a single exit port, the separation factor is useless for describing the separation performance. A better measure is the ratio of product to reactant mole fractions at the top of the reactor. The authors also point out that if a reactor shorter than optimal is employed, the internal discontinuity disappears and the CMCR conversion approaches that of a PFR.

Fish and Carr (1989) obtained detailed solutions of the CMCR model for a reactor with side feed. The configuration they considered is shown in Figure 3.2b. A reversible reaction A $\leftrightarrow$ B and Langmuir isotherms were assumed. They convert the side feed system to a bottom feed one by considering only the top portion of the CMCR. Then their analysis is essentially the same as that used by Cho et al. (1982). Their simulation results were presented in terms of the hodographic or concentration phase plane in Figure 5.8. In this plane, chemical equilibrium for the isothermal case is represented by a straight line with slope $1/K_{eq}$. Curve W satisfies an overall material balance. Fish and Carr introduce curves for the maximum downward flux of A and zero flux of A. They examine several cases using the phase plane

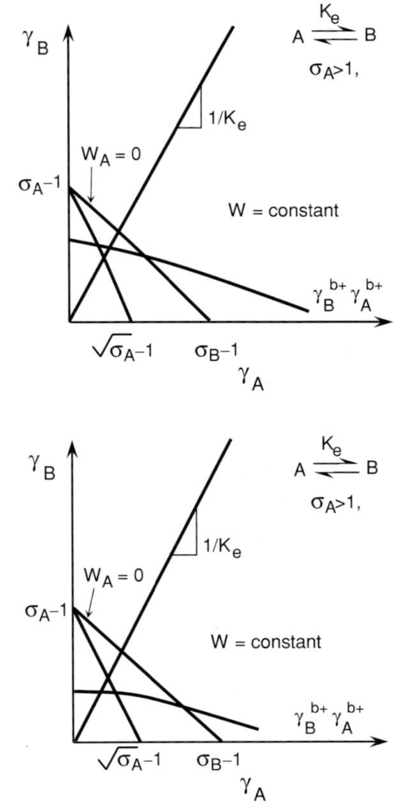

Figure 5.8 Composition phase plane for the top portion of a CMCR showing the equilibrium line, the lines of constant mass flux, zero flux of B and the maximum flux of B. The parts show the effect of increasing the feed rate of A on the position of the constant mass flux line. (Figure adapted from Fish and Carr (1989) with permission. © 1989 Pergamon Press plc.)

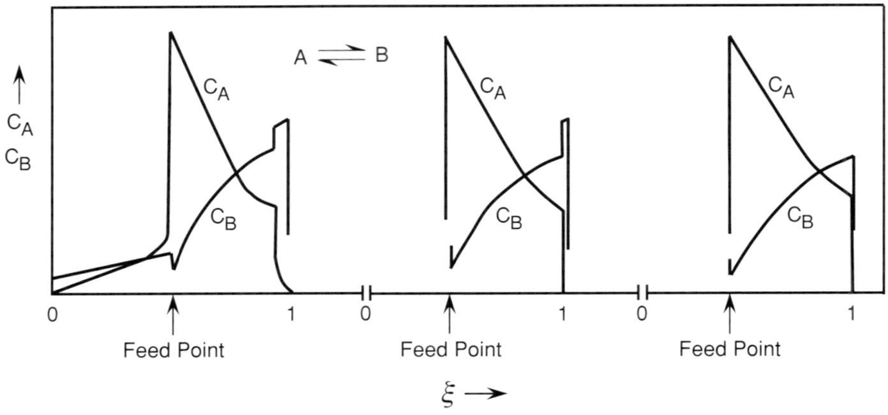

Figure 5.9 Effect on performance of increasing the carrier gas feed rate. The portion of the bed below the feed point is shown in (a). It does not change significantly with the carrier gas feed rate and has been left out of (b) and (c). (Figure adapted from Fish and Carr (1989) with permission. © 1989 Pergamon Press plc.)

representation with the curves just mentioned. When the curve, $W =$ constant, crosses $W_B = 0$ before it crosses the equilibrium line, no reactant leaves at the top of the reactor so that only the product appears at that outlet. This case is shown in both parts of Figure 5.8. The difference between the parts is the feed rate of reactant A. As the feed rate increases, the line representing constant W will eventually intersect $W_B = 0$ after the equilibrium line. Then, the internal discontinuity disappears. Figure 5.8 does not show this case. The crossover situation is associated with the critical height of the top section of the reactor. If the height of this section is greater than the critical height, an internal discontinuity occurs in the column. Fish and Carr point out that the steady state concentration shock, i.e. the discontinuity, moves up toward the top of the reactor as the reactant feed rate increases. Reactor height and feed rate are interchangeable as long as an ideal model is assumed.

The effect of carrier gas feed rate on conversion and the location of the discontinuities is shown in Figure 5.9. At the highest feed rate in Figure 5.9 (right side of the figure), the internal discontinuity has disappeared. Complete conversion of A no longer occurs.

Altshuller (1983) extended the analysis of Carr and Aris and coworkers to an arbitrary adsorption isotherm. He proved two theorems for the CMCR: (1) there exists a parameter value beyond which the behavior of a CMCR is given by that of a fixed bed, (2) the path a discontinuity follows through the CMCR is related to the adsorption isotherm. Altshuller illustrated his results using the Fowler–Guggenheim isotherm. This isotherm introduces a second parameter into the Langmuir isotherm, which results in an inflection point in the isotherm before saturation is reached.

5.3 Experimental studies

The earliest experimental study of a moving-bed chromatographic reactor (Takeuchi and Uraguchi, 1977b) dealt with the catalytic oxidation of carbon monoxide on activated alumina

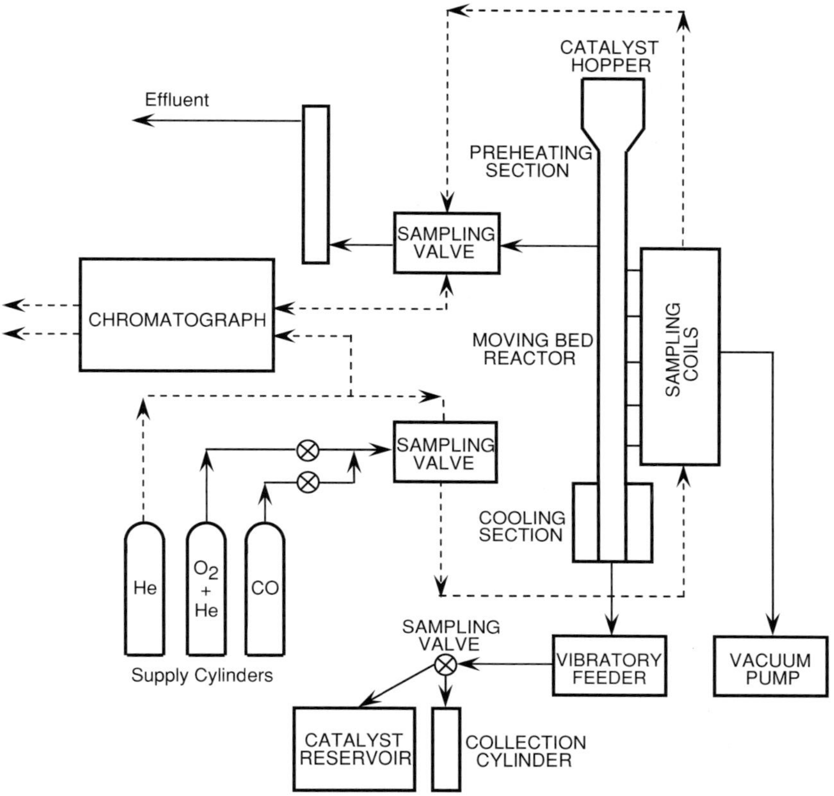

Figure 5.10 Experimental moving-bed chromatographic reactor system. (Figure adapted from Takeuchi and Uraguchi (1977b) with permission. © Society of Chemical Engineers of Japan.)

at a temperature of 614 K. The reactor consisted of a steel tube, 14 mm i.d. by 550 mm, built into the system shown schematically in Figure 5.10. Forty to sixty mesh catalyst/adsorbent particles were used in the moving bed, but were not recirculated. Startup was from a reactor initially filled with an oxygen–helium mixture. The feed entering at the bottom of the reactor contained 5.5 vol.% CO and 19.0 vol.% O_2 in a He carrier gas. Composition at different points in the bed were sampled simultaneously and analyzed by gas chromatography. Typical results are shown in Figure 5.11 as CO and CO_2 concentrations normalized by the feed concentration of CO. The figure also shows model predictions for a CMCR and for a PFR operating under identical feed conditions. For the latter, there is no adsorption so the dimensionless CO_2 concentration is the gas phase concentration. Experimental points fall along the CMCR model. Differences between the models and the divergence of the data points from the PFR model demonstrate that both reaction and separation occurred in the CMCR. It is clear from the figure that CO_2 is being removed from the reactor by adsorption on alumina so that it leaves at the bottom rather than at the top of the moving bed.

This is further illustrated in Figure 5.12a, which shows the CO_2/CO ratio at the top of the reactor as a function of solids velocity in the moving bed. The parameter in the figure is the

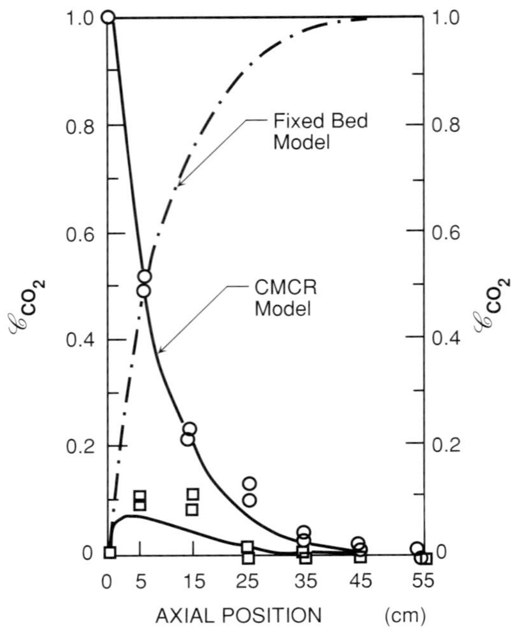

Figure 5.11 Comparison of a PFR model and Takeuchi and Uraguchi's CMCR model with experimental data from the system shown in Figure 5.10 for CO oxidation over activated alumina. (Figure adapted from Takeuchi and Uraguchi (1977b) with permission. © Society of Chemical Engineers of Japan.)

adsorption capacity of the alumina, n, divided by the concentration of CO in the feed. Data points fall roughly on the CMCR model if allowance is made for a nonlinear adsorption isotherm. Without this allowance the agreement between model and experiment is rather poor. Part (b) of the figure plots the normalized concentration of the adsorbate, CO_2, on the solid once again as a function of the solids velocity. Although the figures demonstrate that the CO_2 product is removed at the bottom of the moving-bed reactor as adsorbate on the alumina catalyst, it also shows that the Takeuchi–Uraguchi model is not satisfactory even when a nonlinear isotherm is used.

The objective of the authors was to experimentally test the model they proposed. Their model assumed an isothermal CMCR with a first order irreversible reaction, a linear adsorption isotherm and negligible axial dispersion. Figure 5.12 suggests that this model is oversimplified, perhaps by the use of a linear adsorption isotherm. Of course, no improvement in conversion was observed over a comparable PFR, but the authors did demonstrate a high degree of separation between reactant and product: CO leaves the top of the reactor, while most of the CO_2 formed emerges at the bottom.

A decade later, Fish et al. (1986) undertook CMCR experiments using hydrogenation of mesitylene (MES) to 1,3,5-trimethylcyclohexane (TMC) at 473 K over 0.74 wt% Pt/Al_2O_3. The moving solid phase was a mixture of 15 wt% catalyst and 85 wt% pure alumina. Feed gas, MES in nitrogen, entered at the bottom of the reactor. Objectives of the experiments were to demonstrate that equilibrium conversion can be exceeded, to obtain steady-state

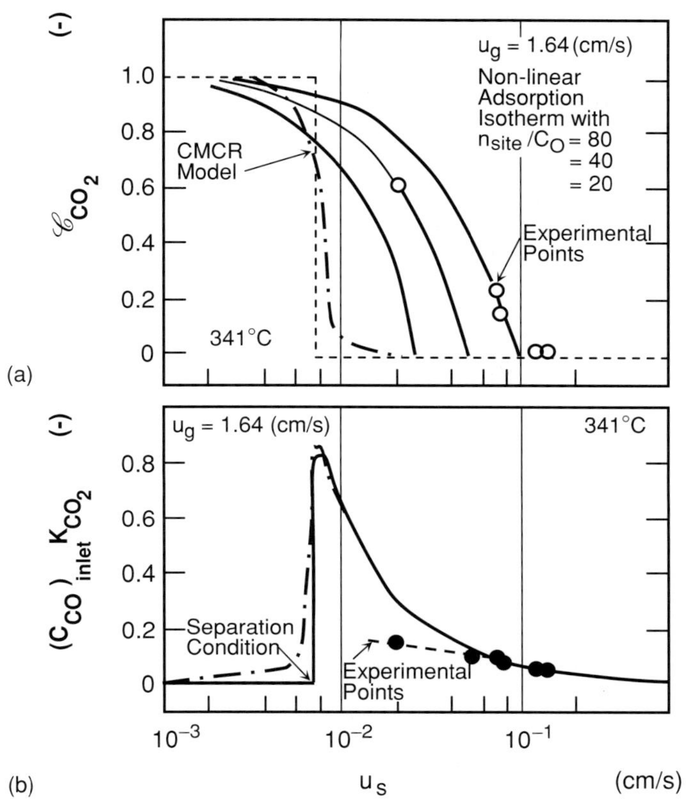

Figure 5.12 Comparison of the Takeuchi and Uraguchi CMCR model with experimental data: (a) dimensionless exit concentration of CO_2 at the top of the reactor as a function of the moving-bed velocity, (b) adsorbate leaving the bottom of the reactor (assuming instantaneous adsorption) as a function of the moving-bed velocity. The solid lines represent the CMCR model, modified for a nonlinear adsorption isotherm. (Figure adapted from Takeuchi and Uraguchi (1977b) with permission. © Society of Chemical Engineers of Japan.)

profiles of reactant and product and to compare the profiles to model predictions. A further objective was to determine the influence of system variables on CMCR performance. Initial experiments showed that careful attention must be given to heating of the solids entering the reactor, close control of the solids flow rate and reliable sampling of the vapor phase within the reactor. The investigators also tried to regenerate the solid phase and recycle it back to the CMCR. For the experiments, a 12.5-mm i.d. tube, about 2 m in length was used. Solid particle size was 32–50 US mesh. Figure 5.13 offers a schematic of the apparatus. Parts (a) and (b) of the figure show the solids preheater and the flow control valve. Not shown in the figure is a regeneration vessel used to remove reactant, MES, product, and TMC, from the solid phase. The thorough instrumentation of the moving bed is clear from the figure. Sampling of temperature and withdrawal of gas phase samples were computer controlled.

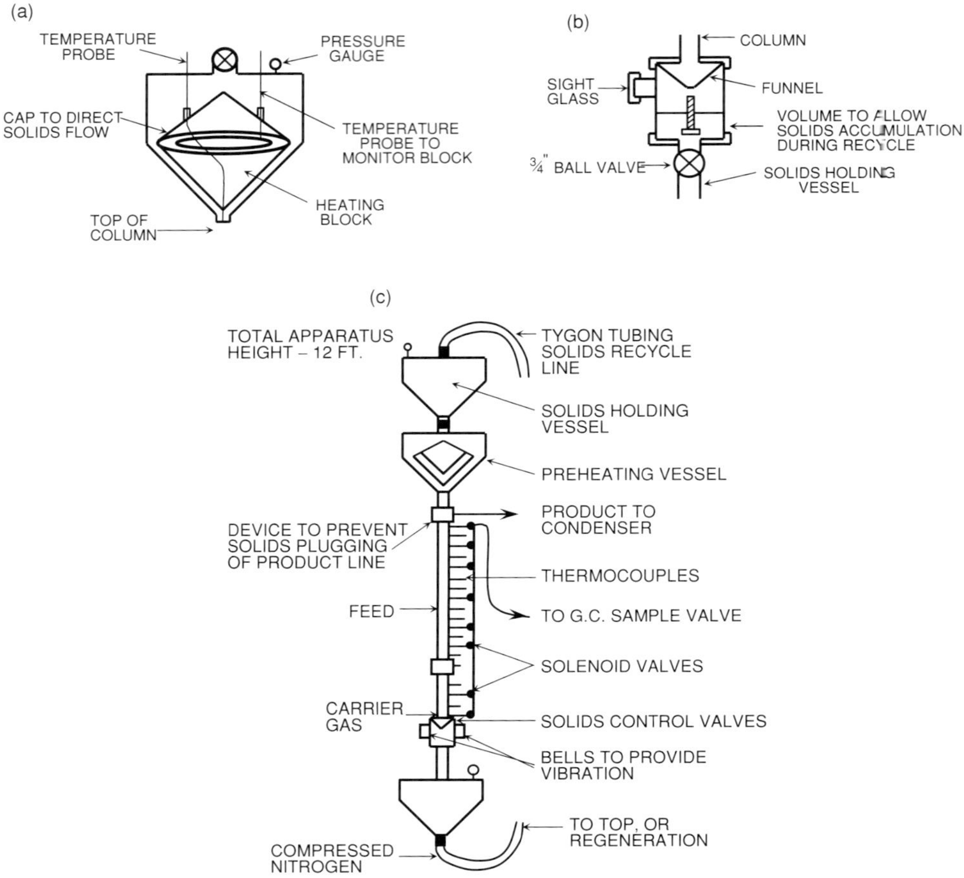

Figure 5.13 Schematic diagram of an experimental CMCR showing (a) the solids preheater, (b) flow control valve and (c) overall setup developed to make the CMCR function satisfactorily. (Figure adapted from Fish et al. (1986) with permission. © 1986 Pergamon Press Ltd.)

Operation of the moving-bed reactor indicated that the adsorption isotherm was convex and not represented by a Langmuir isotherm. Strong adsorption of MES on the alumina caused problems and the authors attempted to deactivate the alumina by treatment with a base. They reported that they were able to obtain pure TMC from their CMCR but that they were not able to exceed equilibrium conversion obtained in a comparable fixed-bed PFR.

Chapter 6
Variations on the Moving-Bed Chromatographic Reactor

6.1 Concept

Periodic or pulse operation of a chromatographic reactor can be avoided by moving the bed of catalyst and adsorbent continuously countercurrent to the flow direction as discussed in the previous chapter. Another alternative is a rotating annular bed of catalyst and adsorbent. This design is widely employed in regenerative heat exchange. Solids in the rotating bed collect heat from a flowing stream of hot fluid located at one point or angle in the annulus and discharge this heat to a stream of cold fluid at a second point or angle. There have been several applications of this design to preparative chromatography (Fox et al., 1969; Scott et al., 1976; Wankat, 1974; Cho et al., 1980a,b; Wardwell et al., 1982). The operation of a rotating annular bed of catalyst and adsorbent is illustrated in Figure 6.1. To simplify the diagram, complete conversion of A in an $A \leftrightarrow B + C$ equilibrium-limited reaction is assumed. Also we assume tailing is negligible so that the product can be collected in a single port. Carrier gas or liquid eluent is injected into the tubes at all but a single point or port. At that port reactant A is fed. Similarly, carrier gas or eluent is drawn off at points along the bottom. However, two points or ports are used to remove the products B and C. Carrier gas or eluent may be withdrawn between these product ports as is shown in the schematic. In this system each tube in the array functions as a CR. Of course, the individual tubes can be stationary and the feed and withdrawal ports can rotate together at a constant velocity. Indeed, if the bed of catalyst and adsorbent can be packed very uniformly, individual tubes can be dispensed with and a single bed may be used. Some lateral dispersion then occurs, but this works only to spread the product elution peaks and might require adjacent ports for product removal. In this arrangement, products are diluted by the carrier gas or eluent. Researchers at the University of Minnesota used a stationary, but single, annular bed with rotating ports (Cho et al., 1980a,b), while Wardwell et al. (1982) employed a single rotating annular bed with stationary ports. Operation of either type of continuous rotating annular bed chromatographic (CRAC) reactor for a biochemical operation using an enzyme catalyst and a solid phase adsorbent is illustrated in Figure 6.2. Note that the reaction products move in helical paths downward through the bed. Separation of these paths depends on the adsorption of the product. The most strongly adsorbed species is retarded and follows the longest path. In biochemical applications conversion of the reactant is often complete so its path would disappear within the bed. A dashed line in the figure shows its continuation. Axial mixing and diffusion in the direction of rotation broaden the path from a line to a band. The angular spread of the product at the bottom of the bed is called the product bandwidth.

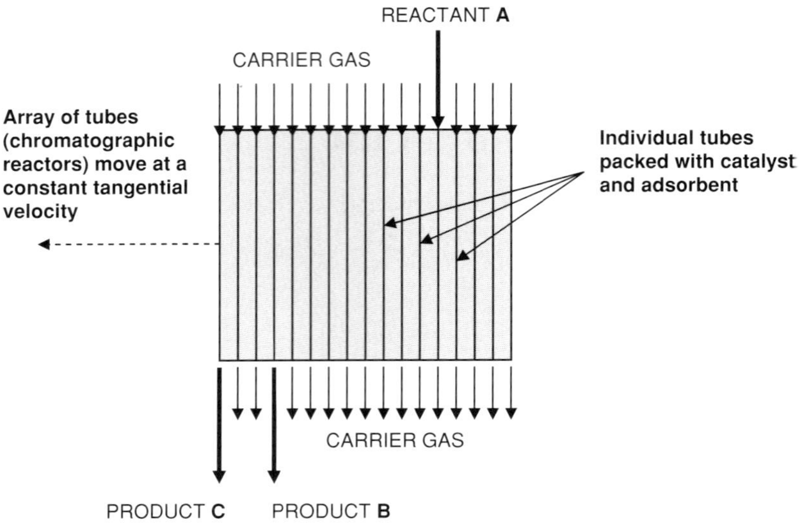

Figure 6.1 Schematic of an array of chromatographic reactors moving at constant velocity pass a fixed reactant and multiple carrier gas feed ports on the top and fixed product and carrier gas removal ports at the bottom.

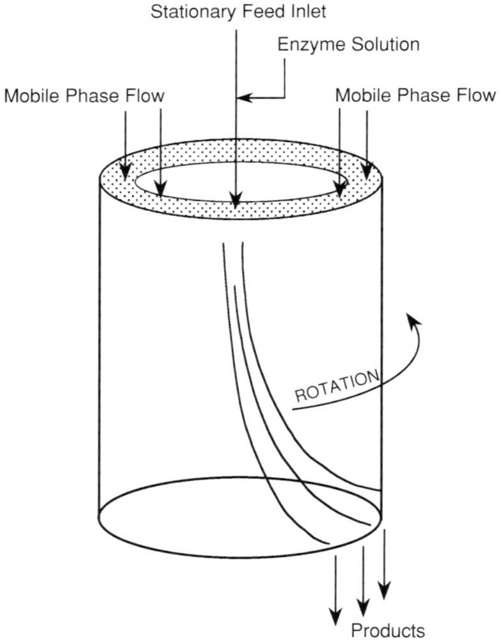

Figure 6.2 Operating principle of a continuous rotating annular chromatographic (CRAC) reactor for an enzyme-catalyzed reaction A → B + C where the product C is more strongly adsorbed than B and the reactant is not adsorbed. Mobile phase is the solvent or eluent. (Figure adapted from Sarmidi and Barker (1993a) with permission. © 1993 SCI.)

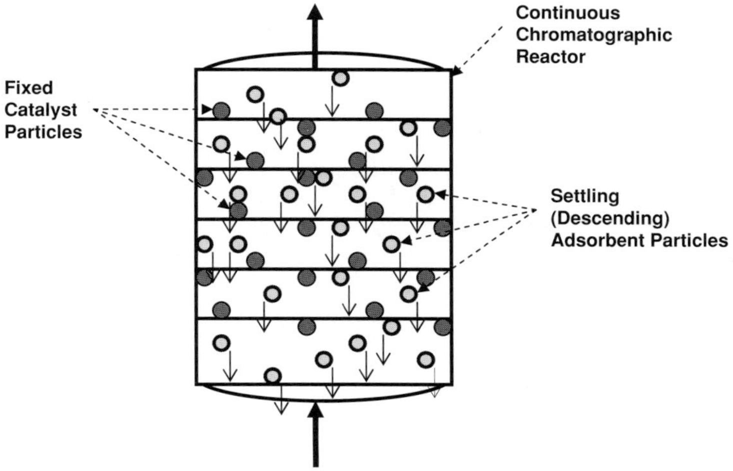

Figure 6.3 Continuous chromatographic reactor with moving adsorbent particles and stationary catalyst particles.

Westerterp and coworkers (Kuczynski et al., 1987b; Westerterp and Kuczynski, 1987a,b) noted that in a CMCR it is just the adsorbent that needs to flow through the reactor. Thus, a continuous moving-bed chromatographic reactor could be constructed with catalyst particles fixed in space and adsorbent and the fluid phase flowing through the fixed bed in opposite directions. Figure 6.3 presents a schematic of such a system. Operation is the same as in the CMCR systems considered in the previous chapter. Consequently, if this system was applied to an equilibrium-limited reaction, such as $A \leftrightarrow B + C$ where the product C is strongly adsorbed compared to B, and A is not adsorbed; the reactant A would be added in the center of the reactor, B would be withdrawn at the top and C at the bottom with the adsorbent. To recover C, a stripping section must be added at the bottom. It is the operation with a stripping section that is shown in Figure 6.4.

To realize the operation illustrated in Figure 6.3, Westerterp and Kuczynski employed large catalyst particles (5 mm i.d. by 5-mm cylindrical pellets) and a fine adsorbent powder ($d_p = 0.09$ mm) that trickled downward past the catalyst and a dumped packing of $7 \times 7 \times 1$-mm Raschig rings. The Raschig rings trapped the large cylindrical pellets as well as increased the bed porosity so as to enable solids flow. The packing also insured a more uniform distribution of the smaller catalyst particles over the reactor cross section. We return to a discussion of the Westerterp and Kuczynski system in Section 6.3.

Another variation of the CMCR was developed by van der Wielen et al. (1990) from a multistage bioreactor that was studied in great detail by Vos (1990) and Vos et al. (1990a,b,c). In their system, the lighter or smaller catalyst particles are fixed or move cocurrent with the fluid while the larger or heavier adsorbent particles move in a countercurrent direction. Their system operated in stages. Each stage is a liquid fluidized bed. Upward flow of liquid segregates catalyst and adsorbent into separate beds in each stage. This is shown in Figure 6.5. Liquid flow is periodically reversed or at least decreased in the earliest variation of their

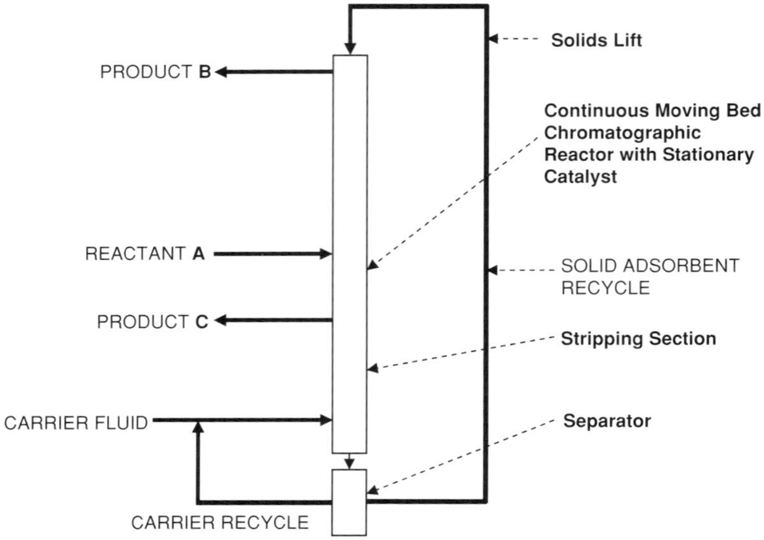

Figure 6.4 Schematic of a continuous chromatographic reactor with moving adsorbent and stationary catalyst applied to an equilibrium-limited reaction $A \leftrightarrow B + C$.

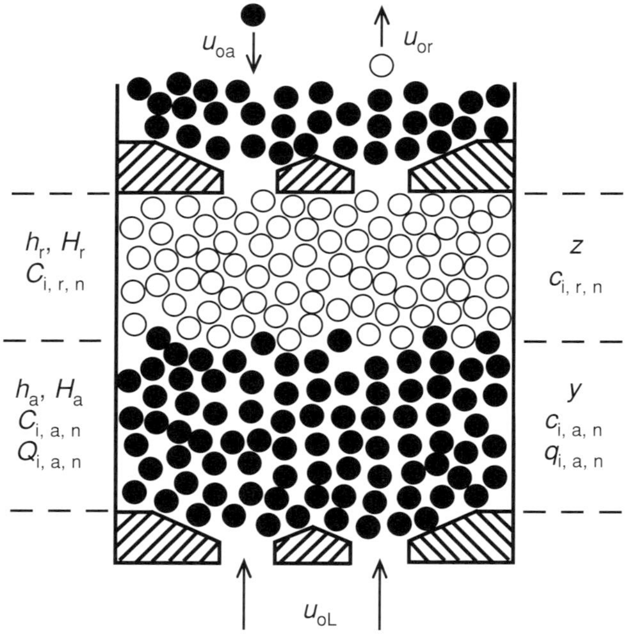

Figure 6.5 Schematic representation of a stage in the moving-bed chromatographic reactor showing flow direction of the phases and variables used in the system model. (Figure adapted from van der Wielen et al. (1990) with permission. © 1990 Pergamon Press plc.)

design. This draws the adsorbent down through a nozzle between stages into the next lower stage. The lighter catalyst is drawn into the adsorbent bed but cannot penetrate the bed so the catalyst is trapped in the stage. When upward fluid flow is restored or increased, adsorbent and catalyst segregate. However, the dense bed of adsorbent in the next upper stage prevents most of the lighter catalyst from moving upward. The catalyst is held in the lower stage. Because of the higher velocity in the nozzles connecting stages, adsorbent cannot move downward. Pulsing thus moves the particles in opposite directions while fluidization of particles with different sizes or densities separates the particles into distinct layers of adsorbent and catalyst. Because of pulsing, the operation is semicontinuous. Figure 6.5 reproduces a schematic of the operation from their earliest paper (van der Wielen et al., 1990).

Refinement of the design through the nozzles connecting stages, the extent of flow reversal and the choice of particle size and density results in a system in which the lighter catalyst particles are fixed in a stage, while the denser adsorbent particles drop downward from stage to stage, countercurrent to the fluid flow. Figure 6.6 from a 1998 paper shows the operation. van der Wielen et al. (1998) comment that the operation resembles the Cloete–Streat contactor used in moving-bed ion-exchange systems (Cloete and Streat, 1963; Dodds et al., 1973; Bennett et al., 1984). Adjustment of the fluid velocity and the design of the nozzle between the stages traps the catalyst while allowing the adsorbent to move. The result is a reactor that functions like the Westerterp and Kuczynski system, except it is periodic while the latter is continuous.

By eliminating the stages, maintaining an L/D of 5 or greater, and adjusting the fluidization parameters, periodic flow reversal can be avoided. Then the denser adsorbent particles rain down through the lighter catalyst particles as a continuous moving bed (van der Wielen et al., 1996). A control nozzle, similar in design to the nozzles used in the pulsed operation (Fig. 6.6) is needed at the top of the reactor to control the flow of adsorbent particles. A dense bed of adsorbent particles must be maintained above that nozzle to prevent carryover of catalyst particles. These particles are held in particulate fluidization. This provides plug

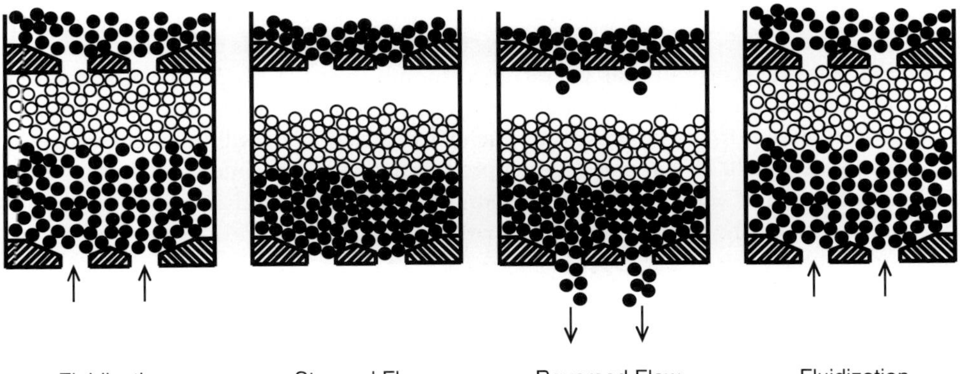

Figure 6.6 Sequence of events in a pulsed, multistage fluidized bed separating reactor with countercurrent movement of adsorbent. (Figure adapted from van der Wielen et al. (1998) with permission. © 1998 AIChE.)

flow characteristics for the liquid phase. Mixing of the lighter particles is limited. However, the operation requires careful attentions to the liquid flow rate and a means of continually adding and removing adsorbents. van der Wielen et al. have applied their continuous moving bed to the preparation of 6-aminopenicillanic acid from penicillin G and have modeled the operation.

6.2 Modeling and design studies

6.2.1 Continuous rotating annular-bed chromatographic (CRAC) reactors

As might be expected from Figure 6.1, the performance of the continuous rotating annular-bed chromatographic reactor system can be described by the system of partial differential equations in Table 3.2 with the simplification for the chromatographic reactor that the solids velocity $U_s = 0$ even though the solids are in angular motion. For each component then, Equation 3.1 may be applied just as in Section 4.1, that is

$$\varepsilon_t \frac{\partial C_i}{\partial t} + \rho_b \frac{\partial q_i}{\partial t} = D_x \frac{\partial^2 C_i}{\partial x^2} - \frac{1}{A}\frac{\partial n_i}{\partial x} - \rho_b r_i \tag{6.1}$$

However, this equation cannot predict the port or angle at which the product can be withdrawn. To do this, Cho et al. (1980a) considered a system at steady state and introduced an angular position variable ς so that the model becomes

$$u\frac{\partial C_i}{\partial z} + W\frac{\partial\left(\varepsilon C_i + \rho_b K_i \left(\frac{C_i}{C_{A0}}\right)^{m_i}\right)}{\partial \varsigma} = r_i \tag{6.2}$$

In this relation, adsorption is assumed to be rapid and is described in terms of the Freundlich isotherm with the exponents for reactants and products <1, u is the superficial fluid velocity and W is the width of the annular bed. Lateral and axial dispersion are neglected.

Cho et al. (1980a) employed the acid catalyzed hydrolysis of methyl formate to validate the above model. This is a homogeneous reaction for which there is an autocatalytic contribution. Activated coconut charcoal served as the adsorbent. Adsorption isotherms were determined experimentally whereas the kinetics of the homogeneous and autocatalytic reactions were drawn from the literature. The experimental work will be discussed in the next section. Figure 6.11 compares model predictions with experimental measurements. Although data points scatter because of the nonhomogeneity of the charcoal bed, the model predicts quite well the concentrations of methanol and formic acid measured at a large number of outlet ports.

A more general model has been proposed by Sarmidi and Barker (1993b) that allows for dispersion in the flow direction and does not assume an adsorption isotherm.

$$\varepsilon\omega\frac{\partial C_i}{\partial \varsigma} + (1-\varepsilon)\rho_{ad}\frac{\partial q_i}{\partial \varsigma} + \varepsilon v\frac{\partial C_i}{\partial z} - D_{ax_i}\frac{\partial^2 C_i}{\partial z^2} = \varepsilon v_i r \tag{6.3}$$

Dispersion in the direction of rotation is often small and so it is frequently neglected. In the above relation, ω = the rotation rate in rpm and the interstitial velocity, $v = Q_{\text{eluent}}/\varepsilon A_{\text{annulus}}$.

The authors apply this model to the enzymatic inversion of sucrose to glucose and fructose. The latter is more strongly adsorbed on the ion-exchange resin chosen as the adsorbent. Rate of sucrose inversion is given by a form of the Michaelis–Menten relation:

$$r = \frac{V_{\text{max}} C_{\text{sucrose}}}{C_{\text{sucrose}} + K_{\text{m}} + \frac{C_{\text{sucrose}}^2}{K_f}} \tag{6.4}$$

Sarmidi and Barker undertook a simulation using this model by assuming the maximum inversion rate $V_{\text{max}} = kC_{\text{enzyme}}$ and linear isotherms for all reaction components. The Dankwerts boundary conditions (see Table 3.7) were used axially and a bed depleted of all components was assumed at $\zeta = 0$. The model was validated experimentally by the authors as will be discussed in Section 6.3.1.

Herbsthofer and Bart (2003) attempted to determine conditions under which a continuous rotating annular chromatographic reactor improves performance for an equilibrium-limited, two-body reaction, $A + B \leftrightarrow C + D$ in which reactant B also serves as the solvent. As noted elsewhere, such systems arise in hydrolysis or in esterification. In their study, the authors compared the CRAC reactor with a PFR followed by a continuous rotating annular chromatographic separator. The separator is identical to the CRAC reactor except that it contains only adsorbent. For good measure, the CRAC reactor was also compared to a double reactor made up of a PFR followed by a CRAC reactor. Restraints were equal reactant feed rates and a product (D) purity of 99.99%. For the comparison, Herbsthofer and Bart assumed (a) 1-D heterogeneous model (Table 3.3) with reaction kinetically controlled but adsorption dominated by a mass transfer resistance. Dispersion was neglected and a linear driving force model was applied to mass transfer to the adsorbent. Reaction kinetics followed a power law model given by the reaction stoichiometry and adsorption obeyed Henry's law. The constants were drawn from the species in the ionic redox reaction: $Ir^{3+} + Fe^{3+} \leftrightarrow Ir^{4+} + Fe^{2+}$. Batch measurements indicated the order of elution times to be $Fe^{3+} < Fe^{2+} < Ir^{3+} << Ir^{4+}$. Note that in this case Fe^{3+} is not an eluent.

For model verifications, simple experiments were conducted in a PFR and in a CRAC reactor using a 1:10 volumetric flow rate of $Ir^{3+}$to Fe^{3+} at a Fe^{3+} feed concentration of 0.15 mol/L and a temperature of 323 K. Equilibrium conversion of Ir^{3+}, attained in the PFR, was 0.86%, while conversion in the CRAC reactor was 4.74%. Samples collected along the bottom of the annulus gave the concentration profiles for $Ir^{4+}$and Fe^{3+} as a function of degrees from the feed point. Satisfactory agreement of the predicted and experimental profiles established that the model used by the researchers was satisfactory.

Using simulation, Herbsthofer and Bart examined the effect of angular velocity on productivity, defined as the molar flow rate of D at a purity of 99.99% divided by the mass of solid functioning as both catalyst and adsorbent. As seen for other moving-bed systems, productivity exhibits a maximum with respect to solids velocity, in this case the rotational speed. For comparison of the CRAC reactor with the PFR + CRAC separator, bed depths were adjusted to assure 99.99% purity of the D product (with respect to C and the educt A). Productivities at the optimal angular velocities were used. For elution times in the sequence $C < A < D$ with a very fast reaction, the researchers found that the CRAC reactor had a higher productivity than the PFR + CRAC separator combination, except at high values of the equilibrium constant as might be expected. With a slow reaction, the CRAC reactor gave only a slightly higher productivity than the PFR + CRAC separator combination up to large values of the equilibrium constant for the reaction. However, the two reactor combination, PFR + CRAC reactor, showed much higher productivity regardless of the equilibrium constant. On the other hand, when the elution time sequence is A < C < D, productivity of the CRAC reactor is always much less than that of the combination when the reaction is slow. When the reaction is very fast, the CRAC reactor shows better productivity than the combination only in a range of equilibrium constants centered about 1.0. For an order of magnitude outside of this range ($K < 0.1$ and $K > 10$), the combination is better or equally productive. Explanation for this is that not all of the catalyst–adsorbent is utilized in the CRAC reactor at any instant and the large requirement of catalyst when the reaction is slow.

The Herbsthofer and Bart study confirms that a separating reactor is not always the best choice even if a good adsorbent is available. Reaction rates and equilibrium must be considered. Their results also suggest that a PFR followed by a CRAC reactor should be considered for some applications. The PFR would be used until a conversion is reached where the reverse reaction becomes important.

6.2.2 Moving bed of adsorbent

Two models have been proposed for a moving bed of adsorbent and a stationary catalyst one for a gas–solid system with a fixed bed of solids and the second for a liquid–solid system in which the catalyst phase has some mobility. The first model was developed for methanol synthesis and neglected interphase mass and heat transfer (Westerterp and Kuczynski, 1987a). It was therefore pseudo-homogeneous and could be described by Equations 3.1 and 3.3 in Table 3.1. Thus, introducing hold up or volume fractions for the catalyst, adsorbent and inert solid phases,

$$(1 - \varepsilon_{\text{cat}} - \varepsilon_{\text{ads}} - \varepsilon_{\text{inert}})\frac{\partial C_i}{\partial t} + \varepsilon_{\text{ads}}\rho_{ad}\frac{\partial q_i}{\partial t}$$
$$= D_x\frac{\partial^2 C_i}{\partial x^2} - \frac{1}{A}\frac{\partial n_i}{\partial x} + U_s\rho_{ad}\frac{\partial q_i}{\partial x} - \sum \varepsilon_{\text{cat}}\rho_c\nu_{i_k}r_k \qquad (6.5)$$

Countercurrent moving-bed chromatographic reactors operate at steady state so the terms on the LHS can be set to zero, unless startup or response to disturbances is to be studied. In the above expression, ν_{i_k} is the stoichiometric coefficient of species i in the kth reaction. In the case of methanol synthesis, there is a side reaction that yields methane. Equation 6.5 must be written for all components in the system, H_2, CO, CH_3OH and CH_4. In their study, Westerterp and Kuczynski neglected methane formation. They did not employ an energy balance. Instead, the authors considered cooling to limit temperature rise in the highly exothermic synthesis, but this was not modeled. Instead two limiting situations were evaluated: wall cooling of the reactor and use of a cooling section in which catalyst was not present, but adsorption equilibrium was maintained. The simulation carried out by Westerterp and Kuczynski used solids hold ups determined experimentally, linear adsorption isotherms and a complex rate expression with inhibition by all components. Rate and adsorption constants, activation energy and heats of adsorption were measured by the authors (Westerterp and Kuczynski, 1987b; Kuczynski et al., 1986, 1987a).

Simulation results showed that complete conversion of CO to methanol could be achieved, but that a large recycle of adsorbent was required. This is because methanol leaves the CMCR only as adsorbate at the bottom of the reactor. A consequence is that there is zero gas flow at the top of the CMCR if a stoichiometric feed containing just H_2 and CO is used. Operating at lower temperature and employing interstage cooling can substantially reduce adsorbent circulation. Of course, an adsorbent with a large capacity for methanol at 470–540 K would reduce the circulation rate significantly.

Trickling or raining of a solid adsorbent through a liquid-fluidized but stationary bed of catalyst, rather than a packed bed, was explored by van der Wielen et al. (1996) for penicillin G conversion. Diffusion is very much slower in liquids than in gases so for a liquid–solid moving-bed system that these investigators discussed, mass transfer must be considered and a pseudo-homogeneous model can no longer be used. Heat effects are small making an energy balance unnecessary. Conversion proceeds over a penicillin acylase, an enzyme, as

$$\text{penicillin G} \rightarrow \text{6-aminopenicillanic acid} + \text{phenylacetic acid} \tag{6.6}$$

In practice, the penicillin acylase is immobilized on a porous solid carrier. The phenylacetic acid exchanges with an anionic resin in the OH form to release water. The exchange shifts the reaction to completion and raises the pH. Phenylacetic acid and penicillin G inhibit conversion. Hydroxide released by the resin on exchange with phenylacetic acid reacts with the acid form of penicillin G. Acidity (pH) is controlled by a potassium phosphate buffer. Thus, there is a homogeneous reaction as well as a heterogeneous one. Equation 3.7 must be expanded to allow for this reaction. van der Wielen et al. (1996) further assume that only mass transfer to and from the adsorbent is important and that there is negligible storage of reactants and products on the immobilized acylase. Thus, for either penicillin G, phenylacetic acid or hydroxide ion,

$$\varepsilon_L \frac{\partial C_i}{\partial t} = \varepsilon_L D_{ax_i} \frac{\partial^2 C_i}{\partial x^2} - \varepsilon_L u_L \frac{\partial C_i}{\partial x} - a_{\text{ads}} J_i + \varepsilon_{\text{cat}} \eta_{\text{overall}_i} \nu_i^{\text{cat}} r^{\text{cat}} - \varepsilon_L \nu_i^{\text{hom}} r^{\text{hom}} \tag{6.7}$$

$$\varepsilon_{\text{ads}} (\rho_b)_{\text{ads}} \frac{\partial q_i}{\partial t} = \varepsilon_{\text{ads}} (\rho_b)_{\text{ads}} D_{ax_{\text{ads}}} \frac{\partial^2 q_i}{\partial x^2} + \varepsilon_{\text{ads}} u_{\text{ads}} \frac{\partial q_i}{\partial x} + a_{\text{ads}} J_i \tag{6.8}$$

Equation 6.8 allows for local mixing of the downward settling adsorbent through a dispersion coefficient. The two-equation dynamic model was used by the authors to examine startup that can often be important in biochemical systems that sometimes exhibit multiple steady states. At steady state, the time derivatives disappear, of course. In these relations, ε is the volume fraction or hold up, a_{ads} is an interfacial area per unit volume of reactor, η is an overall effectiveness factor that includes external mass transfer and J is an interphase flux between the flowing liquid phase and the adsorbent. van der Wielen assumes linear interphase transport so

$$J_i = k_{\text{particle}} \left(C_i - \frac{q_i}{K_i} \right) \tag{6.9}$$

The enzymatic reaction is both reactant and product inhibited and so is described by Michaelis–Menten kinetics. All components of the reaction system influence the rate so the expression is complicated:

$$r = k \frac{1}{1 + \frac{C_{H^+}}{K_{a_{\text{Enz}}}}} \frac{\frac{1}{K_m} \left(C_{\text{Pen}} - \frac{C_{\text{APA}} C_{\text{PA}}}{K_{eq}^{\text{app}}} \right)}{\left(1 + \frac{C_{\text{Pen}}}{K_m} + \frac{C_{\text{PA}}}{K_{i\text{PA}}} \right) \left(1 + \frac{C_{\text{APA}}}{K_{i\text{APA}}} \right)} \tag{6.10}$$

It can be seen that the rate is pH dependent. The adsorption equilibrium constants are also pH functions.

Boundary conditions are simpler than those given by Equations 3.55 and 3.58. The Danckwerts conditions are applied separately to reactants in the fluid phase and adsorbate on the ion-exchange resin at the top and bottom of the reactor. The system of equations is stiff because the rate of the homogeneous reaction is very much faster than the rate of mass transfer and the rate of the enzymatic reaction. To avoid this difficulty, the authors replace the rate term by an equilibrium expression. The investigators used the method of lines to generate a system of ODEs and a fourth-order Runge–Kutta algorithm for integration.

Vos et al. (1990a) discuss a model of a multistage fluidized bed with a solid phase moving downward countercurrent to the liquid flow.

In order to validate the model, van der Wielen et al. (1996) undertook experiments on penicillin G with immobilized acylase. We examine the experimental work later in this chapter.

6.2.3 *Pulsed, multistage fluidized bed with downward moving adsorbent*

van der Wielen et al. (1990) model their staged, countercurrent moving adsorbent and catalyst system assuming that the catalyst and adsorbent in a stage are segregated into zones as is shown in Figure 6.5. Each zone, however, is fully back-mixed. They further assumed that there is no adsorption or desorption as the adsorbent moves from stage to stage. Similarly, there is no reaction on transport of the catalyst in the opposite direction. Because the bottom stage in the chromatographic reactor involves desorption of the product or reactant, this stage was not part of the model. Isothermal operation was assumed, radial gradients neglected, intraparticle diffusion was considered for the catalyst, but not for the adsorbent. Instead, adsorption was assumed to be controlled by mass transfer from the fluid phase to the particle. Thus, a heterogeneous model must be used. This is readily obtained from Equation 6.7 by replacing the flux term by an input–output term to give

$$\varepsilon_L \frac{\partial C_i^j}{\partial t} = u_L\left(C_i^{j+1} - C_i^j\right) - a_{\mathrm{ads}} J_i + \varepsilon_{\mathrm{cat}} \eta_{\mathrm{overall}_i} \nu_i^{\mathrm{cat}} r^{\mathrm{cat}} - \varepsilon_L \nu_i^{\mathrm{hom}} r^{\mathrm{hom}} \qquad (6.11)$$

The terms in Equation 6.11 have been defined above. Stage numbers, j, increase from the top of the reactor. The transport flux to the adsorbent is zero for the reaction zone in each stage and, likewise, the reaction rate is zero for the adsorption zone in the adsorption portion of that stage. The transport flux J_i is given by Equation 6.9, but is simplified by assuming that the adsorbate concentration in the liquid phase is the average concentration leaving the adsorption zone. This means that the liquid phase as well as the adsorbent particles in the zone are assumed to be well mixed. If the reaction rate is fast relative to mass transport, the term containing r^{cat} can be replaced by $a_{\mathrm{cat}} J_i$ where the driving force in Equation 6.9 will now contain the equilibrium concentration of i at the catalyst surface. Material balances for the catalyst phase can be neglected if adsorption on the catalyst is small. Material balances for the adsorbent are required. They can be written as

$$\varepsilon_{\mathrm{ads}}(\rho_b)_{\mathrm{ads}} \frac{dq_i}{dt} = U_{\mathrm{ads}}(\rho_b)_{\mathrm{ads}}(q_i^{j-1} - q_i^j) + a_{\mathrm{ads}} J_i \qquad (6.12)$$

In the equation, U_{ads} is the velocity of the downward moving adsorbent. Boundary conditions are required for each zone but none are required between stages of the chromatographic reactor because of the CSTR assumption.

Analytical solutions to the model exist for linear kinetics but these are not given by the authors. They suggest using a cells-in-series model for each zone with numerical evaluation when the kinetics are nonlinear. van der Wielen et al. (1990) applied the model to the isomerization of glucose with a glucose isomerase catalyst. There is no homogeneous reaction in this system so one of the reaction terms drops out of Equation 6.11. The adsorbent was either an X or Y zeolite depending on whether or not glucose or fructose is preferentially adsorbed. For their simulation, the authors used parameters obtained experimentally by Ruthven and coworkers (Ching and Ruthven, 1988; Ho et al., 1987). Further values came from a thesis (Vos, 1990). For the model parameters assumed, van der Wielen et al. conclude that the overall conversion of glucose and the product purity exceed by 20% those predicted by thermodynamic equilibrium. As might be expected, pronounced improvements in conversion occur with product rather than reactant adsorption. In this case, separation and product purity increase with the number of stages in the reactor. The ratio of liquid and adsorbent velocities, σ, is an important parameter affecting performance. Mass transfer coefficients and liquid phase back-mixing exerted a minor influence on glucose conversion and on selectivity to fructose. An experimental system has been built and is described by the authors but its performance was not given. The authors point out that complete back-mixing of the catalyst and adsorbent phases is unlikely but that this could significantly change system performance.

Design of the pulsed, multistage fluidized bed has been investigated by the van der Wielen team (van der Wielen et al., 1998). They identify segregation of the solids on fluidization, adsorbent flow, the ratio of adsorbent to fluid phase flow, and mixing in the fluid phase as the important design considerations. These were found to depend on the fluid and solid properties, such as catalyst and adsorbent size and density (d_{cat}, d_{ads}, $(\rho_b)_{cat}$, $(\rho_b)_{ads}$), fluid density and viscosity (ρ_L, μ_L) and the Richardson and Zaki indices, n_{cat} and n_{ads}, that characterize the fluidization behavior of the particles (Richardson and Zaki, 1954). They depend also on the volume ratios of catalyst, adsorbent and liquid in the reactor feed streams that determine the volume fractions or hold up, ε_i, in each reactor stage. The pulsing parameters are also important. These are the frequency, f_p, the pulse amplitude or actually the volume withdrawn from the flow in the reverse step, V_p. The pulsing parameters are best understood by examining the schematic of the pulsed, multistage fluidized bed employed by the research team (Fig. 6.7).

Parts of the system are identified in the figure. Pulsing cycle has either two or three parts: fluidization (liquid in up flow with catalyst and adsorbent forming an upper and lower zone in each stage), stop-flow (solids in each zone settle to achieve the highest solids volumetric fraction and adsorbent drops to next lower stage) and reverse flow (adsorbent and fluid sucked down to the next lower stage or to discharge). Parts of this cycle are shown in Figure 6.6. The two-part cycle consists of fluidization and low or reverse flow only. Highest solid flow rates require the three-part cycle. Since highest throughputs mandate high solid flow rates, three-part cycles are used normally.

Parts of the cycle are created by opening the flow control valve after the rotary pump for fluidization (see Fig. 6.7), closing this valve and the valve on the reverse flow chamber for stop-flow, and by opening only that valve for reverse flow. With the pump circulating

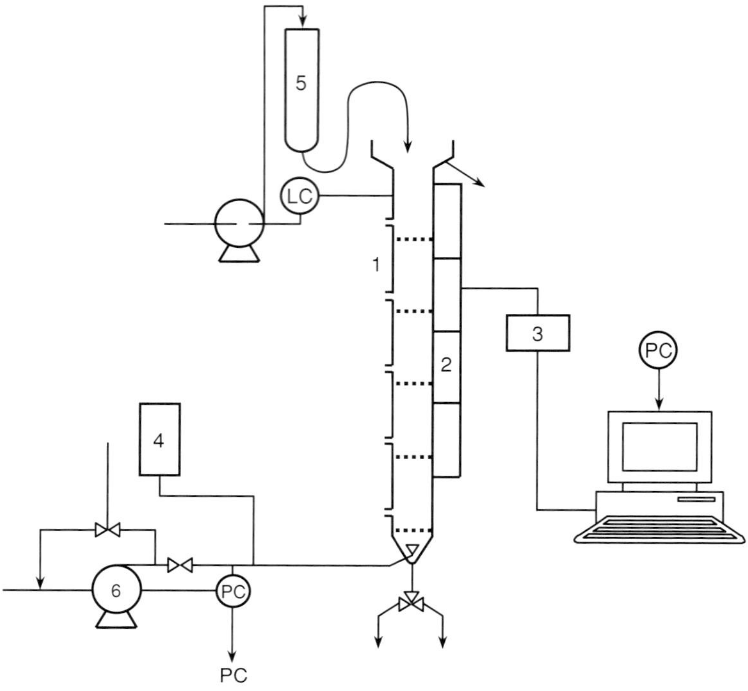

Figure 6.7 Schematic of the experimental unit used by van der Wielen et al. (1998) to investigate the hydrodynamics of the pulsed, multistage, fluidized bed separating reactor. (Figures adapted from van der Wielen et al. (1998) with permission. © 1998 AIChE.)

feedback to its entrance, the reverse flow chamber draws in liquid causing liquid downflow in the multistaged reactor column. A level controller at the top of the column maintains a bed of adsorbent above the topmost perforated plate in the column. The adsorbent in the van der Wielen study had the highest density and the smallest diameter, so it flowed intermittently downward because of the pulsed operation and formed the bottom zone in each stage.

Staging was realized by a perforated plate with conically shaped holes sized at about 5 × the diameter of the adsorbent particle. Hole size and shape in the plates is critical. Design is discussed by Vos et al. (1990c).

Segregation of the catalyst and adsorbent into zones is necessary to maintain the catalyst stationary in the reactor and have only adsorbent flowing countercurrent to the fluid. Value of the segregation function, F_{seg}, indicates if the solid is in the upper layer and whether or not segregation is possible.

$$F_{seg_{ij}} = v_{slip_i}^{\infty}\left(\frac{\rho_i - \rho_B}{\rho_i - \rho_L}\varepsilon_L^{3.8}\right)^{\frac{n_i}{4.8}} - v_{slip_j}^{\infty}\left(\frac{\rho_j - \rho_B}{\rho_j - \rho_L}\varepsilon_L^{3.8}\right)^{\frac{n_j}{4.8}} \tag{6.13}$$

A negative value means that the i particles form the upper zone or layer and the j particles the bottom layer. If F_{seg} is zero, the $i - j$ particles will be well mixed. For the van der Wielen application, $i =$ catalyst. In the above relation $(v_{slip})^{\infty}$ is the slip velocity of an isolated particle in a moving fluid relative to a fixed plane, ρ_i, ρ_j, and ρ_B are the densities of the i and j solid phases and the bulk density of the suspension, respectively, while ε_L is the liquid phase volume fraction under fluidizing conditions. n_i and n_j are the Richardson and Zaki indices for the i and j particles.

A conservative estimate of the adsorbent flow rate for any fluid flow rate and set of pulsing parameters is given by

$$Q_{\text{ads}} = f_p V_p \varepsilon_{\text{ads}}^{\text{max}} \tag{6.14}$$

where $\varepsilon_{\text{ads}}{}^{\text{max}}$ is the volume fraction at the settled or nonfluidized condition in the stop-flow part of the cycle. It is conservative because some adsorbent drops to the lower stage in the stop-flow part. Flow reversal in the three-part cycle reduces the fluid flow. The effective flow rate, $(Q_L)_{\text{effective}}$, is easily estimated,

$$Q_{L_{\text{effective}}} = Q_{L_{\text{fluidization}}} - (1 - \varepsilon_{\text{ads}}^{\text{max}}) f_p V_p \tag{6.15}$$

This flow rate, depth of the catalyst zone and the number of stages allows an estimate of conversion in the reactor–separator.

Volume of the reverse flow chamber, V_p, is limited by the depth of the adsorbent zone or layer in a stage and from stability considerations. It should be just a fraction of the layer volume. Vos et al. (1990c) suggest that it is about 30% of the layer volume. Frequency, f_p, is the reciprocal of the cycle period. The cycle period, τ, is the sum of the durations of the various cycle parts or,

$$\tau = t_{\text{reverseflow}} + t_r + t_{\text{seg}} + t_{\text{sedimentation}} \tag{6.16}$$

In this sum, t_{seg}, is the time needed for the zones or layers to reform after fluidization has begun, while $t_{\text{sedimentation}}$ is the time needed for particles to settle to provide $\varepsilon_{\text{ads}}{}^{\text{max}}$. Generally, the contact time in a stage needed for high conversion is short compared to the other durations. van der Wielen et al. (1997a) develops relations for $t_{\text{sedimentation}}$. In a further paper, estimating methods for t_{seg} are described (van der Wielen et al., 1997b). van der Wielen et al. (1998) claim that $t_{\text{sedimentation}}$ of the dense adsorbent particles through the lighter catalyst particles determines the maximum frequency that can be used.

The relationships given above were investigated experimentally by van der Wielen et al. (1998). An immobilized glucose isomerase with an average diameter of 1.25 mm and a density of 1100 kg/m^3 was the light material, while a palletized CaY zeolite with a particle diameter of 1.27 mm and a density of 1362 kg/m^3 served as the higher density material. Variables studied were frequency, volume of the reverse flow chamber and the liquid fluidizing velocity.

Segregation into solid zones or layer was observed and the layer ordering was correctly predicted by the segregation function calculated from Equation 6.13. Figure 6.8 compares predictions of the solid flow rate with the investigator's experimental measurements. It can be seen that with small flow reversal chambers and higher frequencies the model works reasonably well. Deviations become serious when the chamber volume is increased or the frequency is reduced.

As expected from Equation 6.15, the time average liquid flow through the reactor dropped as the reverse flow volumetric rate, $f_p V_p$, increased. The multistaged reactor exhibited a close approach to plug flow in the absence of pulsing as indicated by a Peclet number of about 50. Liquid phase mixing increased with pulsing and the Peclet number fell exponentially with the volumetric reverse flow rate, $f_p V_p$, reaching values of 2–4 at the pulsing limit. Thus, the

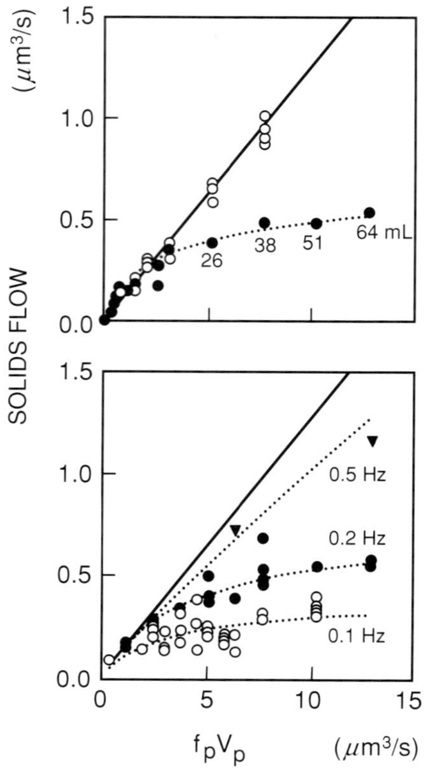

Figure 6.8 Volumetric downflow of the denser particles as a function of the volumetric flow reversal rate: (a) effect of the size of the flow reversal chamber, ○ = $V_p < 15$ mL, ● = $V_p > 15$ mL; (b) effect of pulsing frequency, ○ = 0.1 Hz, ● = 0.2 Hz and = 0.5 Hz. ▼ = Equation 6.14. (Figures adapted from van der Wielen et al. (1998) with permission. © 1998 AIChE.)

liquid in the reactor column was almost fully back-mixed at the operational limit. The key reactor operating parameter, the ratio of adsorbent volumetric downflow to fluid volumetric upflow rates should be the quotient of Equations 6.14 and 6.15. The predicted ratio is the solid line in Figure 6.9. It closely represents the data up to the highest levels of the reverse flow volumetric rate where a negative deviation sets in.

Unfortunately, application of this interesting pulsed, multistage fluidized bed to bioreactions has not been reported.

6.3 Experimental studies

Several different investigators have tested the rotating annular bed concept and established its feasibility. Likewise the trickling solids concept of Westerterp and Kuczynski (1987a) has been demonstrated in both gas–solid and liquid–solid systems. The pulsed multistaged fluidized bed separating reactor appears to have been used by a University of Amsterdam

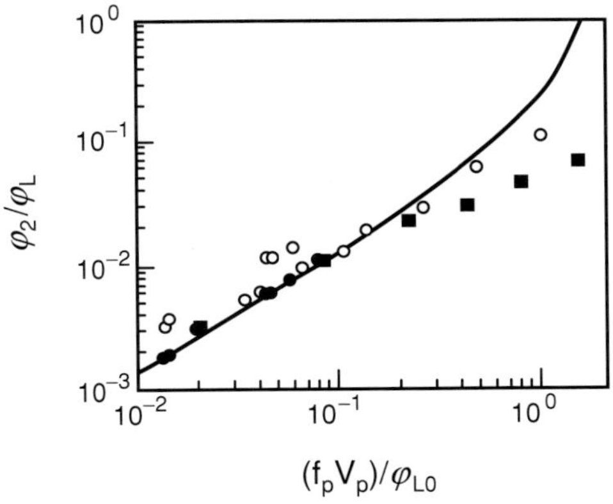

Figure 6.9 Ratio of solid downflow to liquid upflow rates as a function of the volumetric reverse flow rate divided by the liquid upflow rate during the fluidization portion of the cycle. Symbols: ○ = CaY zeolite, ● = immobilized glucose isomerase, ■ = CaY zeolite with $V_p > 15$ mL. (Figure adapted from van der Wielen et al. (1998) with permission. © 1998 AIChE.)

team for almost a decade, but it has not attracted the attention of other investigators. The system does not seem to have been tested experimentally.

6.3.1 Continuous rotating annular chromatographic (CRAC) reactors

The experimental reactor employed by Cho et al. (1980a) for methyl formate hydrolysis appears in Figure 6.10. Outer diameter of the annular vessel was 20.3 cm, the width of the ring-shaped catalyst bed was 19 mm and the depth was 40.6 cm. Feed and withdrawal ports were 1 mm in diameter. Upflow was used and there were 1560 holes (1 mm) in the flange that formed the bottom of the annular bed. Only the reactant injection nozzle rotated. There were 50 withdrawal ports that were stationary.

The annular bed was carefully packed with 60–80 mesh activated coconut charcoal to insure uniformity. Flow rate of the aqueous carrier was varied between 10 and 20 mL/min, while the methyl formate rate was 3% of the carrier liquid. The feed port rotated with an angular velocity set between 0.01 and 0.05 rad/min. Figure 6.11 shows product concentrations measured at the outlet ports during a rotation after steady state was achieved. No methyl formate was detected.

Equilibrium conversion under the conditions used is about 75%. Consequently, the figure indicates that the chromatographic effect and dilution drive the reaction to completion. Lack of a distinct peak for formic acid indicates strong tailing due to adsorption on surface sites of different heats of adsorption. Some tailing is evident for methanol, even though a better recovery of that product is possible. A better choice of adsorbent would have resulted in sharp elution peaks and good product separation.

As mentioned in the previous section, the chromatographic model, Equation 6.2, assuming rapid adsorption and the use of the Freundlich isotherm as well as reaction kinetics from the literature represents the performance of the rotating bed CR quite well. Product tailing and complete conversion of methyl formate were predicted.

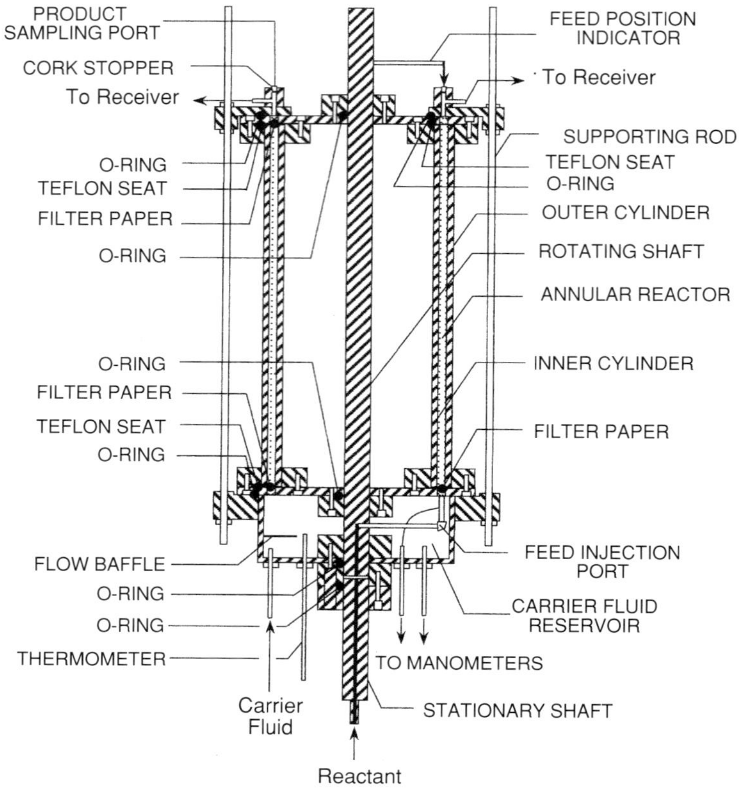

Figure 6.10 Rotating annular-bed chromatographic reactor. (Figure adapted from Cho et al. (1980a) with permission. © 1980 Pergamon Press Ltd.)

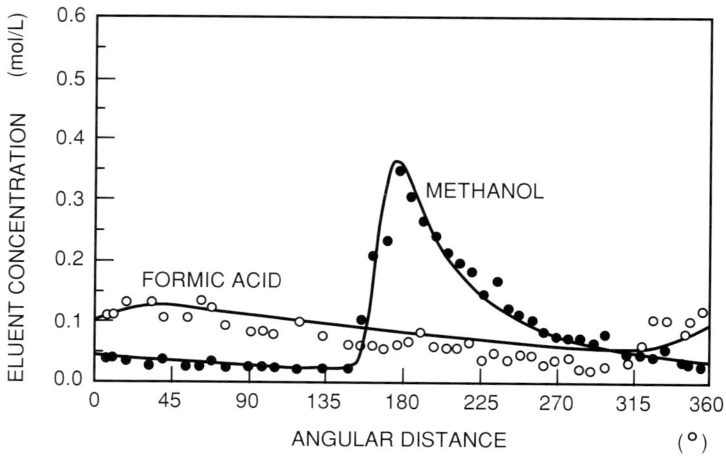

Figure 6.11 Product concentration as a function of sampling point (angular distance) for methyl formate injection at 0° with a concentration of 3.0 mol/L, carrier 1N HCl, $u = 1.04$ cm/min and angular speed of the injection port = 0.03 rad/min, $T = 25$°C. Solid lines are concentrations predicted by the model. (Figure adapted from Cho et al. (1980a) with permission. © 1980 Pergamon Press Ltd.)

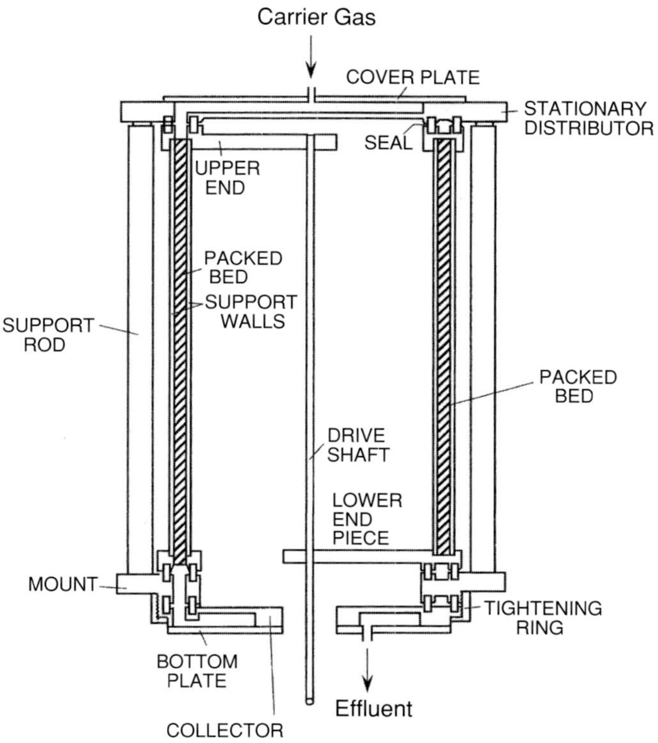

Figure 6.12 Schematic of a rotating annular bed chromatographic reactor used for the catalytic dehydrogenation of cyclohexane. (Figure adapted from Wardwell et al. (1982) with permission. © 1982 American Chemical Society.)

The Cho study was followed quickly by another University of Minnesota project. Wardwell et al. (1982) used a rotating annular bed and studied the dehydrogenation of cyclohexane to benzene over Pt/γ-Al_2O_3 in this system. Figure 6.12 represents a cross section of the reactor they employed. The annular bed was 254 mm in diameter with a height of 457 mm and contained an annular space approximately 7.6-mm wide. This space was filled with 40–60 mesh particles of 0.75 wt% Pt/γ-Al_2O_3 that served as both catalyst and adsorbent. The annular reactor was driven through a variable speed transmission to rotate at 0.4–1.04 rpm. Gas was withdrawn through an annular ring at the bottom that had a single exit. An attempt was made to maintain uniform temperature by operating the system in a thermostatted oven. Samples could be withdrawn from specific points in the annulus so that concentrations of reactants and the products could be measured at different points in the moving bed. Analysis was provided by gas chromatography.

Experimental results at two different temperatures are shown in Figure 6.13. The plots show concentrations on a relative basis at different points in the moving bed. These points are represented in terms of the angular separation between the sampling and feed points. Thus, 0° represents the feed point and 180° is just opposite this point. There is a build up of benzene at positions in the bed opposite the feed point (between 150 and 270°) whereas the

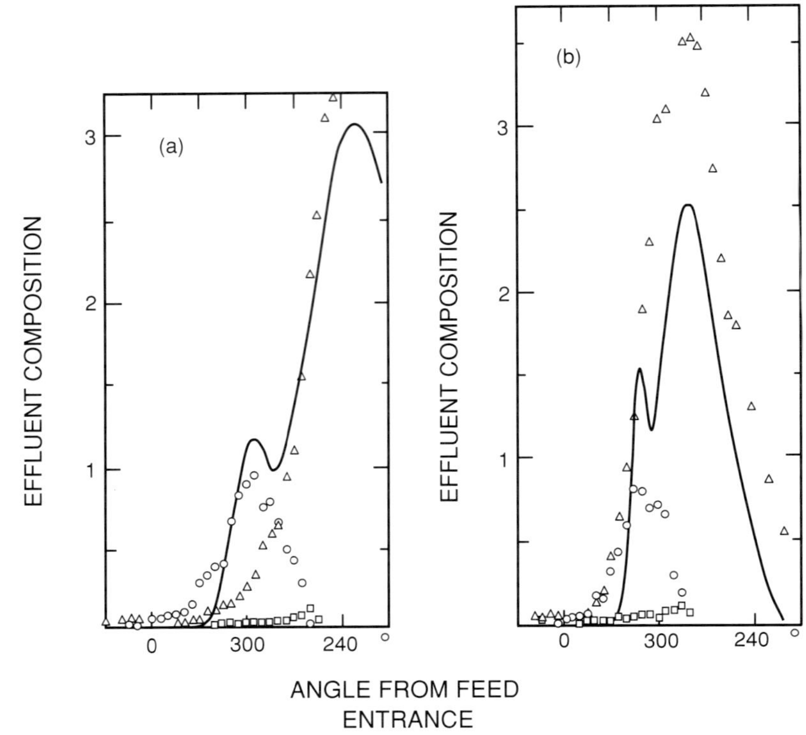

Figure 6.13 Normalized experimental concentrations of reactant (cyclohexane) and product (benzene) as a function of position in a CRAC reactor shown in Figure 6.12: (a) isothermal operation at 477 K, (b) isothermal operation at 500 K. The solid curve is the predicted sum of the normalized benzene and cyclohexane concentrations. (Figure adapted from Wardwell et al. (1982) with permission. © 1982 American Chemical Society.)

peak for cyclohexane is located at about 120° from the feed point. Thus, at 477 K, there is almost complete separation of cyclohexane and benzene. In Figure 6.13a, the flow rate was 2.3 L/min. Increasing the flow rate to 2.9 L/min and the temperature to 500 K brings the benzene and cyclohexane peaks close together so that separation becomes poor. Nevertheless, in both experiments, benzene yields of 87 and 88% were achieved. Equilibrium yields are 13 and 19%, respectively, at the two temperatures used. The figure compares experimental measurements with prediction from simulations that assumed a reactor divided into a two-dimensional array of cells each treated as a back-mixed reactor. Linear isotherms were used but reaction kinetics was nonlinear. It can be seen that the model represented the sum of the benzene and the cyclohexane concentrations reasonably well at 477 K and the lower carrier gas rate, but failed at the higher flow rate and temperature. Under prediction at the higher temperature may be caused by the assumption of isothermality in the model.

The CRAC reactor has been applied to the saccharification of starch by Sarmidi and Barker (1993a) and to another equilibrium-limited biochemical reaction, the enzyme-mediated inversion of sucrose (Sarmidi and Barker, 1993b). A 1.2-cm wide by 140-cm long annulus was used with a stationary feed point and with controllable rotational speed. O-rings between the rotating bed and the stationary housing acted as bearings and also provided a pressure seal. Saccharification used two soluble starches and a tapioca starch with a thermostable, exo-acting α-amylase, maltogenase that hydrolyzes the 1,4-α-glucoside linkages forming maltose and an amylopectin, dextrin, when a branch point is present. CRAC experiments used a mobile phase or eluent rate of 8000 cm^3/h and a rotation rate of 240°/h. Feed to the reactor–separator was a mixture of starch and enzyme at 100°C. The CRAC reactor reduced the amount of enzyme needed considerably over that used in a batch reactor. Conversion and maltose yield depended on the starch used. As might be expected, increasing feed rate decreased conversion and yield. Conversion for the best starch ranged from 68 to 82% with yields of pure maltose from 51 to 78%. These results exceeded greatly those achieved in a batch reactor without adsorbent.

The purpose of the Sarmidi and Barker (1993a,b) study of sucrose inversion was to evaluate the performance of the CRAC reactor and to validate the model they developed for the system. Just as in their saccharification work, the calcium form of a Dowex 50W-X4 resin served as adsorbent. The enzyme, invertase, was introduced with the eluent rather than with the sucrose feed. One of the objects of the study was to examine separation of multiple products so the bottom stationary flange was fitted with 50 collection points. Separation was characterized by the resolution defined as the angular difference between two product peaks divided by the average bandwidth of the peaks. Bandwidth is the angular range in which the product is present. Narrow, widely separated peaks in a chromatograph indicate high resolution. The investigators looked for the effect of enzyme activity and eluent flow rate on resolution. Conversion of sucrose was virtually complete under all conditions examined. Increasing activity and eluent flow rate was found to increase resolution. Essentially pure fructose and glucose were obtained over about half of the bandwidth of each product.

The CRAC model for sucrose inversion discussed in Section 6.2.1 was tested by comparing the predicted product concentrations with the experimental ones. Figure 6.14 shows a typical comparison. The simulation is adequate. However, peak heights are smaller than predicted and the model fails to represent the measured peak asymmetry. Nevertheless, the agreement of model and experiment is better than that seen in Figure 6.13.

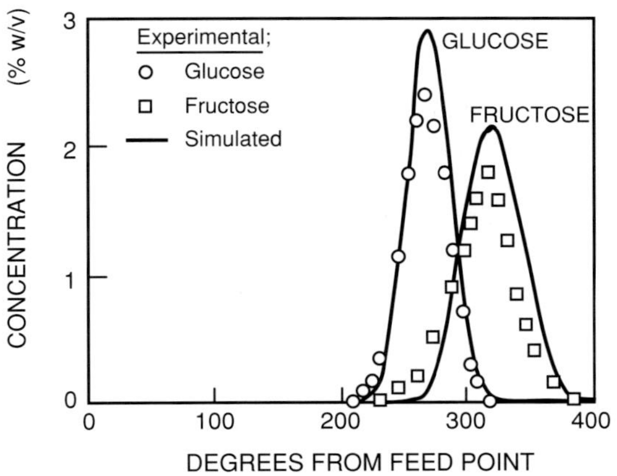

Figure 6.14 Simulated and measured elution profiles for a 135-cm-deep bed for a feed rate of 230 cm^3/h, a sucrose concentration of 25%, eluent flow rate of 8000 cm^3/h, an enzyme activity of 100 units/cm^3, and an angular speed of 240°/h. (Figure adapted from Sarmidi and Barker (1993b) with permission. © 1993 Pergamon Press Ltd.)

6.3.2 Moving bed of adsorbent

Kuczynski et al. (1987b) examined experimentally the physical operation of their trickling solids reactor as well as the application of the concept to methanol synthesis. Kuczynski et al. (1987b) maintained a fixed bed of catalyst, but allowed a solid adsorbent to trickle downward countercurrent to the upward flow of gaseous reactants and the reaction product. Unlike the CMCR systems discussed in Chapter 5, the catalyst operates at steady state. Only the adsorbent experiences a changing composition and temperature environment as it descends through the reactor. Smooth flow of the fine adsorbent powder required the presence of large particles in the bed. Large catalyst particles have small effectiveness factors, however. To overcome this problem, the researchers placed large inert distributors in the packed bed. Figure 6.15 shows the quality of the flow obtained by the authors in a cold flow mockup of their CMCR when Raschig rings were used in the bed.

Kuczynski et al. (1987b) constructed a bench scale reactor that they applied to the synthesis of methanol (MeOH) from CO and H_2 over a Cu/Al_2O_3 catalyst with a silica–alumina adsorbent that adsorbed only MeOH at temperatures above 490 K. Their reactor is depicted in Figure 6.16. Their CMCR was divided into three sections separated by cooling zones that contained no catalyst. Catalyst particle size and the adsorbent used have been discussed earlier. Design of the cooling section and the feeder for the fine adsorbent particles were important problems that required solutions to make the system operable. Schematics of these portions of the system are shown in the figure.

Performance measurements were made with the top of the CMCR open so that some methanol was carried overhead. This always resulted in CO conversions below 100%. Additional experiments were performed with no gas flow overhead so that CO conversion was 100%. Parameters found to influence performance significantly were temperature and the adsorbent–fluid flow ratio, σ, which Westerterp and Kuczynski (1987a) define as $\frac{K_{ad} m_{ad} P}{\rho_{ad} n_g (y_{CO})_0 RT}$. In this ratio, T is a reference temperature (usually the mean reactor temperature), K_{ad} is the adsorption constant for methanol evaluated at T, P is the total pressure, while $(y_{CO})_0$ is the inlet mole fraction of CO and m_{ad} and n_g are the mass and mole flow rates of adsorbent and feed gas, respectively. This ratio, σ, is essentially the phase capacity ratio introduced in the previous chapter as Equation 5.3. Performance at complete CO conversion was expressed as the mean rate of MeOH formation in moles of MeOH/s·kg-cat. For the open CMCR, CO conversion was used. In closed operation, mean rates of MeOH production reached 1.55 mol MeOH/s·kg-cat at 513 K, but with an adsorbent–fluid flow ratio of 22 and a high solid feed rate of 500 kg-adsorbent/s. Lower temperatures and lower adsorbent–fluid flow ratios sharply reduced this mean production rate. With open operation, the adsorbent–fluid flow ratio ranged from about 2 to 10. At temperatures in excess of 510 K, and adsorbent–fluid flow ratios between 4 and 10, conversions in excess of the thermodynamic limit (ca. 70%) were obtained. The highest conversion measured was 91%.

Kuczynski et al. report that their model reproduced experimental results adequately for both open and closed operations of the CMCR. Both theoretical studies and experiment demonstrate not only the attraction of CMCR technology for equilibrium-limited reactions, but also its drawback: high solids circulation rates when reactions occur at elevated temperatures where adsorption capacity is low. The experimental work, in particular, highlighted the operational and design problems associated with moving beds of solids.

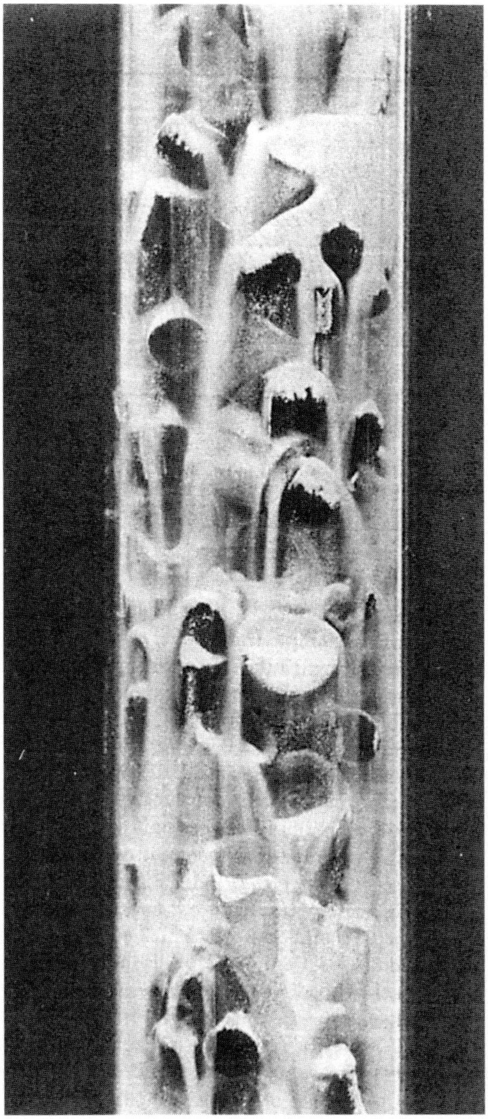

Figure 6.15 Trickle flow of a fine solid adsorbent through larger catalyst particles and a Raschig ring packing under a cold flow condition. (Figure adapted from Westerterp and Kuczynski (1987b) with permission. © 1987 Pergamon Journals Ltd.)

In a follow-up study, Westerterp et al. (1989) replaced the silica–alumina adsorbent by a liquid phase absorbent for methanol. For their investigation, they split the catalyst bed into three sections and contacted the effluent from each section with the liquid phase absorbent that removed methanol and cooled the synthesis gas. The authors reported conversions exceeding the thermodynamic limit for the reaction conditions. Their system, however, is now quite far removed from the periodic systems we are discussing in this review.

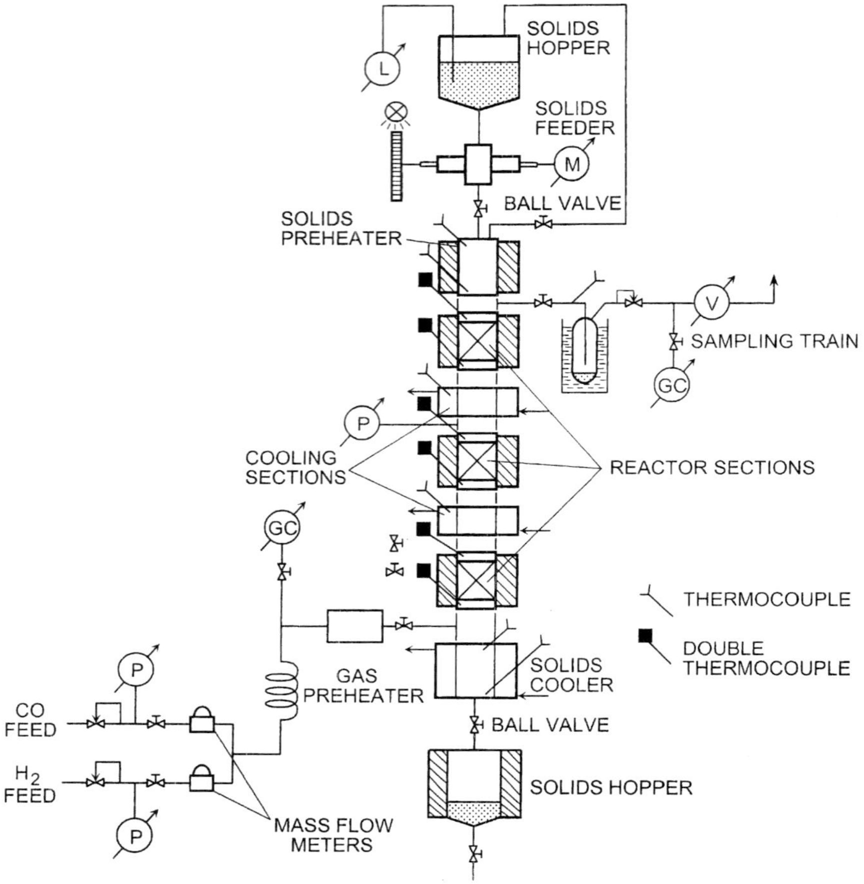

Figure 6.16 Schematic of the high pressure trickling solid fixed-bed catalytic reactor employed for methanol synthesis at 50 MPa and 490–525 K. (Figure adapted from Kuczynski et al. (1987b) with permission. © 1987 by Pergamon Journals Ltd.)

Brun-Tsekhovoi et al. (1986) also examined continuous adsorbent feed but they employed a fluidizing bed of catalyst. Their system is essentially a screening reactor to test if a system is suited to the use of a separating reactor. We discuss their work in Chapter 14.

Zhu et al. (1999) applied the liquid adsorbent proposal of Westerterp et al. (1989) to the synthesis of ethylene oxide over silver catalyst in a slurry reactor where the liquid phase, silicon oil, functioned as an absorbent for ethylene oxide. Like the Brun-Tsekhovoi paper just mentioned, we consider the Zhu et al. (1999) system primarily as a method for screening candidate systems for the possible application of cyclic separating reactors. We review the Zhu work in Chapter 14.

Experiments of van der Wielen et al. (1996) using raining solids in a liquid system employed the deacylation of penicillin G using acylase immobilized in a gelatin/alginate matrix and a strongly basic anion exchange resin, IRA 400, for product separation and pH control.

The separating reactor is schematically similar to the system shown in Figure 6.7. The pulsing system was not used in normal operation. Furthermore, the perforated plates that divided the column into stages were removed except for plates at the top that controlled the feeding of the adsorbent at a uniform rate and one at the bottom for adsorbent removal. The less dense catalyst was fluidized by the up flowing liquid, while the denser adsorbent "rained" or settled downward through the fluidized bed. The system operated at steady state.

Feed for the experiment was an aqueous penicillin G solution of 50 mol/m^3 at pH 8 with a K_2PO_4 buffer solution also at 50 mol/m^3. Adsorbent to liquid fluxes as volume ranged from 0.11 to 0.14.

Hydraulically, the separating reactor functioned satisfactorily. Without the "raining" solid adsorbent, the reactor reduced the penicillin G concentration from 50 to 33 mol/m^3. Using the adsorbent, the enzymatic conversion was driven to completion. Figure 6.17 compares the concentration profile for reactant and products in the liquid fluidized enzymatic reactor without adsorbent with profiles obtained with a downward moving-bed of ion-exchange adsorbent. Comparison of the data points and the model predictions represented by the

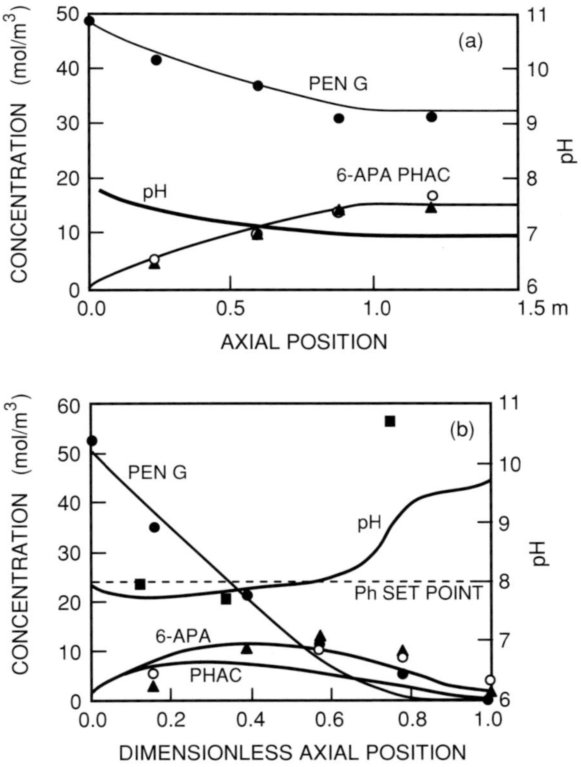

Figure 6.17 Reactant and product concentration versus dimensionless axial position in a fluidized bed enzymatic reactor: (a) without adsorbent, (b) with a countercurrent moving bed of adsorbent at a volumetric solids to liquid ratio = 0.11 at 310 K. Symbols represent experimental measurements. Lines are model predictions. (Figures adapted from van der Wielen et al. (1996) with permission. © 1996 Elsevier Science Ltd.)

curves in the figure show good agreement in (a) in which U_s was set equal to zero. The agreement is poorer in (b), which is the moving-bed operation.

In the absence of ion-exchange resin, conversion of penicillin G lowers the pH. This change together with increasing products concentration inhibits the reaction and drives the conversion rate to zero. When the reactor runs with a descending flow of ion-exchange resin, the pH is held constant in the bottom of the reactor through hydroxide ion release. The buildup of the reaction products in the liquid phase is suppressed by adsorption and ion exchange on the resin. This permits complete conversion of penicillin G. 6-Aminopenicillanic and phenylacetic acid go through maxima in the middle of the reactor reflecting their adsorption on the resin in the upper half of the reactor. Adsorption of products by fresh resin entering the top of the reactor sets up a pH shock front clearly indicated by the pH of 11 data point at the 0.8 position. The front is also predicted by the model, but appears as a steep concentration gradient because of mixing in the bed. The existence of shock fronts and their explanation has been discussed in Chapter 5.

Chapter 7

Simulated Countercurrent Moving-Bed Chromatographic Reactors (SCMCR)

7.1 Concept

Just as the moving-bed chromatographic reactor had its genesis in a purely adsorption process so did the simulated countercurrent moving-bed chromatographic reactor arise from preparative chromatography. Barker and coworkers discuss in several papers a three-section simulated moving-bed chromatographic column (Barker and Deeble, 1975; Barker et al., 1983, 1986; Barker and Ching, 1980). The four-section SCMCR, however, is more frequently used. Liapis and Rippin (1979) demonstrated that dividing a chromatographic column into sections and switching feed between these sections approximates satisfactorily the performance of an ideal moving-bed column with a side point feed. Applications of the simulated moving-bed chromatographic adsorber are thoroughly discussed by Ruthven and Ching (1989).

The extension to systems with both reaction and adsorption appears to have been made first by Zabransky and Anderson (1977) in a US patent. Soon thereafter, Hashimoto et al. (1983) applied the concept to a selectivity problem, the conversion of sucrose into fructose. However, the design they employed differed substantially from the design shown in Figure 3.3 at the beginning of this review. That design, a single packed-bed column with changing inlet and outlet port locations, is similar to the UOP Sorbex separator (Broughton, 1961; Broughton et al., 1970), and has been described by Fish et al. (1988). The Hashimoto system employed multiple beds with separate catalyst and adsorbent.

A chromatographic reactor is effective when reaction products have different adsorptivities. As discussed in Chapter 3, a reactant pulse introduced into a chromatographic reactor consisting of catalyst and adsorbent undergoes reaction accompanied by separation of reactant and products through adsorption. Continuous flow of a carrier or eluent separates the products spatially into a weakly adsorbed component that moves with the carrier and a more strongly adsorbed one that lags behind the carrier flow. The carrier or eluent also regenerates the adsorbents by the desorbing the more strongly adsorbed product. In the simple CR these functions occur at different times after introduction of the feed pulse. In a CMCR, they occur at different axial points in the reactor because the operation is continuous. It is convenient to consider these points as sections of the CMCR. Identical sections are found in a SCMCR as might be expected, but these sections are not fixed in space. They move around the reactor as the feed and withdrawal ports change their location. Figure 7.1 compares the CMCR and SCMCR in terms of these section functions. Adopting nomenclature from separation technology, the adsorbent removes or extracts a product from the flowing fluid phase. This product, the more strongly adsorbed one, is the extract. The weakly adsorbed

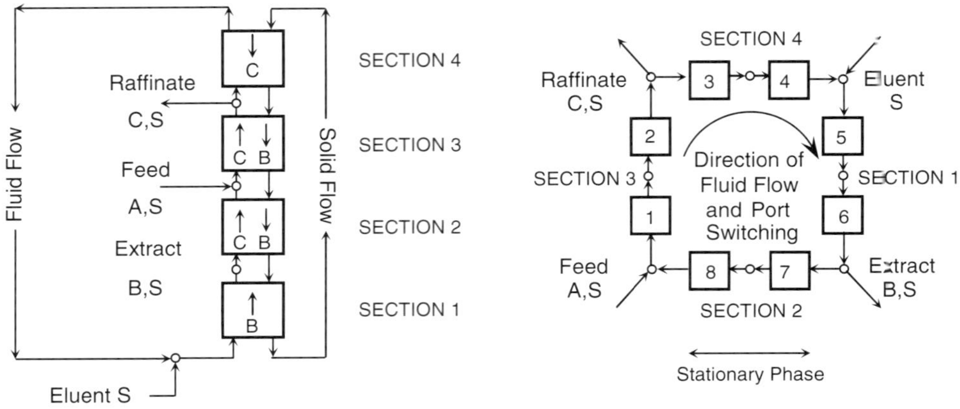

Figure 7.1 Functional comparison of moving-bed and simulated moving-bed chromatographic reactors for an equilibrium-limited reversible reaction. (Figure adapted from Lode et al. (2001) with permission. © 2001 Elsevier Science Ltd.)

product is the raffinate. A section in the SCMCR is made up of two beds or columns in Figure 7.1. There can be more than two beds and the different sections shown on the RHS of Figure 7.1 may not each have the same number of beds. The more beds a section contains, the closer is the approximation to a CMCR. Indeed, Morbidelli and coworkers (Lode et al., 2001) have devised equivalence relations. For a section in an SCMR

$$V_j = n_j V_{\text{bed}} \tag{7.1}$$

$$\tau_s = (1 - \varepsilon_{\text{bed}}) \frac{V_{\text{bed}}}{(Q_s)_{\text{CMCR}}} \tag{7.2}$$

$$(Q_{\text{fluid}})_{\text{SCMCR}} = (Q_{\text{fluid}})_{\text{CMCR}} \left(1 - \varepsilon_{\text{bed}} \frac{V_{\text{bed}}}{\tau_{\text{switch}}} \right) \tag{7.3}$$

In the first equation, V_j is the volume of a section in either a SCMCR or a CMCR, while V_{bed} is the volume of a column in that section of the SCMCR. τ_s is the time between changes of port location in the SCMCR and Q_s is the volumetric rate of solids flow in the equivalent CMCR. Q_{fluid} are the flow rates of the fluid phase in the same section of the CR.

As pointed out in Chapter 3, an SCMCR eliminates the necessity of circulating a solid phase through a reactor vessel and removes the associated problems of introducing and withdrawing this phase, as well as maintaining uniform radial distribution. Attrition of the solid is avoided so there is no need to remove fines. Probably solids inventory can be reduced because collection of solids at the discharge from the bed and hold up at the feed point are no longer needed. These advantages must be weighed against the cost of extra piping and valves and the automated operation that now becomes necessary.

7.2 Isothermal modeling

Multiple packed-bed columns in series are discussed and analyzed by Ray et al. (1990). A model applicable to either single or multiple beds was developed by Ray et al. (1990, 1994) as well as by many other investigators. To simplify the model and its application, Ray assumed rapid mass transfer, isothermal operation and no pressure drop in the bed. The model incorporating these assumptions comes from Table 3.2:

$$\varepsilon_t \frac{\partial C_{ij}}{\partial t} + \rho_b \frac{\partial q_{ij}}{\partial t} = D_{x_{ij}} \frac{\partial^2 C_{ij}}{\partial x^2} - u_{f_j} \frac{\partial C_{ij}}{\partial x} - \rho_b \nu_i r_j \tag{7.4}$$

Of course, the solids flow term has been eliminated. Since the fluid phase is dilute and the reactor isothermal, the molar flux n_i can be replaced by the product of the concentration and the superficial velocity of the fluid, u_f, which is now constant through the bed as well as through the section. The j subscript indicates the jth bed in the multicolumn reactor. Ray et al. (1990) left out the dispersion term because local mixing is usually small and hardly affects reactor performance. More recent investigators, however, leave the term in place because it allows for the minor contribution of local mixing, but also compensates for error introduced by neglecting mass transfer and property changes along the flow path. The SCMCR contains j beds or columns and Equation 7.4 must be written for all reactants. With i independent species, usually reactants, a SCMCR must be described at the minimum by $i \times j$ PDEs. Finding a solution to estimate reactor performance is often a formidible undertaking. The set of equations must be integrated after each switching duration, τ_s. The initial condition at $t = 0$ after a switch is $C_{i_j}(x) = C_{i_{j-1}}(x, \tau_s)$, $q_{i_j}(x) = q_{i_{j-1}}(x, \tau_s)$ where $C_{i_{j-1}}$ and $q_{i_{j-1}}$ are the concentration and the adsorbate loading at the same position, x, in the upstream column just prior to switching the feed and withdrawal points. Other boundary conditions are contained in Table 3.5, bearing in mind that the position at which the boundary conditions apply varies with time. The model just given has been widely used (Kruglov, 1994; Tonkovich and Carr, 1994b; Migliorini et al., 1999a; Lode et al., 2001). Other investigators allow for fluid–solid mass transfer requiring the use of a heterogeneous model, such as given by the PDEs in Table 3.3.

Ray et al. considered a reversible reaction, A $\leftrightarrow$ B, using mesitylene (MES) hydrogenation to 1,3,5-trimethylcyclohexane (TMC) as an example. Key parameters are switching time, τ_s, and a switching velocity, ζ, representing the hypothetical velocity of the solid phase. ζ is defined as $\Delta x/\tau_s$ where Δx is the spacing between inlet and outlet of a bed. With this definition, the switching velocity can be related to the critical CMCR parameter, σ_i, by

$$\sigma_i = \frac{1-\varepsilon}{\varepsilon} N K_{eq} \frac{\zeta}{u_g} \tag{7.5}$$

where N and K_{eq} represent the number of adsorption sites and the equilibrium constant, respectively. By adjusting the gas velocity, u_g, and/or switching velocity, ζ, the condition $\sigma_A < 1$ and $\sigma_B > 1$ necessary to achieve separation and high conversion for a reversible reaction can be obtained.

Ray et al. (1994) used their model to examine the effect of inlet and outlet port location on reactor performance and on the time varying concentration profiles in the catalyst and adsorbent bed. Figure 7.2 shows the evolution of the concentration profiles of the reactant (MES) and the product (TMC) with dimensionless time representing hydrogenation by a simple equilibrium reaction A $\leftrightarrow$ B. The figure assumes a column divided into 20 stages with 8 stages in the central sections between feed and outlet ports and 6 stages above the feed and below the outlet. With respect to Figure 7.1, the SCMCR of Ray et al. has only three sections (Sections 2–4). After 60 units of dimensionless time (corresponding to 300 s), a cyclic stationary profile is achieved. It can be seen that for an appropriate location of the

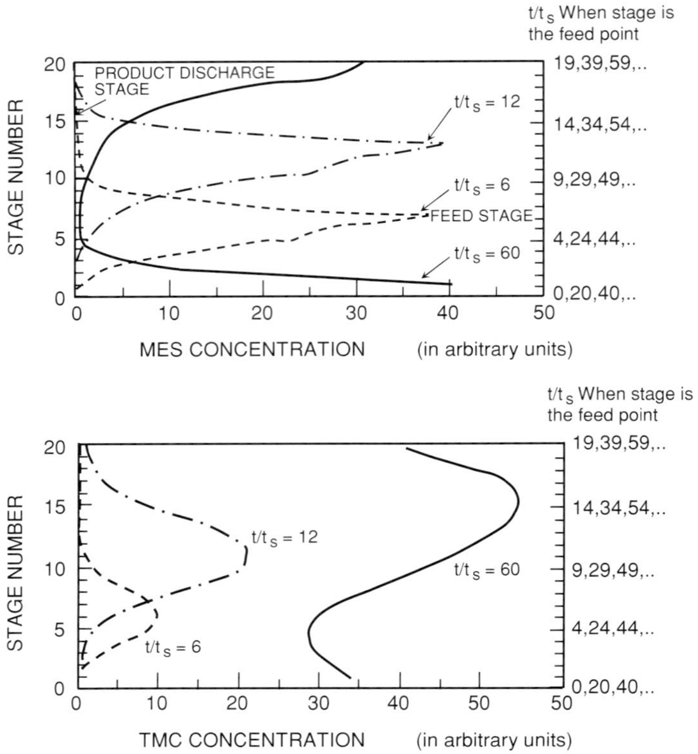

Figure 7.2 Concentration profiles as functions of stage number and normalized time after startup for an SCMCR: The feed and removal points are at stages 7 and 15 ($t/t_s = 7$), stages 13 and 21 ($t/t_s = 12$), stages 1 and 9 ($t/t_s = 60$), respectively. Right-hand axis shows the normalized times at which the stage on the left becomes the feed stage. (Figure adapted from Ray et al. (1994) with permission. © 1994 Pergamon Press Ltd.)

outlet port nearly pure TMC is obtained with almost complete conversion of MES. The figure also shows that a higher concentration of TMC can be realized but with a penalty that separation and conversion are incomplete. The authors point out that for a cyclically stationary profile with withdrawal at stage 9, MES conversion is 97.7% and the product stream contains 98.5% TMC (Fig. 7.2). In contrast to this performance, the equilibrium conversion in a PFR of sufficient length is 62%. Ray et al. explored the effect of increased switching frequency or decreased ζ. They showed that the MES conversion can be increased further to 99.8% and TMC purity to 99.3%. However, if they use a switching frequency of 10 s to ensure that σ_A and σ_B are greater and less than unity, respectively, the gas velocity and the length of the reactor must be changed. If velocity is decreased from 5 to 3, and the length increased 20% to 500, that is increasing residence time, the best performance is obtained using the 11th stage for the feed entry point. Decreasing the switching time reduces conversion while increasing the length raises conversion to 99.9%, but the purity of TMC in the product stream is reduced slightly. Increasing the feed rate holding the other parameters as constant as possible decreases product purity. This also occurs if the concentration of MES

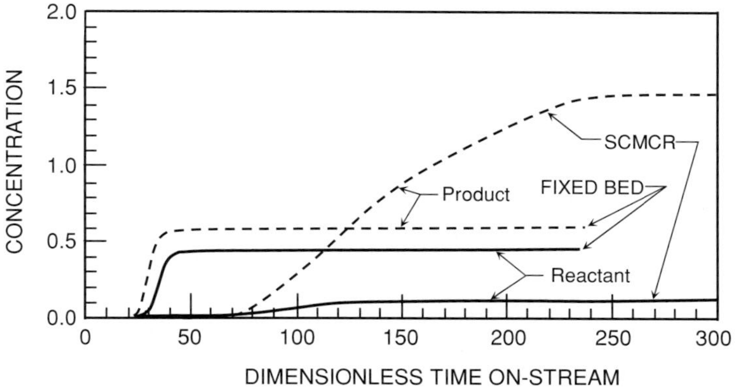

Figure 7.3 Transient concentrations of reactant and product for a fixed-bed reactor (PFR) and for a SCMCR at a switching time of 10 s and a carrier flow rate of 10 cm·s^{-1}. (Figure adapted from Ray et al. (1994) with permission. © 1994 Pergamon Press Ltd.)

in the feed is increased. Although the performance is sensitive to the switching frequency, bed length, feed flow rate and feed concentration, the SCMCR is nevertheless superior to the plug flow fixed-bed reactor as illustrated in Figure 7.3.

The effect of the process parameters on the performance of a SCMCR was examined for a reversible decomposition, A $\leftrightarrow$ B + C, by Fricke et al. (1999). They employed a SCMCR consisting of four sections as illustrated in Figure 7.1 and also in Figure 7.4, taken from their paper. The net fluid flow and the simulated solid flow are shown as solid and dashed lines, respectively. Reactant A is fed into the system between sections 2 and 3. Product B was assumed to be more strongly adsorbed than C. Therefore, C propagates toward the raffinate node and can be obtained at a high concentration between sections 3 and 4 whereas B moves toward the extract node. As an eluent is fed between sections 4 and 1, the total flow rate is greater there than between any other sections. In section 1, high concentration of eluent forces B to desorb where it is then recovered as an extract.

In their model, Fricke et al. considered axial dispersion and interphase mass transfer. They assumed that reaction takes place only in the liquid phase, that is, homogeneous reaction kinetics was employed. Adsorption followed a linear adsorption isotherm. The mass balance employed by the authors is based on Equation 3.7 in Table 3.3 without the convection term

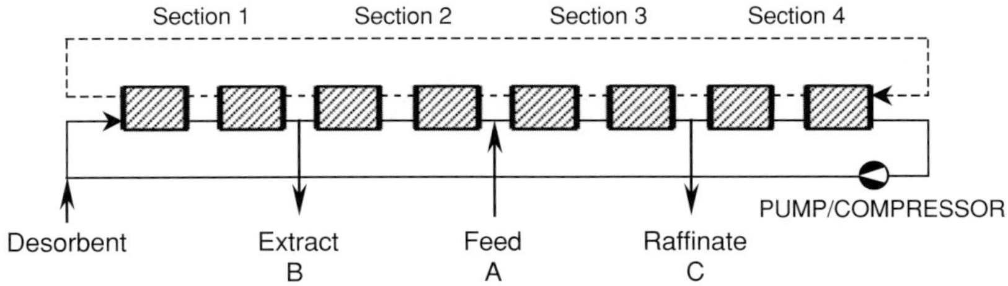

Figure 7.4 Schematic diagram of the multibed SCMCR studied by Fricke et al. (1999). (Figure adapted from Fricke et al. (1999) with permission. © 1999 Elsevier Science Ltd.)

for the adsorbate. Thus,

$$\varepsilon_t \frac{\partial C_{i_j}}{\partial t} = D_{x_{ij}} \frac{\partial^2 C_{i_j}}{\partial x^2} - \frac{1}{A}\frac{\partial n_{i_j}}{\partial x} - k_{m_{ij}} a_m (\rho_b)_{\text{ads}} \left(C_{i_j} - \frac{q_{i_j}}{K_i} \right) - \varepsilon_b \nu_i r_j \tag{7.6}$$

$$\frac{\partial q_{i_j}}{\partial t} = k_{m_{ij}} a_m \left(C_{i_j} - \frac{q_{i_j}}{K_i} \right) \tag{7.7}$$

As above, the j subscript indicates the bed. Note also that the volumetric mass transfer coefficient is based on mass of the adsorbent. Fricke et al. assume that the rate of change of an adsorbed component is given by a linear mass transfer relationship employing an overall mass transfer coefficient for transport between the liquid and the adsorbent surface. The reaction A $\leftrightarrow$ B + C was assumed to be elementary. Boundary conditions are given in Table 3.5. Initial conditions after a switch in port location have been discussed previously.

Performance of the reactor was judged by two criteria: the maximum feed flow rate to achieve a product purity of 99.75% and the specific solvent or desorbent consumption. Their study examined the effects on performance of the linear adsorption constant of the reactant, the separation factor for the products (defined as the ratio of adsorption constants of B and C), the reaction rate constant and the chemical equilibrium constant.

Figure 7.5 shows an example of their calculation results. Clearly, reactant A is present only in sections 2 and 3 so C can be obtained with high purity in the raffinate. Also, nearly pure B can be collected in the extract by introduction of a solvent or desorbent between sections 4 and 1.

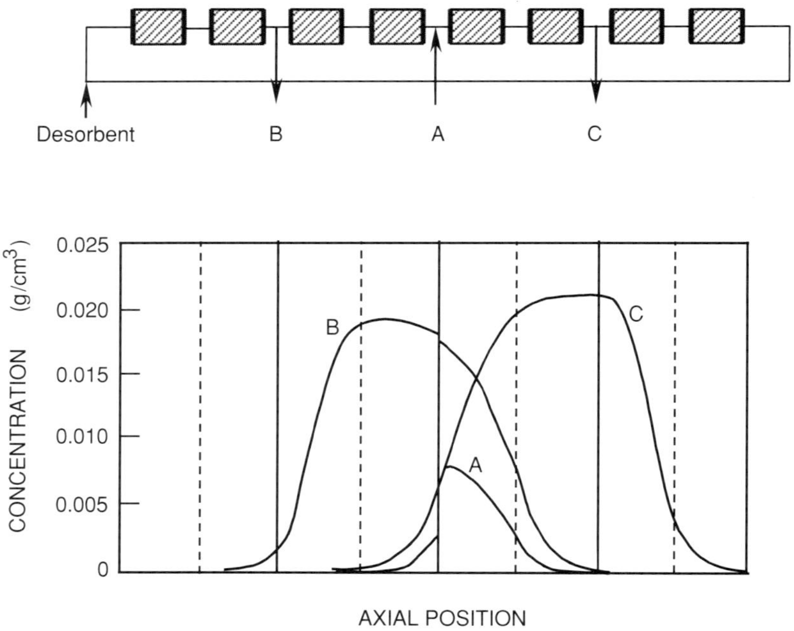

Figure 7.5 Concentration profiles of reactant and products in an SCMCR at the end of a cycle. (Figure adapted from Fricke et al. (1999) with permission. © 1999 Elsevier Science Ltd.)

In order to satisfy the product purity requirement, reactant A must be totally converted in sections 2 and 3. For this to be true, its adsorption equilibrium constant K_A must lie in between K_B and K_C. The throughput of reactant A depends strongly on the overall mass transfer coefficient for A and just weakly on its adsorption constant. Throughput increases with the separation factor. Furthermore, specific consumption of desorbent, volume of desorbent per volume of feed, decreases rapidly as the separation factor approaches 2 but by a separation factor of 4, consumption becomes constant. These results indicate that a high separation factor provides both high purity products and high throughputs. Effects of the reaction rate and the reaction equilibrium constants are similar to that of the separation factor. The observations of Fricke et al. are in good agreement with results of earlier CR and CMCR investigations.

A comprehensive exploration of factors influencing SCMCRs for a $A + B \leftrightarrow C + D$ reaction has been undertaken by a team working at the ETH Zurich (Migliorini et al., 1999a; Lode et al., 2001) using esterification of methanol and acetic acid over an ion-exchange resin as the model reaction. Methanol in this system is both reactant and solvent or eluent. The resin acts as catalyst as well as adsorbent for water. The solid phase exhibits swelling that is component and concentration dependent so that the void fraction shifts constantly within a bed and differs as well between beds making up the SCMCR. Methanol, acetic acid and reaction products have different densities, thus convective mixing driven by density gradients is possible. Migliorini et al. (1999a) based their analysis on the PDE model given earlier, while Lode et al. (2001) noted that SCMCRs and CMCRs perform similarly with respect to key operating parameters and thus used a CMCR model for their work. Both types of chromatographic reactors consist of sections in which regeneration of adsorbent, recovery of extract, reaction and enrichment of raffinate, and enrichment of extract dominate. Using the equivalence relations, Equations 7.1–7.3, simulation results for a CMCR can be applied to an SCMCR.

The ETH Zurich researchers consider the SCMCR system treated by Fricke et al. (1999) shown in Figures 7.1 and 7.4 and not the three-section design examined by Ray et al. (1994). Employing their studies of simulated countercurrent moving-bed chromatographic separators (Storti et al., 1988, 1993; Mazzotti et al., 1997a), they identified sections 2 and 3 in Figure 7.1 as critical for performance, provided sections 1 and 4 are operated to completely recover the raffinate (weakly adsorbed product) and regenerate the adsorbent. This will be the situation normally encountered. In Chapter 5, we observed that performance of a CMCR as a separating system depended on an absorptivity ratio, κ, and a flow rate ratio or a phase capacity ratio σ_i. As might be anticipated, this continues to be true for the SCMCR and the flow rate ratios in sections 2 and 3 control the separation. For the SCMCR, the flow rate ratios are

$$\sigma_j^* = \frac{Q_f \tau_s - V_j(1-\varepsilon_b)\varepsilon_p}{V_j(1-\varepsilon_b)(1-\varepsilon_p)} \tag{7.8}$$

This is a version of Equation 7.5 expressed as volume rather than adsorption capacity. The index j represents a column, that is, a section in Figure 7.4, V_j is the volume of the column and τ_s is the time duration between port changes. Because with complete regeneration of the adsorbent, reaction occurs only in sections 2 and 3, the range of acetic acid conversion and separation between ethyl acetate and water must be represented in the $\sigma_2{}^*$–$\sigma_3{}^*$ plane. Separation performance depends on τ_s, the volumetric flow rate and the bed volume. Feed composition in terms of the acetic acid to ethanol ratio is also important. Ethanol is a reactant as well as the carrier, therefore, pure acetic acid can be fed to the reactor. Diluting the acid with ethanol, however, improves conversion and product purity or rather enlarges the region in the $\sigma_2{}^*$–$\sigma_3{}^*$ plane in which uncontaminated products can be found (Migliorini et al., 1999a). Normalizing Equation 7.4 introduces the Damköhler number. This important parameter is affected by the residence time so that each of the four sections in the SCMCR may have a different number. Fluid flow rates are usually different and the number of beds in each section can be varied even if the total volume of adsorbent and catalyst is kept constant. Once again, since the esterification reaction will usually be confined to sections 2 and 3, it is the residence times or the Damköhler numbers in these sections that matter. Lode et al. (2001) show that if most of the esterification reaction occurs in section 3, increasing the Damköhler number in this section increases conversion, if it is not already complete, and raises productivity.

Migliorini et al. (1999b) and Lode et al. (2001) discuss the design of a SCMCR for equilibrium-limited reactions, emphasizing the similarity of design to a simulated moving-bed chromatographic separator. Both units generally have four zones or sections in which sections 1 and 4 provide regeneration of the adsorbent. For best performance of an SCMCR according to Migliorini, reaction should not occur in these sections. As a consequence, separation and the extent of the equilibrium-limited reaction is governed by behavior in sections 2 and 3 where the flow ratios or Aris numbers, σ_2 and σ_3, and the Damköhler numbers, $(N_{\mathrm{Da}})_2$ and $(N_{\mathrm{Da}})_3$, are important. Neglecting reaction, choice of carrier flow and switching times for beds of a given size should be within a triangle bordered by the diagonal in the σ_2–σ_3 plane if adsorption equilibrium can be represented by a linear isotherm. Within the triangle the more strongly adsorbed component can be obtained at the extract port without contamination by the weakly adsorbed component. Similarly, the weakly adsorbed component is removed in the raffinate port, free of the other component. Complete separation is achieved. Figure 7.6a shows the σ_2–σ_3 plane for this case. The boundaries of the triangle are

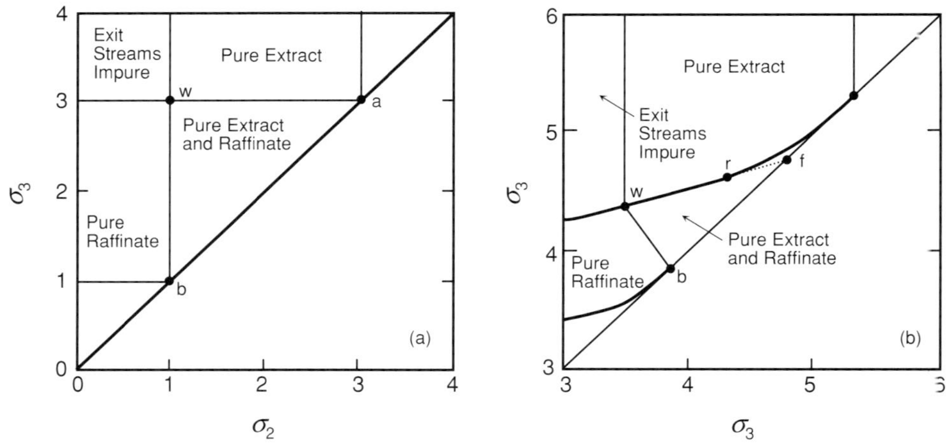

Figure 7.6 Triangle diagrams for the choice of operating conditions in simulated moving-bed chromatographic separators: (a) linear adsorption isotherms, (b) multicomponent Langmuir isotherms for competitive adsorption. (Figure adapted from Mazzotti et al. (1997c) with permission. © 1997 Elsevier Science B.V.

set by the Henry's law constants for the components. If competitive adsorption is present so that a multicomponent Langmuir isotherm is necessary, the shape of the region for optimal choice of flow and switching changes (Fig. 7.6b). Indeed, the shape of the regions in Figure 7.6b depends on the isotherm applicable.

Separation of components or products is preserved, if reaction occurs, as long as the operating conditions are located in the triangle although shape of the triangle in Figure 7.6 will be altered by competitive adsorption of the reactant. Extract and raffinate, however, may be contaminated by the reactant if the reaction does not go to completion. Completion depends upon the Damköhler numbers in sections 2 and 3. Indeed, the region in the σ_2–σ_3 plane for complete conversion is bounded by the diagonal. Fricke and Schmidt-Traub (2003) have studied the change of the complete conversion region with the kinetics of the reaction and the magnitude of the reaction equilibrium constant for both the irreversible and reversible reaction: $A \rightarrow B + C$ where all species adsorb, but adsorption equilibria are given by Henry's law. Figure 7.7 illustrates the effects calculated by these researchers. Migliorini et al. note that an operating point in the triangle furthest from the diagonal offers the highest productivity in terms of the adsorbent for a nonreacting system. Fricke and Schmidt-Traub point out that this is also true for the CMCR and SCMCR.

Design based on the triangle in the σ_2–σ_3 plane should be applicable for reactions of the type $A + B \leftrightarrow C + D$ even when one of the reactants is not the eluent, but one of the reactants will contaminate the products. Migliorini et al. and Lode et al. suggest that design should be undertaken by testing combinations of carrier flow or feed concentration and switching times lying within the triangle, possibly using a search routine. The flow ratios or Aris numbers depend primarily on switching time, carrier flow rate, bed volume and distribution of beds among the SCMCR sections. Lode et al. (2001) undertook a parametric study to estimate the effect on separation, conversion and productivity of varying each of these operating parameters. Migliorini et al. illustrate triangle-based design using a simulation of acetic acid esterification with ethanol and the data of Ching and Lu (1997) on the inversion of sucrose.

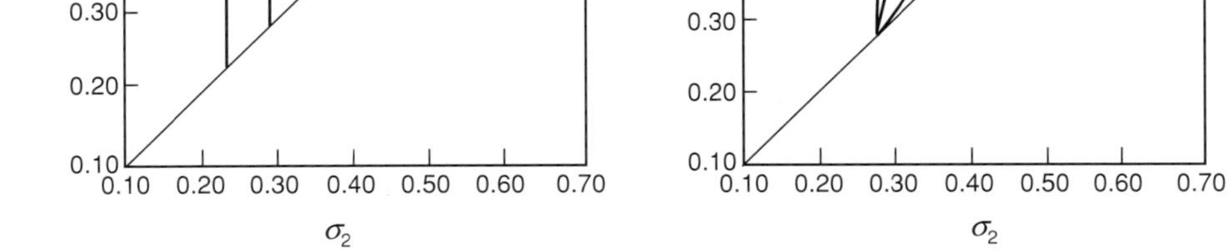

Figure 7.7 Region of complete conversion of the reactant in the σ_2–σ_3 plane in a CMCR or SCMCR for the reaction, $A \rightarrow B + C$, assuming linear adsorption isotherms: (a) effect of the rate constant assuming the reaction is irreversible, (b) effect of the equilibrium constant. (Figure adapted from Fricke and Schmidt-Traub (2003) with permission. © 2003 Elsevier Science B.V.)

Lode et al. (2001) provide experimental verification of the design proposal as we will discuss in Section 7.5.2. Fricke and Schmidt-Traub (2003) derive criteria using a CMCR model that set limiting values for σ_2 and σ_3 for an irreversible reaction A $\rightarrow$ B + C with linear adsorption isotherms. These criteria depend on the Henry's law constant. With a reversible reaction, it appears the flow ratios for a specified conversion can be estimated by a rather complicated iterative procedure. A numerical example suggests that the criteria accurately define suitable choices for the Aris numbers.

Migliorini et al. (1999b) in their consideration of simulated countercurrent moving-bed chromatography for difficult separations point out that in laboratory scale equipment the volume taken up by connections, piping and valving is "dead space" that may account for a significant fraction of the total volume of the preparative chromatographic unit. Dead space affects separation performance in two ways: (1) the measured residence or contact time in the reactor increases, but this does not provide greater conversion or product separation, and (2) mixing in this volume undoes separation provided by the adsorbent. Employing a dispersion model to allow for local mixing in the dead space, Migliorini et al. found that the residence time distortion was the most important effect. Assuming the dead space was fully back-mixed reduced product purity at the take-off ports. Reduction in purity depended strongly on the dead space to total reactor volume ratio. Up to a volume ratio of about 10%, purity decreases were of the order of 1–6%. The residence time problem could be corrected by basing design or analysis on a modified Aris number,

$$(\sigma_j)_{\text{modified}} = \frac{Q_f \tau_s - V_j(1-\varepsilon_b)\varepsilon_p - V_j^{\text{D}}}{V_j(1-\varepsilon_b)(1-\varepsilon_p)} \tag{7.9}$$

where V_j^{D} is the "dead" volume associated with the jth bed or column. It is probably simpler to calculate this dead volume from that associated with the entire laboratory scale unit and divide that value by the number of columns. Although the development was for a separation system, it is probably a good estimate for a SCMCR.

Kawase et al. (1999) investigated production of bisphenol A (BPA) from acetone and phenol in their SCMCR study. An ion-exchange resin catalyzes this liquid phase reaction and functions as the adsorbent. The authors' SCMCR simulation utilized experimental data obtained through batch reactor and fixed-bed breakthrough experiments. The latter were used to obtain adsorption equilibrium constants and mass transfer coefficients. The SCMCR system examined by Kawase et al. had three sections as illustrated in Figure 7.8. Unlike the system used by Fricke et al. (1999), there was no section 1 and so there is no recycle flow from

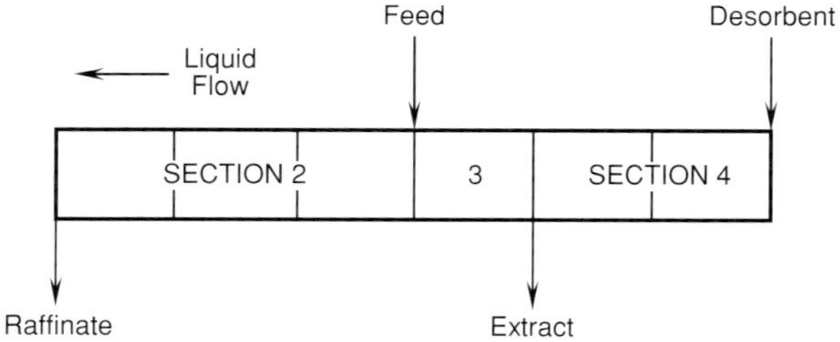

Figure 7.8 Schematic diagram of a simulated moving-bed reactor for production of bisphenol A. (Figure adapted from Kawase et al. (1999) with permission. © 1999 Elsevier Science B.V.)

section 2 to section 4. Recycle could be dispensed with because purification of the desorbent was not necessary due to the weak adsorption of bisphenol A. Phenol served as the solvent. That is, the reaction was conducted in excess phenol. Only acetone is fed between sections 2 and 3. Phenol is fed as desorbent and reactant ahead of section 4.

In their model, Kawase et al. neglected axial dispersion and assumed isothermal operation. Like Fricke et al. (1999), mass transfer between the fluid and solid phases was considered so Equations 7.6 and 7.7 apply. Because there is no homogeneous reaction and the solid phase functions as both adsorbent and catalyst, the rate term must be removed from Equation 7.6 and added to Equation 7.7 in a somewhat different form. The product, BPA, is not adsorbed in their model. A complication, however, is that the adsorption of acetone depends on adsorption of water. A Langmuir isotherm was used for water while for acetone a linear isotherm was assumed with an adsorption constant that depended inversely on the concentration of water. To obtain a rate expression for the reaction, a carbonium ion mechanism was assumed with decomposition of an intermediate species taken to be rate controlling. Under the conditions used, this expression simplifies to give a rate proportional to the concentration of acetone with an inhibition term dependent on the concentration of water. The Kawase et al. model allowed for mass transfer of water and acetone between the liquid phase and the resin surface. Initial and boundary conditions were typical of those used for SCMCRs (see Table 3.5 and the preceding discussion).

Figure 7.9 shows an example of their simulation results. From Figure 7.9a, we can see that strongly adsorbed water is found in all sections whereas BPA and acetone are restricted to sections 2 and 3. Clearly, BPA can be obtained with high purity in the raffinate as shown in Figure 7.9b. At the same time, phenol and water are withdrawn as extract. This stream is much larger than the raffinate stream because it is primarily phenol. Kawase et al. discuss the effect of parameters such as switching time, the number of beds in section 2 and the liquid velocity in section 4 on acetone conversion.

Some years earlier, the Kyoto University group led by Hashimoto considered the esterification of acetic acid and β-phenetyl alcohol using a cationic ion-exchange resin as both catalyst and adsorbent (Kawase et al., 1996). Both simulation and experiments were

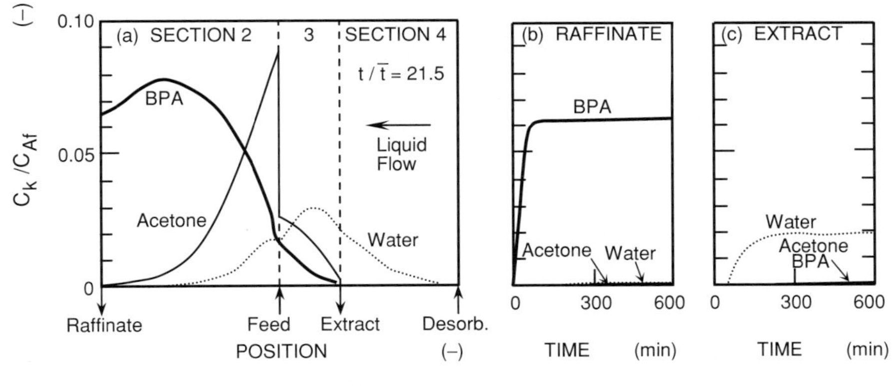

Figure 7.9 Stationary concentration profiles in an SCMCR system (a), and transient concentrations in the raffinate (b) and in the extract (c) for typical operating conditions. (Figure adapted from Kawase et al. (1999) with permission. © 1999 Elsevier Science B.V.)

undertaken. The SCMCR system they used will be considered at the end of the next section. Their simulation employed the model just discussed. Langmuir isotherms were used to describe equilibrium adsorption of reactants and products. A conventional Langmuir–Hinshelwood model was employed for the reaction kinetics assuming that adsorption and reaction steps were all elementary. Kawase et al. conducted a series of experiments in fixed beds to determine the parameters for the isotherms and to measure liquid–solid mass transfer coefficients. Batch experiments determined parameters for the kinetic model. For convenience, we delay discussion of these simulation results until the next section where they are compared to experimental data.

All of the simulation work discussed so far in this section dealt with model validation or with assessing the influence of operating variables on SCMCR performance. These studies are in general agreement. Recent work of Ray and a research team at the National University of Singapore illustrates very nicely the effect of the operating variables (Zhang et al., 2001). The system investigated was the commercially important production of methyl tertiary butyl ether (MTBE) by the reaction of tertiary butyl alcohol (TBA) with methanol (MeOH) over an ion-exchange resin. Methanol also served as the eluent in the system. A four-section SCMCR was considered with two beds or columns in each section (see Fig. 7.1). Flow rate ratios (Equation 7.5) for each section were set so that $\sigma_{\mathrm{MBTE}} < 1$, while $\sigma_{\mathrm{W}} > 1$. Thus, MTBE was the raffinate and water the extract.

For their simulation, Zhang et al. chose to use a feed consisting of 20% TBA in MeOH, raffinate withdrawal was 20% of the total flow at the feed port and solvent introduction after section 3 was 50% of the total feed flow rate. This, with a switching time, $\tau_s = 840$ s, was their base case. For a column length of 25 cm and a feed flow rate of 0.0167 cm^3/s, σ_{MBTE} was 0.223 and $\sigma_{\mathrm{W}} = 2.046$ using experimentally determined adsorption equilibrium constants. The SCMCR was cyclically stationary after 75 switching periods (about 19 cycles) and the predicted concentration profiles for all components except MeOH are shown in Figure 7.10a. The adsorbed product water removed in the extract stream controls the time needed to reach the stationary state. This may also be seen in Figure 7.9.

Our discussion of the Migliorini et al. (1999a) and Lode et al. (2001) contributions pointed out that performance depends critically on the switching time. This is nicely illustrated by comparing Figure 7.10b,c with the base case in Figure 7.10a. Only τ_s is changed in these figures. Reducing τ_s by about 25% lowers the conversion by a small amount, but results in large drops in yield and purity of MTBE in the raffinate. Reduction in residence time causes the lower conversion and yield. The much lower purity occurs because σ_{MTBE} is increased by a smaller τ_s and MTBE is carried into section 4 as Figure 7.10b shows. Increasing τ_s by about 25% should raise conversion. Figure 7.10c shows a small decrease. Explanation for this is that the product separation is poorer allowing the reverse reaction to proceed in section 2. Poorer separation at the raffinate port can be seen in the figure.

Increasing eluent flow rate (Fig. 7.11) strips water from the ion-exchange resin that leaves with the extra solvent at the extract port. Residence time and σ_{MTBE} or σ_{W} are not changed. Because water in sections 2 and 3, where separation occurs, is much lower, TBA conversion and MTBE purity in the raffinate increase. Figure 7.12a shows profiles for a twofold increase in the solvent rate, while profiles for a fourfold increase are given in Figure 7.12b. Although higher solvent usage improves the SCMCR performance with respect to conversion and raffinate purity, Y_{MTBE} does not increase. Of course, recovery of solvent and recycle is an important cost.

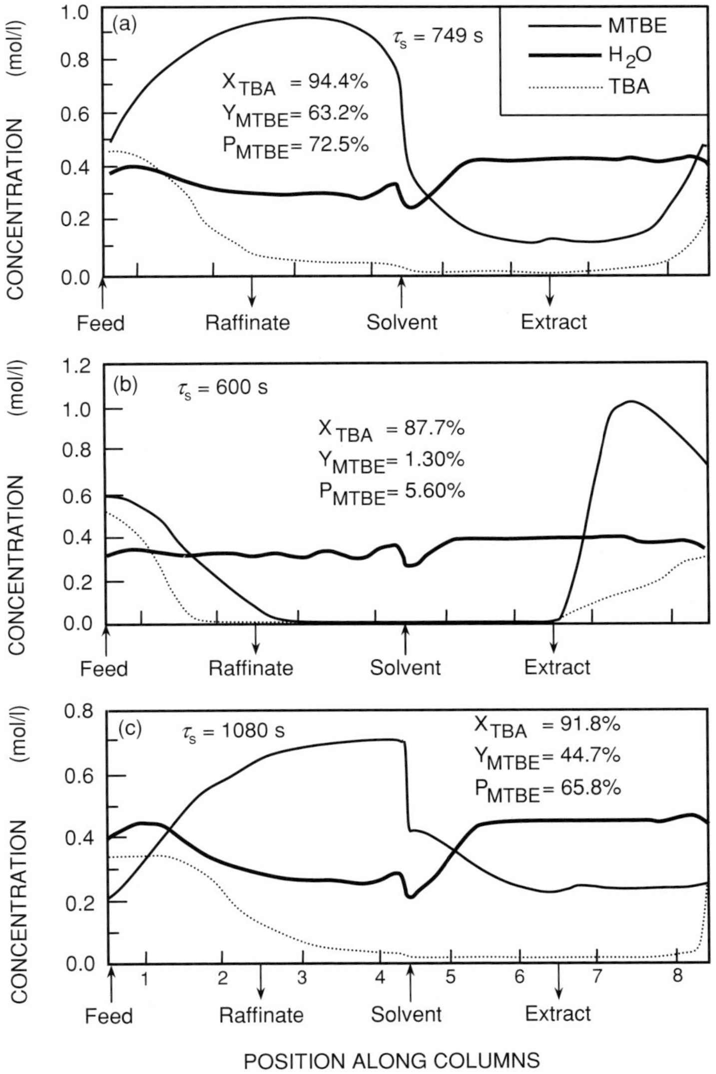

Figure 7.10 Effect of switching time on concentration profiles in a four-section, eight-column SCMCR with an ion-exchange catalyzed reaction between TBA and MeOH to form MTBE: (a) base case, $\tau_s = 749$ s, (b) $\tau_s = 600$ s, (c) $\tau_s = 1080$ s. (Figure adapted from Zhang et al. (2001) with permission. © 2001 American Chemical Society.)

Figure 7.12 examines the effect of increasing the withdrawal rate at the raffinate port. Comparison with Figure 7.11b examines the effect on yield by shifting flow withdrawal to the raffinate port. Indeed, half of the flow is now removed at this port. This changes the flow in sections 3 and 4, improving the separation in these sections and causing more of the reaction to occur in section 2 as the TBA curves in both Figures 7.12a and b illustrate. Both Y_{MTBE} and P_{MTBE} increase significantly. Raising the withdrawal rate at the raffinate port to 75% of

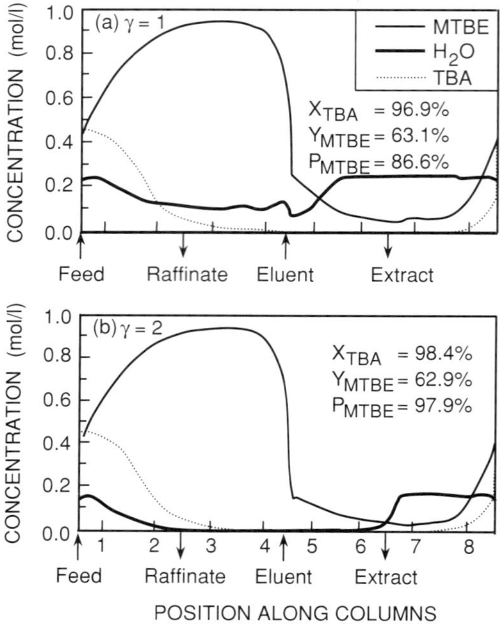

Figure 7.11 Effect of solvent flow rate on concentration profiles for the SCMCR in Figure 7.10. Base case is Figure 7.10a: (a) twofold solvent flow rate, (b) fourfold solvent flow rate. (Figure adapted from Zhang et al. (2001) with permission. © 2001 American Chemical Society.)

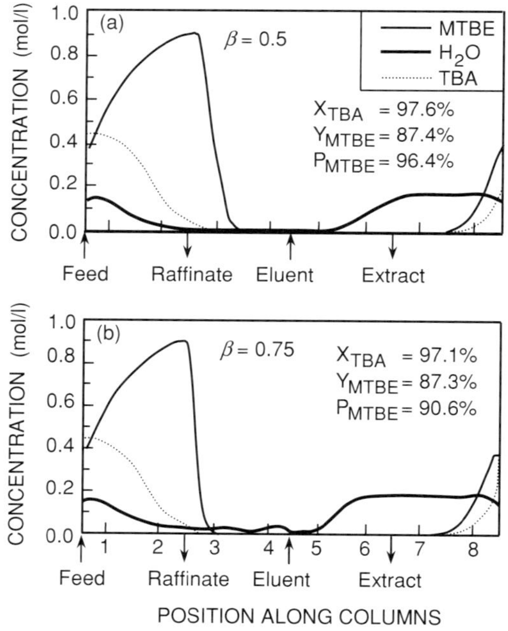

Figure 7.12 Effect of raffinate withdrawal rate on concentration profiles for the SCMCR in Figure 7.10. Base case is Figure 7.11b: (a) withdrawal rate is 50% of feed rate, (b) withdrawal rate is 75% of feed rate. (Figure adapted from Zhang et al. (2001) with permission. © 2001 American Chemical Society.)

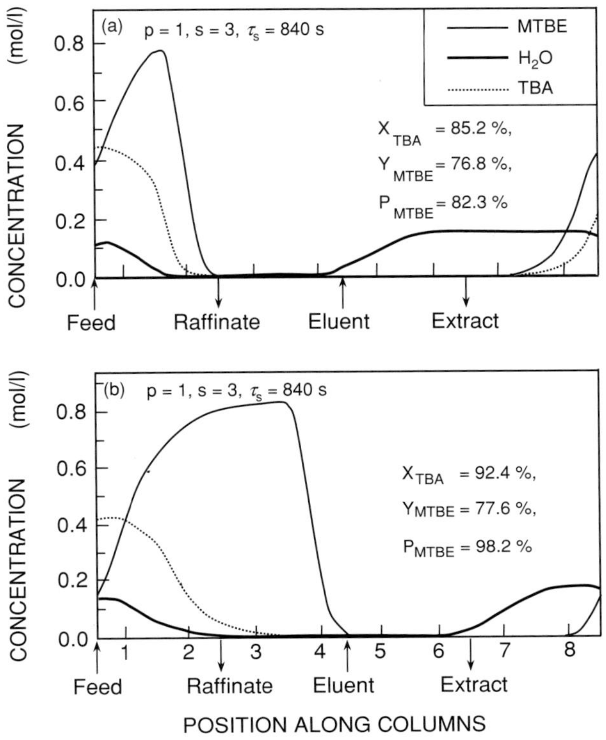

Figure 7.13 Effect of beds (columns) in sections 2 and 3 on concentration profiles for the SCMCR in Figure 7.10. Base case is Figure 7.12a: (a) one bed in section 2 and three in section 3, (b) three beds in section 2 and one in section 3. (Figure adapted from Zhang et al. (2001) with permission. © 2001 American Chemical Society.)

the feed flow causes a slight reduction in performance because the eluent flow is insufficient to sweep all the water from the resin. Thus, water breaks through into sections 2 and 3.

In the next figure, the effect of column location is examined. The base case becomes Figure 7.12a, which gave the best performance. Can performance be still further improved by adding or removing a column in section 2? Almost all the reaction occurs in section 2 according to Figure 7.12. Removing a bed or column from section 2 and adding one to section 1 decreases performance because the residence time in section 2 is reduced by a half and TBA breaks through into section 3 (Fig. 7.13a). This lowers the conversion and the MTBE purity in the raffinate stream. Increasing columns in section 2 by one at the expense of section 1 still reduces performance compared to the base case. The problem is that the residence time in section 1 is too short and water breakthrough into section 2 takes place as Figure 7.13b shows.

The Zhang analysis indicates a complex interaction among the operating variables in a SCMCR just as observed by Lode et al. (2001). This is hardly a surprise as complex interactions are also found in preparative chromatography.

Two contributions have dealt with the optimal design of SCMCRs (Lode et al., 2001; Zhang et al., 2002). Best performance is clearly obtained for a large number of beds or columns and high eluent to feed ratios. Cost considerations, however, limit numbers and ratios that are economically feasible. In some reaction systems, it is difficult to express all design considerations in monetary terms. Safety, for example, is one such consideration. In other systems, insufficient cost data prevents formulation of a single objective function. Often this is the situation in the early stages of process design. When this arises, optimization must be undertaken with multiple variables whose maxima or minima must be sought. There are several ways these variables may be expressed. Zhang et al. (2002) suggests for a reaction $A \leftrightarrow B + C$ that these variable are X_A, conversion of A in the SCMCR, $Y_B{}^*$, yield of the desired product B at the raffinate port, Γ_B, purity of B at that port and $\Xi_B{}^*$, the selectivity to the desired product at the raffinate port. Normally, yield is the product of selectivity and conversion. However, for a SCMCR, yield and selectivity are defined in terms of products collected at the raffinate port while conversion is based on products collected at both extract and raffinate ports. In most of the examples considered so far, water, a product of no value has been the extract. Consequently, only what is recovered at the raffinate port is important. Of course, $Y_B{}^*$ and $\Xi_B{}^*$ will behave much alike when an independent variable is changed provided X_A changes little so it will not always be necessary to consider four functions in an optimization problem. In this formulation, the solvent or extractant to feed ratio is not a dependent variable. It is an independent one and in all cases would be constrained, just as would be the number of beds, bed length and the allocation of beds to the three or four sections of the SCMCR.

An alternate set of dependent variables for the $A \leftrightarrow B + C$ reaction is productivity ϑ_A, and the solvent or eluent requirement:

$$\vartheta_A = \frac{n_A}{V_{\text{reactor}}} \tag{7.10}$$

$$\Delta = \frac{n_{\text{solvent}} + C_{\text{solvent}} Q_f}{n_A} \tag{7.11}$$

where Δ is the eluent requirement in terms of the feed rate, recognizing that solvent may enter after section 3 and with the feed. Lode et al. (2001) use these dependent variables for the case of complete conversion of A and separation of B and C that they considered in their study. For the optimization of a system without complete conversion or product separation, productivity and solvent requirement could be added to X_A, $Y_B{}^*$, Γ_B and $\Xi_B{}^*$, thereby reducing constraints in optimizing a SCMCR.

For a system where complete conversion and product separation is required, Lode et al. (2001) demonstrate that the residence time (Damköhler number) in section 3 and the eluent to key reactant in the feed stream control the size and shape of the complete conversion and separation region in the $\sigma_2{}^*$–$\sigma_3{}^*$ plane. They show also that the solvent ratio is inversely proportional to the switching time and the productivity as defined by Equation 7.10. Productivity, in turn, is inversely proportional to the switching time and the number of column in the SCMCR. Higher productivity through smaller switching times means a higher solvent ratio. There is thus a trade-off between productivity and solvent requirement.

Optimization is iterative. Column or bed length and total number of columns must be specified; then an arrangement of columns into sections must be chosen. If sufficient cost data is available, a single objective function can be fashioned from feed cost, ϑ_A, and Δ. A suitable two-dimensional optimization routine can be applied to find the optimal switching time, τ_s, and feed composition as eluent to feed ratio. Another column arrangement may be then tested to see if the objective function can be reduced further. This is similar to the sequence used by Zhang et al. (2001) that is discussed in Figures 7.10–7.13. Once the optimal arrangement of beds into sections has been identified, the influences of the number of columns and column length can be investigated.

If a single objective function cannot be formulated, the trade-off between productivity and eluent ratio, or, more generally, between conversion, selectivity and purity leads to Pareto-optimal solutions in which pairs of variables, such as τ_s and feed composition give ostensibly acceptable performance. Choice of the pair or set depends on other non-quantifiable information. Zhang et al. (2002) discuss optimization in this case using MTBE synthesis as their example. Two objective functions are considered: yield, Y_{MTBE}, and purity, Γ_{MTBE}. Conversion, X_{TBA}, also important, is treated as a constrained variable rather than an objective function. The authors employed what they referred to as a nondominated sorting algorithm to determine Pareto-optimal sets of variables. A three-dimensional, constrained search was undertaken using τ_s, fraction of the feed flow withdrawn at the raffinate port and the number of columns in section 2 as independent variables. The effect of these variables on the concentration profiles in the SCMR was presented in Figures 7.10, 7.12 and 7.13, respectively. Other variables were kept constant. The sets obtained and the "optimal" values of the independent variables are shown in Figure 7.14. Separate symbols in the plots show the influence of the number of columns, N_{col}. In this figure, p is the number of columns in section 2, and $\beta =$ the fraction of the feed flow withdrawn at the raffinate port. τ_s is also given in the figure although it was not an objective function. Values of other variables in the optimization are given by the authors.

Figure 7.14 illustrates that as long as $Y_{\mathrm{MTBE}} < 0.88$, high values of Γ_{MTBE} and $\Xi_{\mathrm{MTBE}}{}^{*}$ can be achieved. X_{TBA} is greater than 0.98 for all choices of the independent variables within the constraint limits. All of these solutions are satisfactory so that further information is needed to select among them. Withdrawal of raffinate greater than 20% of the feed is necessary to achieve $Y_{\mathrm{MTBE}} > 0.88$, but at this rate water enters the raffinate port causing purity and selectivity to drop drastically. Symbols in the figure show that above six columns in the SCMCR, the number becomes unimportant, that is, the Pareto-optimal solutions are unaffected. However, as the total number increases, the best performance requires adding columns to section 2. When there are just four columns in the four-section SCMCR, section 2 can have just one bed so the achievable performance drops regardless of the independent variables and the conversion constraint cannot be met. Curiously, the optimal switching time does not change with p or with the number of columns. The latter is also true for the raffinate withdrawal rate. Since both affect the residence time, this result indicates that there is a minimal amount of the ion-exchange resin needed to achieve a satisfactory performance. Optimization of yield and purity per unit volume of the SCMCR or purity and the eluent to feed ratio subject to $X_{\mathrm{TBA}} > 0.9$ clearly show the trade-off among these two objective functions in each case.

Simulation of enzymatic systems has also been undertaken. Meurer et al. (1996) explored the enzymatic inversion of sucrose. This reaction has been investigated experimentally as discussed in Section 7.5.3.

A multireaction simulation using a sugar system has been reported by Kawase et al. (2001) as part of an experimental study. These authors considered the formation of lactosucrose from lactose and sucrose. Side reactions are the hydrolysis of sucrose to yield glucose and fructose, and the hydrolysis of lactosucrose, forming fructose and lactose. Transfer of the

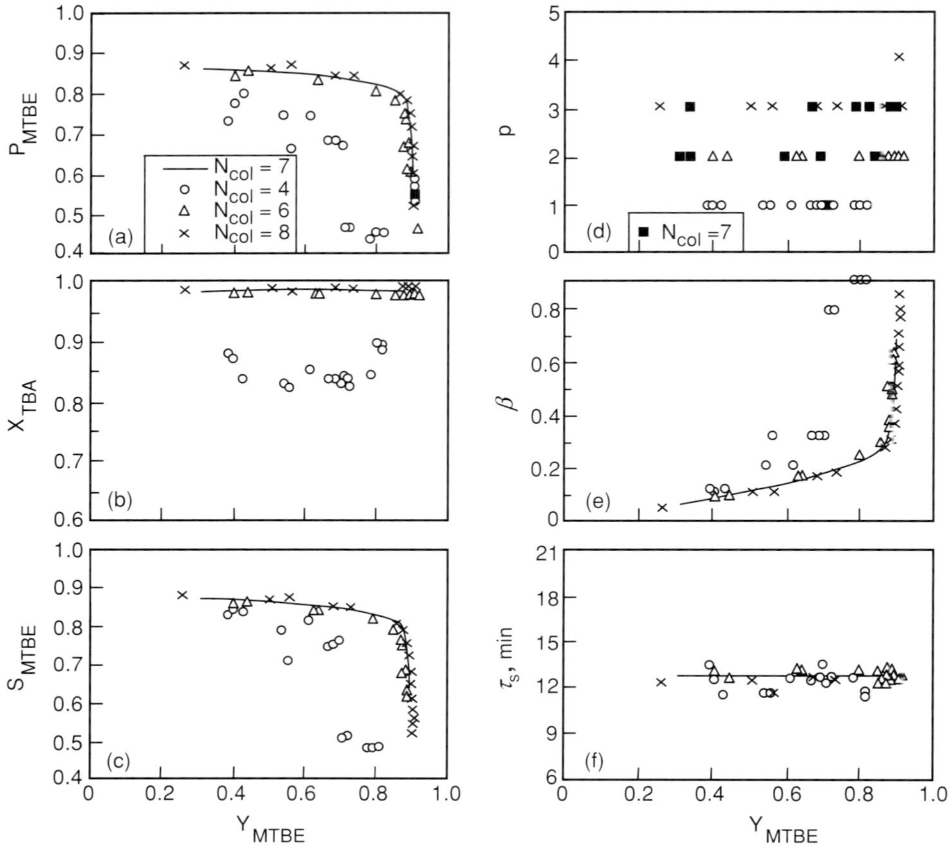

Figure 7.14 Pareto-optimal sets and values of the search variables for optimization of MTBE production in a four-section SCMCR. (Figure adapted from Zhang et al. (2001) with permission. © 2001 American Chemical Society.)

fructosyl moiety of sucrose to lactose is an enzymatic reaction occuring in the liquid phase as are the hydrolysis of sucrose and the lactosucrose product. This product is but weakly adsorbed, whereas fructose and glucose are adsorbed to a much greater extent on the ion-exchange resin. Equilibrium adsorption was assumed so mass transfer from the liquid phase to the solid adsorbent is rate controlling. Isothermality was also assumed. For this system Kawase et al. employed Equations 7.6 and 7.7 but with the assumption of plug flow so the dispersion term was neglected in the former equation. Concentrations C_i^* for the mass transfer driving force term are the mean values in the resin and assumed to be in equilibrium with the adsorbate. Kawase et al. (2001) found from their simulation that application of a SCMCR to lactosucrose formation would substantially increase conversion over that obtainable in a single phase CSTR.

As mentioned in Chapter 3, a cascade of fully back-mixed tanks can be used to represent a chromatographic column. The approach is applicable to an isothermally operated SCMCR as well. Barker and coworkers (Akintoye et al., 1991; Shieh and Barker, 1996) have employed models of this type that they refer to as a theoretical plate model. For example, Shieh and

Barker (1996) developed a dynamic model for the hydrolysis of lactose that assumed reaction only in the fluid or mobile phase and linear adsorption isotherms of the form: $(C_s)_i = (K_{ad})_i C_i$. Thus, for the nth tank in the cascade,

$$(\varepsilon + (1-\varepsilon)(\varepsilon_p + (1-\varepsilon)\rho_{ad} K_{ad_i}))\frac{d(C_i)_n}{dt} = t^*((C_i)_{n-1} - (C_i)_n + \nu_i r \tag{7.12}$$

Hydrolysis of lactose yields glucose and galactose and proceeds through the enzyme lactase derived from aspergillus oryzae. Kinetics follows the Michaelis–Menten model:

$$r = \frac{V_{max} C_L}{K_m + C_L} \tag{7.13}$$

In the model, C_L is the lactose concentration in the mobile phase, V_{max} is the maximum hydrolysis rate and K_m is the Michaelis–Menten constant. Also t^* is the contact time in the reactor divided by the number of stages or tanks assumed. For the hydrolysis, the model contains three equations for each stage or tank, one each for lactose, glucose and galactose. Enzyme lactase is not consumed, although its concentration affects V_{max} in the rate model. Galactose inhibits the hydrolysis so its concentration affects the constant as well. Only an initial condition is needed for this model. It is usually $C_i = (C_s)_i = 0$ for all stages.

Use of cascade models was popular in early simulation studies because integration is often simpler than it is for the one-dimensional pseudo-homogeneous models, which can be hyperbolic PDEs. However, the availability of ODE and PDE integration packages has dramatically lessened the burden of integration and this advantage has disappeared.

7.3 Nonisothermal modeling

Kruglov (1994) used simulation to investigate methanol synthesis from syngas in a SCMCR. Unlike earlier studies, he considered an adiabatic operation so that the heat balance in Table 3.2 and appropriate boundary conditions in Table 3.5 must be introduced. The concept that interested Kruglov was the use of an adsorbent to trap the methanol product so that the reaction is shifted toward a higher conversion. This concept had been applied earlier by Westerterp and coworkers (Kuczynski et al., 1987; Westerterp and Kuczynski, 1987a,b; Westerterp et al., 1989) as already mentioned.

In Kruglov's investigation, a stationary bed was assumed consisting of an intimate mixture of catalyst and adsorbent particles. Operation of the bed is essentially as shown in Figures 3.3, 7.1 and 7.4. For his model, Kruglov allowed for mass transport into the porous adsorbent but he assumed a linear adsorption isotherm. He also assumed uniform initial conditions in the bed and pointed out that initial conditions are not important because the cyclic stationary state is independent of these conditions when bifurcation is not encountered. The model was converted to finite differences for the spatial derivatives and the Thomas algorithm employing a Newton–Raphson procedure handled the sparse matrix that results. Because of the steep gradients that arise, an adaptive grid generation procedure was necessary. Model parameters were developed from known properties of the solids, while kinetic and adsorption parameters were obtained from the literature for the low-pressure catalyst, $Cu/ZnO/Al_2O_3$, and for the adsorption of methanol on silica–alumina.

Typical concentration profiles are given in Figure 7.15. Each plot of concentrations, gas flow rates and temperatures are for different times within a cycle at a cyclic stationary state.

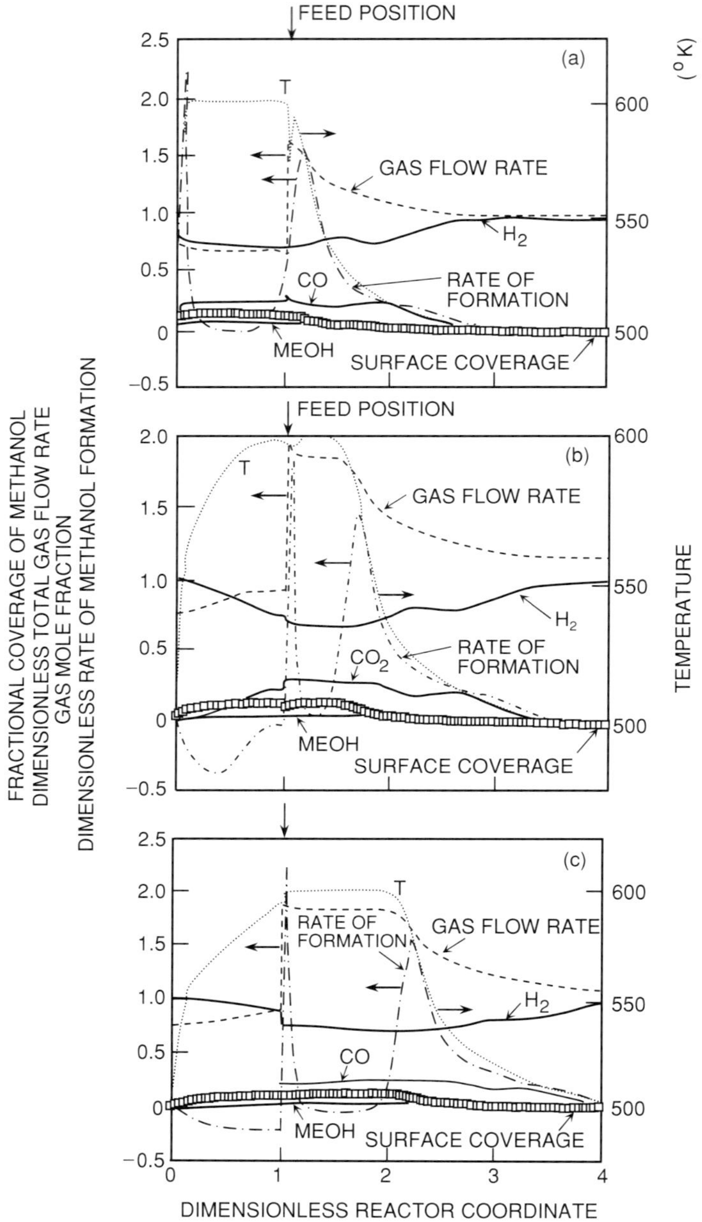

Figure 7.15 Calculated gas phase reactant and product concentrations, catalyst surface coverage, local temperature, total gas flow rate and rate of product formation at three times within a cycle for the adiabatic synthesis of methanol over a low pressure $Cu/ZnO/Al_2O_3$ catalyst using an SCMCR. Parts (a), (b) and (c) show profiles at increasing times within a switching period at a cyclic stationary state. (Figure adapted from Kruglov (1994) with permission. © 1994 Elsevier Science Ltd.)

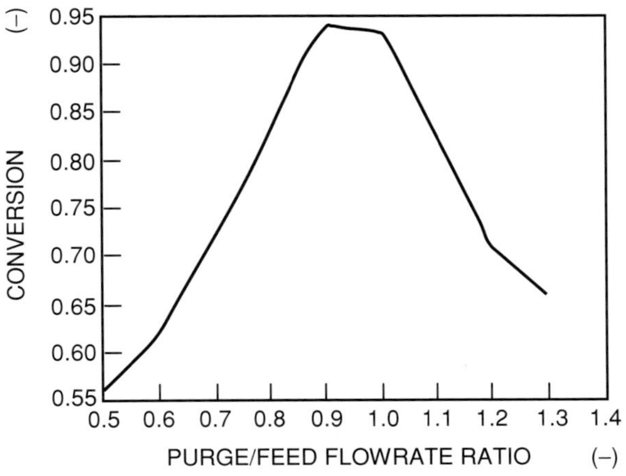

Figure 7.16 Calculated CO conversion to methanol as a function of the carrier gas to feed flow ratio for the conditions given in Figure 7.15. (Figure adapted from Kruglov (1994) with permission. © 1994 Elsevier Science Ltd.)

The stripping bed is to the left of the feed point. Just after the flow has been switched, the stripping bed contains both methanol and CO. With time CO is blown into the section containing the feed point so that very little is left in the stripping bed. Methanol, which is adsorbed, migrates through the bed at a slower velocity and eventually reaches the product removal section of the bed at which point it is recovered. Flow rate of carrier gas through the stripping bed has a strong influence on conversion because it is the carrier gas that sweeps CO from that bed. This is illustrated in Figure 7.16, which plots conversion of CO to methanol as a function of the ratio of eluent to feed flow rate. With a stripping section, conversion approaches 95% between an eluent to feed flow ratio of 0.9 and 1.

7.4 Separate catalyst and adsorbent beds

Kruglov also examined an arrangement in which the catalyst and adsorbent are segregated in the SCMCR by placing each in a separate bed and putting them into a cascade as shown in Figure 7.17. The equations in Table 3.2 need to be altered by dividing each heat and mass balance into two balances, one containing adsorption terms and the other reaction terms. Boundary conditions are not changed in as far as the feed into one bed is the effluent from the upstream one. Kruglov (1994) demonstrates that in this arrangement conversion greater than 98% can be attained if the operation is performed isothermally. For both versions of the simulated SCMCR, the switching frequency has a strong effect on conversion so that an optimum frequency exists. Separation of adsorbent and catalyst was proposed much earlier by Hashimoto et al. (1983). One of the interesting features of a SCMCR is its configurational flexibility compared to a CMCR. Several variations of the archetype in Figure 7.17 have been reported.

Hashimoto et al. (1983) studied the isomerization of glucose to fructose by immobilized glucose isomerase in an SCMCR consisting of 7 reactors (R1–R7) and 16 adsorption

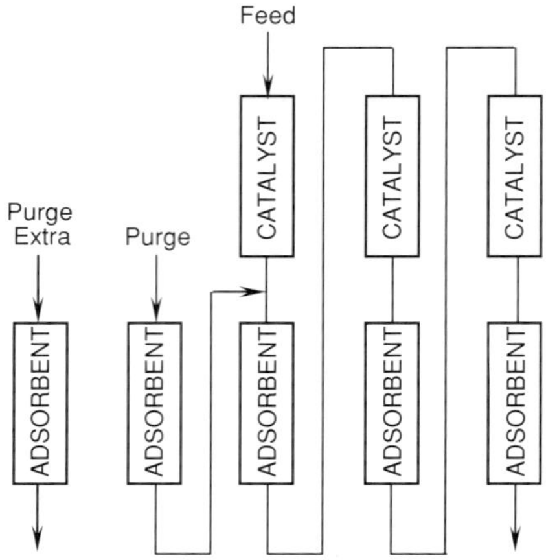

Figure 7.17 SCMCR cascade with separation of adsorbent and catalyst into different beds. (Figure adapted from Kruglov (1994) with permission. © 1994 Elsevier Science Ltd.)

columns (A1–A16). The arrangement of the beds is shown in Figure 7.18. Adsorption columns, A2–A7, were placed in between the reactors, while the other columns were connected consecutively. The reactors are fixed but the adsorption columns "move" in the manner of a SCMCR operation from left to right while the fluid flows from right to left. It is the beds that move; the adsorbent within each bed is fixed. Zone 1 corresponds to section 2 in Figures 7.1 and 7.4, zone 2 to section 3 and zone 3 to section 4. The model of the system, as mentioned above, differs from the material balances in Table 3.2 because the reactors have only a reaction term with no storage on the solid phase whereas the adsorbers have no reaction term. They do have, of course, a term for storage of the adsorbed component. For the adsorber,

$$\varepsilon_b \frac{\partial C_k}{\partial t} = -v_n \frac{\partial C_k}{\partial z} - K_f a_V \left(C_k - C_k^*\right) \tag{7.14}$$

$$(1 - \varepsilon_b)\, m_k \frac{\partial C_k^*}{\partial t} = K_f a_V \left(C_k - C_k^*\right) \tag{7.15}$$

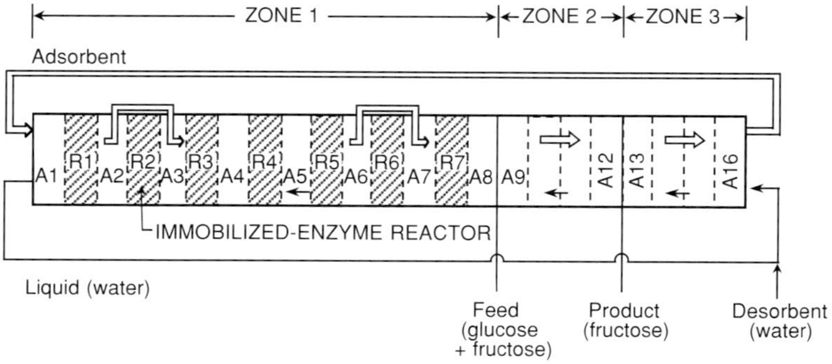

Figure 7.18 Schematic of the operation of an SCMCR with separate beds of catalyst and adsorbent. (Figure adapted from Hashimoto et al. (1983) with permission. © 1983 John Wiley & Sons, Inc.)

Hashimoto et al. proposed an alternate model that assumed the adsorbent particles were moving countercurrent to the liquid flow. In this model, steady state can be assumed so there are no time derivatives. European researchers (Storti et al., 1988; Lode et al., 2001) also used such a model for a combined catalyst and adsorbent bed as mentioned in Section 7.2. Both models were tested against experimental data and both represented the experimental results well. Simulations show that fructose content in excess of 60% can be obtained by a SCMCR using much lower amounts of desorbent than a chromatographic separation system following an enzymatic reactor.

The configuration employed by Tonkovich and Carr (1994b) for their theoretical study of the oxidative coupling of methane (OCM) is another variation of the arrangement shown in Figure 7.18. In their case, separation of adsorber and reactor is essential because adsorbents are not available for the high reaction temperatures employed in OCM. Methane coupling is an example of a consecutive reaction. C_2 products are intermediates between methane and complete oxidation. Rates of methane and C_2 consumption are proportional to O_2 partial pressure so the selectivity to C_2 would be expected to drop as the conversion of methane increases. Thus, at low conversion, high selectivity to C_2 products is obtained. For a PFR, the highest yield of C_2 product reported so far is about 20%. Continuous separation of the C_2 products, to prevent their further oxidation seems to be the most promising route to increasing this yield. The configuration used by Tonkovich and Carr (1994b) is shown in Figure 7.19. Lower temperature adsorption columns separating C_2 products and methane follow the short high temperature reactors, operating near 1000 K. One-pass conversions in these reactors are low so selectivity is high. Initially feed is introduced into the reactor

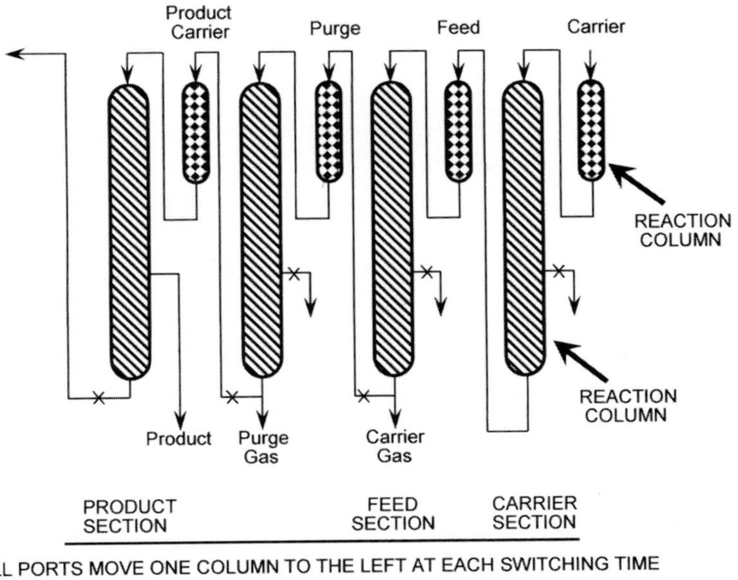

Figure 7.19 Separate adsorbent and catalyst bed SCMCR used to achieve high product yields in the oxidative coupling of methane. Adsorbent beds and reactors operate at widely different temperatures. (Figure adapted from Tonkovich and Carr (1994b) with permission. © 1995 Elsevier Science Ltd.)

second from the right that discharges into the serially connected separation column. Carrier gas goes to the reactor on the far right, passes through both reactor and adsorption column and then mixes with feed and passes through the second reactor. Some adsorbed C_2 is stripped from the column by the carrier gas. However, since methane is weakly adsorbed, it passes through the far right adsorption column sequence more rapidly than C_2, and so it is swept into the second right reactor by the carrier gas. Effluent from the second right adsorption column contains little C_2 and CH_4. It is mainly carrier gas and that gas can be recycled. Carrier gas is recovered at this point. Well before C_2 breakthrough occurs in the adsorption column second on the right, feed is switched to the second left reactor that had been purged in the previous step. Purge continues in this reactor, second on the left, just up to the C_2 breakthrough. Before breakthrough becomes substantial, carrier gas flushes the adsorption column. The effluent is the product stream and contains negligible methane. Product recovery occurs in the far left reactor–adsorption column sequence in Figure 7.19. The reactor that previously was flushed to recover C_2 products continues to receive carrier gas, but it now takes the place of the reactor on the far right and adsorption column effluent goes to the reactor that then receives fresh feed. This effluent stream is rich in CH_4. Methane, thus, cycles through the SCMCR until it is consumed. The sequence of carrier–feed–purge–carrier moves from right to left in the figure at each time step.

In modeling, Tonkovich and Carr used first order, reversible kinetics for the OCM reaction instead of more realistic consecutive reaction kinetics. Solution of the model equation was similar to that used by Kruglov except that integration employed a fourth order Runge–Kutta algorithm. Results were obtained for a cyclic stationary state. Model parameters were based on the experimental work to be discussed later. The authors show calculated concentration profiles as a function of position at different times within a cycle. However, as shown in Figure 7.20, the model did not agree with experiments for switching periods less than 30 s. Calculations predicted maximum C_2 yields of about 60%. The large change of temperature

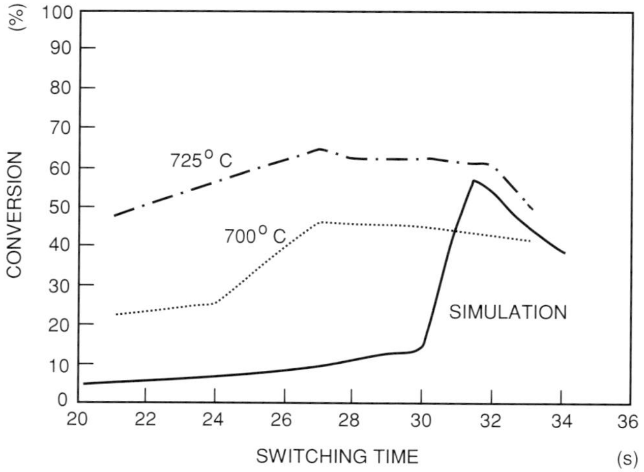

Figure 7.20 Comparison of predicted and experimental methane conversion in the oxidative coupling of methane in a SCMCR. (Figure adapted from Tonkovich and Carr (1994b) with permission. © 1995 Elsevier Science Ltd.)

from the reactor to the adsorption column must result in a high heat loss. It will be necessary to evaluate the energy requirement in a feasibility study of this SCMCR application.

7.5 Experimental studies

7.5.1 Gas–solid systems

Simulation carried out by Ray et al. (1994), discussed earlier, was followed up by an experimental investigation (Ray and Carr, 1995a,b). Catalyst used and reaction temperature in the study were the same as those used by Fish et al. (1986), however, Chromosorb 106 was chosen as the adsorbent instead of alumina. A mixture of 10 wt% catalyst and 90 wt% adsorbent was employed. Their SCMCR consisted of five columns of mixed catalyst–adsorbent as shown in Figure 7.21. The solenoid valves performing the feed, flush and the product removal functions are shown in the figure. Each column had a 13 mm o.d., was 300 mm in length, and contained equal amounts of the catalyst–adsorbent mixture. The columns were wound with heating tape and insulated. Each had separate temperature control to insure the same temperature in each column. With the adsorbent employed, the product, 1,3,5-trimethylcyclohexane (TMC), breaks through the mixed bed before mesitylene (MES). Its retention time was 104 s while that for MES was 392 s. Consequently, switching of the feed, flush and product take-off valves must occur between these two times. Ray and Carr used switching times between 240 and 300 s. Experiments employed 25% excess hydrogen on a volume basis and a feed flow rate of 4.95×10^{-6} mol/s. Nitrogen carrier flow rate was 1.32×10^{-4} mol/s.

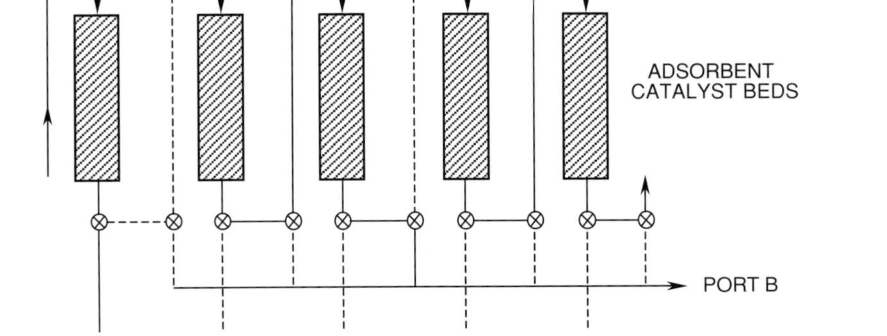

Figure 7.21 Schematic of a multibed, experimental SCMCR for hydrogenation of mesitylene. (Figure adapted from Ray and Carr (1995a) with permission. © 1995 Elsevier Science Ltd.)

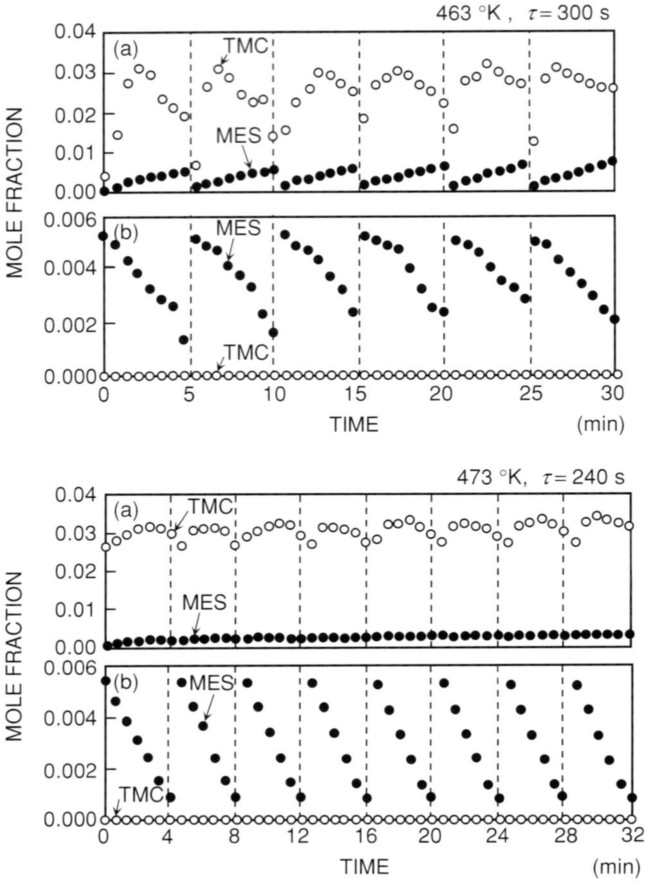

Figure 7.22 Mole fractions of reactant (MES) and product (TMC) leaving port B (a) and port A (b) at 463 K and at 473 K, respectively. (Figure adapted from Ray and Carr (1995a) with permission. © 1995 Elsevier Science Ltd.)

Figure 7.22 shows the mole fraction of MES and TMC leaving port B and port A of the previous figure. Measurements were taken after 1 h of continuous cyclic operation and appear to represent a cyclic stationary state. Port B is located after the column receiving feed while port A is located after the column that is being purged by the N_2 carrier gas. That port is located two columns behind the feed column. It is evident from these figures that the product and reactant separation is excellent. TMC is mainly in the stream leaving through port B while MES exits port A with negligible amount of TMC. Temperature and switching time are important. Figure 7.22 used a switching time of 300 s and a column temperature of 463 K in the upper part while the bottom part of the figure employed a switching time of 240 s and a column temperature of 473 K. The conversion at 463 K was 79%. MES recovery in port A is dictated by its adsorption isotherm. Because of adsorption, not all MES is removed by purging. Raising the temperature by 10 K, however, allows nearly complete removal of MES.

Table 7.1 Comparison of predicted and experimental performance of a CMCR and a SCMCR for mesitylene hydrogenation at 473 K

Reactor type	MES conversion (%)	TMC purity in effluent stream (%)	Reference
CMCR (experimental)	88	95	Fish and Carr (1989)
CMCR (simulation)	97	100	Fish and Carr (1989)
SCMCR (experimental)	83	96	Ray and Carr (1995)
SCMCR (simulation)	97	98	Ray et al. (1994)

This improves separation and raises the conversion of MES to over 82%. Increasing the flow rate of the purge gas improves MES recovery and raises the conversion further; however, TMC purity is decreased. If an infinitely long isothermal packed bed is used at 473 K with identical operating conditions, MES conversion is 40%. At 463 K it is 62%. Ray and Carr's experiments demonstrate that a SCMCR provides higher product purity and higher conversion than can be achieved in a packed bed PFR.

A comparison of experimental and simulation results was given by Ray and Carr and is shown in Table 7.1. The agreement of the simulated values is not surprising as a SCMCR should represent closely a CMCR. The lack of agreement, however, between prediction and experiment is disconcerting. The difference between 100% purity and 95% purity is very large and suggests model inadequacy. The CMCR model assumed steady state, but this may not have been achieved in the experiment. The difference is smaller for the SCMCR. Some of this is due to the SCMCR model that assumed 20 stages rather than the 5 stages actually used. As Ray and Carr observe, a source of the disagreement between experiment and simulation is the inadequacy of the Langmuir isotherm for MES adsorption. MES adsorption occurs on the Chromosorb adsorbent as well as on the alumina used as support for the platinum catalyst.

Because of the high temperatures needed for the oxidative coupling of methane over an oxide catalyst, the SCMCR system used by Ray and Carr could not be used for the coupling reaction. Consequently, Tonkovich et al. (1993) and Tonkovich and Carr (1994a) separated adsorbent and catalyst. They also altered the cycle by adding make-up feed and a second purge step. The adsorbent bed operated at 373 K while the reactor functioned at between 823 and 1048 K. The modified SCMCR is shown in Figure 7.19 and discussed in our section on theoretical studies. The operation of this system is rather complex because of two purges and a make-up feed. It is illustrated schematically in Figure 7.23.

For their experiments, Tonkovich and Carr used pure Sm_2O_3. This was packed in a 13 mm o.d. by 500-mm quartz tube. A reactor column contained 70 mg of catalyst held in place by quartz wool. Each reactor was mounted in a tube furnace. The adsorbent beds used activated charcoal packed into 7 mm o.d. by 76-mm or 110-mm-long steel tubes. The longer tube was placed after the shorter one. This permitted take off within an adsorbent bed during flushing with a product and carrier gas mixture as illustrated in Figure 7.23. N_2 carrier gas flow at 100 mL/min was used in all of the experiments. At this flow rate the breakthrough time for methane in the reactor and two adsorption columns was 34 s whereas the breakthrough time for ethane and ethylene was greater than 225 s. Thus, the switching time had to be less than 34 s. However, since products were removed after the first adsorption column, the switching

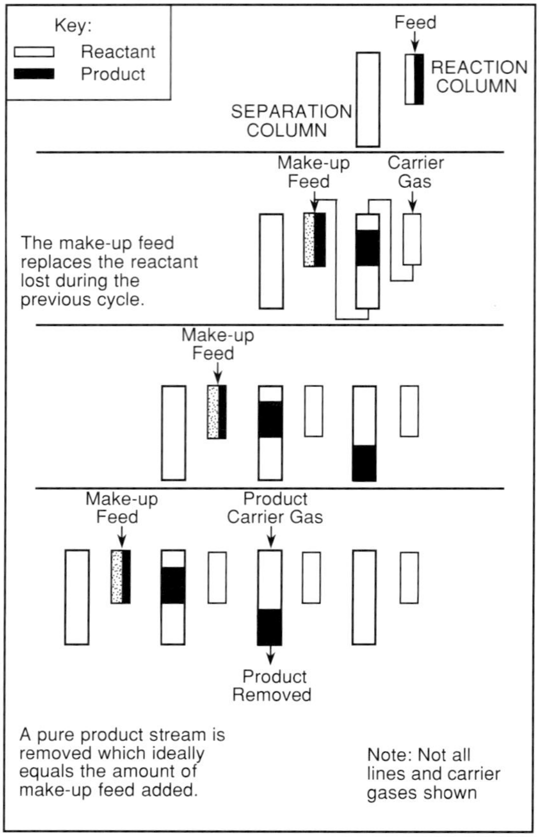

Figure 7.23 Operation of a multibed SCMCR for the oxidative coupling of methane. (Figure adapted from Tonkovich and Carr (1994a) with permission. © 1995 by Elsevier Science Ltd.)

time had to be smaller still. Experiments were performed using switching times from 20 to 33 s. A methane to oxygen feed ratio of 50:1 was selected. The initial methane flow rate was 18 mL/min. In subsequent switches the make-up flow rate was dropped to between 1.4 and 5.1 mL/min. Methane to oxygen ratio was then held between 2:1 and 3:1. This was done to compensate for methane lost per cycle and represents 10–15% of the original flow rate. Reactor temperature was an experimental variable.

Methane conversion per reactor varied between 2 and 3%. Switching time was important. Conversion increased with switching time up to an optimum time of 26–28 s at 973 K. At 998 K, the optimum switching time lies between 24 and 27 s. Selectivity to C_2 products was initially about 100% at switching times of about 20 s but drops sharply as the switching time increases. The yield rises to a maximum between 26 and 28 s at both temperatures. In Figure 7.24 the effect of temperature on selectivity and conversion is shown at close to the optimum switching time. It can be seen that the optimum temperature for OCM lies between 973 and 1048 K. Conversion at the upper temperature reaches 65%, while the selectivity approaches 80%. The C_2 product contains up to 25% ethylene.

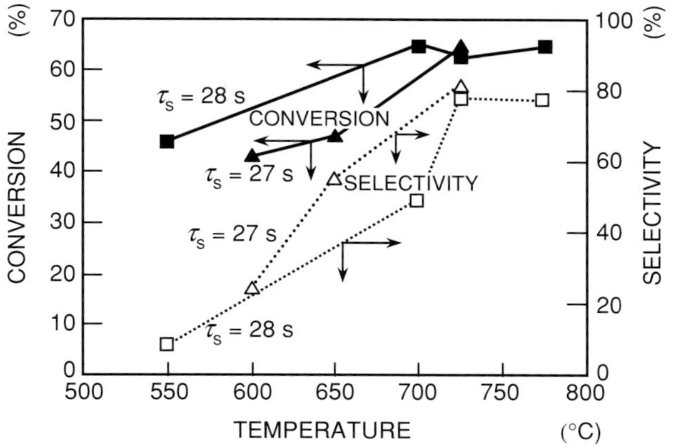

Figure 7.24 Temperature effect on a modified SCMCR performance for oxidative coupling of methane over Sm_2O_3. (Figure adapted from Tonkovich and Carr (1994a) with permission. © 1995 by Elsevier Science Ltd.)

The effect of the CH_4 to O_2 ratio on the modified SCMCR performance is given in Figure 7.25. It can be seen that the ratio has little effect on selectivity and conversion below a ratio of 2.6. Above this value both conversion and selectivity increase. On the other hand, for high conversions in a PFR, high temperatures and low methane to oxygen ratios are needed. Under these conditions the products are mainly carbon oxides and selectivity to C_2 products is poor. As the make-up CH_4:O_2 ratio varies the average methane to oxygen feed ratio in the SCMCR changes and will no longer be about 50:1 as it was in the previous figure.

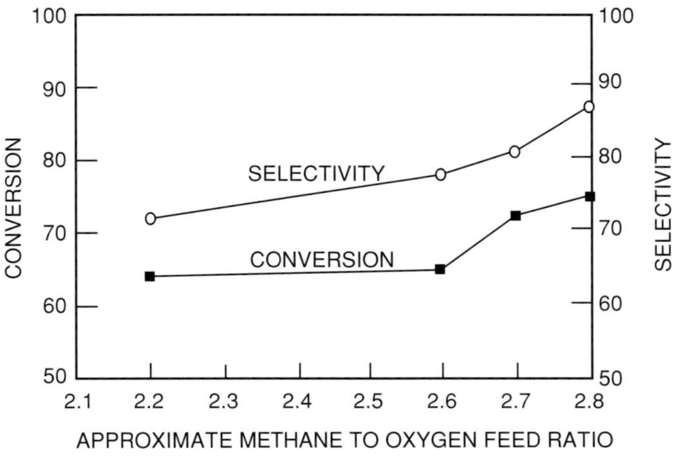

Figure 7.25 Effect of methane to oxygen ratio on conversion of methane and C_2 selectivity for a modified SCMCR. (Figure adapted from Tonkovich and Carr (1994a) with permission. © 1995 by Elsevier Science Ltd.)

Tonkovich and Carr point out that the target for a commercial oxidative coupling process has been a yield of 30% of C_2 products. Figure 7.25 indicates that at 1048 K yields approaching 50% can be achieved in the modified SCMCR. This would be an exciting result except for the huge energy penalty accrued through the large temperature difference between reactor and adsorber. At 1048 K this difference is 675 K. High temperature can be achieved only by burning fuel. Cooling of hot reactor effluent to 373 K would provide some energy recovery through a waste heat boiler, but the Tonkovich–Carr process certainly consumes large amounts of energy per kg of C_2 produced. Furthermore the equipment cost needed to accomplish heating and cooling with five reactor and adsorbent stages would make the capital cost of such a SCMCR process prohibitive. Consequently, realization of the modified SCMCR technology for OCM will require an adsorbent operating at a temperature close to that of the reactor. Nonetheless, the experimental work demonstrates promise for SCMCR technology, even though its application to oxidative coupling seems inappropriate. Electrically heated wire catalysts also provide high conversion and high selectivity for OCM but this design too suffers from high energy loss.

Partial oxidation of methane to methanol, like OCM or oxidation to synthesis gas, is an alternative process for the conversion of natural gas to higher value products. Like the OCM application, the use of a SCMCR for the partial oxidation is an attempt to improve a multiple reaction system in which selectivity is of primary importance. Reactions that are thought to make up this system are

$$\begin{aligned} CH_4 + (1/2)O_2 &\rightarrow CH_3OH \\ CH_3OH + O_2 &\rightarrow CO + 2H_2O \\ CH_4 + (3/2)O_2 &\rightarrow CO + 2H_2O \\ CO + (1/2)O_2 &\rightarrow CO_2. \end{aligned}$$

They probably proceed through free radical generation, propagation and termination. The generation step is promoted by a variety of catalysts, but also takes place homogeneously. Both consecutive and parallel reactions are involved.

Bjorklund and Carr (2002) describe an experimental investigation of noncatalytic methane partial oxidation employing a three-zone CMCR consisting of a single open-tube reactor and three packed bed adsorbers. Figure 7.26 shows a schematic of the system used. Ovals on either end of the adsorption beds indicate switching valves and the numbers within the ovals shows open valves in each step of the three-step cycle. The reactor was a fused silica-lined steel tube, 1 mm i.d. by 1800 mm. Constant temperature was insured by wrapping the tube with a heating tape. Steel tubes, 5.3 mm i.d. by 1800 mm, packed with an 80/100 mesh supelcoport served as adsorbers. This liquid-phase material impregnated onto a solid support had attractive desorption characteristics and was chosen over activated carbon by the investigators. Retention order of the system species on the adsorbent was: $H_2O > MeOH \gg CO \approx CO_2 > CH_4$. The ratio of product and reactant breakthrough times at the adsorbent temperature (375 K) and flow rates employed in the SCMCR was $CH_3OH:CH_4 \approx 2.5:1$, quite close to the ideal ratio of 2:1 for good separation.

Experiments using the empty reactor alone indicated that at 100 atm and $CH_4:O_2 = 16:1$ the reaction commenced at 735 K and gave the highest MeOH yield. At the initiation temperature, selectivity to MeOH was about 50% and the CH_4 conversion was 6%. Consequently, the 16:1 ratio was fed to the SCMCR, shown in Figure 7.26, in the first step. Thereafter the

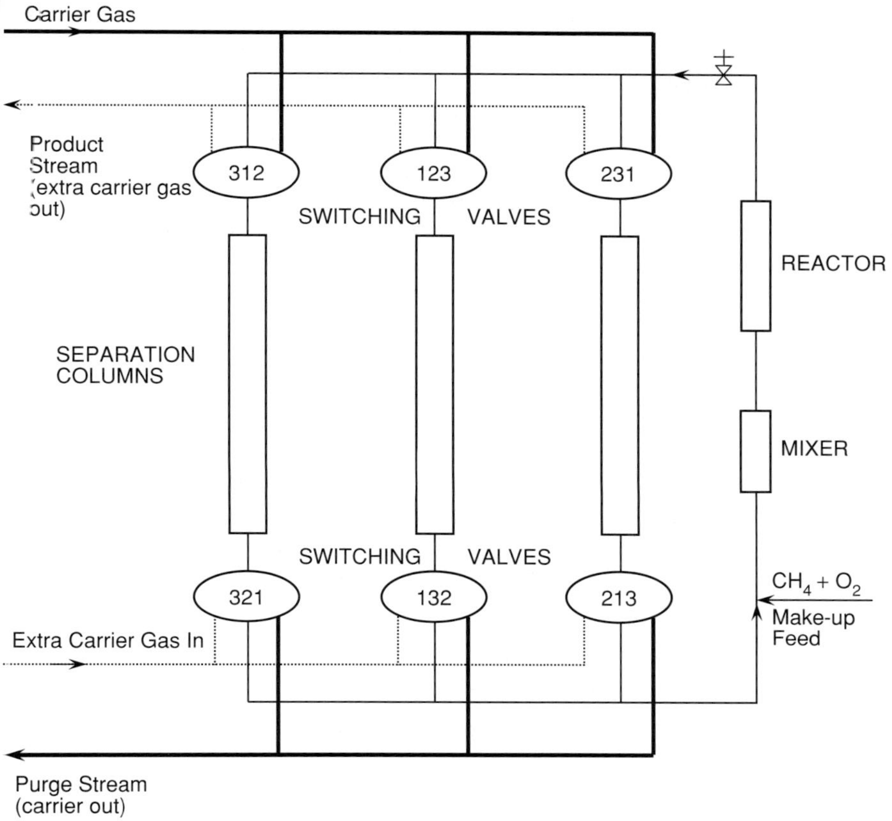

Figure 7.26 Schematic of a modified SCMCR used for partial oxidation of methane. (Figure adapted from Bjorklund and Carr (2002) with permission. © 2002 American Chemical Society.)

make-up feed was at the stoichiometric ratio of 2:1. The latter ratio was found to be optimal experimentally. The switching time, τ_s, was 440 s, also found to be an optimal time.

In the first cycle a 16:1 mixture was fed to the reactor, operated at 750 K for the best performance. Effluent from the reactor went to the first adsorber shown at the LHS of Figure 7.26. Helium carrier gas went to the third adsorber from the left, while additional He, designated as "extra" in the figure, passed into the second bed. Gas leaving the third adsorber mixed with the feed going to the reactor. Gas leaving the second bed became the product stream. Just before breakthrough of CH_4 from the first adsorber, all valves labeled "1" in the ovals were closed and the valves labeled "2" were opened. Reactor effluent now entered the middle bed. The first bed was fed with carrier gas and "extra" carrier gas went to the third bed. Gas leaving the first adsorber mixed with the feed, now at 2:1 CH_4:O_2, and passed on to the reactor. Effluent from the second or middle bed goes to the purge stream and that from the third bed is the product stream. Just before methane breakthrough from the middle bed, the valves are switched and those valves labeled "3" are opened. Effluent from the reactor now flows to the third adsorber on the RHS of Figure 7.26. Other flows can be worked out from the figure. The cycle begins again just before CH_4 breakthrough from the right-hand bed. The feed continues to be 2:1 CH_4:O_2 and this eventually results in a 16:1 mixture entering the reactor.

With the optimal switching time and feed ratio to the reactor, CH_4 conversion and selectivity to MeOH reached 50%, giving a MeOH yield of 25%. Higher yields are desirable. Because the reaction is noncatalytic and selectivity was observed to decrease with increasing temperature, Bjorklund and Carr concluded that only conversion could be used to increase the yield. The problem with their SCMCR system was found to be loss of methane in the product and purge streams. This can be seen in Figure 7.27, which shows the evolution of the CH_4 and MeOH profiles in the adsorber after startup. The first three valve switches shown encompass 1760 s or almost 30 min of operation. The cyclic stationary state is reached after 60 min on stream.

In Figure 7.27, the upper and lower portions give concentration versus position for CH_4 and MeOH with the adsorber assumed to be a single bed beginning on the LHS of Figure 7.26 and ending on the RHS. After the third cycle, the shape of the profiles changes little. Position of the profiles, however, changes in each switching period. Dispersion in the bed alters the front of the profiles from vertical lines to steep slopes.

Bjorklund and Carr (2002) comment that a four-zone SCMCR, which would be possible by lowering the adsorber temperature to increase the MeOH breakthrough time, could reduce CH_4 loss. They suggest yields approaching 50% are attainable.

7.5.2 Liquid–solid systems

Kawase et al. (1996) undertook an experimental study of liquid phase esterification in a SCMCR. They used β-phenetyl acetate synthesis over an ion-exchange resin (Amberlyst 15) that functioned as both catalyst and adsorbent. Water, which adsorbs on the resin more strongly than any other species, is produced together with the ester. 1,4-Dioxane was used as a desorbent or eluent. SCMCR experiments were carried out in an apparatus illustrated in Figure 7.28 that is similar to the system modeled by Fricke et al. (see Fig. 7.4). There are four sections separated by two inlet and two outlet streams.

Figure 7.29 compares model predictions of the stationary cyclic concentration profile with experimental results after 7.7 h of operation. Numerical simulation agrees well with the experimental results. The model was discussed in the section on theoretical studies.

The figure shows that the two products, ester and water, exist in sections 2 and 3 but the amount of water is quite low in section 2. Reactants are found mainly near the feed point. By the point where raffinate is withdrawn, both water and alcohol are virtually absent. Likewise, at the extract point, only water is present. Figure 7.30 shows the time changes of component concentrations in the raffinate and extract streams after start-up. A cyclic stationary state is attained after about 5 h. There is good agreement between simulation and experimental observations. Excluding solvent, purity of the ester in the raffinate stream was 99%. Overall conversion was 99% experimentally, considerably exceeding the equilibrium conversion of 63%. Kawase et al. demonstrate that the SCMCR performance depends critically on the relative flow rates of the extract and raffinate streams. Performance is also sensitive to the temperature. If this is too low, catalyst activity decreases and reactants accumulate in the system seriously affecting purity of the ester obtained in the raffinate.

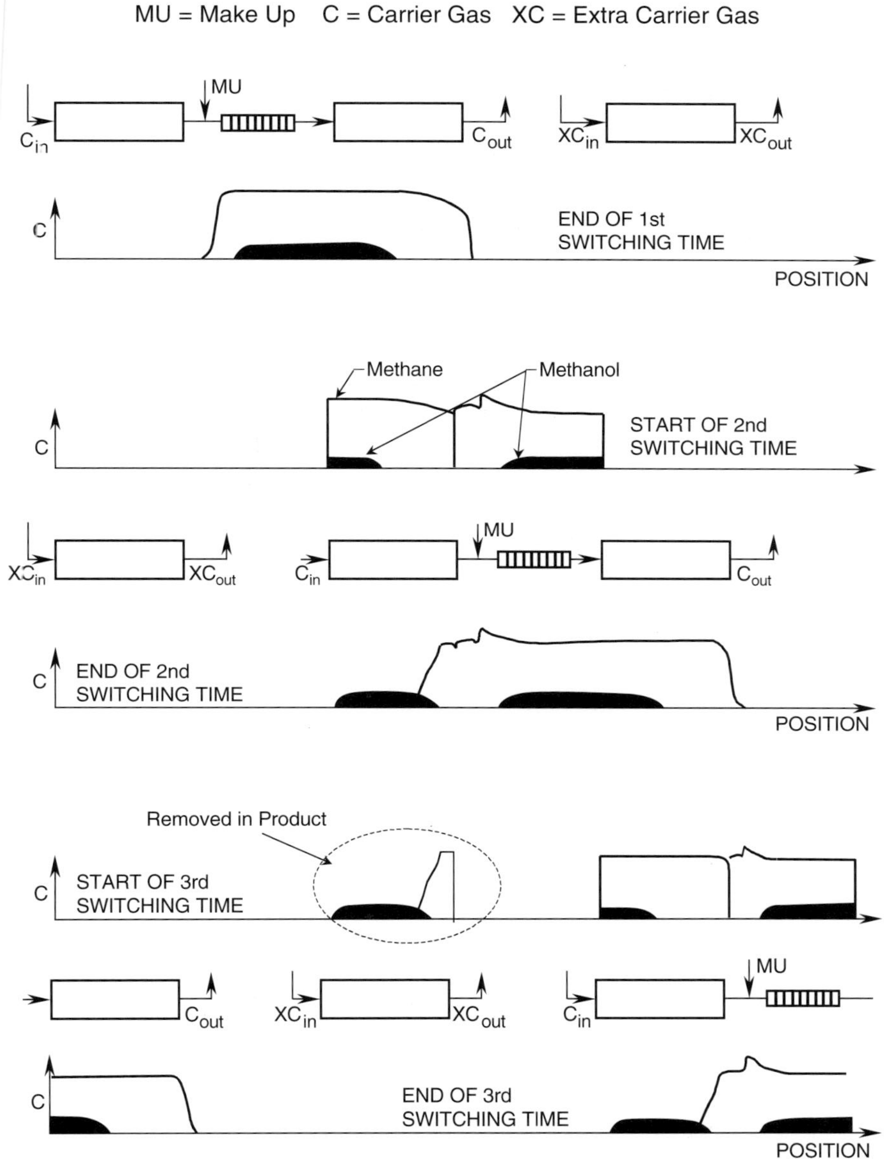

Figure 7.27 Development of the methane and methanol profiles in the modified SCMCR of Figure 7.26 from startup through the first cycle. Middle schematic in each group shows the location of inputs and takeoffs. (Figure adapted from Bjorklund and Carr (2002) with permission. © 2002 American Chemical Society.)

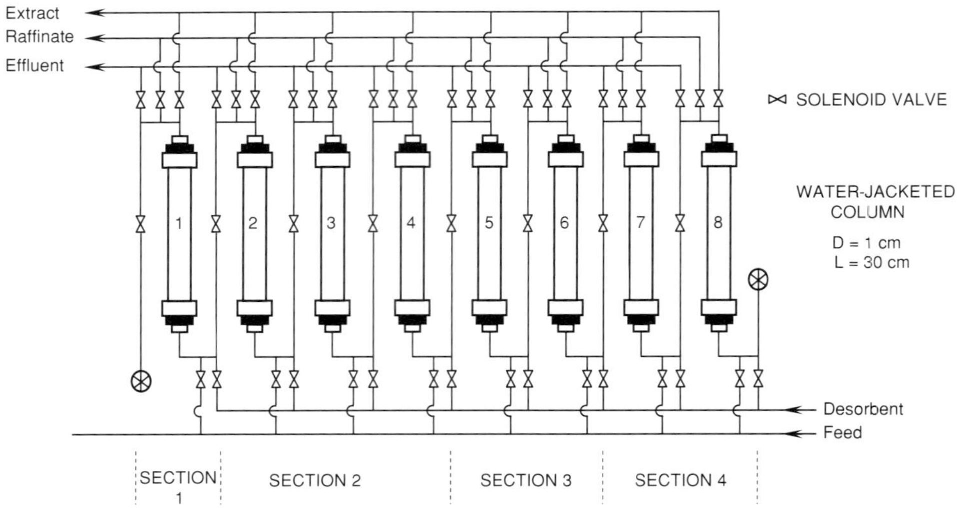

Figure 7.28 Schematic of an experimental SCMCR used for esterification of acetic acid and β-phenetyl alcohol. (Figure adapted from Kawase et al. (1996) with permission. © 1996 Elsevier Science Ltd.)

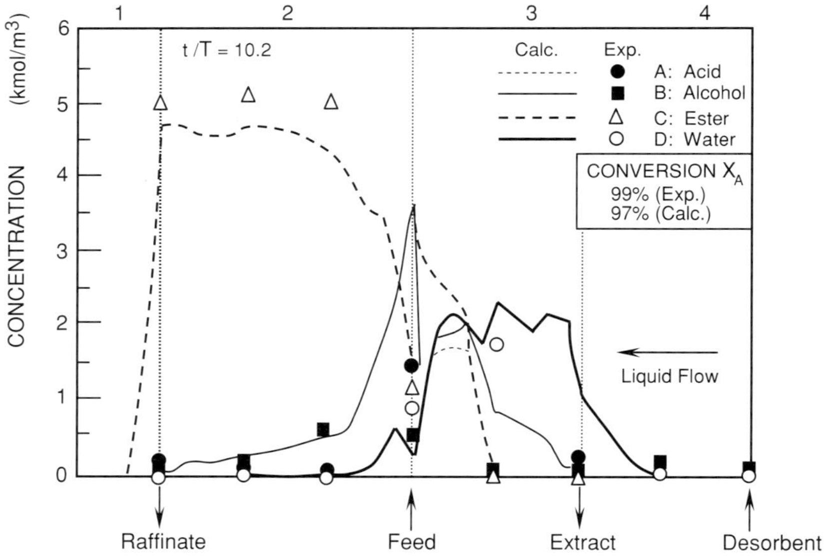

Figure 7.29 Predicted and experimental stationary concentration profiles in an SCMCR system applied to esterification of acetic acid and β-phenetyl alcohol under optimal operating conditions. (Figure adapted from Kawase et al. (1996) with permission. © 1996 Elsevier Science Ltd.)

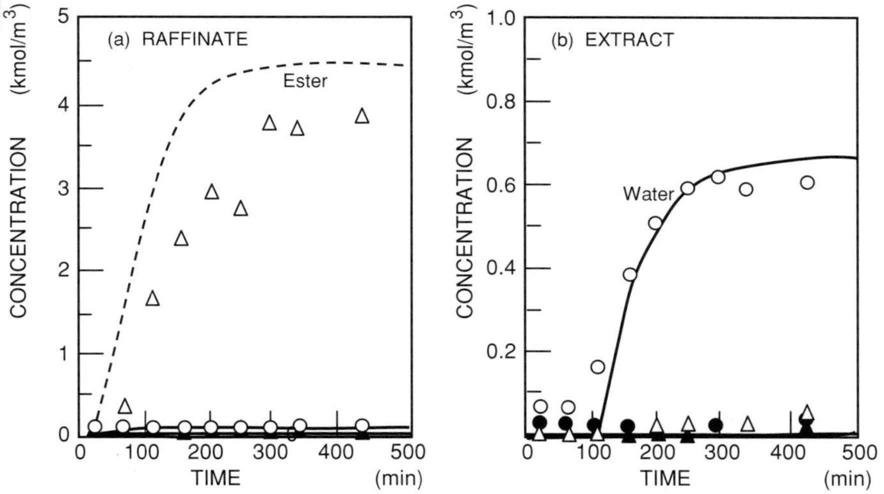

Figure 7.30 Predicted and experimental transient concentrations in the raffinate stream (a), and in the extract stream (b), under optimal operating conditions. (Figure adapted from Kawase et al. (1996) with permission. © 1996 Elsevier Science Ltd.)

Acetic acid esterification was also used by Mazzotti et al. (1996) to demonstrate that a suitably operated SCMCR can achieve complete conversion of the key reactant and complete separation of the reaction products. These investigators used ethanol as the second reactant as well as the solvent or eluent. A commercial strongly acidic cation-exchange resin (Amberlyst 15) was employed as a combined esterification catalyst and product adsorbent. Water is adsorbed by the cross-linked polystyrene–divinylbenzene resin, functionalized by a sulfonic group. Water adsorption on ion-exchange resins was discussed in Chapter 4. An equilibrium distribution of water between the resin, taken to be a polymer phase, and a solution is assumed. Activities for the solution are predicted by the UNIFAC group contribution method, while an extended Flory–Huggins model is used for the polymer phase. This model requires swelling data, which must be acquired experimentally.

For their three-section SCMCR experiments, Mazzotti et al. used eight columns, 7 cm in length, with each column packed with 3.1 g of dry resin. There were five columns in section 3, one in section 2 and two in section 1. Port movement employed a rotary valve consisting of a stationary upper disk connected to the inlet and outlet lines and a rotating lower disk that gave connections to the columns making up the SCMCR. All the columns were held at constant temperature in a thermostatted bath. Operating with $\tau_s = 6$ min, an HAc feed rate of 15 g/h and a 20:1 volumetric ratio of EtOH to HAc, complete conversion of acid was obtained with only ester and EtOH in the raffinate. Mole fraction of ethyl acetate in that stream was 0.10. Mole fraction of water in the extract stream was 0.05 and there was no ester in that stream. Increasing the HAc feed rate to 19 g/h and dropping eluent/feed ratio to 15:1 brought the conversion to 90%. The raffinate stream was water free and the ester mole fraction increased to about 0.17, but ester broke through into the extract stream.

The Mazzotti work (Mazzotti et al., 1996) serves as a "road map" for developing a SCMCR system once a catalytic reaction and a potential adsorbent have been identified. Batch or CSTR experiments provide a kinetic model and a reaction rate constant for esterification.

Further experiments using the individual components and the adsorbent produce equilibrium data for reactants and products. Step change experiments with adsorbent and catalyst yield mass transfer parameters. Prior to the SCMCR experiments, breakthrough runs are carried out in upflow by adding an equimolar mixture of EtOH and HAc at a constant rate to a packed bed of resin that had been saturated with EtOH. The composition at the bed outlet is followed with time until the outlet composition is constant. A regeneration experiment, representing section 4 in the SCMCR, begins with the bed at the end of the reaction experiment and feeds EtOH in downflow. The outlet compositions are again monitored until no further ester leaves the reactor. A comparison with model predictions verifies model and parameters and identifies anomalities such as density-driven flow. Finally, the CR model can be folded into a SCMCR model. This final model, then, is employed for design.

Commercial systems for simulated countercurrent moving-bed chromatographic separations are now being marketed. Lode et al. (2001) employed Model C-920 made by Advanced Separation Technologies Inc. in their study of the esterification of HAc and MeOH employing a sulfonated polystyrene–divinylbenzene ion-exchange resin. Model C-920 consists of ten 2.4 cm i.d. by 30-cm glass columns. The inlet and outlet of each column is connected to a single valve head with stationary and rotating parts. The stationary part is connected to the columns so that the feed and withdrawal have a 3-2-3-2 configuration. MeOH, both reactant and solvent in the Lode study, was fed continuously to the first section of the four-section SCMCR, while HAc was injected in closely controlled amounts between sections 2 and 3 using a precision piston pump. Extract and raffinate withdrawal, after sections 1 and 3 respectively, employed peristaltic pumps.

Like the earlier Mazzotti study, Lode characterized the Amberlyst ion-exchange resin they employed experimentally but instead of using a resin-swelling model and activity coefficients predicted from theory, they used a multicomponent Langmuir model and evaluated model parameters from equilibrium measurements. Kinetics of the esterification reaction was obtained using a well-stirred batch reactor. Adequacy of the parameters of their chromatographic column models was established by step-change breakthrough experiments as just described for the Mazzotti work. We return to these experiments in Chapter 14.

A limited number of parametric experiments were undertaken by Lode et al. (2001) using Model C-920. For a specified eluent (MeOH) rate, when the HAc feed rate was kept small and a relatively fast switching time (24 min) was used, 100% conversion of acid is obtained and separation of the ester and water products is excellent (see Fig. 7.31). However, plant productivity is low, about 1.3 tonnes of ester/day/m^3 of resin and the MeOH/ester ratio is 7.3 by weight. Increasing the eluent flow improves product separation at full conversion, but dilutes the products. Increasing the acid feed rate, holding the remaining variables roughly constant, more than doubles productivity and reduces the MeOH/ester ratio by about a factor of 3. However, conversion drops to below 90% and separation of products as well as acid and ester is significantly poorer.

One of Lode's experiments was undertaken with fluid downflow in the columns in place of upflow. The purpose was to test the effect of mixing caused by density differences on performance. The density ratio of MeOH and HAc, for example, is about 0.75 at 25°C. The mixing problem has been discussed in Chapter 4. Lode observed only a small effect of flow direction and concluded that incomplete conversion, meaning the streams in various columns had smaller density differences, was the explanation. Like the Mazzotti study, just described, the paper of Lode et al. (2001) can be used as guide for development of an SCMCR system.

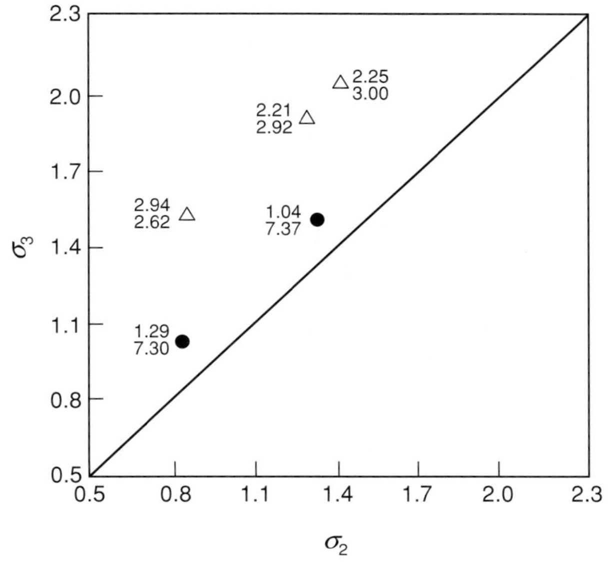

Figure 7.31 Plot of performance of an SCMCR with a 3-2-3-2 column configuration for acetic acid esterification with methanol over an Amberlyst 15 ion-exchange resin: ● = complete acid conversion and product separation, Δ = incomplete acid conversion and product separation; upper number = productivity in tonnes/m^3 resin/day, bottom number = eluent/feed ratio as kg methanol/kg acetic acid. (Figure adapted from Lode et al. (2001) with permission. © 2001 Elsevier Science Ltd.)

In Section 7.2, we discussed the use of a triangle in the σ_2–σ_3 plane for the design of an SCMCR. Complete conversion as well as product separation can be achieved within a triangle bounded on one side by the diagonal of the plane. Lode et al. suggest that optimal operating conditions can be found by testing different operating conditions along a line through the plane parallel to the diagonal. Figure 7.31 shows experimental implementation of this type of a search. Runs performed at $\sigma_2 = 0.81$ and $\sigma_3 = 1.03$, and at $\sigma_2 = 1.30$ and $\sigma_3 = 1.52$ resulted in complete conversion of HAc as well as separation of the ester and water products. These flow ratios have been corrected for "dead space". Numbers next to the data points give the productivity in tonnes/m^3 resin/day (upper number) and eluent/feed ratio in kg MeOH/kg HAc (lower number). Additional runs along a line further removed from the diagonal are also shown. For these points, conversion is less than 100% and product separation is incomplete. Water occurs in the raffinate and some ester may be in the extract. However, productivity is higher and the eluent requirement is substantially less. Thus, complete conversion incurs a cost of higher resin and eluent costs as well as a higher cost for separating product from the eluent.

7.5.3 Biochemical systems

Hashimoto et al. (1983) demonstrated experimentally that high-fructose syrups are attainable from a glucose feed just as predicted by simulation studies discussed in Section 7.2. The rotating system of adsorption columns and stationary enzymatic reactor columns used in

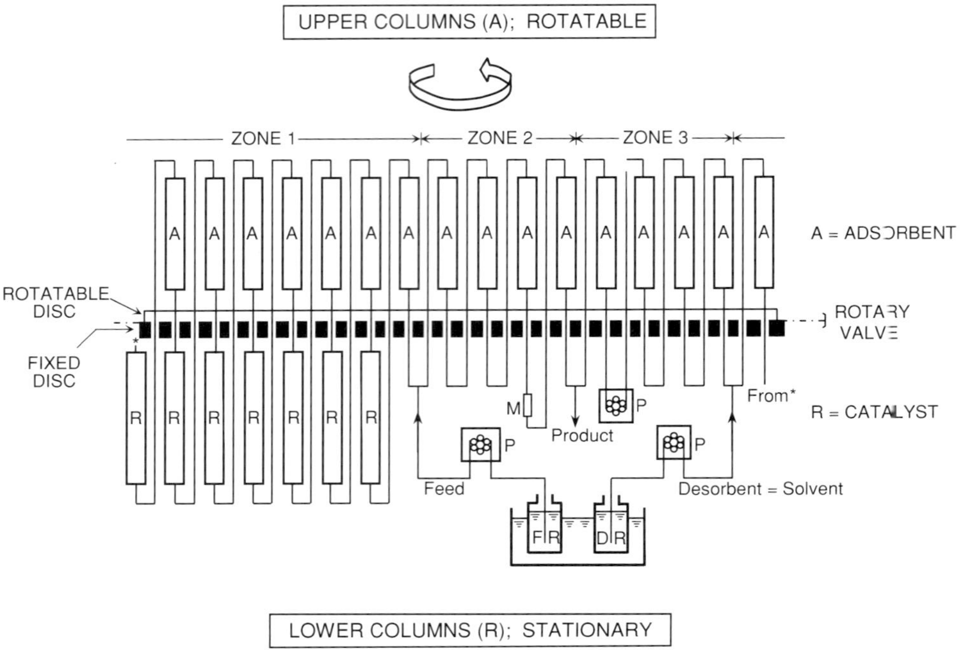

Figure 7.32 Experimental SCMCR with rotating adsorbent columns and stationary reactor columns for glucose isomerization. (Figure adapted from Hashimoto et al. (1983) with permission. © 1983 John Wiley & Sons, Inc.)

their SCMCR experiments is shown in Figure 7.32. As the figure indicates, there are roughly three times as many adsorption columns as enzymatic reactor columns. This is explained by the three functions performed in the system: (1) nearly complete conversion of glucose to fructose which is trapped by the adsorbent, (2) separation of glucose and fructose by their different adsorptivity on Y zeolite, (3) desorption of the mixed fructose–glucose product. These functions are indicated in the figure by zones 1–3. Connections between adsorption columns and reactor columns, or between adsorption columns and the feed or desorbent sources, or between adsorption columns in series takes place through a combination of rotating and fixed discs as shown in the figure. The reactor columns are stationary while the adsorbent columns mounted on the rotating disc move counter-clockwise. At each switching interval the adsorbent column moves one step. Thus the final adsorbent column in zone 1 moves into zone 2 while the final column in zone 2 moves into zone 3 and the final column in zone 3 moves into zone 1. This carries adsorbents heavily loaded with the fructose–glucose mixture into zone 2 where the adsorbent is washed countercurrent with a water stream containing fructose initially at the product concentration. By the end of zone 2 the adsorbent contains a high ratio of fructose to glucose. The adsorbent moves to zone 3 where glucose and eventually fructose are stripped from the adsorbent by countercurrent eluent flow. The stripped adsorbent returns to zone 1.

Hashimoto et al. employed an immobilized glucose isomerase as a catalytic substrate for the conversion of glucose to fructose. The adsorbent was the Ca^{2+} form of Y zeolite. All

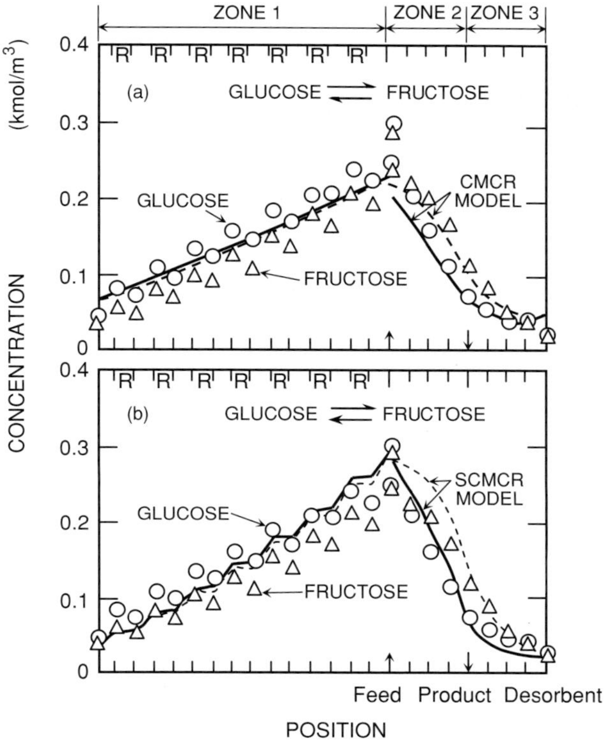

Figure 7.33 Experimental and predicted composition profiles in the modified SCMCR for glucose isomerization: (a) CMCR model, (b) periodic switching model. (Figure adapted from Hashimoto et al. (1983) with permission. © 1983 John Wiley & Sons, Inc.)

columns had the same diameter, 13.8 mm, and adsorption columns were 102 mm in length. Two lengths of the reactor columns were employed, 102 and 180 mm. Isothermality was maintained by circulating thermostatted water at 323 K through the jackets of the columns. The stepwise rotation of the upper disc was in 22.5° steps and switching times were 120 or 180 s.

Because their simulation demonstrated that high conversion of glucose was possible and the fructose fraction was in the desired range, the objective of Hashimoto et al. was to experimentally verify the model developed. For this purpose, they varied flow rates of feed, eluent and product withdrawal. Concentrations of fructose and glucose in the feed stream were also changed. Figure 7.33 shows the variation of glucose and fructose in the sequence of reactor and adsorbent columns in zone 1 as well as changes of these concentrations in zones 2 and 3. The figure demonstrates the conversion of glucose in zone 1 where the concentration goes from about 0.3 in the feed to 0.05 kmol/m^3 in the stream leaving the last reactor. Separation of glucose and fructose is shown as well. They are about equal at 0.3 in the feed, whereas in the stream leaving the adsorbent bed the product, fructose, is 0.12 and glucose is 0.07 kmol/m^3. Adsorbates are further depleted in zone 3 and reach about 0.4 kmol/m^3 in the stream leaving the last adsorbent column in the zone.

Part (b) of Figure 7.33 compares experiments to a detailed model assuming intermittent movement of the beds, namely periodic switching, whereas in (a) experimental data are compared to an approximate model assuming a bed of the adsorbent continuously moving countercurrent to the liquid flow. As mentioned earlier, the CMCR model is a steady state model. Part (b) demonstrates that the model successfully represents fructose and glucose concentrations in zones 2 and 3. Data exhibit considerable variation in zone 1 with the model predictions falling between the data limits. The model represents discontinuities evident at the feed point and at the point of introduction of eluent reasonably well. It is also seen that the product composition is close to that given by the model. The CMCR model, much easier to handle calculationally, does not disclose discontinuities but does give a good representation of the composition of the SCMCR product.

Hashimoto et al. (1983) go on to demonstrate that under comparable conditions the modified SCMCR can achieve the target fructose content of about 55% with a significantly lower eluent to feed ratio than a fixed-bed enzymatic reactor followed by a fixed-bed adsorber with recycle of unconverted glucose. This means that the energy requirement needed to concentrate the fructose syrup is significantly reduced.

Quite independent of the Japanese research, Barker and coworkers in England investigated the enzymatic conversion of sugars using the calcium form of an ion-exchange resin as the adsorbent (Barker et al., 1987a,b; Barker and Ganetsos, 1988, Akintoye et al., 1990, 1991; Barker et al., 1992a,b). A 12-column, preparative scale unit, built originally for chromatographic separations, was used for their studies. Each column was 5.4 cm in diameter and had a length of 75 cm. Six on–off valves were attached to each column, three at each end. These valves were connected to one of the inlet or outlet streams (raffinate, extract and stream to the succeeding column) and were timer-operated. A water purge was introduced just upstream of the eluent and enzyme feed point to desorb fructose from the resin. This operation is shown schematically in Figure 7.34. At each switching time, the inlet and outlet streams move one column in the clockwise direction.

Test reactions were sucrose inversion using an invertase enzyme (Akintoye et al., 1990, 1991), biosynthesis of a polyglucose, dextran, employing the enzyme dextransucrase and

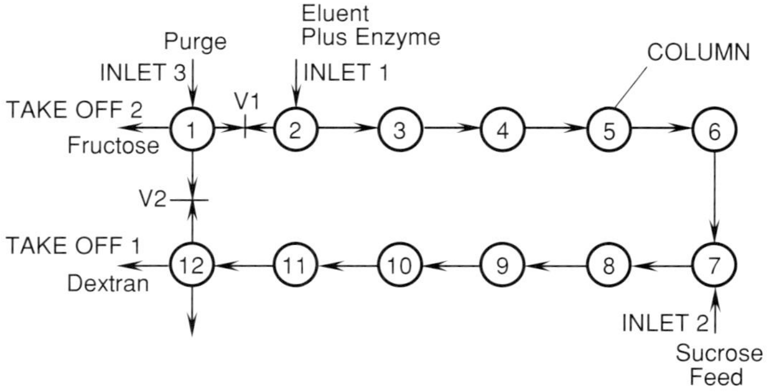

Figure 7.34 Twelve-column SCMCR operating scheme employed in the biosynthesis of dextran from sucrose. V1 and V2 are on–off valves used to isolate column 1 in order to strip fructose from the resin using a separate purge stream. Numbers represent columns in the three-section SCMCR. (Figure adapted from Barker et al. (1992c) with permission. © 1993 International Adsorption Society.)

the saccharification of a modified starch using maltogenase (Barker et al., 1992a,b). In all cases the adsorbent was the calcium form of either a Purolite ion-exchange resin (PCR 563) or a Korela resin (V07C). The inversion reaction is apparently inhibited by the educt or substrate for the sucrose system in the presence of products. Dextran production appears to be equilibrium limited. Fructose, a product of both sugar reactions, is preferentially adsorbed by the resin. The enzymatic reaction occurs in the aqueous phase so the enzymes, at low concentrations, are added to the SCMCR with the water eluent.

Akintoye et al. (1990, 1991) found that sucrose inversion was readily carried to completion at 25°C in their SCMCR system even for concentrated solutions. Separation of the glucose and fructose products was virtually complete with glucose in the raffinate at throughputs of 16 kg sucrose/m^3 resin/h. Enzyme usage was just 20% of that consumed in a stirred, batch reactor in the absence of adsorbent.

Biosynthesis of dextran was less successful. Although initially, complete conversion of sucrose was obtained at a pH of 5.2, at 25°C with $\tau_s = 30$ min and a dextran product free of fructose was recovered, the separation of dextran and fructose deteriorated after 50 h on stream even at low sucrose throughputs. Levan, a polyglucose, was found in the dextran and glucose was present in the fructose extract. The problem was traced to displacement of Ca^{2+} from the resin by small amounts of K^+ and Na^+ in the enzyme. Periodically regenerating the resin with calcium hydroxide permitted much longer run times, but incomplete separation persisted as Figure 7.35 demonstrates. In the figure, FRP is the extract port and it can be seen that the extract is fructose uncontaminated by glucose or sucrose. Dextran, however, was contaminated by levan, glucose and a small amount of fructose as shown by the concentrations in the raffinate port (DRP). There is no sucrose in either port so sucrose conversion is 100%.

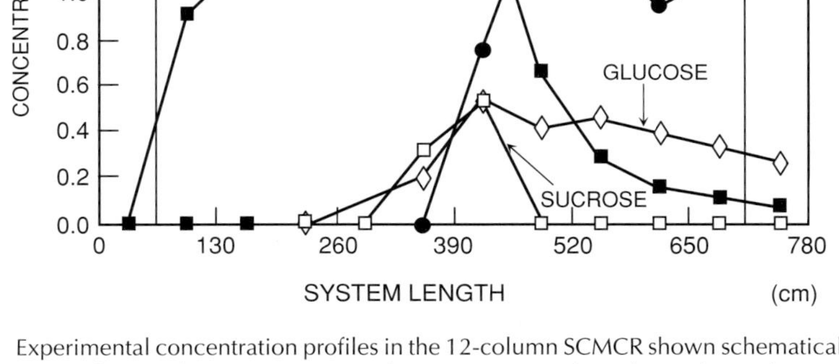

Figure 7.35 Experimental concentration profiles in the 12-column SCMCR shown schematically in Figure 7.28 for dextran production from sucrose using dextransucrase and a calcium form of an ion-exchange resin. Resin periodically regenerated, $\tau_s = 30$ min and a fed rate of 16.2 g sucrose/h. (Figure adapted from Barker et al. (1992c) with permission. © 1993 International Adsorption Society.)

For the production of maltose and dextrin from modified starch, conversion of 60% was reached at a feed rate of 116 g starch/h with maltose purities of 96%. Enzyme consumption was substantially reduced. There was no deterioration of performance with time so Barker et al. (1992b) concluded that this application of an SCMCR was successful. Further results are reported by Shieh and Barker (1995). These investigators used an exo-acting α-amylase, maltonase, that can hydrolyse the maltotriose of starch into maltose and glucose. The calcium exchanged resin preferentially adsorbs maltose. Shieh and Barker examined experimentally the effect of the eluent flow rate (containing the diluted maltonase), the switching time, τ_s, feed concentration and enzyme activity on the maltose yield, starch conversion and product purity. They observed strong interaction among these variables. Increasing eluent flow rate changes σ in all sections of the SCMCR and reduces residence times. This decreases conversion only slightly because more enzyme is fed to the reactor. Maltose purity improves. Raising the starch concentration decreased the enzyme to starch ratio and thereby the conversion. A larger effect on conversion and purity resulted from the higher viscosity of the liquid phase that reduced mixing of the enzyme with starch. Pressure drop increased and this effectively sets a limit on the starch concentration that can be used. Increasing the switching time has a similar effect to that of the eluent flow rate, starch conversion is lowered, maltose purity increases but the dextrin purity decreases as expected.

Shieh and Barker (1995) compared their results for the SCMCR described above to studies by Sarmidi and Barker (1993a,b) on maltose production from starch using a CRAC reactor. We discussed this work in Section 6.3.1 of the previous chapter. For essentially the same operating conditions, enzyme and adsorbent, the SCMCR used a considerably smaller amount of enzyme for an even larger starch conversion. This can be attributed to the depth of bed used: 7.8 m compared to 1.35 m for the CRAC reactor. The advantage of the CRAC reactor is that multiple draw-offs are easily made so that purer products can be obtained. The maltose obtained from the SCMCR contained small amounts of a glucose by-product.

Hydrolysis of lactose by the enzyme lactase obtained from the organism *Aspergillus oryzae* to produce glucose and galactose was also studied by Shieh and Barker (1996). Galactose strongly inhibits the hydrolysis reaction. The investigators used the 12-column SCMCR illustrated in Figure 7.34. The lactose feed and the enzyme were introduced together and a second inlet fed the mobile phase, deionized water. Galactose is strongly adsorbed by the resin and a purge eluent stream, also deionized water, was used to desorb the product from the resin. For this purpose, a column of the system was isolated from the remaining columns. Shieh and Barker also explored the influence of eluent flow rate, the switching time, τ_s, feed concentration and enzyme activity on the glucose yield, lactose conversion and product purity. Their findings are difficult to summarize succinctly. As might be expected there is strong interaction among these operating variables. Enzyme activity had perhaps the least effect on performance. Adsorption of galactose, the lactase inhibitor, drove the lactose conversion to completion even with a low activity enzyme. SCMCR performance was governed, thus, by the separation requirement. This is strongly dependent on switching time, which controls the ratio of adsorbent to mobile phase flow rates. Eluent flow rate is also important. The SCMCR systems reduced the enzyme required by 1/3 compared to the conventional batch reactor used for the hydrolysis. Glucose purities reached 99.9%. Galactose purities were lower and the product was quite dilute.

Shieh and Barker (1996) also conducted the hydrolysis in a single pulse chromatographic reactor (CR) and reexamined the effect of eluent flow rate, feed concentration, pulse size and enzyme activity on CR performance. They compared experimental results for the two chromatographic systems and observed SCMCR throughputs 2.4- to 5-fold those in a CR for the same glucose recovery. Much less enzyme was also required. The comparison demonstrates a much more efficient use of the adsorbent bed in a SCMCR as we have noted in Chapter 4.

A CSTR-cascade model, discussed in Section 7.2, was compared to the SCMCR results as an exercise in model validation. The model provided a reasonably good representation of the component profiles in a column at a point in the cycle. However, it badly overestimated the product purities at the draw-off points. Shieh and Barker attribute this model failure to the effect of the enzyme on the adsorption equilibrium constant in the linear model they assumed.

Hashimoto's team (Kawase et al., 2001), unlike the Barker study of dextran formation, intentionally took up the application of a SCMCR to a multicomponent, multireaction system: production of the commercially interesting lactosucrose from lactose and sucrose. Single pulse experiments indicated that a substantial improvement in throughput and yield over use of a batch reactor could be obtained. Lactosucrose formation is catalyzed by the enzyme β-fructofuranosidase. However, this enzyme also catalyzes the hydrolysis of sucrose to yield glucose and fructose as well as the hydrolysis of the lactosucrose product to produce lactose and fructose. Glucose is also produced in the primary reaction. All reactions are reversible. Use of an adsorbent specific to glucose would increase lactose conversion by forcing the primary reaction. It would also raise the fructose level in the system, thereby discouraging hydrolysis of the lactosucrose product.

Kawase et al. carried out their experiments with a cationic exchange resin (Amberlite CR-1310) in the sodium form. Adsorption equilibrium expressed as Henry's Law constants were in the sequence fructose $>$ glucose $>$ lactose $\approx$ sucrose $>$ lactosucrose. Experimental runs were made at 50°C using water with a small amount of NaCl as the eluent. Their SCMCR consisted of 12 columns of 12 mm i.d. by 190 mm, arranged in four sections with two columns in section 1, four in each of sections 2 and 3 and two in section 4. Several experiments used a 2-6-2-2 arrangement. Figure 7.36 presents a schematic of their experimental unit. Note that the enzyme is introduced with the lactose–sucrose feed stream. It is eluted with the raffinate where, in the absence of glucose, hydrolysis of the lactosucrose reduces both lactose conversion and product yield. Additional hydrolysis before the fluid phase is removed further decreases performance. The authors mention that the use of an immobilized enzyme would greatly enhance performance by avoiding product degradation in the liquid phase. Feed and withdrawal rates were carefully adjusted so that lactosucrose and glucose were separated in sections 2 and 3.

Using a switching time, $\tau_s = 10$ min, a cyclic stationary state was attained in about 7.5 cycles or 900 min. High lactosucrose yields approaching 60%, at sucrose conversions in excess of 50% were readily obtained with the 2-6-2-2 column arrangement and there was complete separation of the primary reaction products, lactosucrose and glucose as Figure 7.37 demonstrates. Lactosucrose yields drop off rapidly in a batch reactor without an adsorbent when sucrose conversion exceeds 50% because lactosucrose formation approaches equilibrium and hydrolysis reactions then dominate.

Figure 7.37 shows the distribution of educt and product as a function of column position for a cyclic stationary state. It can be seen that raffinate is essentially lactosucrose with small sucrose and lactose impurities. The extract is predominately glucose with some fructose and unconverted sucrose.

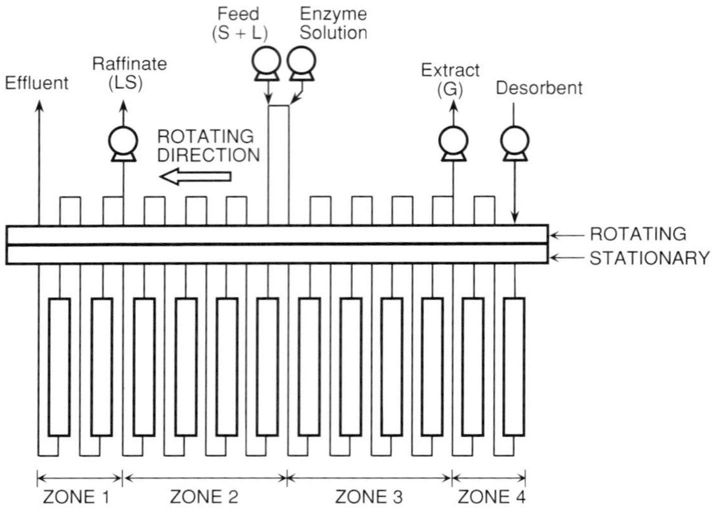

Figure 7.36 Schematic of the 12-column SCMCR showing the operation of the rotary valving. (Figure adapted from Kawase et al. (2001) with permission. © 2001 Elsevier Science Ltd.)

If one product adsorbs and the second product does not, product separation is quite simple and a simulated countercurrent moving-bed reactor–adsorber can consist of just two zones: a reaction–adsorption zone and a regeneration zone. Such a system, an extreme type of SCMCR, has been studied experimentally by Mensah and Carta (1999). It can also be considered to be a form of pressure swing reactor in which desorption is carried out by flushing with a carrier rather than by reducing pressure. Consequently, it falls in between the

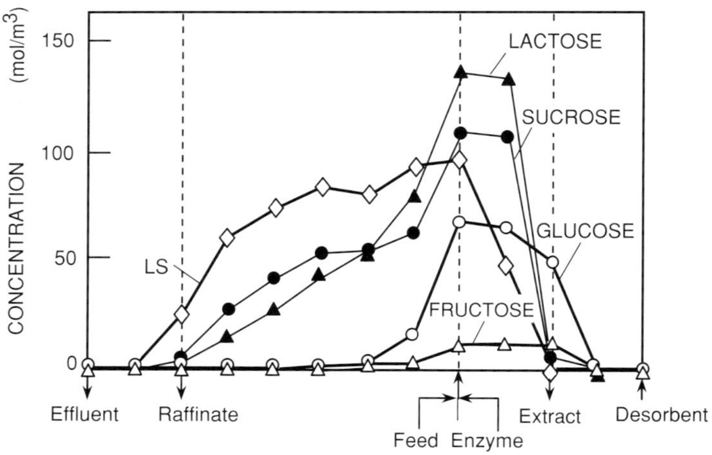

Figure 7.37 Concentration profiles in a 12-column SCMCR used for the formation of lactosucrose from sucrose at $(C_{\text{sucrose}})_0 = 500$ mol/m^3 and lactose at $(C_{\text{lactose}})_0 = 530$ mol/m^3, and $\tau_s = 10$ min. Feed concentration of the β-fructofuranosidase was 100 enzyme activity units/m^3. (Figure adapted from Kawase et al. (2001) with permission. © 2001 Elsevier Science Ltd.)

two classes of cyclically operated separating reactors considered in this monograph. Rather than discuss the interesting Mensah and Carta work in two places, we will examine their contribution in Chapter 12.

7.6 Nonseparation applications

Moving beds can be used for heat trapping. In the beginning of Chapter 5 we rather briefly described a two-section annular rotating bed, the Ljungstrom heat exchanger, that was developed for this application. The rotating bed swings past a downflowing hot stream in the first sector of the device, then through a sealing sector where there is no flow. In the next sector the bed passes through an upflow of cold gas and again into a sealing sector. Heat is captured from the first sector heating the solids, while in the third sector, heat is released to the gas phase and the solids cool. The annular rotating bed functions as a heat trapping and heat recovery device.

Such an operation could be carried out in downward moving beds of solids with lift devices to return the solids to the top of the bed. A simpler exploitation of a bed of solids for heat trapping is to periodically reverse the direction of flow through the bed. Periodic flow reversal has been combined with solid-catalyzed reaction to realize a low capital cost autothermal reactor that is often referred to as a reverse flow reactor (Matros, 1985). We will consider reverse flow operation when we examine pressure and temperature swing separating reactors in Chapters 9–12. With short cycle periods, a problem of the reverse flow reactor is that unconverted reactant at the front of the reactor is expelled on a flow-direction switch. This is wash-out and, of course, it lowers the time-average conversion.

A moving bed of solids avoids the washout problem. For reasons similar to the development of the SCMCR, a multiple bed with rotating inlet and outlet ports has been proposed. Matros (1985) refers to the system as a ring reactor. van den Bussche and Froment (1996) call it a STAR reactor and have investigated the reactor's application to methanol synthesis.

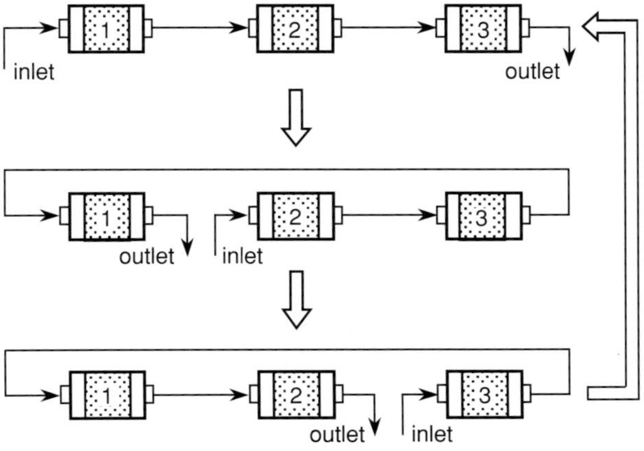

Figure 7.38 Working principle of a simulated moving-bed reactor (SMBR). (Figure adapted from Velardi and Barresi (2002) with permission. © 2002 Elsevier Science Ltd.)

Brinkmann et al. (1999) explore the use of a simulated moving-bed reactor (SMBR) for VOC destruction. Velardi and Barresi (2002) returned to methanol synthesis and compared the performance of a three-bed SMBR to that of a reverse flow reactor. Operation of their system is illustrated by Figure 7.38.

As might be expected, Velardi and Barresi (2002) find that at long cycle times, there is little or no difference between the two reactors. For short cycle times, however, conversions are higher for an SMBR, there is less variation of outlet temperature and, indeed, temperatures through the reactor are more uniform. The latter contributes to improved conversion for the equilibrium-limited exothermic synthesis reaction. The investigators observed complex cyclic behavior at intermediate cycle times and in some cases reaction extinction.

Chapter 8

Chromatographic Reactors: Overview, Assessment, Challenges and Possibilities

8.1 Overview and assessment

Papers discussed in Part I demonstrate that chromatographic reactors of all types are able to force equilibrium-limited or product-inhibited reactions toward completion. They are also able to improve selectivity when multiple reactions occur, but this property needs further study, particularly experimental study. Chromatographic reactors, too, can provide good separation between products or between reactants and products if conditions and properties are suitable. Indeed, literature discussed in the preceding chapters indicates that for biochemical systems at least the challenge in design is to achieve high purity products that minimize the need for further downstream separation. The conditions for which chromatographic reactors are not better than conventional ones are known as well: short space times where conversions are low, when one reactant is strongly adsorbed and when just small differences between adsorption equilibrium constants exist. Although strong adsorption of a product can improve performance, there are limits. Eluent requirement and productivity of the catalyst–adsorbent bed drop if adsorption is too strong.

Creators of the now large chromatographic reactor literature have been remarkably inventive. The single pulse of feed in continuously flowing eluent passing through a packed bed has evolved into moving beds of solids or solid adsorbents raining through a fixed bed of catalyst. Solids movement has cleverly utilized fluidized beds employing differential settling velocities to move the adsorbent while keeping the catalyst stationary. Rotating annular beds or stationary annular beds and rotating inlet and outlet ports have been shown to be another means of converting a discontinuous operation into a continuous one. Stationary beds of solids and switching inlet and outlet ports has seen several innovations to adapt the simulated countercurrent moving-bed design to different applications. Mixed beds of catalyst and adsorbent, matched separate beds of catalyst and adsorbent and a single catalyst bed with multiple adsorbent beds have been examined in Chapter 7. Single- and two-point introduction of eluent as well as rotating an adsorbent bed out of circulatory flow for off-line regeneration have all been explored. Nevertheless, opportunities for innovation remain. Some will be mentioned at the end of this chapter.

Chromatographic reactors have been applied to many chemical systems. Table 8.1 summarizes these.

Table 8.1 Chemical and biochemical reaction systems investigated

Reaction system	Type of chromatographic reactor used
Chemical reaction system	
Dehydrogenation of cyclohexane	CR, CRAC
Dehydrogenation of *n*-butane	CR
Hydrolysis of methyl formate	CR, CRAC, SCMCR
Ammonia synthesis	CR
Dehydrogenation of ethane	CR
CO oxidation	CR, CMCR
Dehydroisomerization of *n*-butane	CR
Esterification of acetic acid	CR, SCMCR
MTBE synthesis	SCMCR
Diethyl acetal synthesis	CR
Triacetine synthesis	CR
Hydrogenation of mesitylene	CMCR, SCMCR
Redox reaction between Ir^{3+} and Fe^{3+}	CR, CRAC
Methanol synthesis	CMCR, SCMCR
Aromatics alkylation	SCMCR
Oxidative coupling of methane	SCMCR
Synthesis of β-phenetyl acetate	SCMCR
Synthesis of Bisphenol A	SCMCR
Partial oxidation of methane	SCMCR
Biochemical reaction system	
Saccharification of starch	CRAC, SCMCR
Enzymatic diol esterification	CR
Enantioselective esterification	CR
Esterification of racemic naproxen	CR
Glucose isomerization	CMCR, SCMCR
Deacylation of Penicillan G	CMCR
Production of dextran	SCMCR
Hydrolysis of lactose	CR, SCMCR
Enzymatic esterification	SCMCR
Lactosucrose preparation	SCMCR

8.1.1 The chromatographic reactor (CR)

The single-pulse chromatographic reactor has had a historical role in the development of separating reactors utilizing adsorption. This type of chromatographic reactor, discussed in Chapters 4 and 14, continues to have an important role as a fast, low-cost tool for testing catalyst and adsorbent combinations or for screening potential reactions for CMCR or SCMCR applications. Screening is particularly important for biochemical applications where there are possibilities of adsorbent–enzyme interactions, such as adsorption and denaturing of the enzyme or fouling of the adsorbent.

With CR models, if the model restrictions are respected, it is possible to extract kinetic and adsorption data from pulse measurements as Hattori and Murakami (1968) and Sardin and Villermaux (1979) have demonstrated. Pulse size and contact time are critical conditions. Pulse durations must be short and reactants must be dilute to avoid adsorbent saturation or highly nonlinear adsorption isotherms. Nevertheless, measurement requirements restrict dilution and pulse duration. Accurate measurements are needed if rate and isotherm data are to be obtained from the effluent composition. As tools for catalytic kinetics, CMCRs or SCMCRs are less suitable simply because they are more difficult to design, equipment is expensive and time-consuming to build and they are more difficult to operate. They may require many cycles to reach a cyclic stationary state needed for reliable measurements.

Multiple pulse CRs are also useful as a study tool because they can suggest the range of operating variable for CMCRs and SCMCRs by easily performed and interpreted experiments such as varying pulse frequency, pulse duration and reactant concentration. Although the multiple pulse CR offers no more information than that drawn from a single pulse CR, it has the advantage that experimental chromatograms demonstrate visually conversions and separations achievable. Thus, the multiple pulse CR effectively presents the potential advantages of separating reactors to management or investors.

8.1.2 The countercurrent moving-bed chromatographic reactor (CMCR)

Countercurrent moving-bed chromatographic reactors (CMCRs) are continuous reactor–separators, but the adsorbent and often the catalyst undergo a periodically changing environment. Several variants have been considered in Chapters 5 and 6. Falling beds of a combined catalyst–adsorbent as well as finely divided adsorbent trickling through large, fixed catalyst particles have been investigated, but they pose severe operating problems resulting from radially uneven flow and particle attrition. Even though the nonuniform flow problem probably becomes less serious as reactor diameter increases, the attrition problem will mandate catalyst and adsorbent particles possessing either high mechanical strength or low cost to reduce replacement expenses. On the other hand, a CMCR has several advantages over SCMCR or CRAC alternatives. Consider a reaction with a strongly adsorbed product or reaction systems that foul either the adsorbent or the catalyst. Continuous removal of the solid phase would permit recovery of a strongly adsorbed product through an external stripper, operated perhaps at a higher temperature or with a different eluent. Fouled or coked solids could be regenerated and recycled back to the CMCR. Continuous catalyst–adsorbent removal and makeup are simple when solid must be withdrawn from the bottom of the reactor and carried to the top. Solids transport could be done mechanically or via a pneumatic lift. Hoppers would be necessary to provide holdup. A CMCR with continuous recycle of the solid phase would function like fluidized bed catalytic cracking or butane partial oxidation. Such systems, however, are a subject for future research. There is no mention in the research literature of operation of a CMCR with adsorbent–catalyst recycle although Fish et al. (1986) discuss a bench-scale design of such a system.

There appears to be little current interest in the CMCR for gas–solid systems. This neglect, however, may be unwarranted. In the final section dealing with research opportunities, we suggest an approach that could make the CMCR alternative more attractive.

Beds of moving catalyst or adsorbent particles seem to be much more practical for liquid–solid systems. The multistaged fluidized beds pioneered by Dutch investigators (Vos et al., 1990a,b,c and van der Wielen et al., 1996, 1997a,b, 1998) and discussed in Chapter 6 seems well suited to isothermal biochemical reactions. These operate with immobilized enzyme and adsorbent settling downward against a rising liquid phase. Pulsing the mobile phase flow allows particles to descend through holes in plates enclosing the stages. By changing size and density of the immobilized enzyme and the adsorbent, the multiple beds can be operated so that only the adsorbent descends from stage to stage. Problem-free mechanical operation of these systems has been shown for laboratory-scale equipment, but it will be necessary to demonstrate operation on a much larger scale before commercialization can be considered. It appears to be a promising technology that deserves further study.

With just two exit ports in a CMCR, separation is limited when multiple reactions occur. This is not a problem in a consecutive reaction system as long as an adsorbent can be found for the sought after product. It could be a problem when more than one reaction product has value. It may be possible to introduce an intermediate take-off port to exploit a concentration build up in a CMCR, but this option has not been explored. Consequently, CMCRs studied so far are restricted to single reactions or to rather simple consecutive reaction schemes.

Another potential application of a CMCR is when there is a solid reactant. Such a reactant could be fed to the system along with catalyst and/or adsorbent. This possibility also has not been considered in the literature. However, there are several potential applications: incineration of solid waste with toxic gaseous products, fuel gasification or calcination reactions that have equilibrium limitations. In the former case, adding adsorbent at the top of the incinerator would convert the unit into a CMCR. Possibly, a CMCR with a lime adsorbent could provide in situ CO_2 capture in coal gasification to produce synthesis gas.

The applications just mentioned represent highly nonisothermal operations. A further such application might use the moving bed or raining solids as a heat source for an endothermic reaction or, possibly, as a heat sink to remove heat from an exothermic reaction. We believe that the nonisothermal operation of CMCRs deserves study along with the mechanical problems of flow distribution and solids feeding and removal. Experimental work is expensive, of course, but nonisothermal operation and the use of moving solids as a heat source or sink is amenable to computational study. So far this use of CMCRs has not been examined in the research literature.

8.1.3 Continuous rotating annular-bed chromatographic reactor (CRAC)

The continuous, rotating, annular-bed chromatographic reactor (CRAC), considered in Sections 6.2.1 and 6.3.1, has two different designs: one in which the solid phase is contained in an annular basket that rotates past fixed inlet and outlet ports. The second, perhaps somewhat simpler design, places the inlet and outlet ports on plates that rotate past an annular bed that is now fixed in place. Both designs offer many outlet ports. These can be grouped together to capture a particular product. Thus, several relatively pure products may be obtained when there are multiple reactions and more than one product has value. This is a unique feature of the CRAC reactor–separator. Curiously, research on the use of the CRAC system for multiple reactions does not seem to exist.

Widespread use of fairly large rotating heat exchangers means that the mechanical design of CRAC systems is well understood. Furthermore, CRAC separators have found commercial use in preparative chromatography. Thus, the CRAC reactor–separator must be deemed a technology ready for commercial application. Nevertheless, it seems unlikely that the technology will be practical for systems with very large volumetric capacities because of problems with uniform gas distribution over a large cross section and mechanical problems of rotating a large mass of solids if that option is chosen. There is a further problem of sealing for gaseous systems. This could be severe for reactions that require temperatures above 600 K. These limitations will have to be spelled out by mechanical designers. We believe that CRAC reactors will find most of their applications in liquid phase processes, probably low-temperature biochemical ones, in which sealing is less of a problem. Because of growing distribution and mechanical problems as the annular diameter increases, most CRAC applications will likely be for small-scale production. This again points toward its use for biochemical reactions. It is no surprise that much of the literature discussed in Chapter 6 deals with such applications.

CRAC experiments reported up to now, have been carried out on thin annular beds with widths on the order of a centimeter. A commercial unit would have a diameter on the order of a meter or more and the bed width would be certainly a half-meter. Annular or radial dispersion has not been a problem with thin beds, but perhaps it becomes more so in wide beds where differences in the centrifugal forces that the fluid phase experiences at the inner and outer walls could be significant at higher rotational speeds. This question needs some study. Segmenting the bed with dividers or baffles running in the flow direction would be possible if radial or annular dispersion is serious. Heating or cooling fluids could be routed through these baffles offering a means of temperature control.

8.1.4 Simulated countercurrent moving-bed chromatographic reactors (SCMCR)

The literature reviewed in Chapter 7 has expanded rapidly since 1990, suggesting many fruitful applications of simulated countercurrent moving-bed chromatographic reactors (SCMCRs) in industry. Interest in SCMCRs has been driven by two factors: commercial success of the Sorbex separation system described in papers by Broughton (1968), Broughton et al. (1970) and De Rosset et al. (1976) and the adoption of preparative chromatography, largely based on simulated moving-bed systems, for preparing high value chiro or enantio materials. It has been evident from past studies that the biochemical reactions necessary to produce starting materials for separation are better undertaken in a chromatographic reactor than in the conventional batch reactor because product inhibition is often encountered. Inhibition can be sharply reduced by product adsorption. Indeed, a raft of research studies suggest that product-inhibited enzymatic reactions can be easily forced to completion in an SCMCR so that design is dictated by product separation considerations rather than by conversion requirements.

An important advantage of an SCMCR over a CMCR is flexibility. It is straightforward to arrange mixed bed columns or to separate catalyst and adsorbent and use different size beds for each. These can be connected in different ways as the work of Hashimoto et al. (1983) illustrates. Furthermore, different cycle structures can be used so that multiple purges are possible as Tonkovich and Carr proposed in their 1994 papers.

Because an SCMCR uses fixed beds, design and scale-up depends upon our understanding of fluid dynamics in packing. This indeed is quite well developed so that scale-up poses no problem.

There is no "standard" design for a SCMCR, but four-section reactor–separators are more frequently encountered than three-section ones. The difficulty of recovering the more strongly adsorbed product governs this choice. The number of columns in each section range from 1 to 4 in most applications. Since columns are of equal size, it is the depth of bed needed in each section that is important. Research has shown that just two columns are adequate to represent the countercurrent moving bed of solids model. Studies have been carried out on mixed catalyst and adsorbent systems, separate beds of catalyst and adsorbent and, for the biotechnology area, an adsorbent bed only, with an enzyme in the fluid or mobile phase. Separating catalyst and adsorbent allows more flexibility as different temperatures can be used for the catalyst bed and for the adsorbent bed as demonstrated by Tonkovich et al. (1994a,b). Piping can be designed to isolate the reactors in the desorption portion of a cycle. On the other hand, combining catalyst and adsorbent reduces the catalyst requirement. This would be a consideration if the catalyst is expensive.

Just as for its moving bed of solids counterpart, SCMCRs have normally just two outlet ports and so are restricted to single reactions with just two products or rather simple multiple reactions in which a desired terminal or intermediate product is adsorbed so that it leaves in the extract port. In principle, another outlet port can be added, even a further eluent port in section 3 or 4, so a product of intermediate adsorption could be removed and the mobile phase flows adjusted by addition of eluent. Thus, an SCMCR could be made suitable for a two-reaction system with three products. How this might be done and the effect on performance is yet to be tackled by researchers.

Mechanical design of SCMCRs is based on the principles of flow through packed beds of randomly distributed solids. Uniform flow distribution is a problem only with shallow beds or, potentially, with a wide distribution of particle sizes. Rapid switching of valves, often cited as a problem with cyclic operation (see Silveston, 1998), is not one for SCMCRs because switching times are usually of the order of tens of minutes or more. Even with shorter switching times of feed and exit port valves, motor-driven fast-acting valves can be employed as long as fluid temperatures do not exceed 600–650 K and pressures are below about 10 bar. Beyond this limit, large valves can be built but they are expensive. Multiple valving required between vessels adds dead time and this dead time has been shown by Migliorini et al. (1999b) to affect separation performance.

Much of the theoretical and experimental research on SCMCRs has assumed isothermal operation. Because the imminent applications seem to be biochemical and thus liquid phase reactions, this is neither surprising nor negligent. Enzymatic reactions have small heats of reaction and heats of adsorption are low. For gas phase reactions, however, this is not the case. Thus, attention needs to be paid to nonisothermal operation of an SCMCR and to heat removal for temperature control. Cooling in the piping and valving between beds needs to be examined along with heat losses from individual beds. This is a problem because of time-varying concentrations, heat release or demand, and temperatures in each bed. Even for liquid systems, some theoretical studies would be worthwhile to see if a problem exists.

The volume of research on SCMCRs makes this type of chromatographic reactor likely to be the first that enjoys widespread industrial use. Indeed, applications appear to have begun. Applications of the CRAC version lag, but it too should soon see industrial application. By way of contrast to the CRAC system, the SCMCR does not seem to be size limited so it would appear to be the choice for large volume applications. Size will probably be limited by cooling or heating requirement, but not only because there are severe problems in heating or cooling large packed beds. The main problem is that the bed operation is characterized by moving reaction and adsorption fronts. The heating or cooling demand, thus, varies with time and position in the bed. This problem may dictate near isothermal operation, which could force a limit on size and production capacity of a single bed. Fronts are stationary in a CRAC system, but in one version, in which the annular bed rotates, heat transfer to and from a bed in motion will be needed.

Because of its simplicity in terms of involving just a packed bed and valving, the SCMCR is a suitable system for down scaling or miniaturization to almost microreactor size. A miniature SCMCR could conceivably serve as an on-board hydrogen generator for fuel cells through the steam reforming of methanol or dimethyl ether. These are fuels that are handled and transported as liquids.

A limitation of SCMCRs and indeed of CMCRs is the availability of adsorbent materials. Conventional adsorbents function normally at low temperatures. Consequently, the SCMCR technology is probably limited to reactions proceeding at the temperatures between 300 and 550 K. Although Tonkovich and Carr (1994a,b) and Bjorklund and Carr (2002) have demonstrated SCMCR operation at temperatures above 800 K, cooling and heating of the adsorbent over a temperature difference of 500°C was used and the cost this imposes cannot be justified economically.

8.2 Modeling

For the last 30 years virtually every chromatographic reactor study published has discussed a reactor model. Most have included experimental model validation. CR modeling, consequently, is a mature subject. Challenging systems, such as those employing ion-exchange resins that swell or shrink depending upon the water or alcohol content of the fluid phase, have been successfully modeled.

For isothermal systems considered in the literature, one-dimensional pseudo-homogenous models have been found to be satisfactory for relatively slow reactions, namely most liquid phase reactions. With reactions that proceed rapidly, one-dimensional heterogenous models may be needed. Recalling that adsorption rates should exceed or at least match reaction rates in an effective separating reactor, rapid reactions can mean that the adsorption step becomes mass transfer limited so that transport between phases must be considered. Nevertheless, even if mass transfer is considered for the fluid–adsorbent system, validation work in the literature suggests that apart from adding an effectiveness factor the catalyst–fluid system can be treated as a pseudo-homogeneous one.

Experimental investigations of chromatographic reactors for model validation abound in the literature, but there are relatively few available on the moving-bed systems (Wardwell et al., 1982; Takeuchi and Uraguchi, 1977b; Fish et al., 1986). In general, experimental work substantiates theoretical analysis. One result, though, is disturbing. The CMCR models of Wardwell et al. (1982) and Takeuchi and Uraguchi (1977b) did not fit all experimental results adequately. Models discussed in the literature are isothermal and many use linear isotherms. They are, thus, restricted to very dilute systems. Probably the model failures for CMCRs arise because the systems studied have not been dilute enough to meet the isothermal–linear isotherm assumption. There are other explanations. Wardwell et al. used a cell-in-series model that adds dispersion to the CMCR, whereas Takeuchi and Uraguchi assumed plug flow for both the fluid and solid phases. It is doubtful that solids move downward in a tube as a plug. Thus, an incorrect allowance for dispersion may have led to the failure of the models.

Considering that the isothermal reactor is an artifact of the research laboratory, the challenge in CR modeling is nonisothermal operation, particularly when heat loss must be considered. Many of the reactions considered in the literature, such as mesitylene hydrogenation, have substantial heats of reactions. Some components may also exhibit large heats of adsorption. The packed-bed literature (Froment and Bischoff, 1990; Westerterp et al., 1993) suggests that one-dimensional models are adequate for steady state operation. Will this be the case for transient operation? With the exception of the moving-bed systems considered in Chapters 5 and 6, CRs are periodic. Furthermore, their operation can switch from endothermic to exothermic within a cycle. Modeling such systems is a largely uncharted subject. Prior to any difficult experimental study, simulation should be undertaken to see under what conditions one-dimensional pseudo-homogeneous models fail to represent CR behavior. We suggest that this be undertaken first for a relatively simple system such as a single pulse CR with a hot feed and a constant wall temperature. A study would compare three models: a two-dimensional, a one-dimensional and perhaps an isothermal one-dimensional model. Of course, allowance should be made for the heat of reaction, heat effects on adsorption and desorption and temperature effects on reaction rate and the adsorption isotherm. Initially, heat transfer parameters, such as effective conductivity and wall heat transfer constants, would be drawn from the packed-bed literature. Heterogeneous versus pseudo-homogeneous is not an issue. Pseudo-homogeneous models work well for energy balances.

Indeed, if temperature variations within the catalyst–adsorbent bed are shown through simulation to have a significant effect on performance, further experimental study will be important. Temperature variation in CMCR and SCMCR systems has been tackled by simulation just once (Kruglov, 1994) and then only for an adiabatic reactor. No experimental studies have been reported. If temperature variation has an important effect on SCMCR performance, the matter of energy supply or removal will have to be examined. Initially, this would be investigated through modeling different heat supply or removal alternatives.

Another issue in modeling is solution methods for representing the cyclic stationary state of SCMCR systems. Integration of the model's equation over time to reach this state often consumes vast amounts of a large computer's CPU time and is prohibitively expensive for desk top PCs. "Shooting" methods, discussed in Chapter 9 for pressure swing reactors, have been found to be capable of sharply reducing CPU time but other investigators dispute this claim. Solution methods are of critical importance in modeling so the use of "shooting" techniques needs further exploration. Likely, the choice of model parameters, the complexity of the kinetic model and the adsorption isotherm determine whether or not these techniques are efficient and which among the several possibilities should be used.

A further problem, though perhaps a secondary one, that has not been examined in the literature is that steady-state kinetic models are routinely used in simulating chromatographic reactors. But if we look at each catalyst particle, that particle is exposed to an unsteady-state environment. Transient behavior on the particle level has been neglected. Under unsteady conditions, intraparticle diffusion can be significant even when the Thiele modulus is small as it is in the case of fine particles (Cho, 1983; Cho and West, 1986; Aida et al., 1997). This question can and should be handled through analytical studies. A start has been made on more realistic and sophisticated models, such as in the recently published papers of Mazzotti et al. (1997a,b,c), but a useful effort would be to compare performance predictions from simplified and sophisticated models. If, indeed, model simplification leads to the prediction of unrealistic benefits, more experimental studies of CMCR systems will become necessary to establish what simplifications are permissible.

8.3 Design

The matter of design does not arise for single and multiple pulse CRs because these reactors are interesting only for exploratory studies. Size, namely length and diameter, of these systems are often set by convenience or by experiment. Diameter influences the quantity of reactant needed, while length depends on product and reactant retention times.

With the exception of the CRAC system, the development of CMCRs has not advanced sufficiently to focus attention on design. As mentioned above, the continuous rotating annular bed chromatographic reactor or CRAC reactor has emerged as a commercial reactor. The literature, however, contains no specific design procedure for the system. As evident in Chapter 6, performance is characterized by the separation achieved. Thus, it is likely that a design procedure will follow the procedure for a separatory system with the provision that retention time is long enough for the desired conversion or yield. It is retention time and loading as the volumetric flow per square meter of bed surface per second that will determine bed depth. Bed width will then be set by the loading and the specified production in moles of product per unit time. Angular velocity of the bed or the feed and withdrawal ports will be 2π radians divided by the retention time. The latter is an experimentally determined quantity. What then is needed is a guide to or bounds on loading. This needs exploration.

By way of contrast to CRAC systems, design of SCMCRs has received attention in the literature. Guidelines for choice of an SCMCR for an application are nicely summarized by Fricke et al. (1999):

1. Mass transfer should not be rate limiting.
2. Educts should be significantly more weakly adsorbed than any product. This is particularly important when mass transfer is a rate-limiting step.
3. Reaction rates for single reactions should exceed $1 \times 10^{-2}\ s^{-1}$.
4. Reaction equilibrium constant should exceed 0.01 for an equilibrium-limited, single reaction.
5. Separation factor, ratio of adsorption equilibrium constants, for the reaction products should be <0.5 or >2.

Design of an SCMCR consists of several steps: choice of mode either mixed catalyst and adsorbent or separate beds is the first step. This is followed by the selection of total column length and its subdivision into separate beds or columns, then the number of sections is chosen, usually four, and individual columns assigned to each section. Total bed length generally determines conversion. Product purity is often specified so it is the recovery of a product at this purity that must be assured in the design. Key to recovery is switching time. This sets the equivalent flow rate of solids and thus the σ for the equivalent moving-bed reactor. Complete separation occurs when the flow ratios in section 2 and section 3 on either side of the feed point are within a region bordered by a diagonal in the $\sigma_3 - \sigma_2$ plane and limiting values of these variables where σ_3 and σ_2 are the flow ratios or also ratios of switch time, τ_s, to contact time, $t_j^* = V_j/Q_j$, in the j section. A much larger region above the diagonal provides either pure extract or pure raffinate. Regeneration of the adsorbent requires that the switch time exceed the contact time in section 1, while in section 4 the switch time must be less than the contact time. SCMCR design follows the design of a separating system closely, except when the reaction proceeds slowly. Then, the loading rate must be reduced to reach the desired conversion or yield and the productivity of the unit suffers. A design procedure for multiple reactions when an intermediate product must be recovered is not discussed in the literature.

8.4 Research needs

Modeling of chromatographic reactors is now a mature art. Improvements are still possible, but will come through improved isotherm models that cover a wider range of concentrations and allow for competitive adsorption. Better reaction rate models would be useful too. Further study of "shooting" methods for calculating performance of SCMCRs at the cyclic stationary state should be undertaken. Integration of the sets of coupled, nonlinear partial differential equations that form the model of a SCMCR remains a challenge because of the CPU time needed. "Shooting" methods have been shown to reduce the CPU requirement in some cases, but apparently this is not universally true.

Simulation needs are generally well served by existing models. However, a problem that needs to be addressed is nonisothermal operation of chromatographic reactors as heat effects in gaseous systems are appreciable. An isothermal assumption has been made in almost all of the theoretical papers. This assumption is valid for dilute systems but would not be true for concentration that would be encountered in commercial processes. Further analytical study of the effect of temperature variation in a CMCR or a SCMCR is needed.

Although there have been no experimental reports of catalyst deactivation or loss of adsorption capacity with time in chromatographic reactors, it is a problem that can be expected. Capacity loss has been observed in temperature swing reactors (Chapters 10 and 11). There is then a need to determine how gradual loss of catalyst activity or adsorptive capacity will affect the performance of all types of chromatographic reactors. A modeling investigation should be adequate to define the problem and indicate strategies that can be used to mitigate capacity loss or deactivation effects. Isobaric operation has been assumed in all studies reported to date. Although this assumption is valid for many chromatographic reactors, it is questionable for SCMCRs that may contain as many as 16 packed-bed columns connected through several on–off valves at each column. It would be instructive to redo some of the simulations published in the past allowing now for pressure drop in the flow circuit. Migliorini et al. (1999b) have shown that "dead" volume between columns effects performance. A pressure drop effect on performance should also occur.

Tools for design of CRAC and SCMCR systems are incomplete. The literature does not have information on suitable loadings for CRAC reactors and a procedure for SCMCR design with multiple reactions is not available.

Despite the large number of system studied (see Table 8.1), there remains a need to explore additional systems in order to reinforce understanding and enhance confidence in modeling. A particular need is to examine systems of multiple reactions where selectivity is a primary concern. Table 3.1 shows just two numerical studies and two experimental ones.

8.5 Research opportunities

Unlike "needs" discussed above, opportunities represent new applications of chromatographic reactors or means of improving performance. "Needs" are to enhance understanding.

8.5.1 *Improving SCMCR performance*

The close relationship between a stimulated countercurrent moving-bed chromatographic (SCMC) separator and an SCMCR make it likely that innovations for the separator will work as well for the reactor. Several novel separator operations have been proposed and studied. Variation of eluent composition has been used in analytical chromatography for many years as a means of shortening elution times. This technique was first studied for preparative chromatography by Nicoud et al. (1995), who referred to the operation as a gradient method. Maintaining a constant solvent composition, in contrast, was called an isocratic operation. Jensen et al. (2000) suggested a bang-bang modulation, using one solvent for the feed and a better solvent for the adsorbate as the eluent. A thorough analysis of this operation has been published by Antos and Seidel-Morgenstern (2001). They examined the separation of cyclopentanone and cycloheptanone over a silica gel. Solvent was *n*-hexane with the addition of ethyl acetate to modify the solubility of the two species separated. The separation was modeled and the predicted effect of running *n*-hexane with the feed and pure ethyl acetate as the eluent on the performance of the four-zone SCMC was verified experimentally. When flows were adjusted so that the σ_2 and σ_3 for the simulation fell about in the triangle in the σ_2– σ_3 plane for complete separation, the mean composition of ethyl acetate was 15 mol%. This condition tripled the SCMC separator productivity with just a small loss in extract purity over an isocratic operation using 15 mol% ethyl acetate in *n*-hexane. Experiments on a laboratory scale using the compositions and flow employed in the simulation gave productivity, product purities, and modifier concentration in the raffinate and extract streams that were virtually identical to those found in the simulation.

Increased throughput possible using a gradient method, that is, different compositions of the solvent with the feed and that used as eluent, will probably decrease conversion and perhaps worsen product separation. Nevertheless, there may be some advantage accrued by applying the method to an SCMCR and this deserves investigation. First step would be a numerical study, but eventually experiments will be needed.

Variation of feed composition and flow rates within a time step can also improve SCMC separation performance (Zang and Wankat, 2002a,b; Zhang et al., 2003). Reasoning for utilizing flow variation is nicely illustrated by Figure 8.1, which shows the variation of the extract purity and yield with time in an SCMC separator, independent of the eluent. Purity and yield are high early in the switching period because of the adsorbate concentration upstream of the extract port. Product removed from the separator is the volumetric mean of the flow withdrawn during the switching period. Consequently, if the extract flow rate is augmented early in the switching period when the adsorbate concentration is high and reduced when it is low, so that the mean equals the flow under the constant flow condition, extract purity should increase. On the other hand, if extract purity meets specification, the quantity of adsorbent could be reduced, raising productivity. Alternatively, the amount of

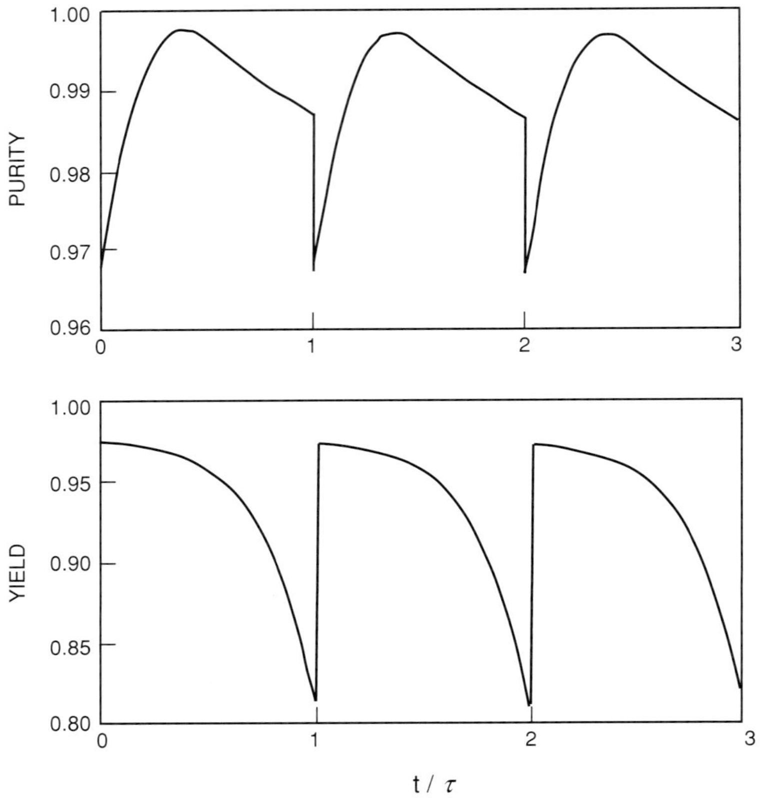

Figure 8.1 Variation of extract purity and yield with time at the extract port over three switching periods at the cyclic stationary state for an SCMC separator at constant flow rate. (Figure adapted from Kloppenburg and Gilles (1999) with permission. © 1999 by Wiley-VCH Verlag GmbH.)

eluent used might be lowered, raising the throughput/eluent volume ratio. Kloppenburg and Gilles (1999) have examined extract flow variation by simulation for the separation of p-xylene and m-xylene in an eight-column SCMC separator using toluene as the solvent. They demonstrated an 11% reduction in the eluent requirement by changing the extract flow rate three times out of a possible four changes during the switching period. The duration of each flow rate interval was equal. Throughput, and thus productivity, was maintained as was extract and raffinate product purity. Of course, feed rate, eluent and raffinate flow rates changed together with the extract flow rate. From an examination of the simulation results, it seems likely that a simpler flow variation, two changes of extract flow rate during a switching period, would accrue much of the reduction in eluent requirement that Kloppenburg and Gilles identified.

Zang and Wankat (2002a,b), Zhang et al. (2003) and Schramm et al. (2003c) also considered flow variation within a switching period. They observed that much of the benefits of changing the flows of all inlet and outlet streams, as just discussed, can be achieved by stepwise variation of just the feed stream or the feed and one of the product withdrawal

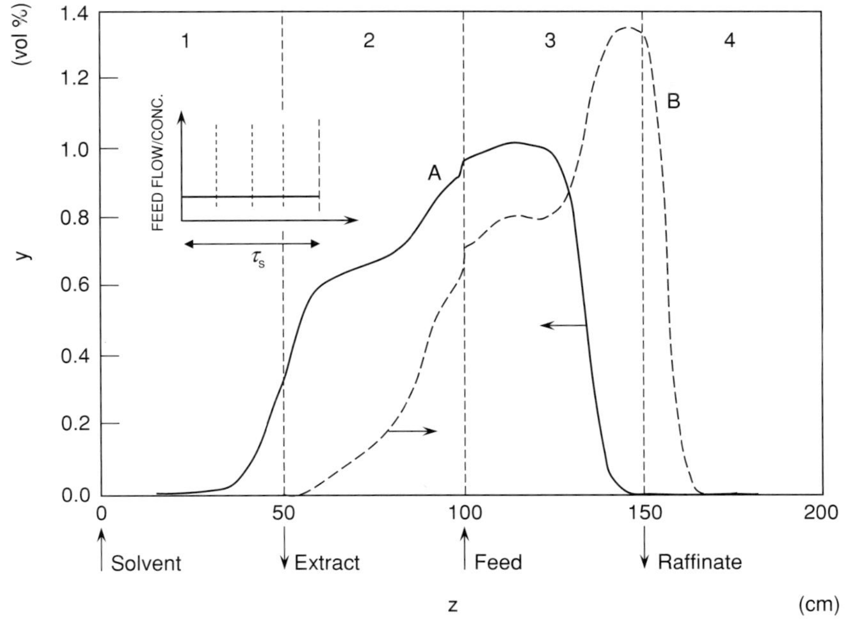

Figure 8.2 Extract and raffinate profiles at the midpoint of the switching time for a four-section SCMC separator with continuous, uniform flows at all inlet and outlet ports. (Figure adapted from Schramm et al. (2003c) with permission. © 2003 by Elsevier Ltd.)

streams. We illustrate this observation with the help of Figure 8.2. In this figure, showing concentration profiles of the extract (A) and the raffinate (B) versus position in the bed at the midpoint of a switching interval, the profiles are moving from left to right with the fluid flow. On each switch, the zone moves to the left simulating the movement of the adsorbent. Thus, the profile at the end of the interval in zone 3 becomes the profile in zone 2 at the start of the next interval. These initial profiles migrate to the right across the feed point or node during the interval. Feed, of course, creates a concentration change as may be seen in the figure.

In order to increase the purity of the raffinate, the feed must be manipulated to slow the velocity of the extract (A) front. Conversely, the purity of the extract can be raised by accelerating the movement of the raffinate (B) front (see the arrows in the figure). These profiles behave like shock waves in sections 3 and 4 and as spreading waves in sections 1 and 2 as we have discussed in Chapters 5 and 7. The velocity of a shock wave depends on concentration of the wave species, while the velocity of each point in the spreading wave likewise depends on concentration at that point for the latter wave. Consequently, manipulating feed flow rate will affect wave velocity. Velocity change can be reinforced by manipulating the withdrawal rate at either the extract or raffinate port. Alternatively, the concentration of the feed can be manipulated to alter front velocity. Reducing, for example, the concentration in the feed at the beginning of the switching interval reduces the concentration of A in section 3 thereby decelerating the velocity of the extract front. Velocity change is illustrated in Figure 8.3. Slowing of the extract front in section 3 causes the purity

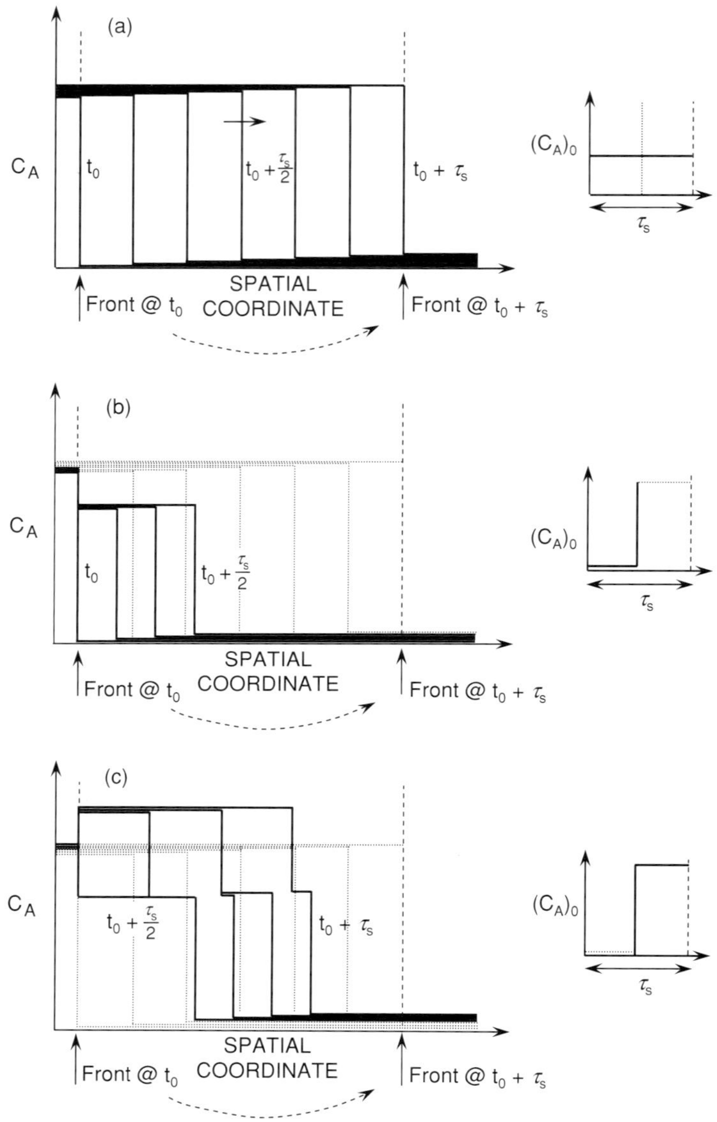

Figure 8.3 Effect of concentration on the velocity of an ideal shock front in section 3 of an SCMC separator: (a) front velocity with uniform concentration, (b) front velocity in the first half of the switching time interval when concentration is reduced by dropping feed concentration, (c) front velocity in the second half of the interval when concentration is increased by raising feed concentration. Feed concentration manipulations shown on the RHS. Dashed lines show section 3 concentrations and front velocities for (a). (Figure adapted from Schramm et al. (2003c) with permission. © 2003 Elsevier Ltd.)

of B in the withdrawal at the raffinate port to increase. Maintaining a constant feed rate to the SCMC separator means, however, that the feed concentration must be increased in the remaining portion of the switching interval. Raising the concentration now increases front velocity, but the higher velocity front only catches up with the slower moving front from the previous portion of the interval. As a result, the extract front does not have time to reach the raffinate port. Note that front velocities are affected only in section 3. No changes occur in section 2 so the feed manipulation does not alter the extract purity.

Concentrations in zone 3 can also be manipulated by the feed flow rate. Altering the flow rate causes flow changes throughout the SCMC separator. Manipulating withdrawal rates at the raffinate port mitigates such changes.

Schramm et al. (2003c) carried out simulations of feed manipulation using as an example the separation of cyclopentanone and cycloheptanone over silica gel with a *n*-hexane and ethyl acetate mixture as solvent. The SCMC separator consisted of four zones with two columns in each zone. Langmuir isotherms were assumed and dispersion included in their models (thus, the slopes in Fig. 8.2). These simulations demonstrated that if product purity is constrained to 98%, feed flow or feed concentration manipulation in a two-part cycle shown in Figure 8.3 increases productivity and reduces eluent demand substantially. Feed concentration manipulations are also examined in Schramm et al. (2002, 2003a,b).

These manipulations are quite likely to yield separation benefits in an SCMCR. Nevertheless, they need to be examined in a reaction application as mentioned above in reference to eluent manipulation. They are not applicable, it seems, to CRAC reactors or CMCRs.

The literature on chromatographic reactors has considered only isotropic catalyst and adsorbent beds. It seems likely that beds exhibiting axially varying ratios of adsorbent and catalyst, adsorbents with varying heats of adsorption or catalysts of changing activity could enhance conversion or selectivity further. This possibility is suggested by the work of Cote et al. (1999) using layered beds of catalysts with different activities. Shi et al. (1996) used layers of different catalyst to improve selectivity for multiple reactions. Aida et al. (1999) demonstrated that different arrangements of layered catalyst beds under periodic operation altered conversion in a reactant-inhibited system.

8.5.2 New applications

Reactors utilizing trapping may have some interesting applications in conventional processes. Consider the removal of trace amounts of CO from a hydrogen stream obtained by steam reforming of a hydrocarbon. One process for this is to selectively oxidize CO over a supported Pt catalyst. Reaction occurs through the adsorbed species. The catalyst strongly adsorbs CO, whereas the other reactant, O_2, is just weakly adsorbed. Hydrogen is only slightly adsorbed.

As now carried out using a packed bed of catalyst, oxygen and the raw hydrogen stream containing a small amount of CO are fed cocurrently to a plug flow reactor. Order of the reaction is positive with respect to O_2 and negative with respect to CO. As O_2 is consumed by reaction, the rate drops, although this loss is partially compensated for by the drop in the CO concentration. Toward the end of the reactor, H_2 oxidation commences. CO conversion without a significant loss of H_2 could be achieved by using a CMCR or SCMCR in place of a packed bed. Figure 8.4 sketches a CMCR application and shows the expected profiles of oxygen, adsorbed CO (as CO(s)), and hydrogen. The raw H_2 stream enters at the middle of

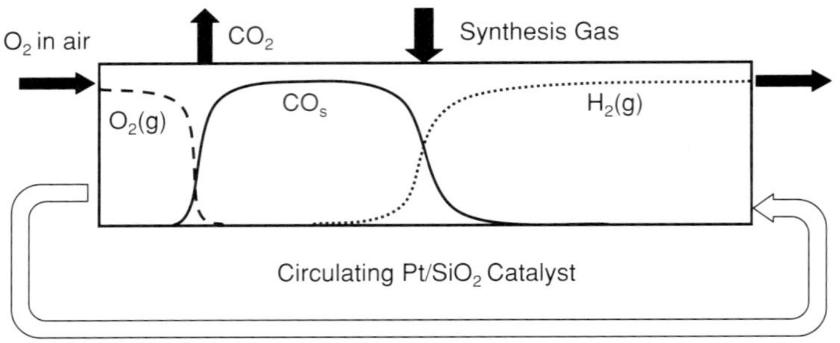

Figure 8.4 Schematic of a possible CMCR application to purifying hydrogen from steam reforming by selective oxidation of CO. Curves within the rectangle representing the moving bed show schematically the expected concentration profiles of hydrogen, oxygen and adsorbed CO in the moving bed.

the bed and flows to the right against a moving bed of catalyst that functions as an adsorbent for CO. Oxygen enters from the left and reacts with CO adsorbed on the catalyst, releasing CO_2. A product gas rich in CO_2 with traces of O_2 and H_2 is withdrawn between the feed nodes. The hydrogen product is very pure with just ppm levels of CO, O_2 and CO_2. Excess oxygen can be introduced to lessen the amount of catalyst needed.

8.5.3 Moving-bed design

The problem of radial maldistribution of moving solids might be resolved by employing structured packing in place of dumped beds used by Kuczynski et al. (1987b). There might be as well a reduction in attrition. Extruded ceramic or metal monoliths could be attractive as packing for moving beds of solids. These structures have a regular grid spacing with uniform, straight channels that should provide well-distributed downward flow of fine, dense catalyst–adsorbent particles against a rising gas stream. Grid dimensions range from several millimeters to a centimeter. Thus, fine particles of millimeter to submillimeter size would be required. Although flow in each channel may be nonuniform, there should be no radial maldistribution over the cross section of the monolith. Choking, of course, will limit throughput. Utilization of monoliths deserves attention. Attrition remains a problem, but this can be attacked through the development of hard shell catalysts or adsorbents now used in some circulating fluidized bed processes. Monoliths may also be applicable to a liquid–solid system. Larger cell dimensions would be needed to accommodate the larger particles required for a reasonable slip velocity in an up flowing liquid phase.

Use of monoliths with catalytic walls is a further possibility. In this alternative, the catalyst phase is fixed in place and only adsorbent flows countercurrent to the rising fluid phase. Monolithic structures with the catalyst phase deposited on the monolith walls are now being intensively studied in several research laboratories (e.g. Nijhuis et al., 2001; Garcia-Bordeje et al., 2002; de Deugd et al., 2003).

Part III
Swing Reactors

Chapter 9
Pressure Swing Reactors

9.1 Introduction to swing reactors

Separating reactors employing adsorption can utilize abrupt pressure or temperature changes for regeneration. Chromatographic reactors employ an eluent for this purpose. Just as the chromatographic reactors considered in Chapters 3–8 are derived from preparative chromatography, the swing reactors, either pressure swing (PSR) or temperature swing (TSR), are drawn from adsorption systems: pressure and temperature swing adsorbers. Both classes of separating reactors are basically cyclic in operation. Work on the swing reactor–separators is less advanced than that on chromatographic reactors, even though their separation counterparts have a well-established technology. Extending the application of this technology to reaction systems appears to be straightforward if swing reactors can improve yields or selectivity or can decrease capital and operating costs when both reaction and separation are necessary. Thus, the question addressed by most researchers has been whether or not carrying out reactions in swing systems improves either separation performance or reaction conversion and selectivity.

Pressure swing reactors were the first to be considered (Vaporciyan and Kadlec, 1987, 1989). Table 9.1 lists many of the publications on swing reactors. Studies of periodically pulsed reactors are also listed. In these reactors, carrier gas flushes the catalyst–adsorbent bed to recover product rather than reducing pressure. Periodically pulsed reactors fall within some middle ground between chromatographic and swing reactors. They rely on carrier gas for reactant–product separation, but some downstream separation will be normally required. Periodically pulsed reactors also appear to be attractive devices for screening potential applications of PSRs.

Perhaps the most important incentive for exploring pressure swing reactors is that they can operate at lower temperatures for equilibrium-limited endothermic reactions than those usually employed. Dehydrogenation is a good example of such a reaction. Lower temperatures could reduce the importance of secondary reactions and would certainly lower capital and/or operating costs. On the other hand, for equilibrium-limited exothermic reactions with large activation energies, a pressure swing reactor might increase reaction yield without requiring a reduction in temperature.

9.2 Concepts and types

There are two design alternatives for pressure swing systems. A single catalyst–adsorbent bed can be employed and feed, delivery or exhaust flow can be cycled through several steps.

Table 9.1 Literature on pressure or temperature swing separating reactors

Reference	Type of study	Reaction considered	Remarks
Pressure swing reactors			
Vaporciyan and Kadlec (1987)	Calculational	General fast reversible (reaction equilibrium)	Reaction may improve separation, separating factor reversal
Vaporciyan and Kadlec (1989)	Calculational and experimental	CO oxidation, general multiple reactions	No increased conversion, selectivity improvement, separating factor reversal
Alpay et al. (1993)	Calculational	Dehydrogenation of methyl cyclohexane	Studied computation methods
Alpay et al. (1994)	Calculational	Reversible reactions and methyl cyclohexane dehydrogenation	Higher conversion except when total moles increase by reaction
Chatsiriwech et al. (1994)	Calculational	General reversible	Yield improvement, higher conversion
Kirkby and Morgan (1994)	Calculational	Irreversible isomerization	Separating factor reversal
Lu and Rodrigues (1994)	Calculational	General reversible reaction	Improved conversion over a PFR
Carvill et al. (1996)	Experimental	Reverse water gas shift	Increased conversion and high purity CO
Yongsunthon and Alpay (1998a,b)	Calculational	Endothermic reaction (dehydrogenation of methyl cyclohexane)	Demonstrated that heat effects influence performance
Cheng et al. (1998)	Calculational	General reversible dissociation reaction	Optimization of the control variables
Kodde et al. (2000)	Calculational	Sequential reaction with a common reactant	Enhancement of selectivity depends strongly on adsorption
Sheikh et al. (2001)	Calculational and experimental	Dehydrogenation of 1-butene	Demonstrated improved product purity but no increase in conversion
Gomes and Yee (2002)	Calculational and experimental	Propene disproportionation	Demonstrated improved product purity, conversion, recovery, yield and productivity
van Noorden et al. (2003a)	Calculational	General reversible dissociation reaction	Cycle optimization by a numerical gradient method
Al-Juhani and Loughlin (2003)	Calculational	Isomerization of *n*-C_5	Investigated separate catalyst and adsorbent beds
Ding and Alpay (2000a)	Calculational and experimental	Hydrogen by steam reforming of methane + water gas shift	Model validation and investigation of the effect of operating conditions
Nonisothermal pressure swing reactors			
Xiu et al. (2002a)	Calculational	As above	Intraparticle diffusion and reaction adequately represented by an effectiveness factor

(Continued)

Table 9.1 Literature on pressure or temperature swing separating reactors *(Continued)*

Reference	Type of study	Reaction considered	Remarks
Xiu et al. (2002b)	Calculational	As above	Use of a four-step process with a reactive purge
Xiu et al. (2003a,b)	Calculational	As above	Use of a four-step process with a reactive purge in a temperature zoned reactor
Temperature swing reactors			
Han and Harrison (1994)	Experimental	Water gas shift	Almost quantitative shift to H_2 and greater than 99% removal of carbon oxides
Han and Harrison (1997)	Experimental	Water gas shift	Dolomite CO_2 acceptor can be used over many cycles, but with declining activity
Yongsunthon and Alpay (1998b)	Calculational	General reversible reaction	Demonstrated improved conversion with lower energy demand
Balasubramanian et al. (1999)	Experimental	Steam reforming of methane	Obtained >95% H_2 in single bed process
Yongsunthon and Alpay (1999)	Calculational	Dehydrogenation of methyl cyclohexane	Demonstrated that higher purity can be obtained by staging and use of combined serial and parallel flow
Yongsunthon and Alpay (2000)	Calculational	Dehydrogenation of methyl cyclohexane	See above. Further improvement can be obtained by reversing flow direction in regeneration
Lopez Ortiz and Harrison (2001)	Experimental	Steam reforming of methane	Obtained >95% H_2 in single bed process with only a moderate decline in dolomite capacity
Elsner et al. (2002a,b)	Calculational and experimental	Claus reaction and HCN synthesis	Sulfur discharge restrictions can be met in a 2-step process. Application to HCN synthesis ineffective
Menge et al. (2003)	Experimental	Claus reaction and HCN synthesis	Side reaction and coadsorption problems
Elsner et al. (2003)	Experimental	Claus reaction and HCN synthesis	Conditions for successful application of temperature swing system
Harrison and Peng (2003)	Experimental	Steam reforming of methane	Obtained >95% H_2 and less than 30 ppm CO with a dolomite CO_2 acceptor
Periodically pulsed reactors			
Brun-Tsekhovoi et al. (1986)	Experimental	Steam reforming of methane	Increased conversion through CO_2 capture and high purity H_2
Goto et al. (1993)	Experimental	Cyclohexane dehydrogenation	Used Ca/Ni alloy as H_2 adsorbent
Goto et al. (1995)	Experimental	Dehydrogenative aromatization of *n*-hexane	Mg-based metal alloys used as H_2 adsorbents

(Continued)

Table 9.1 Literature on pressure or temperature swing separating reactors *(Continued)*

Reference	Type of study	Reaction considered	Remarks
Goto et al. (1996)	Experimental	Dehydrogenative aromatization of *n*-hexane	Used Ti and Ta metals as H_2 adsorbent
Sheikh et al. (1998)	Calculational	General reversible reaction	Circulation of adsorbent through a CSTR
Mensah et al. (1998a)	Calculational and experimental	Liquid phase esterification	Data collection for reactor model
Mensah et al. (1998b)	Calculational and experimental	Liquid phase esterification	Data collection for reactor model
Štěpánek et al. (1999)	Calculational	Production of nicotinamide	Optimal design method for a monolith reactor-separator
Mensah and Carta (1999)	Calculational and experimental	Liquid phase esterification	Model validation and study of parameter effects
Aida et al. (2002)	Calculational and experimental	NO reduction by CO	Investigation of the effect of adsorbent on NO conversion

A continuous moving-bed system can also be used with two vessels: one for reaction and adsorptive separation operated at high pressure and a second for adsorbent regeneration at lower pressure. A more practical alternative for continuous operation would be a bed in the shape of a rotating wheel so that the catalyst and adsorbent sequentially pass the feed, exhaust and delivery connections. In principle, the rotating bed is identical to the discontinuously operated fixed bed, that is, the set of equation describing the two systems are the same. In this review, we discuss just the fixed bed alternative and restrict our considerations to just a single bed. Only this case has been considered in the literature.

The pressure swing reactor (PSR) was developed from pressure swing adsorption (PSA) by simply adding catalyst to the adsorbent. Consequently, PSR designs follow those used for adsorption systems. Figure 9.1 illustrates two PSA designs that have been applied to reaction systems. In (a), the reactor is completely filled with catalyst and adsorbent, while in (b) a void volume follows the bed. This is often a portion of the reactor at the discharge end empty of catalyst and adsorbent, but it could be an attached empty vessel. This variation, pioneered by Kadlec and coworkers (Turnock and Kadlec, 1971; Kowler and Kadlec, 1972), finds use in rapid PSA cycles where the void volume provides purge and some backfill. It is referred to as a reflux volume. Shallow beds and cycle times well below 1 min are used. We will refer to the reflux volume variation with a short cycle time as a rapid-cycle pressure swing adsorber or reactor. With deeper beds built without a reflux volume, cycle times of several minutes can be used. Purge must then be provided from an external source.

In recent years, further variations have been introduced: the reactor bed can be divided into zones and maintained at different temperatures by heating or cooling the wall for that zone. Also, stream withdrawal or purge can be carried out between zones.

A pressure swing adsorber can be operated in several different ways. Figure 9.2 shows the basic pressure swing cycles employed: (a) two steps, (b) three steps with a delivery step, (c) four steps with added purge or (d) with added backfill and (e) six steps with two purges and

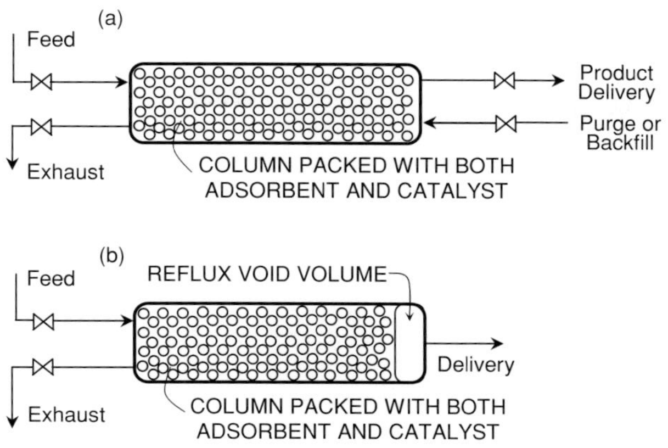

Figure 9.1 Connections and bed arrangements for pressure swing adsorbers and reactors: (a) standard design, (b) with reflux volume. (Adapted from Vaporciyan and Kadlec (1987) with permission. © 1987 AIChE.)

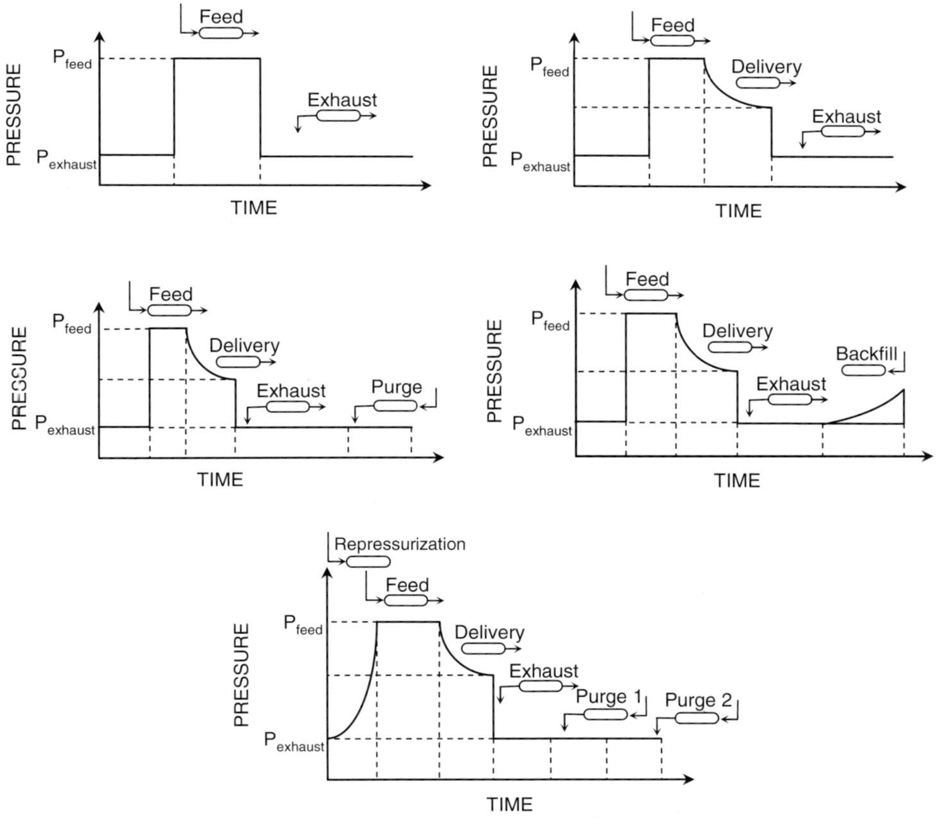

Figure 9.2 Basic operational cycles for pressure swing adsorbers and reactors. (Adapted from Vaporciyan and Kadlec (1987) with permission. © 1987 AIChE.)

repressurization. All of these cycles have been applied to PSRs. For industrial operations, at least two PSA units are used in parallel to generate quasi-continuous flow of the feed, product (discharge) and exhaust streams. There are many variations of the basic cycles shown in Figure 9.2. The number of PSA units used and their arrangement in a network has also many variations. As of 1994, there were about 2000 patents in the PSA literature on cycles, networks and operating conditions.

The pressure swing vessel in Figure 9.1 has one inlet for the feed. This inlet is held at the highest operating pressure in the system. Normally the adsorbent bed is vertically oriented with an inlet at the bottom of the bed. The product or discharge stream is removed through an outlet usually at the top of the bed, but in any case located opposite to the inlet. This stream is withdrawn at an intermediate pressure. The second outlet providing the exhaust stream is located at the feed end of the bed and is held at the lowest pressure used in the cycle.

The basic PSA or PSR cycle consists of a pressurization or feed step (see Fig. 9.2b) when feed is introduced to the adsorbent bed, followed by a discharge step, sometimes called a delivery, in which the pressure is reduced to an intermediate value. During these two parts of a cycle, the discharge port is open. The third and final step is exhaust or depressurization, when the feed side outlet is opened and the discharge port is closed. Pressure in the outlet is maintained constant at the lowest pressure in the cycle and bed pressure falls to this level by the end of the step. Step duration varies with particle size of adsorbent, bed length and adsorption properties of the mixture to be separated. Usually the discharge step is the longest. The basic cycle is implemented when a reflux volume is used. This volume provides a back-flow purge during the exhaust step. If the exhaust port is closed during the step, backfill can also take place. Reaction products can be collected at either or both of the discharge and exhaust ports.

A purge cycle adds a fourth step following the depressurization. As can be seen in Figure 9.2c, the discharge port is opened and product is allowed to flow back into the bed. The product flow in Figure 9.1 then moves from right to left providing a flush of the catalyst and adsorbent. The exhaust port remains open. Purge cycles improve separation. They are used in rapid pressure swing systems, but at the cost of either a smaller throughput or a longer cycle period if the same feed amount is maintained.

Figure 9.2d illustrates a backfill cycle. In this four-step operation, the exhaust port is closed after the depressurization and the product stream from the discharge port partially repressurizes the adsorbent bed. The cycle is employed to improve separation. With both four-step cycles, it is assumed that product is discharged into a well-mixed vessel. This vessel provides the purge or backfill. The stream thus has constant composition and is usually assumed to be at constant pressure.

A variant of the backfill cycle employs product or a nonadsorbing feed component to repressurize the packed bed. In this case, shown in Figure 9.2e, the feed in step 1 forces the repressurizing gas out of the reactor so that there is no change in pressure during that step.

More detailed descriptions of pressure swing adsorption, parametric pumping as well as performance data on various adsorptive separations are to be found in books by Ruthven (1984), Wankat (1986), Yang (1987), and Suzuki (1990). Rapid cycles seem to be most suited to reaction applications. These generally employ short, low aspect ratio beds of fine particles and cycle periods of the order of 1–2 s. Performance of these rapid-cycling systems for adsorptive separation is discussed by Jones and Keller (1981), Pritchard and Simpson (1986) and Hart and Thomas (1991).

9.3 General models for pressure swing reactors

Tables 3.2 and 3.3 summarize the equations used to model PSRs. They assume that a single solid acts both as catalyst and adsorbent. A one-dimensional, pseudo-homogeneous model (Table 3.2) is widely used because it has been demonstrated in the packed-bed reactor literature that such a model describes reactor behavior adequately at the limits of adiabatic or isothermal operation. Often partial pressures are used in material balances in place of concentration because adsorption equilibria are stated in terms of that variable. Gas properties are assumed to be independent of composition and temperature, and mixtures behave ideally. Catalyst and adsorbent are taken to be well mixed and uniformly distributed in the reactor. All inlet/outlet effects are neglected. In Tables 3.2 and 3.3, a phenomenological model has replaced the momentum balance. This is the Ergun equation in the table, but in some studies Darcy's law is applied. Equation 3.4 reduces to Darcy's law if J_k is set to zero. If adsorption equilibrium is assumed, only one mass balance per species is necessary. However, a solid phase balance for each species must be added if adsorption or mass transfer rates are considered.

Equilibrium adsorption can be modeled by the Langmuir isotherm, if heats of reaction and adsorption are modest so that the system can be treated as isothermal. Equation 9.1 is the form for multicomponent adsorption on a single type of adsorption site.

$$q_i^* = \frac{q_{s_i}\alpha_i P_i}{1 + \sum_{j=1}^{N_c} \alpha_j P_j} \tag{9.1}$$

where q_i^* is the equilibrium amount of component i adsorbed and $(q_s)_i$ is maximum amount adsorbed or the density of adsorption sites. Many of the simulations considered in this review assume a linear isotherm such as given by Equation 9.2.

$$q_i^* = q_{s_i}\alpha_i P_i = m_i P_i \tag{9.2}$$

Theoretical studies generally assume just a single reaction occurs, but any rate expression for that reaction is possible. Then,

$$r_i = \nu_i \left(\frac{r_k}{\nu_k}\right) = \nu_i r \tag{9.3}$$

where r_k is the general form of the rate expression written for the kth reactant.

$$r_k = k_k \cdot \mathrm{f}(y_i, y_j) - k_{-k} \cdot \mathrm{g}(y_i, y_j) \tag{9.4}$$

The functions $\mathrm{f}(y_i, y_j)$ and $\mathrm{g}(y_i, y_j)$ represent arbitrary functions of the component mole fraction and can include additional temperature-dependent constants. If both the forward and reverse reactions are second order overall, Equation 9.4 becomes $r_k = k_k \cdot \mathrm{f}(y_i y_j) - k_{-k} \cdot \mathrm{g}(y_i y_j)$. The rate constants, k_k and k_{-k} are temperature function through the Arrhenius expression. Constants embodied in the rate expression, $f(y_i, y_j)$ and $g(y_i, y_j)$ are usually adsorption equilibrium constants. Their temperature dependence is given by the van't Hoff expression. If adsorption rates must be considered,

$$\frac{\partial q_i}{\partial t} = k_{\text{ads}_i}\left(q_i^* - q_i\right) \tag{9.5}$$

Mass transfer, if controlling, is also model by a linear driving force (LDF) model (Equation 9.5), however, the driving force is then expressed as concentration $(C_i - C_i^*)$ and the mass transfer coefficient, $(k_{\text{m}})_i$, replaces the adsorption rate constant, $(k_{\text{ads}})_i$.

In models used by Vaporciyan and Kadlec (1987, 1989), Darcy's law and a linear adsorption isotherm are assumed. Velocity in Equations 3.1–3.3 in Table 3.2 is replaced by pressure gradients and the isotherm is inserted for the dq/dt term. More recent modeling studies use the Ergun equation, explicitly model velocity variation and allow for nonlinear adsorption equilibrium. These variations will be introduced later in this chapter.

Tables 3.2 and 3.3 do not give a model for the reflux volume in Figure 9.1b. The appropriate model is for an empty, well-mixed vessel with a variable inlet flow set by the pressure gradient at the end of the catalyst–adsorbent bed and an outlet flow dependent on the product discharge policy used. For the simulations reviewed, this flow was assumed constant.

Most of the simulations discussed in the literature use dimensionless forms of the model. The choice of dimensionless variables and thus the dimensionless groups that arise are not the same as those given earlier in Table 3.4. A typical dimensionless model (Vaporciyan and Kadlec, 1989) for a pressure swing reactor with a reflux volume is given in Table 9.2. This model is written in terms of partial pressure. In Equation 9.6 isothermal, plug flow operation, a linear adsorption isotherm, and Darcy law are assumed. The isotherm is substituted into the component mass balance to generate the term γ_i. Reaction kinetics are given by a power law model with integer exponents and each component i is involved in just a single reaction. The dependent variable for this isothermal model is a partial pressure rather than a concentration.

Table 9.3 defines the dimensionless variables, parameters and groups appearing above. Time has been normalized by cycle duration and axial position by bed length. Durations of the parts of a cycle are defined by f as cycle fractions. Definition of the Damköhler number depends on the kinetic model. In Table 9.3, a power law model has been assumed. There are separate definitions for forward and reverse reactions. If multiple reactions were considered, each reaction step would have a different definition of this number. We have designated one of the important dimensionless groups, N_{Ru}, to honor Professor Douglas Ruthven who has

Table 9.2 Dimensionless model for a PSR with a reflux volume

Component mass balance for catalyst–adsorbent bed	$\frac{\partial}{\partial\lambda}\left[Z_i\frac{\partial Z}{\partial\lambda}\right] = N_{\text{Da}_f}\prod_{i=1}^{N} Z_i^{\beta_i} - N_{\text{Da}_r}\prod_{i=1}^{N} Z_i^{\alpha_i} + \left(\frac{\mu L^2}{\tau P_f K}\right)\gamma_i\frac{\partial Z_i}{\partial\theta}$ for $i = 1, 2, \ldots, N$	(9.6)
Component mass balance for the reflux volume	$\frac{\partial Z_{R_i}}{\partial\theta} = -N_{\text{Ru}}\, f_R Z_i \left.\frac{\partial Z}{\partial\lambda}\right\vert_{\lambda=1} - N_{\text{Ka}}$ for $i = 1, 2, \ldots, N$	(9.7)
Total mass balance	$Z = \sum_{i=1}^{N} Z_i$	(9.8)

Table 9.3 Definition of the dimensionless variables, parameters and groups in Equations 9.6–9.8

Partial pressure (concentration)	$Z_i = \dfrac{P_i}{P_f}$
Axial position	$\lambda = \dfrac{z}{L}$
Time	$\theta = \dfrac{t}{\tau}$
Partial cycle duration	
Feed	$f_f = \dfrac{t_f}{\tau}$
Discharge	$f_d = \dfrac{t_d}{\tau}$
Exhaust	$f_e = \dfrac{t_e}{\tau}$
Damköhler number for the forward reaction of order β	$N_{\mathrm{Da}_f} = k_j P_f^{\beta-1} \rho_c \mathrm{RT}\tau$
Damköhler number for the reverse reaction of order α	$N_{\mathrm{Da}_r} = \dfrac{k_j}{K_{\mathrm{eq}_j}} P_f^{\alpha-1} \rho_c \mathrm{RT}\tau$
Pressure ratio	$\dfrac{P_e}{P_f}$
Kadlec number	$N_{\mathrm{Ka}} = \dfrac{n_d \mathrm{RT}\tau}{P_f V_R}$
Ruthven number	$N_{\mathrm{Ru}} = \dfrac{\tau P_f k}{\mu L^2}$
Volume ratio	$\dfrac{1}{f_R} = \dfrac{V_R}{A_x L}$
Capacity	$\gamma_i = \varepsilon + \rho_A \mathrm{RT}\alpha_i$
Conversion	$\dfrac{n_{i_0} - n_{i_{\mathrm{exit}}}}{n_{i_0}}$
Separation factor (i/j)	$\dfrac{(y_i/y_j)_{\mathrm{discharge}}}{(y_i/y_j)_{\mathrm{exhaust}}}$

been a pioneer in the study and modeling of pressure swing adsorbers. Another group, N_{Ka}, arises when a reflux volume is incorporated into a PSR or PSA. This number honors Professor Robert Kadlec who considered the use of a reflux volume in a reactor–separator. Conversion and yield measure the performance of PSRs. Separation achieved in either a PSA or PSR is given by a separating factor. This dimensionless group is defined in the table. Separation can also be described by the yield of a product removed in either the discharge or exhaust port.

Boundary and initial conditions complete the models. The former are required for the feed port and for the discharge and exhaust ports. When an inlet or outlet from the bed is closed, boundary conditions are not required. However, if there is no open inlet or outlet at an end of the reactor bed, an end of bed boundary condition must be specified. These are the standard zero flux conditions, namely,

$$\frac{dT}{dz} = \frac{dP}{dz} = \frac{dy_i}{dz} = 0 \tag{9.9}$$

Table 9.4 summarizes these conditions. Note that these sets are independent of the models employed for adsorption and reaction, and also independent of reaction stoichiometry.

Phenomenological models for reaction and adsorption vary between investigators so these will be introduced as necessary when we discuss specific contributions.

9.4 Computational considerations

Periodic operation of reactors results in moving reaction and temperature fronts. These fronts manifest abrupt changes in concentration and temperature that lead to unstable behavior in many common integration routines. Such instability can be managed only by employing inordinately small space and time steps. In the presence of both adsorption and reaction, the instability problem appears to be exacerbated. Vaporciyan and Kadlec (1987) found that integration routines that functioned satisfactorily with PSA systems became unstable when an equilibrium reaction was introduced. Consequently, several investigators have devoted attention to identifying efficient computation methods (Alpay et al., 1993; Chatsiriwech et al., 1992, 1994; Cheng et al., 1998; Van Noorden et al., 2002).

In the earliest modeling study of a PSR, Vaporciyan and Kadlec (1987) encountered instabilities using finite differences to convert their PDE model to a set of ODEs that they then tried to integrate with a Newton–Raphson method. These authors resorted to cell discretization and then solved the resulting system of nonlinear ODEs by a Runge–Kutta–Fehlberg procedure.

A systematic study of computation procedures was undertaken by Alpay et al. (1993). Spatial discretization to convert the reactor model into a set of ordinary differential equations (ODEs) was undertaken using orthogonal collocation (OC), orthogonal collocation on finite elements (OCFE), and by assuming the reactor to consist of well mixed cells-in-series (CIS). Double collocation on finite elements (DOCFE) to convert the PDEs to a set of algebraic equations was also tested. For the collocations, Lagrangian interpolation polynomials were employed. Roots of the Legendre polynomials were used to choose the collocation points. In the finite element application of OC, the reactor was divided into elements and OC applied to each element. This permitted lower order interpolation polynomials to be used. However, additional equations must be added to the set to insure continuity of the solutions and the first derivatives of the polynomials at the element boundaries. With double collocation, time as well as space is discretized. For spatial discretization, the reactor was divided into elements and the OCFE procedure followed. Just orthogonal collocation was used for time. The large system of ODEs containing some nonlinear terms was stiff so integration employed a variable order, variable step size Gear's method, while for the large system of algebraic equations produced by the DOCFE procedure, Powell's hybrid method of solution was implemented.

The four solution methods were tested for the computer CPU time needed to achieve convergence to two decimal places at a cyclic stationary state using a simplified model of a two-step cycle for the separation of air in a short-cycle pressure swing adsorber. For this rather simple system, the OC method required between 10 and 15 collocation points to bring convergence error to about 1%, while three elements with just 3 collocation points were adequate for the OCFE procedure. A similar result was found for DOCFE. At least

Table 9.4 Boundary and initial conditions for pressure swing reactors

Bed initial condition for all part cycles	$\zeta(z, \tau_p)^- = \zeta(z, \tau_p)^+$	(9.10)
Also once a cyclic stationary state is achieved	$\zeta(z, t) = \zeta(z, t + \tau)$ where ζ is any dependent variable (y_i, P, T, for example)	(9.11)
Inlet port		
Pressurization	$P = P_f$	(9.12)
	$D_z \dfrac{d}{dz}(Cy_i) = \dfrac{n}{A_x}(y_i - y_{i_f})$	(9.13)
	$\lambda_z \dfrac{dT}{dz} = \dfrac{nc_{p_g}}{A_x}(T - T_f)$	(9.14)
Exhaust/purge/backfill (usually closed)	$dP/dz = dy_i/dz = dT/dz = 0$	(9.15)
Discharge port/purge or backfill inlet port		
Pressurization/delay/discharge	$dP/dz = dy_i/dz = dT/dz = 0$	(9.16)
	And $n = C_v\sqrt{C(P_{z=L} - P_d)}$	(9.17)
	Or $n = n_d$ (constant delivery)	(9.18)
	$\dfrac{d}{dz}(Cy_i) = \dfrac{dT}{dz} = 0$	(9.19)
Exhaust (two operations are possible)	(1) The discharge port is open and conditions given by either Equations 9.16–9.19 apply (2) the discharge port is closed and the condition given by Equation 9.15 applies	
Purge	$P = P_p$	(9.20)
	And $n = C_v\sqrt{C(P_p - P_{z=L})}$	(9.21)
	Or $n = n_p$ (constant delivery)	(9.22)
	Also $D_z \dfrac{d}{dz}(Cy_i) = \dfrac{n}{A_x}(y_i - y_{i_p})$ where $y_{i_p} = y_{i_{\text{mix}}}$	(9.23)
	$\lambda_z \dfrac{dT}{dz} = \dfrac{nc_{p_g}}{A_x}(T - T_p)$ where $T_p = (T)_{\text{mix}}$	(9.24)
Backfill	These are the same as for the purge step, except P_b replaces P_p and n_b replaces n_p	
Exhaust (depressurization) port		
Feed/delay/discharge	Discharge port is shut and the end of bed boundary conditions must be used. (Equation 9.15)	
Exhaust	$P = P_v$	(9.25)
	$n = C_v\sqrt{C(P_{z=0} - P_v)}$	(9.26)
	And $\dfrac{d}{dz}(Cy_i) = \dfrac{dT}{dz} = 0$	(9.27)
Backfill	Discharge port is closed and end of bed conditions apply. (Equation 9.15)	
End of packed bed (with reflux void volume)		
Pressurization/discharge	$\dfrac{dP}{dz} = \dfrac{d}{dz}(Cy_i) = \dfrac{dT}{dz} = 0$	(9.28)
Exhaust	$D_z \dfrac{d}{dz}(Cy_i) = \dfrac{n}{A_x}(y_i - y_{i_R})$ where $y_{i_R} = y_{i_{\text{mix}}}$	(9.29)
	$\lambda_z \dfrac{dT}{dz} = \dfrac{nc_{p_g}}{A_x}(T - T_R)$ where $T_R = (T)_{\text{mix}}$	(9.30)

50 cells were needed using CIS. In terms of CPU usage, all the collocation methods were better than CIS. Alpay et al. concluded that the OCFE method is significantly better than the other two collocation methods with respect to CPU usage.

The accuracy of the OCFE procedure was ascertained by comparing the oxygen concentration with an analytical solution for a drastically simplified model of a two-step PSA cycle separating air. Excellent agreement with the analytical solution was obtained. A further test of the OCFE procedure was carried out for an equilibrium reaction, $A \Leftrightarrow B + 3C$, where just species B adsorbs. In the absence of axial dispersion and pressure gradients, with first order kinetics and a linear adsorption isotherm, the model for an isothermal, two-step cycle can be solved by the method of characteristics (Chatsiriwech et al., 1992, 1994). With modifications to the OCFE procedure to adopt it to the absence of dispersion and pressure gradients, good agreement between numerical and analytical concentration profiles in the catalyst–adsorbent bed was found, except where the analytical solution predicted discontinuities.

Alpay et al. (1993) tested the adequacy of their model rather than the computational method by comparing experimental measurements to the predicted delivered purity of the oxygen in air separation via a two-step PSA system. The prediction was satisfactory in spite of scatter in the experimental points. An interesting result of this test was that for adsorbent particle diameters below 0.4 mm, the assumption of instantaneous local equilibrium gave results similar to those predicted by diffusion control in the presence of equilibrium employing a lumped parameter diffusion model. Both assumptions gave good agreement with experimental measurements. Apparently, diffusion can be neglected for fine particles. For particles greater than 0.4 mm, diffusion is important and must be considered by any PSR model.

Cheng et al. (1998) were concerned with optimization of the pressure swing variables such as feed pressure, feed flow rate and step durations as well as estimating the cyclic steady state for the operation. They considered a two-step cycle and a reversible dissociation reaction. For estimating cyclic steady states, a comparison was made between discretizing the spatial variables only and solving a system of nonlinear dynamic equations with discretization of both temporal and spatial variables and solving a large system of nonlinear algebraic equations. For the first case, discretization employed the OCFE method, while for the second case DOCFE was used. With the parameters used, Cheng et al. observed that almost 5000 cycles were needed when starting from feed conditions throughout a reactor and assuming initially zero rate of adsorption. Fourth order orthogonal collocation on eight elements provided satisfactory accuracy. Using DOCFE, a hundred iterations were needed and satisfactory results were found for third order collocation using 16 temporal elements. Optimization of the two-step cycle was able to increase the yield by about 30% over an equilibrium fixed bed process using the same system parameters. Feed flow rate and step durations exhibited optimum values, while yield increased monotonically with feed pressure.

Engineering of pressure swing reactors requires, in the first instance, knowledge of the cyclic stationary state, that is the behavior with time within a cycle after cycle to cycle variations at a point in space have disappeared. Optimization of the manipulated or control variable, such as duration of the feed or depressurization steps, or the exhaust pressure, will almost always be undertaken with reference to this state. Transient behavior on startup or on a change in feed rate or composition is also important, but this behavior will be explored usually only after a design has been chosen, that is, after vessels have been sized and the control variables selected, at least tentatively. The simulation techniques discussed above often require a large number of time steps until the cyclic stationary state is reached and therefore require a large computational investment. Of course, they do have the advantage in indicating the time it takes to reach the stationary state and what excursions of the dependent variables may occur during the transition to that state. Nevertheless, the problem of rapidly estimating cyclic stationary states has attracted much attention. The problem arises not only for pressure swing reactors but also for chromatographic systems, particularly for SCMCRs. It has been of concern for all types of periodic processes so it is not surprising that a large research literature has arisen.

Techniques for determining the cyclic stationary states of all types of periodic processes employ "shooting" methods to produce Poincaré maps. The condition of a cyclic stationary state is $y(z, t) = y(z, t + \tau)$ where y is a dependent variable at a spatial point z. A "shooting" method begins at an initial condition, a fixed point, and adjusts y incrementally until the condition is fulfilled. When the increment approaches zero to a preset amount then the fixed point has been determined. This can be done from the eigenvalues of the Jacobean of the state, e.g. model, equations at the fixed point. Once a starting point or guessed value of the fixed point is identified, the increments Δy in each iteration can be generated from the Jacobean on a period 1 map of the state equations in the Picard method. Thus, in vector notation

$$\Delta y_i \approx -(\mathbf{J}(y_i) - \mathbf{I})^{-1}(\mathrm{f}(y_i) - y_i) \tag{9.31}$$

where $\mathbf{J}$ is the Jacobean of the period 1 map of the state equation at y_i. The method is discussed by Croft and Levan (1994), for example. A Newton method can also be used and is a more popular approach. This uses the state equations directly rather than their map, but also requires the evaluation of Jacobeans (Croft and Levan, 1994). An interesting recent development uses a hybrid of these methods (van Noorden et al., 2002, 2003a,b). The method uses the observation that in many periodic operations most of the eigenvalues of the Jacobean of the map of the state equations are small and can be neglected in determining the increment Δy. Consequently, a greatly reduced set of eigenvalues needs to be evaluated. This much smaller number can be used to construct a subspace of the discretized state space with matrices representing the orthogonal basis of the subspace and its orthogonal complement. The increment can be decomposed into components in the subspace and its complement. The van Noorden et al. papers formulate the relationships and offer algorithms for implementation of their hybrid "shooting" method.

In an extension of their computational study, van Noorden et al. (2003a) showed that their hybrid "shooting" method could be applied advantageously to optimization of time-dependent variables of periodic processes employing a first order gradient approach (Bryson and Ho, 1975). Time dependent variables are those involved in the operating cycle, such as purge duration or feed rate. Optimization involves a function that defines a performance level or index that is to be maximized, such as yield or product purity, or to be minimized, such as energy consumption per unit of production. Often constraints exist that can be formulated as a mathematical relation. Optimization of the time-dependent variables, usually referred to as optimal control, involves solution of the adjoint equations that embed the performance index and the constraints as well as solution of the state equations. In their paper, van

Noorden et al. (2003a) show how their "shooting" method can be applied to optimization through the adjoint equations. Many of the matrices generated for estimation of the fixed points can be applied directly to increments for the control variables. The authors develop relationships and provide algorithms.

The first order gradient Newton–Picard optimization method was demonstrated by van Noorden et al. using the two-step, isothermal PSR system dealt with by Cheng et al. (1998) and mentioned briefly several paragraphs earlier. The performance index to be maximized was the yield of the weakly adsorbed product, while the purity of the product was constrained as a minimum value. Control variables were feed pressure, duration of the feed step and the cycle period. Sequential quadratic programming for the two-step cycle resulted in a set of control variables that led to a 0.12% yield. Use of a first order gradient with the van Noorden Newton–Picard "shooting" method resulted in a variable pressure during the feed step and caused the yield to increase threefold to 0.34%. The shape of the feed pressure profile suggested that a four-step cycle would be better than the cycle selected by Cheng et al. (1998). That cycle consisted of a short pressurization through the discharge, feed for a short duration, followed by no feed or pressurization for a much longer duration. The cycle terminated with a relatively long depressurization step.

9.5 Isothermal modeling studies

In the first theoretical study of a pressure swing reactor, Vaporciyan and Kadlec (1987) examined fast reactions under conditions of adsorption equilibria. Because reaction equilibrium is attained as the feed enters the reactor and is maintained at all points inside this vessel, pressure swing operation cannot force the reaction to exceed its equilibrium limit. It seems then that the authors were interested in how the presence of reaction affected the separation performance of a pressure swing adsorber (PSA).

For their study, Vaporciyan and Kadlec chose a rapid cycling pressure swing reactor (RPSR) with a reflux volume as illustrated in Figure 9.1b. The system operated with the discharge port continuously open and assumed withdrawal at a constant rate. Turnock and Kadlec (1971) showed that for cycle periods of the order of seconds a nearly constant discharge flow is possible. The cycle examined is illustrated in Figure 9.2. Four reaction stoichiometries were considered: (1) A $\leftrightarrow$ C, (2) A $\leftrightarrow$ 2C, (3) 2A $\leftrightarrow$ C, (4) A $\leftrightarrow$ (1/2)B + C. In the latter three cases, changes in the number of moles on reaction means that the equilibrium partial pressures are affected by reactor pressure.

Several simplifications were made in the model shown in Table 9.2 because of the assumptions introduced. The rate term in Equation 9.1 was deleted and because isothermal operation was assumed the heat balance, Equation 9.3, was not needed. Deleting the rate term requires the addition of an equilibrium constraint:

$$\prod_{i=1}^{N} y_i = K_{eq} P^{-\Delta\nu} \tag{9.32}$$

where $\Delta\nu$ is the sum of the stoichiometric coefficients for the reaction. An additional group of material balances must be added for the reflux volume. The dimensionless form for the reflux volume balance appears as Equation 9.7 in Table 9.2. Boundary conditions for

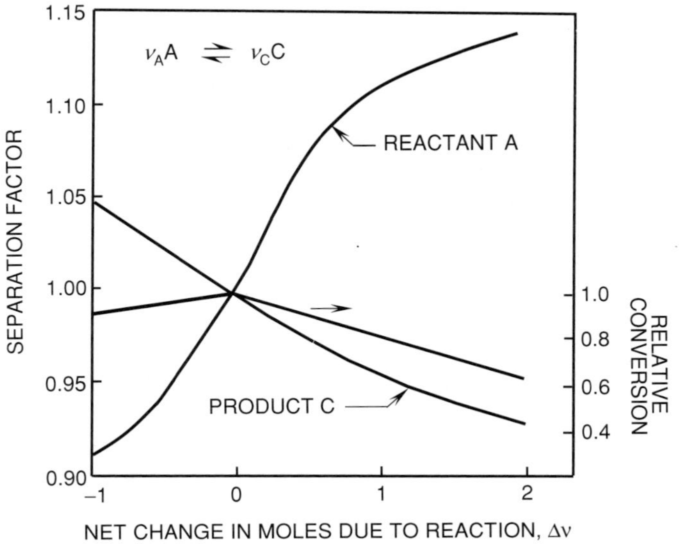

Figure 9.3 Effect of reaction stoichiometry on separations achieved in an RPSR when reaction equilibrium is attained. (Figure adapted from Vaporciyan and Kadlec (1987) with permission. © 1987 AIChE.)

a PSR with a reflux volume now apply. These are given in Table 9.4. The dimensionless form of the model, Equation 9.6, also changes as the sink terms containing the Damköhler numbers vanish. The equilibrium constraint, however, generates a new dimensionless group, $K_{eq} P^{-\Delta\nu}$. Problems integrating the model due to the equilibrium constraint have been discussed in the previous section. Vaporciyan and Kadlec used a cells-in-series method and observed that six cells were adequate. The approach to a cyclic stationary state was slow from an empty reactor initial condition. About 80–100 cycles were required.

Simulations were undertaken for a feed consisting of A without diluent. A small discharge rate, 0.00163 mol/s, was used and assumed to be constant throughout the cycle. For most cases, the exhaust pressure was 1/3 of the feed pressure. Duration of the feed and exhaust portions of the cycle were the same, while the duration of the delivery portion of the cycle was set to zero. The dimensionless equilibrium constant changed with the reaction, but was kept between 1 and 10. For the single product reaction, adsorption of the reactant was tenfold that of the product, whereas for the fourth reaction, A ↔ (1/2)B + C, both reactant and product C were strongly adsorbed relative to product B.

Figure 9.3 shows, as expected, that the equilibrium conversion of A is not exceeded in a RPSR when reaction rates are so fast that equilibrium is achieved at the entrance to the reactor. Furthermore, without a change in the total number of moles during reaction, no separation is achieved. Discharge and exhaust compositions in the RPSR become different when the total number of moles in the reaction changes. This is a consequence of the pressure dependence of the equilibrium composition. The exhaust is always at a lower pressure than the discharge. Thus, when there is an increase in total moles, equilibrium conversion of A is less and the reactant becomes concentrated in the discharge stream as Figure 9.3 illustrates. This behavior differs from that of a RPSA in which the strongly adsorbed reactant would appear in the exhaust stream. Of course, the separation factor, defined for a PSR as the ratio of a component's mole fraction in the discharge and the exhaust, depends on both adsorptive capacity and the equilibrium constant for any reaction stoichiometry. Increasing the equilibrium constant, all other parameters held constant, augments conversion but decreases the throughput for a constant discharge rate.

Separation factors for the reactants increase if there is an increase in total moles, but decreases if there is a decrease. This means that the reactant is concentrated in the exhaust when there is a decrease in total moles. On the other hand, as the equilibrium constant increases, the separation factor for the product C tends to unity.

For the reaction with two products with one product and the reactant relatively strongly adsorbed, B is concentrated in the discharge and C in the exhaust as would be expected for a RPSA, but the separation factors are independent of the equilibrium constant.

Vaporciyan and Kadlec also examined the influence of design parameters, such as cycle period and the feed fraction of a cycle, on separation. Some of their results are shown in Figure 9.4. According to these investigators there is a restricted range for the cycle period. If the period is too short, pressure attenuation in a packed bed of finely sized catalyst and adsorbent is so great that the specified discharge flow rate cannot be attained. This problem also arises with long cycle periods because then the constant discharge flow rate cannot be maintained because of dropping pressure in the bed. The feasible range depends on design variables such as permeability of the packing and the fraction of the bed taken up by reflux volume.

Within the feasible range, the authors observe that the affect of cycle period on reactant conversion is small regardless of reaction stoichiometry. However, a reaction, in which the total number of moles decreases, will show a decrease in conversion as the cycle period increases. The opposite is seen if total moles increase in the reaction. Throughput for a constant product discharge also depends on the cycle period. If total moles increase, larger cycle periods diminish the throughput. These effects arise primarily from the influence of pressure on the equilibrium conversion.

Figure 9.4a illustrates the dependence of the separation factor on cycle period. This factor, defined earlier and given in Table 9.4, measures the separation achieved in a PSR. It can be seen that with very rapid cycling, the separation factor goes to unity so that no separation is achieved. In this condition, the pressures in the reflux volume and at the feed end of the bed are about the same. However, for the parameters used in the simulation, once the cycle period exceeds 2 s, a good separation between the products is obtained. Just as in a RPSA, the weakly adsorbed product concentrates in the discharge. The figure shows that separation of the reactant is poor. For cycle periods greater than 2–4 s, separation becomes independent of period. For reactions with a single product, separation is poor, although some separation of the reactant is achieved. Figure 9.4a shows results for a single product reaction with an increase in total moles. Behavior is similar when there is a net decrease in total moles.

Range of the feed fraction of a cycle is limited at the low end because of the constant product discharge requirement. At a feed fraction of unity, the reactor behaves as a plug flow reactor (PFR), of course. Within the possible range, feed fraction has only a small effect on conversion and it is opposite to the effect of cycle period. When moles are created in the reaction, conversion decreases with increasing feed fraction. However, when the total moles decrease, increasing the feed fraction raises the conversion, but just slightly. Once again, this

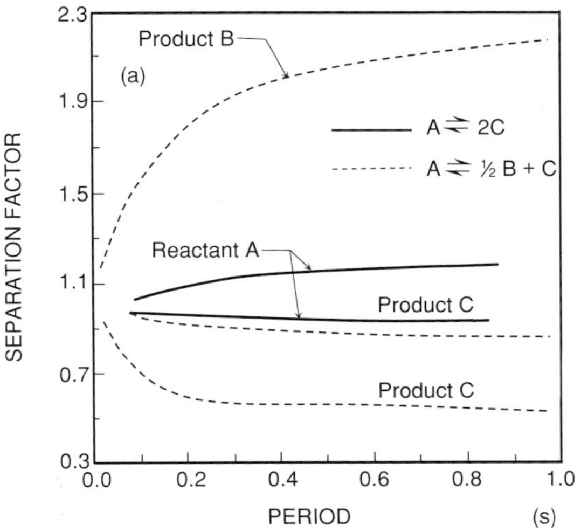

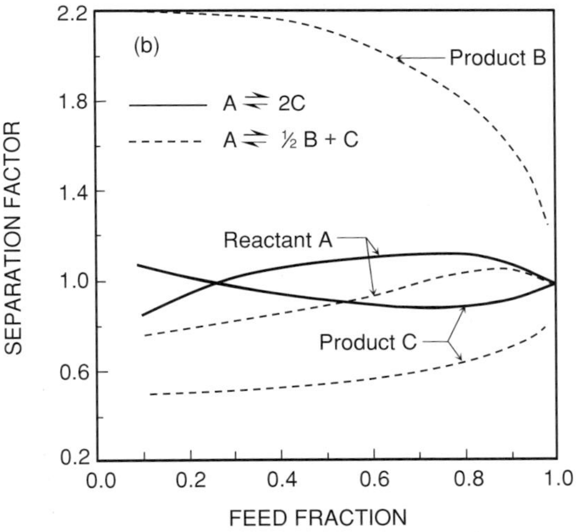

Figure 9.4 Effect of design parameters on separations achieved in an RPSR in which reaction equilibrium is attained: (a) cycle period, (b) feed fraction. (Figure adapted from Vaporciyan and Kadlec (1987) with permission. © 1987 AIChE.)

behavior arises from the pressure dependency of the equilibrium conversion and the effect of feed fraction at constant cycle period on pressure in the catalyst–adsorbent bed.

An unexpected discovery in the authors' simulation is a reversal in the separation factor as feed fraction is varied. This is shown in Figure 9.4b. For the reaction, A $\leftrightarrow$ 2C, below $f_f = 0.15$, product C is concentrated in the discharge, while reactant A dominates in the exhaust. Between $f_f = 0.2$ and 0.25, there is no separation, but as the feed fraction increases further A is concentrated in the discharge and C in the exhaust. Thereafter the separation factor for A goes through a maximum before it becomes unity at $f_f = 1$ where the pressure swing reactor behaves as a PFR. The separation factor for C goes through a minimum, but this occurs at a higher feed fraction than that for the maximum in A.

Separation reversals are not observed for pressure swing adsorbers with a reflux volume. Explanation of the reversal draws on the pressure effect on the equilibrium composition and the trend of pressure at the bed outlets with the feed fraction. Thus, at low feed fractions (<0.2), the bed pressure at the exhaust is lower than the pressure in the reflux volume at the discharge end of the bed. Equilibrium conversion is therefore lower at the exhaust and the reactant is enriched there. As feed fraction augments, the pressure at the exhaust tends to rise and approaches the pressure at the discharge. This increases conversion, augmenting C in the exhaust.

For the reaction with two products only one of which is strongly adsorbed, the RPSR acts like a RPSA, but low feed fractions enhance separation of the products over that found in a simple adsorber. Just as with cycle period, the effect of feed fraction on reactant–product separation is small. Indeed, reactant is not well separated in the RPSR.

Simulation results obtained by Vaporciyan and Kadlec are generally applicable to other pressure swing cycles. They demonstrate that for fast reactions in which equilibrium is established almost instantaneously, conversions greater than the equilibrium limit cannot be attained. The presence of reaction in the adsorbent bed can change the separation performance of the pressure swing system in terms of the exit steams in which reactants and products appear. However, the extent of separation is not greatly altered. Reaction in the bed does introduce a new phenomenon: reversal of the separation as the feed fraction of the cycle is varied. The authors point out that RPSRs have a behavior limit: the equilibrium pressure swing reactor when rates are very fast.

A second paper on RPSRs by Vaporciyan and Kadlec (1989) discusses both numerical simulations and experimental measurements. The system is the same as in their first paper: a mixed catalyst–adsorbent bed containing a reflux volume at the discharge end. The cycle is also the same. Vaporciyan and Kadlec's numerical studies were exploratory. Their objective was to examine the performance and behavior of an isothermal RPSR for several reaction classes as well as to test their model by comparing simulation with experiment. The authors used the dimensionless model given by Equations 9.6 and 9.7 and the boundary and initial conditions stated in Table 9.4. Further simplifications came from assuming Darcy's Law governed flow through the packed bed and that linear adsorption isotherms were applicable.

Vaporciyan and Kadlec remark that dimensional analysis, examination of the dimensionless groups appearing in Table 9.3, delivers useful insight into the role of design parameters in PSR performance. They observe that the length of the reflux volume, L_R, appears squared in the denominator of the Ruthven number, N_{Ru}, for the reflux volume while the cycle period, τ, is in the numerator. Cycle period also appears along with the delivery molar flow rate, n_d, in the numerator of the Kadlec number, N_{Ka}. Decreasing reflux volume length requires a much larger reduction in τ to maintain the Ruthven number constant. However, to keep the Kadlec number constant, n_d must increase strongly. This means that reducing the size or length of the reflux volume by a modest amount can accommodate an increase in the delivery rate and thus bed throughput, while maintaining the contribution of the reflux volume to separation performance and yield unchanged. However, the Ruthven number for the reactor bed increases so its performance must change. With irreversible reactions, the Damköhler number, N_{Da}, shows that increasing catalyst activity can reduce temperature. Through the Ruthven number, higher activity appears to provide higher throughput while

Table 9.5 Physical characteristics of an RPSR and model parameters used in simulations (Vaporciyan and Kadlec, 1989)

Cross-sectional area	3.45 cm^2
Reflux volume	50 cm^3
Packed-bed length	157.5 cm
Packed-bed permeability	48.6 cm^2·cp/atm·s
Gas viscosity	<0.0228 cp
Mass of catalyst	229.6 g
Δ	0.174 mm
δ	0.163 mm
Platinum	0.5 wt%
Pt dispersion	13.7%
A	10 040 cm^2-Pt/g-cat
K_1	1.485 × 10^8 mol·K/cm^2-Pt·s·atm^2
K_2	5,484.5 L/atm
E	15 000 cal/mol
H	2000 cal/mol
Mass of adsorbent	143.1 g
γ_{CO}	8.4
γ_{O_2}	1.9
γ_{CO_2}	51.4
γ_{N_2}	2.7

maintaining the same separation performance and yield. The ratio of bed permeability and gas viscosity appear in the N_{Ru} so that dimensionless analysis can predict how the bed length must be changed to maintain reactor performance as the gas properties and bed structure (particle size) are changed. Restrictions apply. The bed length must be long enough so that there is sufficient pressure drop in the bed to maintain a constant delivery rate through the discharge port.

Two irreversible reactions: A $\rightarrow$ (1/2)B + (1/2)C, A $\rightarrow$ B, and a simultaneous reaction C $\leftarrow$ A $\rightarrow$ B were studied by simulation for a RPSR with bed dimensions, catalyst–adsorbent properties and rate constants representative of the authors' experimental work. These model parameters are summarized in Table 9.5. For the simulations, cycle period was 5 s and the delivery duration was set to zero so that the cycle had just feed and exhaust steps. These were of equal duration. Feed pressure was set at 1.68 bar, while the ratio of exhaust to feed pressure was 0.595. Capacity enhancements through adsorption were 10 for A, 36 for B and 1 for C. Diluent made up 95% of the feed by volume and was adsorbed with a capacity enhancement factor of 3.7. The Ruthven number was set at 2.1, the Kadlec number at 0.74 and the ratio of packed bed to reflux volume was 10.9.

Vaporciyan and Kadlec found that absorption of B relative to the other products did not increase the yield of B relative to a PFR containing an equal amount of catalyst and operating at the same feed rate. If adsorption of A dominates, conversion of A relative to a PFR can actually decrease because adsorption of A at the entrance to the bed reduces the residence time of A in the reactor. Damköhler number or in other words rate of reaction does not effect the conversion or yield of B in the RPSR relative to a PFR, but it does effect the separation. Figure 9.5 shows that when N_{Da} increases a separation inversion occurs. At low rates of reaction, products B and C are separated from A. Both appear mainly in the discharge, but

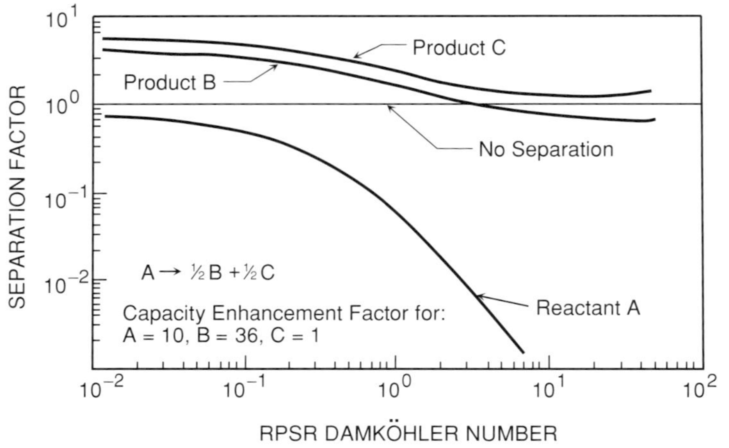

Figure 9.5 Effect of rate of an irreversible reaction on separation in an RPSR. (Figure adapted from Vaporciyan and Kadlec (1989) with permission. © 1989 AIChE.)

at very different concentrations. Reactant A is poorly separated. As N_{Da} increases, reactant A concentrates in the exhaust stream and is separated from the carrier gas. At $N_{Da} = 10$, product B is recovered in the exhaust. C is the dominant reaction species in the discharge stream, but this stream is primarily the carrier gas. Although Figure 9.5 gives simulation results for the two-product reaction, the separation behavior is similar for the other two reactions with the exception of course that there is no C product in the second reaction. The authors conclude from this simulation that the presence of an irreversible reaction enhances importantly the reactant–product or the product–product separation achievable in a RPSR. Throughput or capacity of the reactor is not strongly influenced by the presence of a reaction.

The same three reactions were studied for a dimensionless equilibrium constant = 1.0 and the same set of bed and operating parameters. Effect of rate of reaction (as N_{Da}), cycle period and feed fraction of the cycle on yield and separation were investigated. Adsorption of reactant influences yield and separation strongly. If the reactant is not adsorbed but a product is, conversions exceeding those obtained in a PFR are possible for single, reversible reactions. Conversion of A goes through a maximum and near this maximum it exceeds conversion in a comparable PFR. The enhancement, observe Vaporciyan and Kadlec, is solely a consequence of a shift in equilibrium through adsorption of a product. Separation of reactant and product or products from one another, if two or more are formed, also depends on N_{Da}. At low N_{Da}, the separation behavior is like that discussed in the previous paragraph, whereas at high N_{Da}, separation follows the pattern discussed earlier for a reaction that instantaneously attains equilibrium. Vaporciyan and Kadlec remark that in general the separation performance follows that exhibited by an RPSA.

Interesting selectivity behavior occurs between the high and low values of N_{Da} for the simultaneous reaction, $C \leftarrow A \rightarrow B$. Expressing the effect of N_{Da} as conversion, selectivity to the more strongly adsorbed product B increases from unity to a maximum as conversion increases, which for the parameters used in the simulation was about 30% greater than obtainable in a PFR. The maximum selectivity to B is achieved at about the conversion

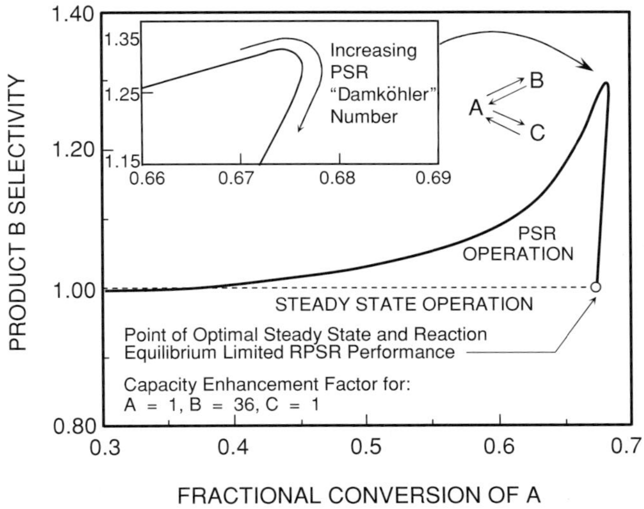

Figure 9.6 Variation of selectivity with rate of consumption of A in an RPSR for a parallel, reversible reaction. (Figure adapted from Vaporciyan and Kadlec (1989) with permission. © 1989 AIChE.)

corresponding to chemical equilibrium. Increasing N_{Da} further leads to a sharp decrease in selectivity to B as shown in Figure 9.6, and at high N_{Da} equilibrium occurs at the reactor entrance and selectivity drops to unity.

Vaporciyan and Kadlec (1989) observe that design parameters affect selectivity and some exhibit optimal values. The pressure ratio, P_e/P_f, causes selectivity to B to increase as the ratio decreases. A decrease from 0.595 to 0.149 increases selectivity from 1.3 to 1.6. Conversion of A also increases from 67 to 70%. Cycle period and the feed fraction of a cycle are important design parameters that are limited in practice by the assumption that discharge flow remains constant over a cycle. Nevertheless, within this limit, each of these parameters influence selectivity and exhibit optima. Figure 9.7a shows that decreasing the cycle period to about 3 s from the 5 s used in Figure 9.6 increases the selectivity to1.34, while Figure 9.7b indicates that increasing the feed fraction of the cycle to close to 0.9 from 0.5 raises the selectivity to about 1.47.

An interesting variation to a PSR design was introduced by Lee and Kadlec (1989) who also investigated the parallel equilibrium reaction system discussed above in a three-step pressure swing cycle. Instead of an intimately mixed bed of adsorbent and catalyst, they considered use of three adsorbents, each specific for a reactant or a product. Catalyst and the adsorbents were distributed into one to three sections of a four-zone reactor as indicated in Figure 9.8. The zones were of variable length. The idea was to force the concentration profile of the components so as to maximize the purity of one of the products, in this case product B, in the discharge or product stream. This could be accomplished by a profile, which maximized the partial pressure of the desired product (B) at both ends of the bed, while minimizing the partial pressure of undesired product (C) at these locations. Lee and Kadlec undertook a search of bed structure and found that the catalyst should be placed only in the first 10% of the bed. The adsorbent for A should be uniformly distributed in the remaining

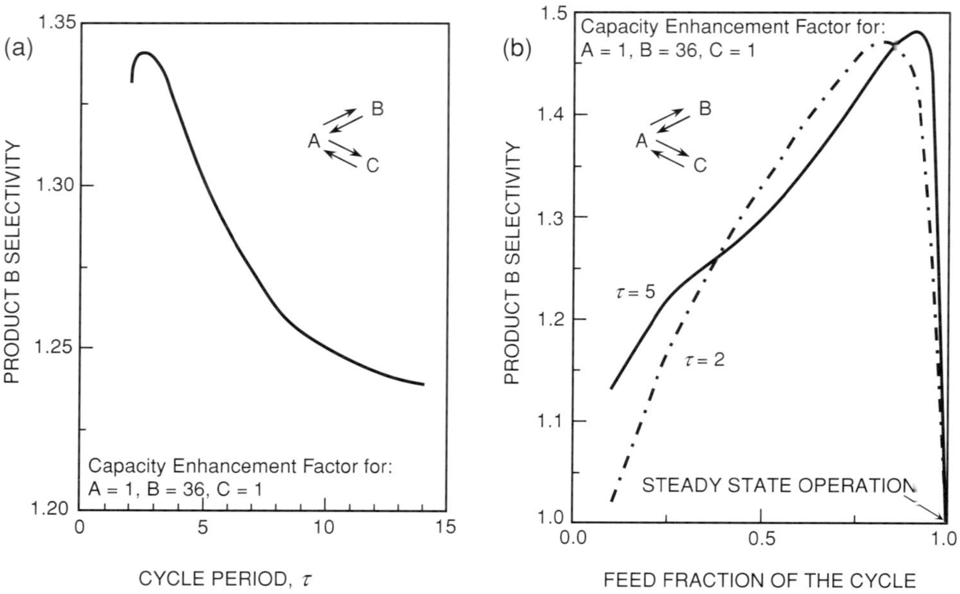

Figure 9.7 Effect of RPSR design parameters on selectivity with parallel reversible reactions: (a) cycle period, (b) feed fraction. (Figure adapted from Vaporciyan and Kadlec (1989) with permission. © 1989 Air Products and Chemicals, Inc.)

90%, whereas the adsorbents for B and C should be distributed uniformly throughout the reactor. However, the amounts of each of these adsorbents differed so that the capacity for component C greatly exceeded the adsorption capacity for component B. Nevertheless the Lee and Kadlec's results indicated that distribution of catalyst and adsorbents had a relatively small effect on conversion and selectivity to B in the product and exhaust streams. Perhaps this explains why the interesting concept has not been studied further.

European researchers (Chatsiriwech et al., 1992, 1994; Alpay et al., 1993, 1994; Cheng et al., 1998; Yongsunthon and Alpay, 1998a,b, 1999, 2000; Xiu et al., 2002a,b, 2003a,b) have taken up research on PSRs in the last decade. Reactions considered in the earlier work have been arbitrary ones of the A → B type such as isomerization and disproportionation with no change in the number of moles by reaction. They also examined a dissociation reaction in which the total number of moles increased. Dehydrogenation of methylcyclohexane (MCH) to toluene is an example of such a reaction. For each mole of toluene, 3 moles of H_2 are formed. To undertake their simulation of PSR performance, Alpay et al. (1994) measured dehydrogenation rates in a continuous plug flow reactor packed with a 5 wt % Pt on Al_2O_3 catalyst powder. Rates were fitted to a first order model and used to estimate the rate constant of the forward reaction for this equilibrium-limited dehydrogenation. Adsorption measurements were carried out for toluene, H_2 and MCH on molecular sieves, various clays, activated alumina and activated carbon. Molecular sieves were the best adsorbents, but significant adsorption of toluene took place on the clays. With the exception of a 13X sieve, adsorption of toluene was about an order of magnitude greater than that of MCH. Hydrogen, effectively, was not adsorbed. Adsorption data were well fitted by linear isotherms

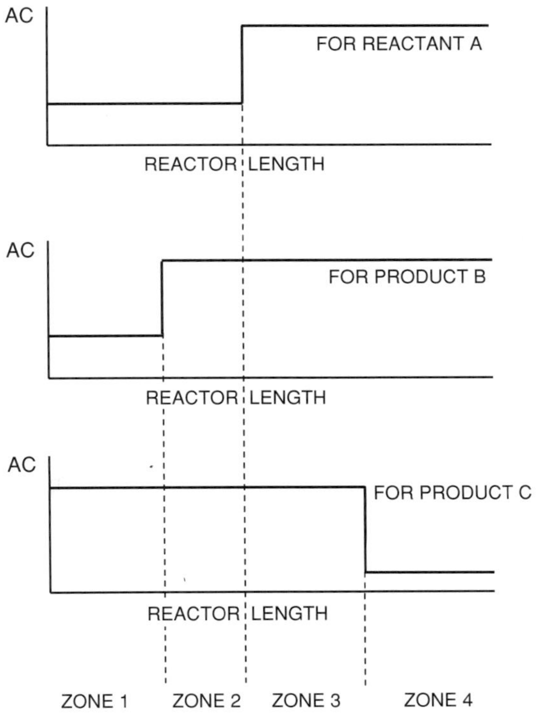

Figure 9.8 Multi-zone reactor. (Figure adapted from Lee and Kadlec (1989) with permission. © 1989 AIChE.)

and adsorption equilibrium constants were evaluated from these isotherms. Adsorption equilibrium constants used for the authors' simulation of MCH dehydrogenation were in the range of values measured experimentally for 3A and 4A sieves and the clays, however, much smaller values of the adsorption equilibrium constant than those measured were employed to allow for total pressure and catalyst dilution. For the simulation of isomerization and disproportionation reactions, rate constants were adjusted so that conversions were comparable to those obtained in their MCH dehydrogenation measurements. It appears that Alpay et al. took care to use parameters in their model that were representative of real reaction systems.

A simple pressure swing system with a four-step cycle (Fig. 9.2c) in which a portion of the product stream is used to purge the PSR bed was simulated. The reactor was filled with equal amounts of well-mixed catalyst and adsorbent. Composition of the purge was constant at the time average composition of the product stream. To restrict the variables considered in their study, the authors used just a single feed composition (5 vol% reactant) and feed pressure (10 bar). Total cycle time for most runs was 24 s for the 1-m-long fixed bed. The simple cycle consisted of 6 s of pressurization, 12 s of discharge and 6 s of exhaust or depressurization. For the purge option, the feed step was held to 12 s, but the other two steps and the purge were 4 s in duration. The authors used a longer cycle time, 36 s, for their simulation of MCH dehydrogenation. Discharge lasted 18 s, while the other steps in the purge cycle were set at 6 s each. Exhaust pressure was 5 bar during depressurization. For the other simulations, it was held to 1 bar.

Equations 3.1 and 3.4 in Table 3.2 give the one-dimensional, isothermal model used by the English researchers (e.g. Alpay et al., 1993). Catalyst and adsorbent are separate phases. For convenience, these relations are restated below in terms of an axial dimension z.

$$\frac{\partial}{\partial t}(\varepsilon_t C_i + (\rho_b)_{\text{ads}} q_i) = \frac{\partial}{\partial z} D_z \frac{\partial C_i}{\partial z} - \frac{\partial u C_i}{\partial z} - (\rho_b)_{\text{cat}} r_i \tag{9.33}$$

$$\frac{dP}{dz} = -J_v u - J_k u^2 \tag{9.34}$$

Of course, Equation 9.33 must be applied to each reactive component. Equation 9.34 is the Ergun relation. It must be invoked if a significant pressure drop is anticipated. Adsorption was either assumed to be instantaneous and mass transfer and intraparticle diffusion were neglected or a linear driving force model was used to represent the combined resistance of adsorption and mass transfer. In the latter case, the model was written as

$$\frac{\partial q_i}{\partial t} = k_i (q_i^* - q_i) \tag{9.35}$$

The coefficient k_i is usually estimated from the effective diffusivity of the component i in the porous adsorbent. Various relations have been proposed for this coefficient (Alpay et al., 1993; Ding and Alpay, 2000b). Reaction rates are usually assumed to be kinetically controlled. Of course, adsorption and reaction occur on different particles. In most earlier works, a linear adsorption isotherm was assumed, while the reaction rate model was first order overall. Adsorption of the reactant and one of the products, if two were formed, were assumed to be negligible. Boundary conditions for the model are stated in Table 9.4. Simplifications allowed the investigators to use the method of characteristics to reduce the model to a set of ODEs for integration by a Runge–Kutta–Merson routine.

Simulated yields for an isomerization, A $\leftrightarrow$ B, are compared as a function of the forward rate constant for a PSR and a PFR in Figure 9.9. Yields for the PFR are at 10 bar, the feed pressure to the pressure swing reactor. They are thus the maximum yields for a PSR. At low reaction rates where the reactor operates well away from equilibrium, a PSR underperforms the PFR because the reverse reaction does not influence conversion. On the other hand, for a high rate constant, equilibrium is reached at the front of the bed. The gas phase concentration of B in the remainder of the bed is sufficient to make the reverse reaction important so that the yield of the PSR approaches the equilibrium yield just as Vaporciyan and Kadlec (1987) demonstrated earlier. Nevertheless, there is a significant improvement in yield at $k_f = 1 \times 10^{-4}$ where the PFR reaches the equilibrium yield. The improvement depends strongly on the magnitude of the adsorption constant. Indeed, even at $k_f = 1 \times 10^{-3}$, where Figure 9.9 shows little improvement over the equilibrium yield, increasing the adsorption constant by a factor of about 4 brings about a 20% increase in yield.

With high rate constants, the PSR behaves as a pressure swing adsorber (PSA). Isomerization product B is obtained essentially in the exhaust during depressurization. This behavior is illustrated in Figure 9.9b. At $k_f = 1 \times 10^{-3}$, 92% of B appears in the exhaust stream. Using a purge cycle further improves the separation of reactant and product as is well known in PSA practice. This can be seen in (b) also. The significance of the figure is that for a high separation factor near complete recovery of product in the exhaust can be attained at much

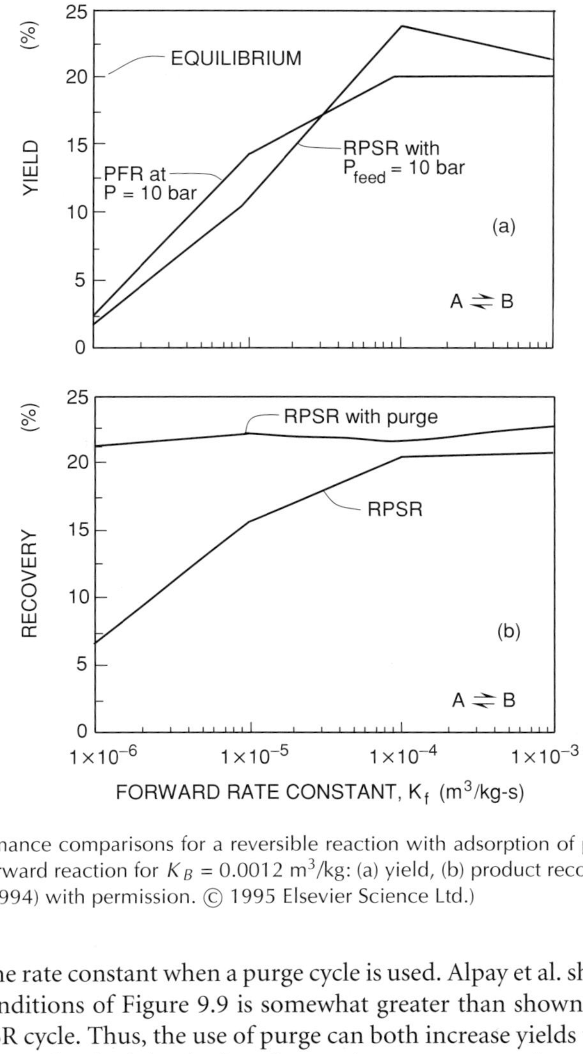

Figure 9.9 Performance comparisons for a reversible reaction with adsorption of product as a function of the rate of the forward reaction for K_B = 0.0012 m^3/kg: (a) yield, (b) product recovery. (Figure adapted from Alpay et al. (1994) with permission. © 1995 Elsevier Science Ltd.)

lower values of the rate constant when a purge cycle is used. Alpay et al. show that yield with purge for the conditions of Figure 9.9 is somewhat greater than shown in Figure 9.9a for the three-step PSR cycle. Thus, the use of purge can both increase yields under equilibrium conditions and provide a high level of product and reactant separation.

A limit exists for the adsorption constant in the purge cycle according to Alpay et al. (1994). At high values, the isomerization product is so strongly adsorbed that total recovery in the exhaust and purge steps is sharply reduced. Product yield falls below the yield at equilibrium. Alpay et al. also demonstrate that the yield for the PSR depends on the purge volume.

Simulation of a disproportionation reaction, $2A \leftrightarrow B + C$, in PSR and PFR reactors for rate and adsorption constants similar to those used in the isomerization study gave the general behavior shown in Figure 9.9. An increase in yield, if any, depended upon the magnitude of the adsorption constant for product B and also on the amount of purge as measured by moles of purge per cycle. Just as with adsorption constants, there is an optimum amount of purge. In the simulation carried out by the authors, the amount of product per cycle was

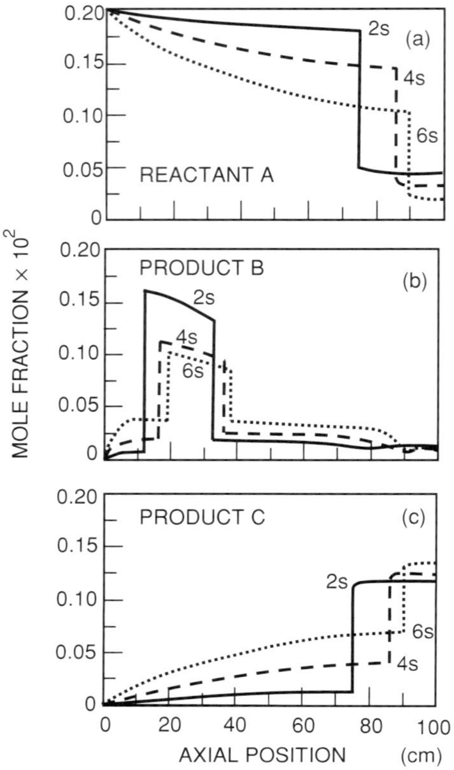

Figure 9.10 Reactant and product concentration profiles in a PSR at different times within a cycle once a cyclic steady state is attained. (Figure adapted from Alpay et al. (1994) with permission. © 1995 Elsevier Science Ltd.)

held constant. Consequently, increasing the purge amount increases the amount of feed. Eventually decreasing contact time in the bed overcomes the yield improvement from the greater purge amount.

In the range of the rate constant where the yield of product B exceeds the equilibrium yield, the improvement is nicely explained by Figure 9.10. This figure shows the gas phase mole fractions of reactant and products after pressurization commences. As the nonadsorbed reactant fills the bed of mixed catalyst and adsorbent, the adsorbed product, B, becomes trapped at the feed end of the bed. Product C, which is not adsorbed, moves through the bed and collects at the discharge end. Its mole fraction is low in the front of the reactor thus minimizing the extent of the reverse reaction. The shape of the product C profile comes about because the PSR operates on a purge cycle in which pressurization follows purge. The purge gas is largely C so that during this step, reactant A is forced backwards. Once the purge ends, product C is forced out of the reactor by the entering feed consisting of reactant A.

Figure 9.10 also explains the separation between products B and C achieved by the pressure swing operation. The adsorbed product B, trapped in the feed end of the bed, is recovered from this end during the exhaust and purge steps provided the total cycle time is kept short and the exhaust step is significantly longer than the pressurization one. Good separation of

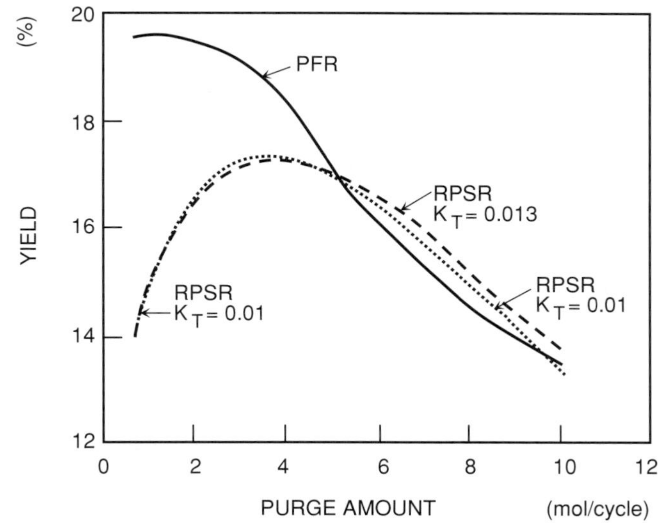

Figure 9.11 Predicted influence of the purge amount and adsorption equilibrium constant on PSR yield for methylcyclohexane dehydrogenation. (Figure adapted from Alpay et al. (1994) with permission. © 1995 Elsevier Science Ltd.)

reactant A and product B is not achieved in either the simple or purge cycles. Reactant A is recovered with product B in the exhaust stream. Some improvement in this separation may be achieved by switching the exhaust between collecting vessels. Initially the exhaust during depressurization will be mainly A. Toward the end of the step, product B will predominate.

Methylcyclohexane dehydrogenation in a PSR was simulated for a feed pressure of 10 bar and an exhaust at 5 bar. As mentioned earlier, equilibrium yields are functions of pressure. Because of an increase of moles in the reaction, yield changes inversely with pressure. Yields for a pressure swing reactor were less than the maximum PFR yield at 5 bar for all values of the forward rate constant in contrast to the situation in Figure 9.9a.

Optimal values of the adsorption constant and amount of purge were observed for the PSR yield. Figure 9.11 shows the effect of purge amount on toluene yield for two values of the adsorption constant. It can be seen that the PSR yield slightly exceeds the PFR yield at high purge volumes, but these yields are below those at the optimum purge. They are well below the equilibrium or maximum yield of about 19.5% for a PFR operating at 5 bar assuming the smallest PSR purge volume. The decrease in the PFR yield with increasing purge arises from the simulation condition of a fixed amount of the toluene product regardless of the purge and the further condition of identical feed rates for comparison of reactor performance. To meet the first condition, the feed rate must increase in the PSR as the amount of purge increases. The second condition requires a feed rate increase for the PFR. As the feed rate increases, equilibrium is no longer attained at the end of the plug flow reactor.

The study of Alpay et al. (1994) suggests that a pressure swing reactor may not be the best choice for equilibrium reactions for which an increase in moles through reaction causes the equilibrium yield to decrease with increasing pressure. Further simulations of methylcyclohexane dehydrogenation showed that if the PSR is operated at lower pressures (feed pressure of 2 bar and an exhaust pressure of 1 bar), PSR yields are always below the PFR yield at the same purge volumes regardless of the adsorption constant and the purge amount. Nevertheless, a PSR can exceed PFR yields for dehydrogenation reactions in which 1 mole of H_2 is formed for each mole of product, e.g. ethane dehydrogenation to ethene. For this reaction, Figure 9.9a and b illustrates the PSR behavior.

An important contribution of Alpay's theoretical study was to demonstrate that the interaction of the PSR operating variables with reaction properties is complex. Indeed, because of this, there are situations in which a PSR will not be an attractive reactor choice.

Chatsiriwech et al. (1994) in a closely related work examined a reversible reaction, $2A \leftrightarrow B + C$, without a change in total moles in a three-step PSR cycle and in a cycle with purge. Their model was identical to the one used by Alpay et al. (1994), although their model parameters were somewhat different. Figure 9.9a represents the yield behavior they observed. Improvement over the equilibrium yield and the forward rate at which this occurred was found to depend on the adsorption constant for the adsorbing component, product B. If adsorption is weak, yields are always below those obtained in a PFR operating at the feed pressure. Increasing product adsorption increases the yield improvement. The effect of purge amount was studied. As in Figure 9.11, an optimal purge that maximized the yield of product B was found. This optimum and the improvement over the equilibrium yield depended strongly on K_B. The optimal purge and yield improvement was small for low values of the constant. The importance of the adsorption constant carries through to the separation factors as might be expected. Low factors or high separation (see Table 9.3) arise with strong adsorption. The fractional duration of the delivery and purge steps in the purge cycle could be varied widely with only a small effect on yield improvement and the optimal purge. The amount of product discharged in a cycle also exerted a strong influence on yield and separation. As this amount increases, yields drop and can decrease below the equilibrium yield. The separation factor increases toward unity where no separation is achieved. The effect of product amount was further explored by Kirkby and Morgan (1994), albeit for an irreversible reaction.

Dehydrogenation was also examined by Sheikh et al. (2001) for a two-step RPSR. The stoichiometry considered is $A \leftrightarrow B + C$ as typified by 1-butene dehydrogenation. The authors assumed that the kinetic of both forward and back reactions are elementary and that adsorption equilibrium followed a linear isotherm. Model parameters were experimentally determined (Sheikh et al., 2001). Two modes of operations were considered cocurrent and countercurrent. In the former, depressurization occurred only through the product port while in the latter both product and exhaust port were used. The simulation showed a substantial concentration of the desired B component (butadiene) in the product stream for countercurrent depressurization and a significant yield increase compared to a PFR containing only catalyst and an inert solid diluent operating at identical inlet conditions. Cocurrent depressurization increased conversion of A dramatically, but at the expense of the separation of reactant and product species.

Kirkby and Morgan (1994) examined an irreversible isomerization, $A \rightarrow B$, for three pressure swing cycles. Irreversible reactions were the subject of a 1987 paper by Vaporciyan and Kadlec that we have already discussed. The objective of Kirkby and Morgan in their paper was similar. Three pressure swing cycles were considered: a three-step cycle (Fig. 9.2b), a purge cycle (Fig. 9.2c) and a backfill cycle (Fig. 9.2d). The one-dimensional, isothermal model used by Alpay et al. (1994) was assumed. The reaction was taken to be first order. Solution of the model PDEs utilized the method of characteristics.

The authors' simulation considered the type of cycle, the amount of product per cycle, that is the amount delivered at the discharge end of the reactor, and for the purge cycle, the amount of purge. Purge had a constant composition, namely that of the well-mixed product stream leaving the discharge port. For the backfill cycle, the backfill pressure was considered along with the product amount. The backfill composition was also that of the well-mixed product stream. The catalyst and adsorbent filled reactor had a 1 m^2 cross section and was 1 m in length. It had a total void fraction of 0.75. Duration of each part of a cycle was 30 s, so for the three-step PSR, cycle period was 90 s. Feed pressure was 1 MPa while exhaust pressure was 0.1 MPa. Reactant mole fraction in the feed was just 5×10^{-3}. For the simulation, the adsorption constant for product B was set at 1×10^{-3} mol/kg/Pa, reactant A was nonadsorbing and the rate constant was 1×10^{-5} mol/s Pa m^3.

Kirkby and Morgan remark that for a three-step PSA cycle, product B would be concentrated in the exhaust rather than in the discharge. However, when an irreversible reaction is possible, an inversion occurs as the amount of product drawn off per cycle decreases. At very low product draw-off through the discharge port, product B is concentrated at the discharge end resulting in a large separation factor. As draw-off increases, product B becomes concentrated in the exhaust. The mole fraction of A in the exhaust as well as in the product stream approaches that in the feed. The inversion is shown in Figure 9.12. Vaporciyan and Kadlec (1989) also observed inversion but it arose under variation of a different design parameter. Explanation for this behavior in the Kirkby and Morgan study is that at low draw-off, product B is forced by the feed toward the discharge end of the reactor.

When a purge cycle is used, an inversion is once again observed as product draw-off through the discharge port increases. With 40 mol of product/cycle, inversion occurs when

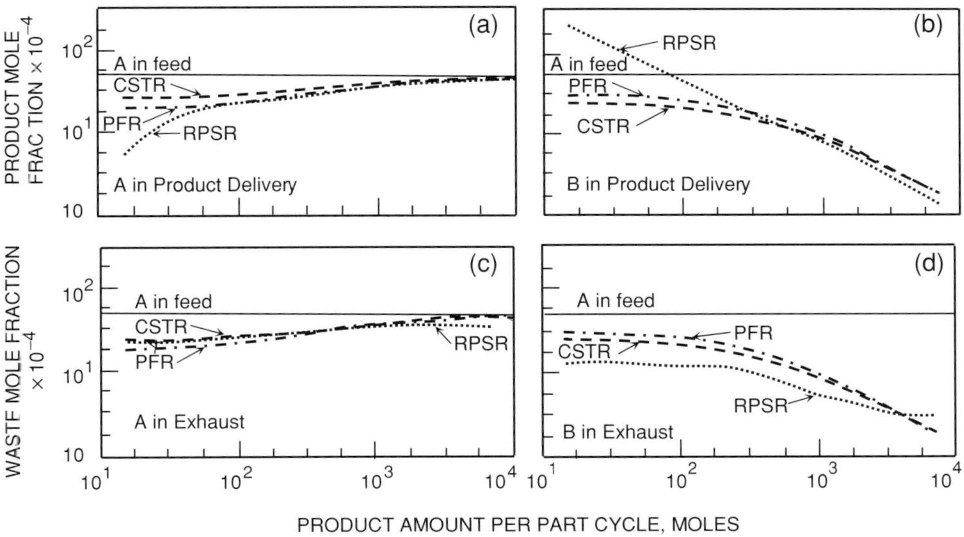

Figure 9.12 Exit concentrations of reactant and product as a function of the product draw-off: (a) reactant A in the discharge, (b) reactant A in the exhaust, (c) product B in the discharge and (d) product B in the exhaust. (Figure adapted from Kirkby and Morgan (1994) with permission. © 1994 Institution of Chemical Engineers (U.K.).)

the purge amount reaches about 1 mol/cycle. Above that purge amount, product B is largely found in the exhaust. Moreover, carrier gas becomes concentrated in the discharge stream. Smaller purge amounts provide a good separation between reactant and product with the latter found mainly in the discharge stream. As the amount of product increases, separation becomes poorer. Inversion is still observed but a greater purge amount is needed for it to occur.

The backfill cycle results in separations similar to those for the purge cycle. Inversion occurs, but it is found at a low level of backfill pressure. For example, at a product draw-off of 40 mol/cycle, inversion occurs at a backfill pressure of 0.2 MPa. Once the backfill pressure reaches 0.4 MPa, the PSR effectively separates reactant and product from the nonadsorbing carrier gas.

The Kirkby and Morgan study shows that PSR performance with respect to operating parameters is largely the same regardless of the mode of PSR operation. The parameters, however, affect conditions for separation inversion.

Another type of equilibrium-limited disproportionation or olefin metathesis reaction, $2C_3^{=} \leftrightarrow C_2^{=} + C_4^{=}$, was studied by Gomes and Yee (2002) through simulation as well as experimentation. We consider the experimental portion further on. A four-step isothermal pressure swing cycle was examined and the authors assumed isobaric operation in each step. The PSR model the authors adopted is given by Equation 9.33. An experimentally determined second order rate expression for the metathesis reaction was used (Gomes and Fuller, 1996), while adsorption was assumed to be an equilibrium process that followed a multicomponent Langmuir isotherm. However, mass transfer was taken to be rate controlling for adsorption. In the system, 2-butene was strongly adsorbed, compared to ethene. Thus, ethene is recovered primarily in the product stream and is four times as concentrated as 2-butene. An appendix in the Gomes and Yee paper show the collocation equations employed by the authors for their simulation.

Gomes and Yee (2002) used simulation to explore the effect of cycle period and reactant feed rate on PSR performance. Just the effect of period is shown in Figure 9.13.

The effect of the feed rate to the unit was also investigated. Behavior similar to that shown in Figure 9.13 was found, however, a threefold change in feed rate did not greatly alter either purity or recovery from those shown in the figure. Gomes and Yee explained this effect by noting that increasing the flow rate increases propene entering the reactor so that the adsorption of additional propene displaces butene making the separation of ethene and butene less effective.

Separating catalyst and adsorbent into separate beds was examined by Lee and Kadlec (1986) as a means of improving selectivity for a consecutive reaction as discussed earlier. Al-Juhani and Loughlin (2003) have examined separate beds as a way of reducing the capital outlay for equilibrium-limited reactions for which recycle of unreacted feed is necessary. These investigators considered the Union Carbide "Total Isomerization Process" for light straight run naphthas, which uses an isomerization reactor followed by pressure swing adsorbers. Hydrogen is used in the process to prevent catalyst deactivation. Al-Juhani and Loughlin place the catalyst, 0.5 wt% Pd on Y zeolite, in the front section of a single bed and adsorbent, 5A zeolite, in the rear section. A four-step cycle was chosen with (1) bed pressurization with feed, (2) feed step with recovery of the isomerate product in H_2, (3) depressurization producing a waste stream of mixed reactant, product and H_2, and (4) a low pressure purge using H_2. The isomerate product from the high pressure feed step is flashed to recover H_2, which is recycled. Figure 9.14 shows the process proposed.

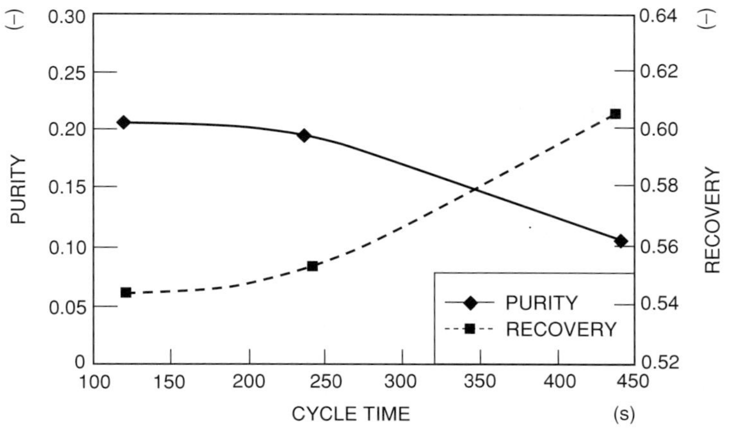

Figure 9.13 Ethene product purity and total recovery as a function of PSR cycle time. (Figure adapted from Gomes and Yee (2002) with permission. © 2002 Elsevier Science Ltd.)

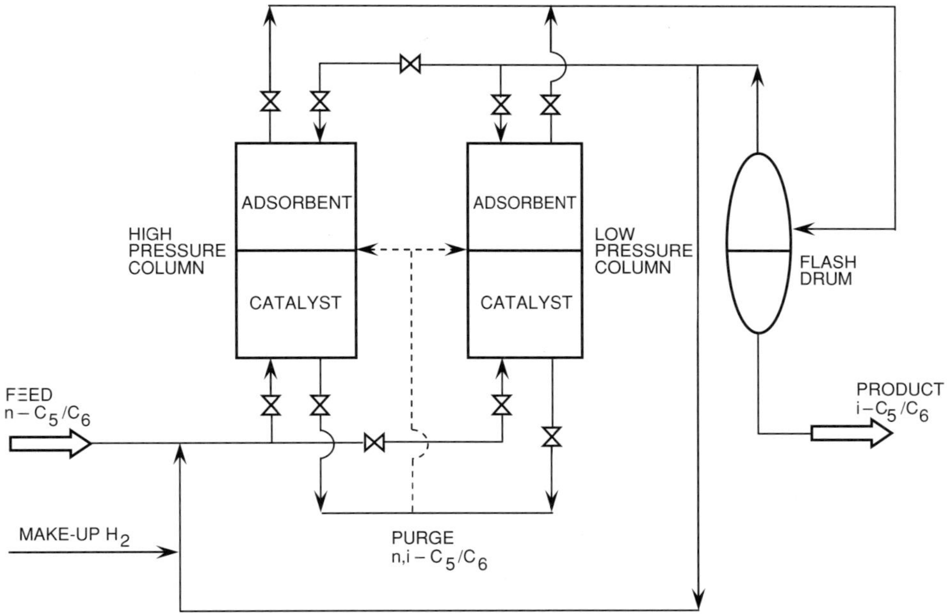

Figure 9.14 Process flowsheet for *n*-paraffin isomerization using a segregated catalyst–adsorbent bed and pressure swing operation. Optional purge with a paraffin stream shown as dashed lines. (Figure adapted from Al-Juhani and Loughlin (2003) with permission. © 2003 Kluwer Academic Publishers.)

Al-Juhani and Loughlin (2003) explored the influence of catalyst activity and temperature on system performance by means of an isothermal simulation. Of particular interest was the optimal fraction of the bed occupied by catalyst. Literature models were used for the isomerization kinetics and adsorption equilibria. They assumed that adsorption was mass transfer controlled and that both external film and internal diffusional resistance was important. Isomerization, however, was kinetically controlled. The model given in Table 9.2 was employed with allowance for axial dispersion and for volume change that altered velocity in the bed. The Danckwerts boundary condition was assumed to apply at the junction of the catalyst and adsorbent beds. A linear pressure gradient was assumed for the depressurization and repressurization steps, otherwise the bed operated isobarically. The study considered a 20-cm-deep bed and feed and discharge pressures of 15 and 2 bar, respectively. The temperature range encompassed 506–533 K. Orthogonal collocation on finite elements was used to integrate the model.

As might be expected, increasing catalyst activity or adjusting space time or bed length to raise the Damköhler number increases paraffin conversion. For the conditions chosen by the authors, reaction equilibrium is attained at about $N_{Da} = 3$. Because the isomerization reaction is modestly exothermic, increasing temperature for $N_{Da} = 3$ drives down conversion. For the 20-cm bed, reaction equilibrium is reached at $N_{Da} = 2.6$ and this gives the highest paraffin conversion. Paraffin concentration profiles are shown in Figure 9.15 at the end of each step after a cyclic stationary state has been reached for a bed evenly split between catalyst and adsorbent. The PSR product is withdrawn at the end of the reactor, $\xi = 1$, in the feed or adsorption step. It is evident that this product contains a negligible amount of paraffin. The figure shows that the paraffin concentration in the discharge has about the

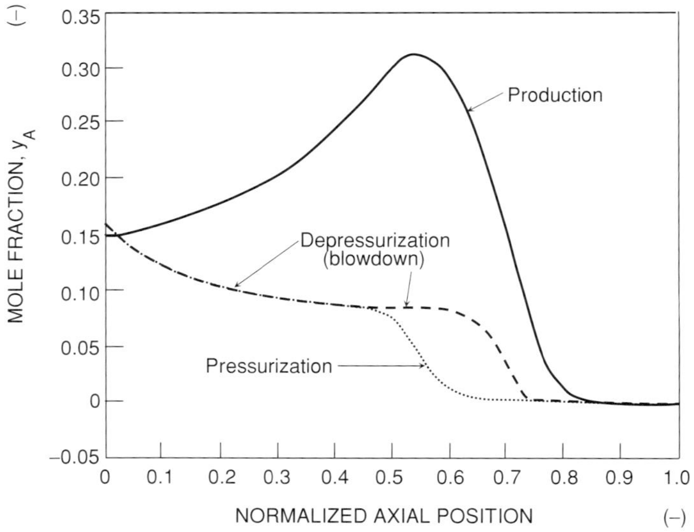

Figure 9.15 n-C_5 profiles at the end of each cycle step in the segregated catalyst–adsorbent bed of Figure 9.14. Simulation is for equal volumes of catalyst and adsorbent, $N_{Da} = 2.6$, $T = 506$ K, a 20-cm deep bed and $\tau = 23$ min. Cyclic stationary state is assumed. (Figure adapted from Al-Juhani and Loughlin (2003) with permission. © 2003 Kluwer Academic Publishers.)

concentration of the feed. This is too rich a stream to vent and Al-Juhani and Loughlin suggest that this stream should be used to purge the catalyst bed during the low-pressure step. The dashed lines in Figure 9.14 show this proposed process modification.

9.6 Nonisothermal modeling studies

Most reactions proceed with a large consumption or release of heat; adsorption processes too exhibit important heat effects. Only in laboratory scale reactors can these reaction and adsorption heats be balanced by external heating or cooling so that isothermal conditions are realized. In vessels whose diameter exceeds 4.5 cm, radial as well as axial temperature variations arise. Nevertheless, nonisothermal operation of a PSR has been addressed only recently (Yongsunthon and Alpay, 1998a, 1999). According to isothermal analysis of PSRs, product adsorption reduces the rate of the reverse reaction thereby raising conversion as equilibrium is approached. With endothermic reactions, Yongsunthon and Alpay (1998) note that adding an adsorbent to a reacting system changes the temperature conversion trajectory favorably with respect to conversion. This is illustrated in Figure 9.16. based on the analysis of a well-mixed, adiabatic flow reactor in which the catalysis is stationary, but a fluid feed and a solid adsorbent are added continuously.

Defining conversion to include reactant trapped by the adsorbent gives the following expression:

$$X_{\mathrm{A}} = 1 - \frac{C_{\mathrm{A}}(1 + (\varphi_{\mathrm{A}}))}{C_{\mathrm{A}_0}(1 + (\varphi_{\mathrm{A}_0}))} \tag{9.36}$$

In this definition, φ_{A} is the volumetric capacity of the adsorbent for reactant, A, per unit volume of fluid in the reactor. The subscript 0 indicates a feed or initial condition.

For a well-mixed continuous flow reactor, leaving out the convective and diffusive terms and the time derivatives in the material and heat balances of Table 3.2 and combining them gives the conversion-temperature trajectory

$$X = -\frac{1}{(\Delta H_r)_f}\left(\frac{(\gamma)_f(T - T_0)}{Q_0 C_{\mathrm{A}_0}} + \frac{(\varphi_{\mathrm{A}})_f}{1 + (\varphi_{\mathrm{A}})_f}\Delta H_{a_{\mathrm{A}}}\right) \tag{9.37}$$

whose slope is

$$\frac{dX}{dT} = -\frac{1}{(\Delta H_r)_f}\left(\frac{(\gamma)_f}{Q_0 C_{\mathrm{A}_0}} + \frac{(\varphi_{\mathrm{A}})_f}{1 + (\varphi_{\mathrm{A}})_f}\frac{d\Theta}{dT}\right) \tag{9.38}$$

where $(\Delta H_r)_f = \Delta H_r + \sum_i \frac{\nu_{\mathrm{i}}}{|\nu_{\mathrm{A}}|}\frac{(\varphi_i)_f}{1 + (\varphi_i)_f}\Delta H_{a_i}$ is an effective heat of reaction for the batch reactor, $\Theta_i = \frac{(\varphi_i)_f}{1 + (\varphi_i)_f{}^2}\frac{\Delta H_{a_i}}{T}(1 + \frac{\Delta H_{a_i}}{RT})$ is a dimensionless adsorbent capacity, and $(\gamma)_f = \rho_s(C_p)_s + \varepsilon\sum_i (C_p)_g M_i C_i$ is an effective thermal capacity. If there is no reactant entering with the adsorbent, conversion in Equation 9.36 becomes

$$X_A = \frac{n_{\mathrm{A}_0} - n_{\mathrm{A}} - m_{\mathrm{A}} q_{\mathrm{A}}}{n_{\mathrm{A}_0}} = 1 - \frac{n_{\mathrm{A}}(1 + (\varphi_{\mathrm{A}})_f)}{n_{\mathrm{A}_0}} \tag{9.39}$$

where the f subscript indicates a flowing adsorbent. It is Equation 9.38 that is shown in Figure 9.16. The increase in slope of the trajectory is evident. Thus, the reactor temperature

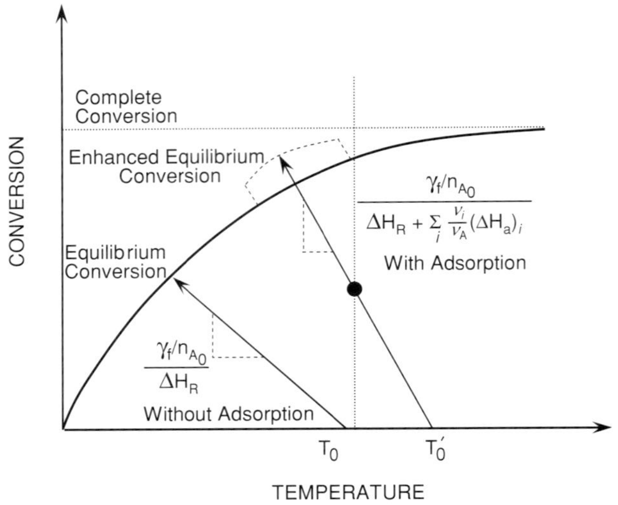

Figure 9.16 Comparison of the conversion–temperature trajectories for an A → Products reaction with or without an adsorbent for the reaction species. An adiabatic steady state reactor fed continuously with reactant and adsorbent is assumed. (Figure adapted from Yongsunthon and Alpay (1998a,b) with permission. © 1998 Elsevier Science Ltd.)

is greater for the same degree of conversion and the rate will proceed faster. Conversely, if the reaction is exothermic, nonisothermal operation may decrease conversion with respect to that predicted for an isothermal PSR.

A nonisothermal PSR model was invoked by Ding and Alpay (2000a) for the steam reforming of methane in the presence of an adsorbent for CO_2. The objective of their investigation was the validation of Xu and Froment (1989) model for the reforming kinetics over Ni catalysts. Ding and Alpay employed a pseudo-homogeneous, one-dimensional model such as given above (Equation 9.33) for their mass balance. For their energy balance to allow for heat loss through the reactor walls, they used

$$C_{p_g}\frac{\partial(\rho_g T)}{\partial t} + \rho_b C_{p_s}\frac{\partial T}{\partial t} = k_z\frac{\partial^2 T}{\partial z^2} - C_{p_g}\frac{\partial(u\rho_g T)}{\partial z} - (\rho_b)_{\text{ads}}\sum_{i=1}^{N}\Delta H_{a_i}\frac{\partial q_i}{\partial t}$$
$$- (\rho_b)_{\text{cat}}\sum_{i=1}^{N_R}\Delta H_j \eta_j R_j + \frac{4h_W}{d_R}(T_W - T) \tag{9.40}$$

In this balance, the index j indicates the jth reaction, R_j is the rate of that reaction based on catalyst weight, η_j is the effectiveness factor and ΔH_j the heat effect associated with the reaction. The overall heat transfer coefficient for heat loss through the reactor wall is h_W and T_W is the wall temperature, while d_R is the reactor diameter. A linear driving force model was employed for the rate of adsorption in Equations 9.33 and 9.40. The appropriate boundary conditions from Table 9.4 were used. For the simulation, the 22-cm reactor modeled was divided into 100 elements and a second order collocation, implemented using gPROMS, was employed for solution of the system of equations. Simulation and experiment are compared in the next section. The kinetics proposed by Xu and Froment (1989a,b) for steam reforming of methane at steady state are

$$R_{\mathrm{I}} = \frac{k_1}{P_{\mathrm{H_2}}^{2.5}\varsigma^2}\left(P_{\mathrm{CH_4}}P_{\mathrm{H_2O}} - \frac{P_{\mathrm{H_2}}^3 P_{\mathrm{CO}}}{K_1}\right) \tag{9.41}$$

$$R_{\mathrm{II}} = \frac{k_2}{P_{\mathrm{H_2}}^{3.5}\varsigma^2}\left(P_{\mathrm{CH_4}}P_{\mathrm{H_2O}}^2 - \frac{P_{\mathrm{H_2}}^4 P_{\mathrm{CO}}}{K_2}\right) \tag{9.42}$$

$$R_{\mathrm{III}} = \frac{k_3}{P_{\mathrm{H_2}}\varsigma^2}\left(P_{\mathrm{CH_4}}P_{\mathrm{H_2O}} - \frac{P_{\mathrm{H_2}} P_{\mathrm{CO}}}{K_3}\right) \tag{9.43}$$

where $\varsigma = 1 + K_{\mathrm{CO}}P_{\mathrm{CO}} + K_{\mathrm{H_2}}P_{\mathrm{H_2}} + K_{\mathrm{CH_4}}P_{\mathrm{CH_4}} + K_{\mathrm{H_2O}}P_{\mathrm{H_2O}}/P_{\mathrm{H_2}}$. The rate constant, k_j, and the equilibrium constants, K_j and K_i, are temperature functions that are summarized in the Xu and Froment papers. These kinetics have been widely used, but are without validation for dynamic processes. Adsorption data for the hydrotalcite was obtained in a separate study and fitted to a Langmuir isotherm (Ding and Alpay, 2000b).

Probably the most comprehensive theoretical investigation of pressure swing reactors has been contributed by Rodrigues' University of Porto team (Xiu et al., 2002a,b, 2003a,b) who considered hydrogen production via the steam reforming of methane. In their first paper, they examined the five-step cycle used experimentally by Carvill et al. (1996) and later by Hufton et al. (1999) involving (1) feed and production, (2) depressurization or blowdown, (3) low pressure, countercurrent purge with a reactant (CH_4), (4) low pressure, countercurrent purge with product (H_2), and (5) countercurrent repressurization with product (H_2). Feed, purge and repressurization temperatures were identical as was the wall temperature of the reactor bed. For their model of this PSR operation and, indeed, for other systems considered in later papers, Xiu et al. employed Equations 9.33–9.35 and 9.40. Their treatment followed Ding and Alpay (2000a) except that they assumed adsorption was governed by the rate of mass transfer and used a linear driving force model to represent that rate. Thus, their model was a heterogeneous one. They employed the kinetic model proposed by Xu and Froment (1989) for steam reforming given above as Equations 9.41–9.43. Different adsorption equilibria relations were used in the presence or absence of steam in the gas phase. Unlike Ding and Alpay, Xiu et al. (2002a) simulated the entire PSR cycle.

For their simulation of the five-step PSR steam-reforming process, Xiu et al. chose a 2-m reactor length, step durations of 1000 s for t_1, 900 s for t_3 and 100 s for $t_2 = t_4 = t_5$. They assumed a 6:1 steam to CH_4 ratio, $T_f = T_p = T_W = 723$ K, $P_f = 445.7$ kPa, $P_e = 125.7$ kPa, volumetrically equal amounts of catalyst and adsorbent, but with $(\rho_b)_{\mathrm{cat}} = 249$ kg/m^3 and $(\rho_b)_{\mathrm{ads}} = 498$ kg/m^3. Velocities at the two entrance ports for the start of the individual steps were $(u_1)_0 = 0.05$ m/s, $(u_3)_0 = (u_4)_0 = 0.2$ m/s. These parameters are reasonably representative of the Hufton et al. (1999) experiments discussed at the end of the next section. Figure 9.17 shows profiles of pressure, temperature and superficial velocities within the 2-m reactor at the end of the cycle steps for the 1st to 15th cycles. Variations disappear after the second or third cycles except for step 1 where they persist for more than five cycles. By the eighth cycle, the average H_2 mole fraction in the product has become constant at about 0.81. The simulation showed that CO_2 is present mainly in the first half of reactor. Concentrations in the gas phase are highest at the end of the second step and drop continuously in the third to fifth steps as might be expected. At the end of the feed step, the front 2/3 of the adsorbent

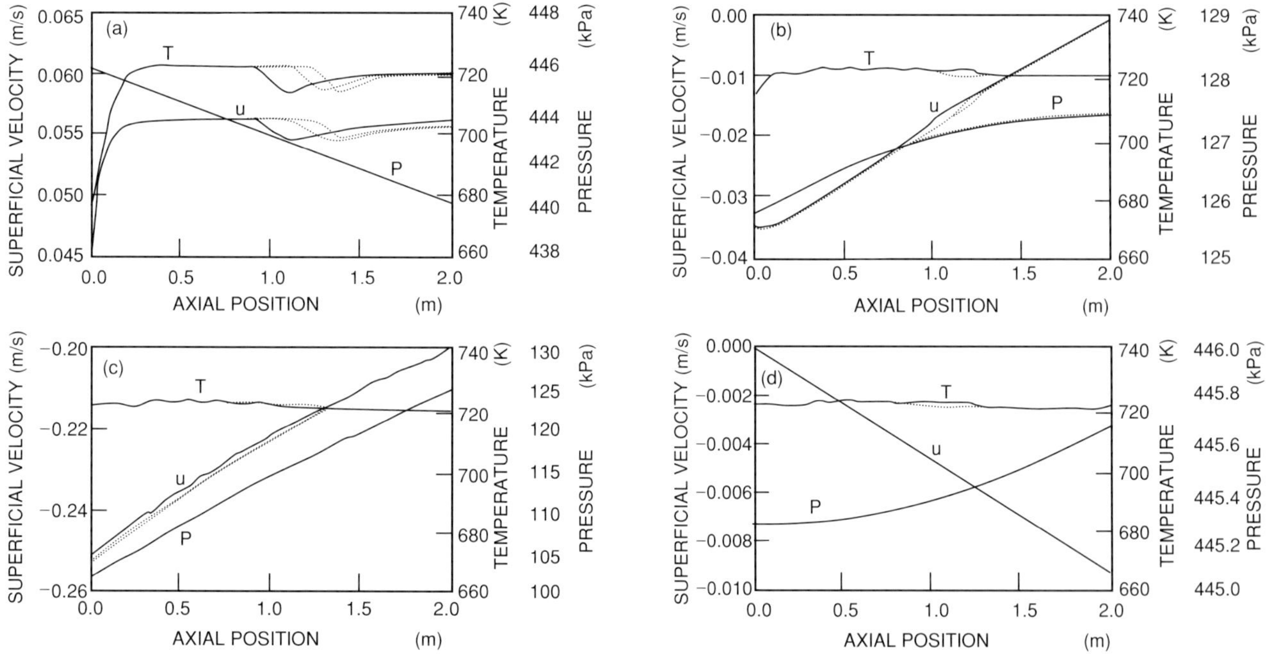

Figure 9.17 Pressure, temperature and superficial velocity profiles in a five-step PSR system running the steam reforming of methane. Operating conditions are stated in the text. Solid lines are profiles at the end of the 1st cycle, dashed lines are profiles after the 2nd, 5th and 15th cycles: (a) at the end of step 1, (b) at the end of step 2, (c) at the end of steps 3 and 4, (d) at the end of step 5. (Figure adapted from Xiu et al. (2002b) with permission. © 2002 AIChE.)

bed is saturated with CO_2. Some CO_2 is present in the front half of the reactor even after the two purge steps.

Simulation also disclosed that for a set of operating conditions increasing the packing density, the adsorbent to catalyst ratio and the residence time enhances both purity and conversion. Residence time can be increased by using a longer reactor, but Xiu et al. point out that this makes regeneration of the adsorbent difficult because of pressure loss or, alternatively, lengthens the regeneration time. Xiu et al. used their simulation to represent the experimental results of Ding and Alpay (2000a) and Hufton et al. (1999) that are discussed in the following section. They report good agreement with these experiments.

In an extension of the simulation just discussed, Xiu et al. (2003a) investigated the effect of catalyst and adsorbent particle size on PSR performance. The authors, however, considered just the feed or production step and not the entire cycle. They added to the model discussed above a material balance on a shell within the spherical catalyst or adsorbent particle assuming that catalytic sites are isotropically distributed. Thus,

$$\varepsilon_{\text{ads}} \frac{\partial C_i}{\partial t} + (1 - \varepsilon_{\text{ads}})(\rho_b)_{\text{ads}} \frac{\partial q_i}{\partial t} = \frac{1}{r_p^2} \frac{\partial}{\partial r_p} \left((D_i)_{\text{eff}_{\text{ads}}} r_p^2 \frac{\partial C_i}{\partial r_p} \right) \tag{9.44}$$

$$\varepsilon_{\text{cat}} \frac{\partial C_i}{\partial t} = \frac{1}{r_p^2} \frac{\partial}{\partial r_p} \left((D_i)_{\text{eff}_{\text{cat}}} r_p^2 \frac{\partial C_i}{\partial r_p} \right) + (\rho_b)_{\text{cat}} \sum_{j=1}^{n} \nu_{ij} R_j \tag{9.45}$$

In these relations, the concentrations, C_i, within the porous particles will rarely be the same in the catalyst and the adsorbent. Also $(d_p)_{\text{cat}} \neq (d_p)_{\text{ads}}$ so that the particle radius, r_p, in the two models will be different. The effective diffusivities, $(D_i)_{\text{eff}_{\text{cat}}}$ and $(D_i)_{\text{eff}_{\text{ads}}}$, depend on pore structure and the bulk diffusivities. Formulas for their calculation are readily available (Froment and Bischoff, 1990; Westerterp et al., 1993). Xiu et al. assume rapid gas–solid heat transfer and isothermal particles. They also modify their bed material balance to include a storage term for component i within the porous catalyst and adsorbent particles. Evaluation of $\partial q_i/\partial t$ and $\sum_i \sum_j \nu_{ij} R_j$ as well as the storage term in the bed balances require the solution of Equations 9.44 and 9.45 for C_i and q_i. From these profiles as a function of radius, average values of C_i, $\partial q_i/\partial t$ and R_j can be obtained by integrating over the sphere. These average values may then be used in the bed heat and material balances, e.g. in Equations 9.33 and 9.40. Alternatively, effectiveness factors may be estimated from the ratios of the average values and the values calculated for the external surface of a particle.

For their simulations, Xiu et al. (2003a) chose to use a 2-m-long bed and 3-mm spherical particles, with $\varepsilon_p = 0.5$, $\varepsilon_b = 0.47$, and twice the amount of adsorbent as catalyst. Operating conditions chosen for the feed step were those used in their earlier study discussed above. Mass transfer was found to be important as Figure 9.18 illustrates. Note that the authors did not calculate effectiveness factors for adsorption. There is justification for this. Once product or reactant enters the adsorbent particle, it is no longer available for reaction even if it is not adsorbed.

Reaction rates in Figure 9.18a are always high at the entrance of the reactor. After 100 s on stream, they fall as methane is consumed. Adsorption rates are low initially because little CO_2 is available. They increase downstream so that by $z = 0.1$ they are comparable to rates of reaction. Adsorption becomes important then in maintaining the rates of reaction. This can be seen as a decrease in the slope of the rates around $z = 0.1$. Adsorbent at the entrance

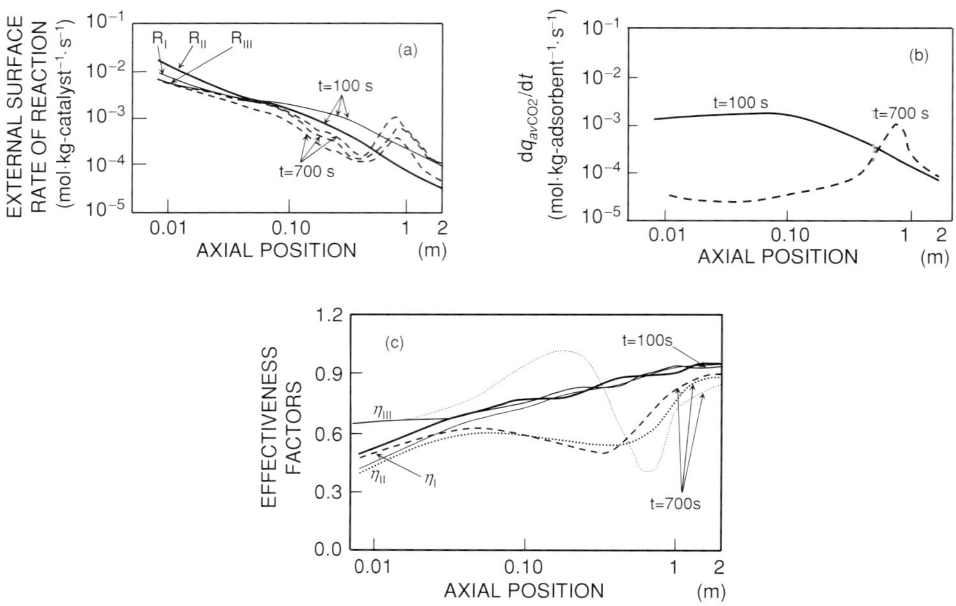

Figure 9.18 Variation of external surface rates of reaction (a), adsorption (b) and particle effectiveness factors (c) with axial position and time after the start of the feed step for steam reforming of methane in a PSR. Operating conditions are the same as in Figure 9.17. Subscript I refers to the reaction given by Equation 9.41, II to Equation 9.42 and III to Equation 9.43. (Figure adapted from Xiu et al. (2003a) with permission. © 2003 Elsevier Science B.V.)

is saturated with CO_2, but this does not affect reaction rate. After 700 s on stream, reaction rates at the entrance remain high, but now drop by $z = 0.1$. Figure 9.18 shows adsorption rates are low at this point as the adsorbent is approaching saturation. The adsorption front has moved downstream to $z = 1$. The rate increases there and again becomes comparable to the reaction rates. Adsorption is now contributing to methane conversion. These time and position variation of rates causes an erratic variation of effectiveness factors that is evident in Figure 9.18c. This figure shows these factors are well below unity so intraparticle diffusion and external surface mass transfer are important. Effectiveness factors for reforming ranged from about 0.4 to about 0.95 depending on position in the reactor, whereas those for the water gas shift ranged from 0.7 to 0.95. The lower values occurred at the front end of the reactor where rates are high. Consequently, although concentration gradient exists in the catalyst particles, the effect on rate of reaction per weight of the catalyst is important only in the first 10% of bed length. Because rate is high in this region, intraparticle diffusion does not have a large effect on the length of the reactor needed to meet the H_2 purity or the CO constraint on the product gas. Conversion and productivity are affected though.

Because of the computing burden imposed by considering intraparticle diffusion, Xiu et al. (2003a) recommend assuming constant effectiveness factors for the rate expressions. They suggest calculating these factors at the beginning and end of the feed step and using an average for a simulation. They demonstrate that the error introduced by this approximation is quite small.

9.7 Experimental studies

Unfortunately, experimental work has always lagged behind theoretical studies of pressure swing reactors. Vaporciyan and Kadlec (1989) studied the oxidation of dilute CO by O_2 in a carrier gas of N_2 (>95 vol%) over an eggshell Pt catalyst in the presence of a molecular sieve adsorbent. A rapid pressure swing configuration was used with a reflux volume following the catalyst–adsorbent bed. A simulation of experimental results was undertaken assuming that the oxidation reaction was irreversible and was inhibited by CO. No increase in conversion was predicted vis-à-vis a PFR of the same catalyst content, but substantial alterations of the separation behavior of a RPSA with the same adsorbent were projected. We reviewed this work in the previous section.

The experiments undertaken by Vaporciyan and Kadlec (1989) used a mixed bed of a Pt/Al_2O_3 catalyst and much smaller particles of a zeolite 5A adsorbent. By maintaining the CO concentration at about 1.6 vol%, oxidation could be treated as isothermal and at a reaction temperature of 430 K it was irreversible. The investigators employed a 2.1 cm i.d. tubular reactor with an empty reflux volume at the discharge end. The size of the reflux volume could be changed between experiments. Using a shell-impregnated alumina support and adsorbent particles of 32/48 mesh reduced diffusion interference. Small particles are chosen generally for rapid pressure swing adsorbers operating between 0.1 and 0.5 Hz. A schematic of the Vaporciyan–Kadlec apparatus is shown in Figure 9.19. Reactor dimensions and adsorbent–catalyst properties are summarized in Table 9.5.

Objectives of the Vaporciyan and Kadlec study were to understand the behavior of the periodic process and to establish what advantages accrue from this operation mode. To these

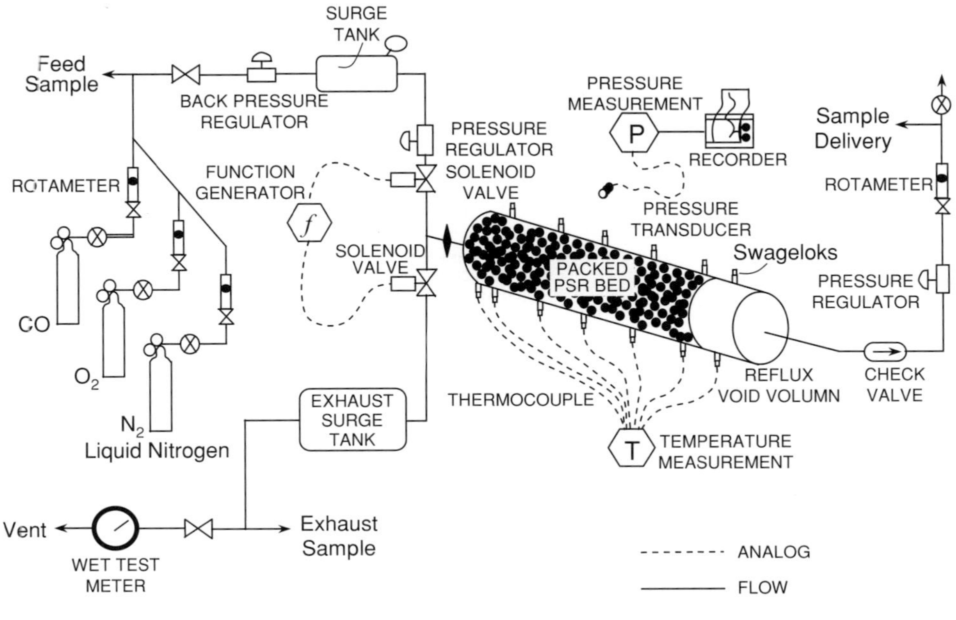

Figure 9.19 Experimental scale rapid pressure swing reactor (RPSR). (Figure adapted from Vaporciyan and Kadlec (1989) with permission. © 1989 AIChE.)

ends, the authors investigated the effect of oxygen and CO concentrations, delivery flow rate, the feed pressure, the ratio of reflux to packed-bed volumes and the cycle parameters. An exhaustive study of the variables was not undertaken.

An effect on the conversion of CO was observed as is evident in Figure 9.20a. As the CO concentration in the feed increases in a PFR, the rate of CO_2 production climbs to a maximum and then sharply decreases. This is a consequence of strong adsorption of CO on

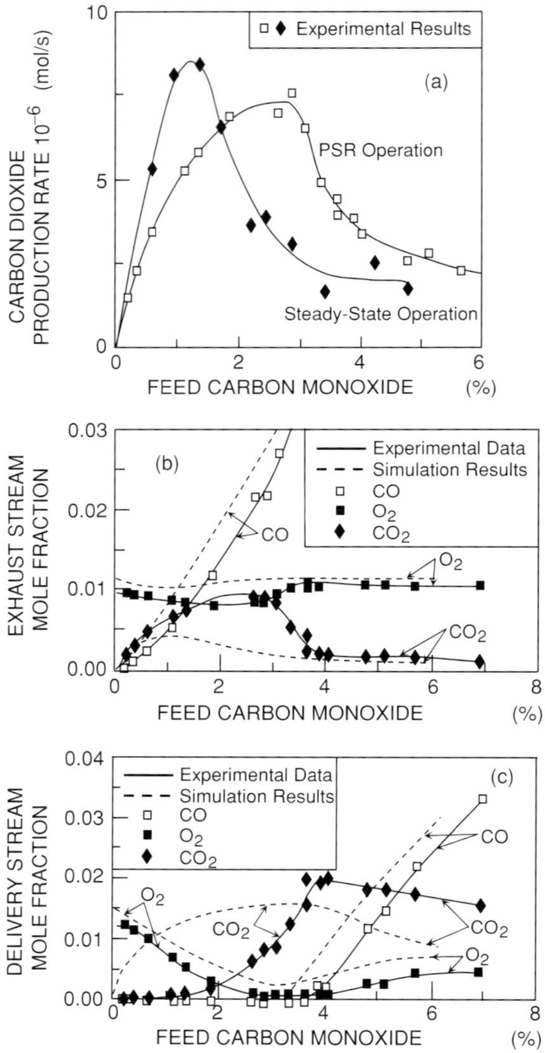

Figure 9.20 Variation of time-average conversion, exhaust and discharge concentrations as a function of CO concentration in the RPSR feed. O_2 concentration was 1.15 vol%. In the figure: (a) CO conversion as CO_2 production rate, (b) exhaust, (c) discharge. (Figure adapted from Vaporciyan and Kadlec (1989) with permission. © 1989 AIChE.)

Pt and reactant inhibition of the oxidation reaction. A maximum in CO_2 production is also observed in a PSR, but the rate maximum is shifted to a significantly higher CO concentration in the feed. Nonetheless, this maximum production rate is always less than that for a PFR containing just catalyst. Vaporciyan and Kadlec explain the shift of the maximum by the change in the CO/O_2 adsorbate ratio on the catalyst surface due to the adsorption of CO on the zeolite in the pressurization step and its release for reaction on depressurization. Thus, periodic operation increases the residence time of CO in the bed. The absence of a global increase in the CO_2 production rate was predicted by the authors' simulation and is not surprising on physical grounds because reactant separation by the adsorbent lowers the reaction rate. Strong adsorption of the CO_2 product can have no effect, as the reaction is irreversible. Lack of a rate increase contrasts with the periodic reactor results of various authors (Silveston, 1998) in which rate enhancements of more than twofold have been observed. Those reactors, however, did not contain an additional adsorbent.

Vaporciyan and Kadlec assess the effect of the system variables on reactant and product separation using a separating factor, defined in their study as the ratio of the time average mole fraction of a system component in the discharge and in the exhaust. Note that this is not the definition used in Table 9.3. They report that the separating factor for oxygen changes from very small values to about 1.3 as O_2 concentration in the feed increases from nearly 0 to about 2 vol%. Thereafter it does not change. At low concentrations, O_2 is enriched in the exhaust stream, while above a concentration of about 2 vol% it is enriched in the discharge stream. This is the inversion effect discussed in the previous section. Above 0.5 vol% O_2 in the feed, CO is not present in the discharge stream. CO_2 behaves inversely to O_2 and attains a separating factor of about 0.1 at 2 vol% O_2. Location of the reaction front as a function of the O_2 and CO concentrations account for the O_2 and CO_2 inversions. Increasing O_2 concentration shifts the reaction front toward the entrance of the bed. CO_2 formed in the front is trapped by the adsorbent. On depressurizing with gas from the reflux volume flowing toward the entrance, CO_2 desorbs and is carried into the exhaust stream. Weakly adsorbed oxygen on the other hand penetrates the reaction front and is present in the discharge stream.

Changing the O_2 concentration has little effect on CO inhibition; it is the CO concentration in the feed that matters. Figure 9.20b and c shows how the increasing concentration of CO alters the distribution of products in the PSR exit streams. Inversion, that is, reversal of product–reactant exit locations can be seen. At low CO concentrations, only O_2 appears in the discharge stream. Just as in a PSA, CO_2 adsorption forces CO_2 to the exhaust and CO is largely consumed. When the CO feed fraction reaches 0.02, reactants and product appear in the exhaust at equal mole fractions. A further increase switches CO_2 to the discharge port as CO displaces CO_2 on the catalyst surface at the front of the reactor. Between 2 and 4 vol% CO in the feed, the discharge stream contains virtually only CO_2. Above 4 vol% CO, CO breakthrough at the discharge port occurs. Reaction inhibition also raises the O_2 mole fraction in the discharge stream leaving the RPSR.

Simulation results are also shown in Figure 9.20b and c. There is a reasonable agreement between measurements and predictions at feed CO concentrations up to 1 vol%. Between 2 and 4 vol%, prediction of the critical CO_2 concentration displays considerable error. Vaporciyan and Kadlec attribute this error to the failure of their simulations to reach a cyclic stationary state. They report, however, temperature variations of up to 5°C along

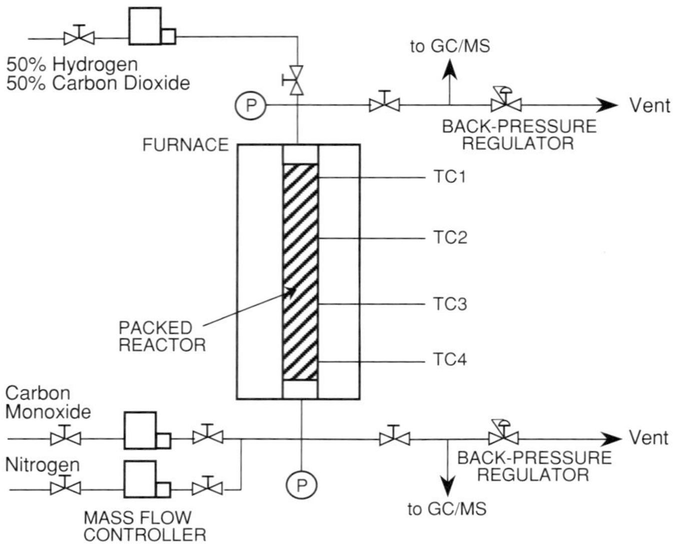

Figure 9.21 Schematic of the bench scale PSR used for CO production from CO_2. (Figure adapted from Carvill et al. (1996) with permission. © 1996 Air Products and Chemicals, Inc.)

their reactor as well as moving temperature fronts. An isothermal model was used in their simulations.

For the irreversible CO oxidation reaction, Vaporciyan and Kadlec's measurements show that the periodic separating reactor operates primarily as a reactant–product separator with no important effect on conversion.

Carvill et al. (1996) describe a six-step PSR cycle illustrated in Figure 9.2e. Feed is introduced into a reactor already at the feed pressure. The discharge port is opened and the desired product, just weakly adsorbed, is removed through this port. This step is followed by depressurization through the open exhaust port with the discharge port closed. The reactor is flushed countercurrent to the feed direction with a nonadsorbing carrier gas in step 3. This is followed by a second purge, but now using the desired product collected at the discharge port. In the fifth and final step, the exhaust port is closed and the product now pressurizes the reactor up to the feed pressure.

This system was demonstrated at a bench scale by the authors for CO production from CO_2 by the reverse water gas shift reaction: $CO_2 + H_2 \leftrightarrow H_2O + CO$. A commercial low-temperature shift catalyst was used along with a NaX zeolite as an adsorbent for water. A 3.8 cm i.d. by 53 cm long tube served as the reactor (see Fig. 9.21) and was packed with 0.5 kg of a 1:1 mixture by weight of catalyst and adsorbent. The experiments were carried out isothermally using an equimolar mixture of CO_2 and H_2 at 523 K and 480 kPa. In the PSR cycle, pressure at the exhaust port in the depressurization and purge steps was held to between 115 and 129 kPa.

Table 9.6 compares the PSR experimental results with an infinitely long PFR operating at 838 K so as to reach the 36% equilibrium conversion of CO_2 obtained in the PSR. At 523 K, the equilibrium conversion of CO_2 to CO is 9.8%. The PSR product was essentially pure CO at 480 kPa. Effluent from the PFR would have to be purified to give 99% CO. Probably

Table 9.6 Performance of the five-step PSR for CO production from CO_2 by the reverse water gas shift reaction with ideal PFR performance as a comparison

	PSR	PFR
Operating temperature (isothermal) (K)	523	838
Operating pressure (kPa)	480	480
CO_2 conversion (%)	36	36
Product composition (vol%)		
CO	99	22
CO_2		39
H_2		39
CO product pressure (kPa)	480	45–90

a PSA unit would be used to deliver the product at 45–90 kPa. Carvill et al. point out that the conversion measured in their reactor depended on bed length relative to the length of the reactive front moving through the reactor. Higher conversion could have been achieved by using a longer reactor.

Figure 9.22 shows the time variation of CO in the discharge during the feed step of the cycle. The CO produced up to 30 min (hatched area in the figure) was CO stored in the bed during the backfill step. The second rectangular area, from 30 to 67 min, is the CO produced in the cycle. After 67 min, reactants break through to the discharge. An experiment in which backfill used either CO_2 or H_2 showed that high-purity CO could not be produced. The Carvill et al. process offers a low temperature, single vessel process for a relatively pure product delivered at a high pressure.

Disadvantages of the process are that it delivers a mixture of reactants and products in the exhaust step and a nonadsorbing carrier gas is required.

Sheikh et al. (2001) undertook an experimental study of dehydrogenation of 1-butene to butadiene using CrO_2/Al_2O_3 as a catalyst and KY zeolite as an adsorbent in an isotropic bed of catalyst and adsorbent with a 1:1 ratio by volume. A 4 mm i.d. by 120 mm tubular reactor was used. The investigators employed a two-step RPSR system with two variations of the depressurization step: (1) depressurization through the exhaust (countercurrent

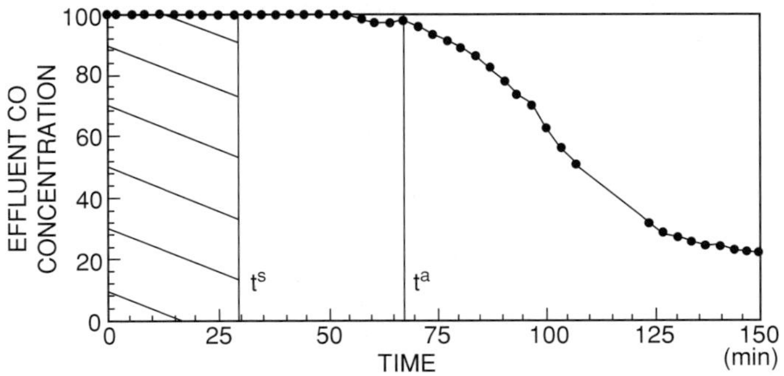

Figure 9.22 Time variation of CO vol% in the PSR discharge for the production of CO from CO_2 using the reverse water gas shift reaction with selective adsorption of water. (Figure adapted from Carvill et al. (1996) with permission. © 1996 Air Products and Chemicals, Inc.)

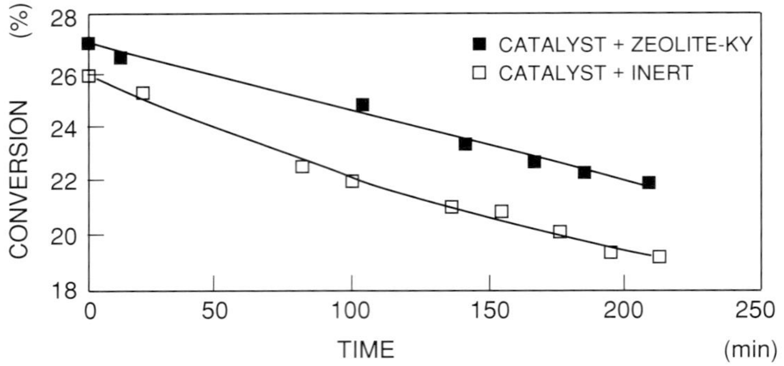

Figure 9.23 Comparison of butene conversion at 763K in a two-step RPSR with conversion in an equivalent PFR for an equilibrium-limited catalytic dehydrogenation as a function of time on stream for a feed pressure of 2 bar, a feed flow rate of 1.5 mL/s with cocurrent depressurization. (Figure adapted from Sheikh et al. (2001) with permission. © 2001 Elsevier Science Ltd.)

mode) or (2) depressurization through the product exit (cocurrent mode). Fast operation was used with equal periods of pressurization and depressurization of 1 s. The only variable in the experiment was the feed rate to the unit. Experiments were also undertaken on a PFR version of the reactor in which the zeolite adsorbent was replaced by a nonadsorbing silica. The experimental results failed to confirm the simulation study that was described earlier. Butene conversion in the countercurrent mode was lower than conversion in an equivalent PFR, but the mole fraction of butadiene in the product (on an inert and H_2 free basis) was only modestly greater. The authors attributed the lack of agreement between simulation and experiment to an error in the adsorption equilibrium measurement. Apparently the KY zeolite adsorbed butene much more strongly than anticipated. Nevertheless, experiments showed that cocurrent depressurization increased butene conversion over that in an equivalent PFR. Furthermore, the rate of catalyst deactivation was lowered as illustrated in Figure 9.23. Catalyst deactivation is due to coke deposition presumably. The Sheikh et al. explanation for the improved performance under cocurrent depressurization is that butadiene is formed in their catalyst–adsorbent system primarily in the latter part of the reactor. The countercurrent mode remixes this product with unreacted butene.

The simulations of Gomes and Yee (2002) were tested experimentally in a continuously operated two-column PSR unit shown schematically in Figure 9.24. The columns had an inside diameter of 3.5 cm and were 7.5 cm in length. The system was computer-controlled through two mass flow controllers and sampling of the product and exhaust streams was automatic. Catalyst was rhenium oxide on γ-Al_2O_3 and the adsorbent was zeolite 13X, which has a high capacity for butene. Equal masses of catalyst and adsorbent were used. Experiments were carried out at a feed pressure of 3 bar and a purge pressure of 1 bar. Purge to feed volumetric ratio (STP) was 0.8 and the feed contained 33.3% propene in helium. The experiments were conducted at ambient temperature.

Fixed-bed experiments under the PSR conditions show insignificant recovery of ethene, however, equilibrium conversion indicates a conversion of 25%. It appears that the catalyst rapidly deactivated for the feed compositions used. PSR operation for a cycle of 240 s resulted in a product stream containing 5% propene whereas butene at the cyclic steady state was 1% in the same stream. Figure 9.25 compares the simulation with the experimental results.

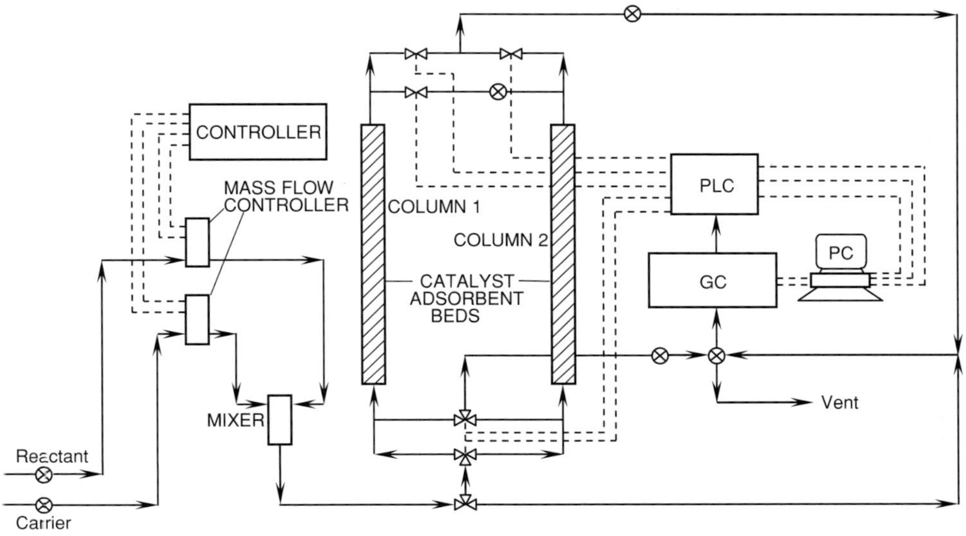

Figure 9.24 Dual column experimental unit for metathesis of propene in a four-step PSR cycle. (Figure adapted from Gomes and Yee (2002) with permission. © 2002 Elsevier Science Ltd.)

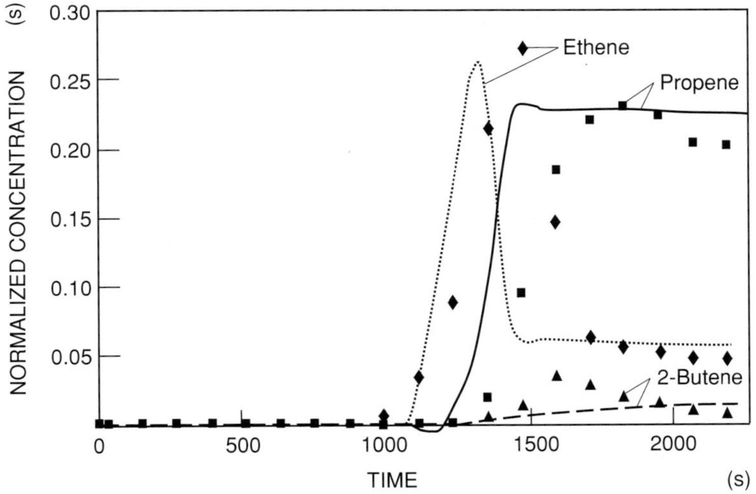

Figure 9.25 Comparison of simulated and experimental compositions for the product stream at a cycle time of 240 s in a four-step PSR. (Figure adapted from Gomes and Yee (2002) with permission. © 2002 Elsevier Science Ltd.)

Both indicate a relatively long time was needed to reach the cyclic stationary state. About 12 cycles are necessary. Due to the weak adsorption of ethene, ethene elutes first and its concentration goes through a maximum during startup. It can be seen from the figure that the simulation represents the experimental behavior adequately.

It is no surprise, of course, that this PSR operation substantially improves reactor performance. Concentration of ethene in the product stream increased by about fivefold, while the yield of ethene rose by about fourfold compared to packed-bed performance in the absence of an adsorbent.

The previous section discussed briefly the simulation of methane steam reforming over a nickel catalyst and a potassium promoted hydrotalcite adsorbent for CO_2 by Ding and Alpay (2000a,b). These researchers also undertook an experimental investigation using a 12.4 mm i.d. by 220 mm tubular reactor packed with 7.2 g of the MgO-$Na_2Si_2O_4$-Ni-NiO catalyst and 14.8 g of the commercially available adsorbent. Additional experiments were undertaken in the same reactor, in which a SiC inert volumetrically replaced the adsorbent. Most runs were performed at a 3:1 steam to CH_4 ratio, 523 K and 455 kPa. A limited study of the effect of the forenamed variables was undertaken. Only the feed or production step of the cycle was studied. However to regenerate the hydrotalcite adsorbent, a PSR cycle was followed. After breakthrough occurred in the feed step, the reactor was depressurized, and then purged with H_2 and steam. Finally, the bed was repressurized with H_2 at 753 K to ensure the catalyst was reduced.

Temperature distribution was measured after methane was introduced to the feed and one measurement well before breakthrough is shown in Figure 9.26. A low temperature wave associated with the endothermic reforming reaction develops and moves downstream in the reactor. Endothermicity and heat loss drop the temperature in the front end of the reactor by more than 60°C. Clearly, most PSR systems will operate nonisothermally.

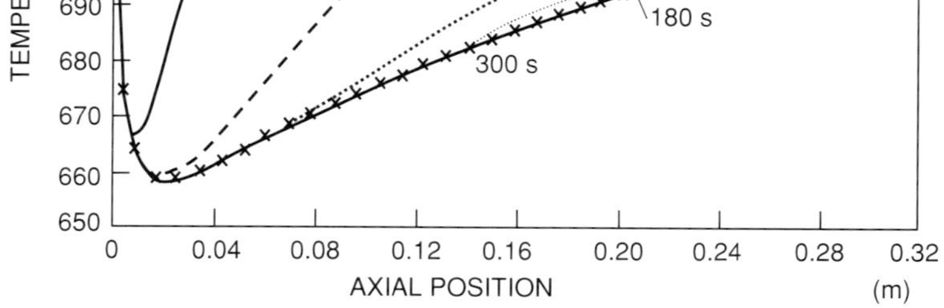

Figure 9.26 Temperature profiles as a function of time after start up for the production step in the steam reforming of CH_4 over a mixed bed of a nickel catalyst and a hydrotalcite adsorbent in a PSR cycle with a feed temperature of 723K, steam/CH_4 ratio of 6:1 and a pressure of 446 kPa. (Figure adapted from Ding and Alpay (2000a) with permission. © 2000 Elsevier Science Ltd.)

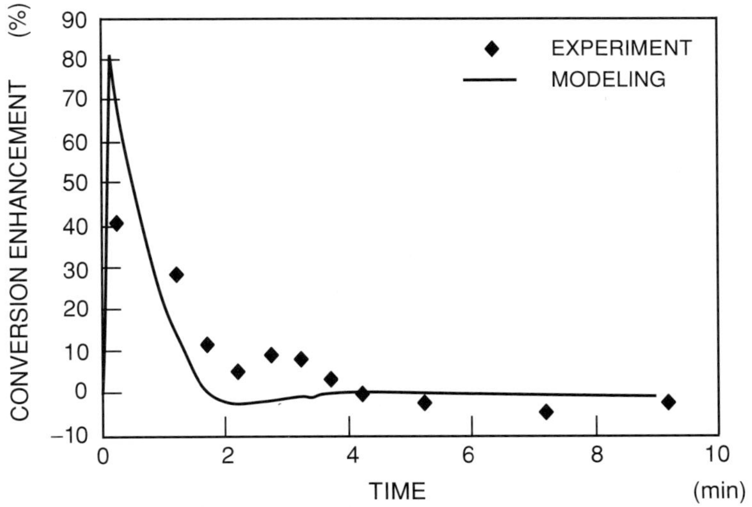

Figure 9.27 Comparison of predicted and measured enhancement of end-of-reactor CH_4 conversion for reactor operation with a Ni catalyst when a CO_2 adsorbent replaces an inert solid diluent in the reactor. Operating conditions are as given in Figure 9.26. (Figure adapted from Ding and Alpay (2000a,b) with permission. © 2000 Elsevier Science Ltd.)

The experiments performed by Ding and Alpay demonstrate that the steady-state kinetic model of Xu and Froment (1989) for steam reforming of methane is adequate for simulation of a PSR operation. The end-of-reactor conversion enhancement predicted by the PSR model was compared to the enhancement measured experimentally. Figure 9.27 presents the comparison as a function of time after startup. Allowing for the inevitable mixing and sampling problems, the agreement is good.

Ding and Alpay observed that conversion enhancement through the use of an adsorbent, i.e. a PSR operation, improves as space time and pressure increase, but drops off as the steam to CH_4 ratio rises. At high ratios, methane conversion is forced to completion so that use of an adsorbent contributes little to the process. Temperature affects equilibrium conversion and kinetics as well as adsorption so that there is an optimal temperature for the mixed catalyst–adsorbent bed.

Production of hydrogen in a PSR system has attracted much recent attention probably because recent commercial hydrogen technology incorporates pressure swing adsorption (PSA). New hydrogen plants reform natural gas at up to about 4.2 MPa, between 1020 and 1170 K using a steam–CH_4 ratio of about 3–6. The product from the reformer is cooled in a waste heat boiler to 570–670 K and then goes to a high-temperature water gas shift reactor where the CO content is forced to 1–4 vol%. A second waste heat boiler reduces the temperature further and a low temperature shift reactor operating at about 525 K drops CO into the ppm range. A chiller-condenser reduces the temperature to below 320 K. The chilled gas is compressed and fed to a PSA battery. This adsorption process provides a high purity H_2 stream (98 to >99.9 % H_2) at a H_2 recovery well above 70%. The waste stream from the separation has a moderate heating value so that, along with some additional natural gas, it supplies the energy demand of the endothermic reforming reaction.

At about the same time as the Ding and Alpay study, Air Products and Chemicals in the United States began an examination of combining catalyst and a CO_2 adsorbent in a system they call "sorption-enhanced reaction process" or "SERP". The objective, it seems, was to reduce the process capital cost.

The work at Air Products has been described in publications by Hufton et al. (1999) and Waldron et al. (2001). The former paper deals primarily with the properties of the K_2CO_3 promoted hydrotalcite developed as a sorbent for CO_2. This material and other similar oxide/hydroxide materials are not strictly adsorbents. The base material undergoes a gas–solid reaction with the formation of a carbonate, which decomposes during the regeneration step. Hufton et al. view these gas–solid reactions as adsorption and desorption of CO_2. They demonstrate that the processes are rapid so that they can be described in terms of an adsorption/desorption equilibria. They found that the equilibrium followed a Langmuir isotherm with $K = 0.393\ \text{kPa}^{-1}$. The material exhibits a capacity of about 0.45 mol CO_2/kg of hydrotalcite at 673 K at a partial pressure of 30 kPa CO_2 after undergoing about 15 adsorption/desorption cycles (see Fig. 9.29). Several experiments were carried out using the equipment shown in Figure 9.28 that was a modification of the Carvill system seen in Figure 9.21. Hufton and coworkers used a 38 mm by 1067 mm tube packed with a uniformly mixed commercial Ni-reforming catalyst and hydrotalcite adsorbent in a 1:1 ratio by weight. Particle size of both catalyst and adsorbent was 3 mm. A five-step cycle was used: (1) a feed and production step using a 6:1 mixture of steam and CH_4 at 480 kPa and 723 K, (2) a depressurization step through the feed end, which was held at ambient pressure, (3) a countercurrent purge of CH_4 or steam at 115 kPa and 723 K, (4) a second countercurrent purge using the H_2 product, followed by (5) repressurization with the H_2 product. Reactor

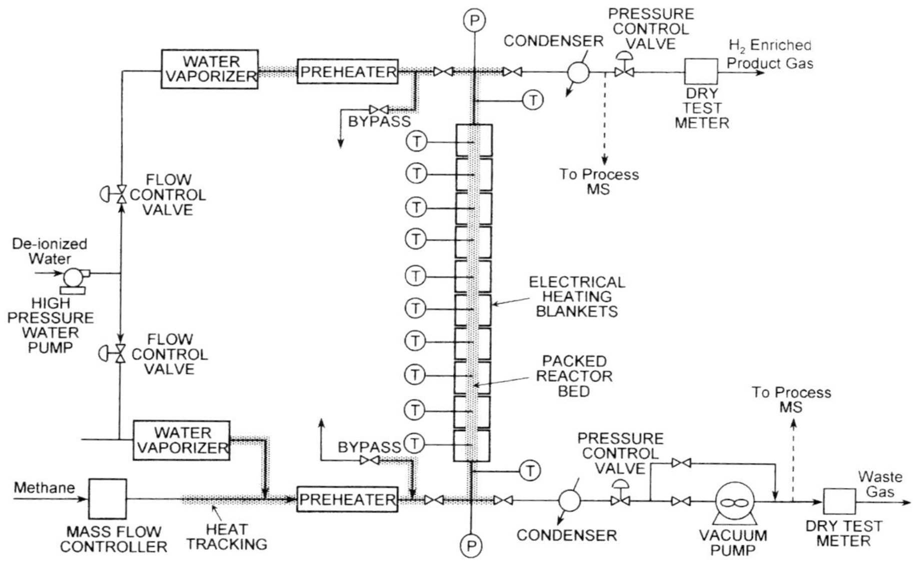

Figure 9.28 Schematic diagram of the miniature pilot plant for hydrogen production from the steam reforming of methane using a four-step isothermal PSR system. (Figure adapted from Waldron et al. (2001) with permission. © 2001 Air Products and Chemicals, Inc.)

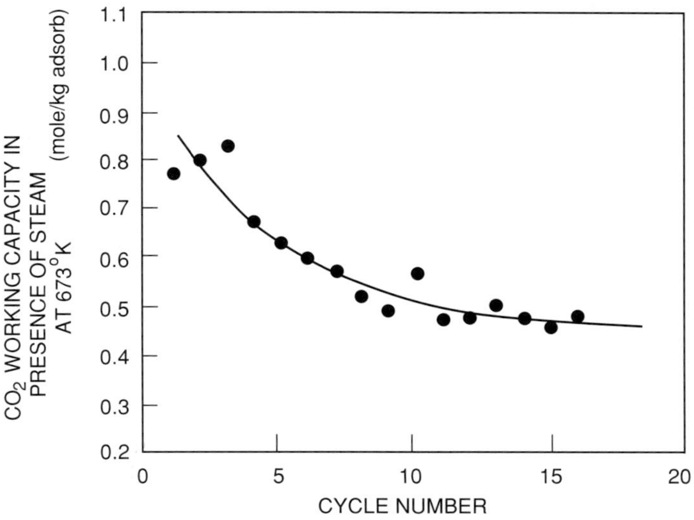

Figure 9.29 Loss of sorbent capacity with successive cycles for steam reforming of CH_4 over a 1:1 mixture of a Ni catalyst and a promoted hydrotalcite adsorbent with a 6:1 steam/CH_4 ratio, $T = 723$K, $P_H = 480$ and $P_L = 101$ kPa. (Figure adapted from Hufton et al. (1999) with permission. © 1999 Air Products and Chemicals, Inc.)

wall was maintained at 723 K, the feed temperature, but the bed did not operate isothermally. The single result reported by Hufton et al. showed a product stream containing 96 vol% H_2 with less than 50 ppm of carbon oxides at a CH_4 conversion of 82%. Productivity was 0.8 mole of H_2/kg catalyst. A packed-bed reactor operating under the same conditions had a conversion of just 28% and the H_2 mole fraction in the product gas was 0.53. Loss of sorbent capacity concerned the Hufton team and was studied. Figure 9.29 shows the loss of capacity with successive cycles. Capacity reaches a plateau after about 15 cycles. Loss of capacity affects productivity and alters cycle period and split. Its effect on product composition is minor.

A second publication (Waldron et al., 2001) discusses a four-step PSR operation involving (1) a feed step using a 6:1 steam/CH_4 mixture at 763 K, (2) a depressurization step through the feed end, maintained at 103 kPa, (3) a low pressure countercurrent purge step using 763 K steam at 69 kPa with the exhaust end maintained at 34 kPa and (4) repressurization using steam. This system was studied experimentally in a small continuously operated pilot unit illustrated schematically in Figure 9.28. The reactor consisted of a heated stainless steel tube 25 mm i.d. by 61 mm. Although operated at various temperatures, only data at 763 K has been reported. The reactor was packed with 0.9 kg of a noble metal/Al_2O_3 catalyst with 1.8 kg of potassium-promoted hydrotalcite as an adsorbent. A feed load for most operations was 38 kg-mol/h/m^2 and the feed pressure ranged between 179 and 455 kPa. Steam flow rate varied between 37 and 45 kg-mol/h/m^2. Step durations varied from run to run but the cycle period ranged between 630 and 1130 s.

Regeneration of the hydrotalcite was carried out by passing superheated steam through the reactor. This provided the energy input for CO_2 desorption. Nevertheless, the process developed by Air Products is a pressure swing one, albeit a nonisothermal one.

Table 9.7 Performance of a four-step nonisothermal PSR experimental unit for hydrogen production (Waldron et al., 2001)

Feed pressure (kPa)	Gas quantities (g mol/kg solid)			Hydrogen product purity (dry basis)				CH_4 conversion to H_2
	Feed	Purge (steam)	H_2 product	H_2 (%)	CH_4 (%)	CO_2 (ppm)	CO (ppm)	
180	0.60	0.75	0.20	88.6	11.4	60	Not detected	59
178	0.76	1.84	0.29	90.0	10.0	20	"	66
181	0.60	1.88	0.25	94.4	5.6	40	"	73
458	0.54	0.77	0.16	88.7	11.3	136	"	54

Through control of step durations and feed rates, the hydrogen purity was maintained above 88% (dry basis) and the concentration of CO_2 was kept below 130 ppm. About 30 cycles (or 5–9 h) were needed to attain a cyclic stationary state. Table 9.7 summarizes the performance of the miniature pilot plant. As can be seen in the table, raising the amount of purge increases the purity to 94% and improves conversion significantly.

As described briefly in Chapter 4, Goto and coworkers examined various dehydrogenation reactions, such as cyclohexane, *n*-hexane and propane to yield aromatic products using hydrogen storage alloys or metals. Most of their experiments were conducted in a periodically pulsed reactor system, that is, an isobaric operation with a cyclic pulse of reactant imposed on a continuous flow of an inert carrier gas. Partial regeneration of the hydrogen storage alloys or metals took place through the carrier gas when no reactant was present in the feed. However, they also examined regeneration under vacuum (Goto et al., 1993, 1995, 1996). In this case, their operation system becomes a pressure swing system. The authors mention that reducing pressure worked well for the regeneration of the alloys or metals for all reaction systems examined but that the conversion of cyclohexane is slightly less than when a carrier gas purge was used. No explanation of the difference between the carrier gas purge and PSR operation was given. Perhaps, some of the heat necessary for H_2 desorption might have been supplied by carrier gas, in this case helium, which has high thermal conductivity despite its low heat capacity.

Chapter 10
Temperature Swing Reactors

10.1 Introduction

Conventional adsorbers for VOCs and other low concentrations of impurities often use temperature regeneration of the adsorbent. In these systems adsorption occurs at a low temperature, the bed is then flushed with a hot carrier gas to strip the adsorbent and to concentrate the adsorbate in the gas phase. This regeneration technique is applicable to periodically operated separation systems but has not been widely used. The problem is that most of the heat is retained by the solid adsorbent so there is a relatively large energy demand to heat and cool the system. Because the temperatures are low in many adsorption systems, the heat expended is difficult to recover so a considerable amount of energy can be wasted.

The use of flow reversal is a technique to reduce energy loss. Flow reversal was developed in Russia during the 1970s by the Boreskov Laboratory of Catalysis (Boreskov et al., 1977; Boreskov and Matros, 1983) as a technique for preheating reactants using waste heat leaving the reactor exit. In this case, just a single reactor is used. If applied to regenerating adsorbers, at least two separate beds seem to be needed to continuously treat a target stream. Application of flow reversal to adsorption was proposed by Agar and Ruppel (1988a,b) to ensure complete NO_x conversion without ammonia slip. However, the deNO_x unit is not a separating reactor. The adsorbent functions as a chemical trap, holding, in this case, ammonia in the reactor. We will briefly consider trapping reactors in Chapter 12 as they are normally operated cyclically.

Energy loss is a major problem in temperature swing reactors (TSR) so that flow reversal is often implemented. Its use conceptually is nicely illustrated by the Claus reaction. In this reaction,

$$2H_2S + SO_2 \leftrightarrow (3/n)S_n + 2H_2O$$

The reaction is carried out usually above 230°C. Several stages must be used in the conventional process in order to achieve target levels of H_2S removal, which are currently 99.5%. An examination of Figures 9.1 or 9.2 in the previous chapter shows that flow reversal is employed in the depressurization or exhaust step as well as in the purge step of a pressure swing reactor cycle. A two-bed, temperature swing separating reactor is shown in Figure 10.1. The bed consists of a mixture of a commercial alumina catalyst and zeolite 3A molecular sieve. The upper part of the figure shows the forward direction of operation. In this, reactants are fed to the front of the hot bed and exit through a condenser that removes molten sulfur by cooling the gas. The gas then is reheated to a temperature exceeding reaction temperature

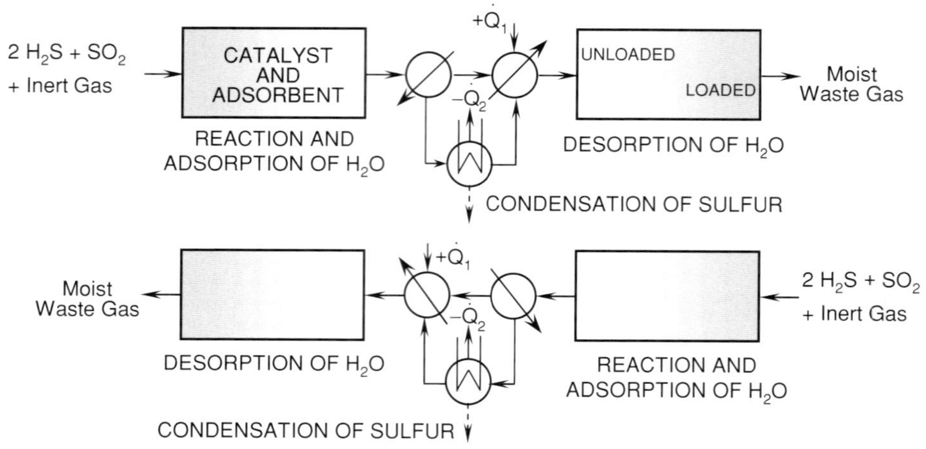

Figure 10.1 Schematic of the operation of a two-bed temperature swing reactor (TSR) for the Claus reaction. (Figure adapted from Elsner et al. (2002a) with permission. © 2002 Elsevier Science Ltd.)

of about 250°C and passes through the second bed. The hot gas strips water vapor from the adsorbent in what we have termed a regenerative purge. When either adsorbent in the first bed is saturated or water is fully stripped from the adsorbent of the second bed, the flow direction switches and then cold reactor feed enters from the right end of the diagram as shown in the figure. Consequently, most of the energy supplied to heat the gas to reaction temperature is recovered from the second bed. The Claus reaction is exothermic so some low-pressure steam can be recovered before the gas flows to the condenser.

Temperature swing with flow reversal illustrated above was not conceived by Elsner et al. (2002a,b). It is routinely used for adsorber regeneration and it was proposed several years earlier by Yongsunthon and Alpay (1999) for separating reactors. They point out that for an endothermic reaction with product adsorption, heating the purge gas in a three-step cycle increases product recovery and purity. With temperature forcing, the separating reactor system can be complicated: many cycles and different arrangements of multiple catalyst–adsorbent beds are possible. We consider these arrangements in the next section.

In TSRs, thermal energy is supplied through the fluid phase for regeneration of the adsorbent, at least on an industrial scale. This permits use of a moving-bed reactor in place of cyclic operation. In the moving-bed design, a mixture of catalyst and adsorbent flows through a vessel fed by reactants where products are formed. The solids then flow through a regenerator in which adsorbate is recovered. Operation of this reactor type was discussed in Chapter 5 for chromatographic applications. Insuring uniform flow of solids across a reactor cross section has been a problem for moving beds. Identical residence time for all solid particles is necessary for an efficient chromatographic separation. With temperature or pressure swing systems, however, uniform residence time is no longer essential so that moving beds may be practical for these classes of reactors. Indeed, a circulating fluidized bed might be used. Harrison and his coworkers at Louisiana State University have proposed a two fluidized bed system for the water gas shift reaction (Han and Harrison, 1994) and the

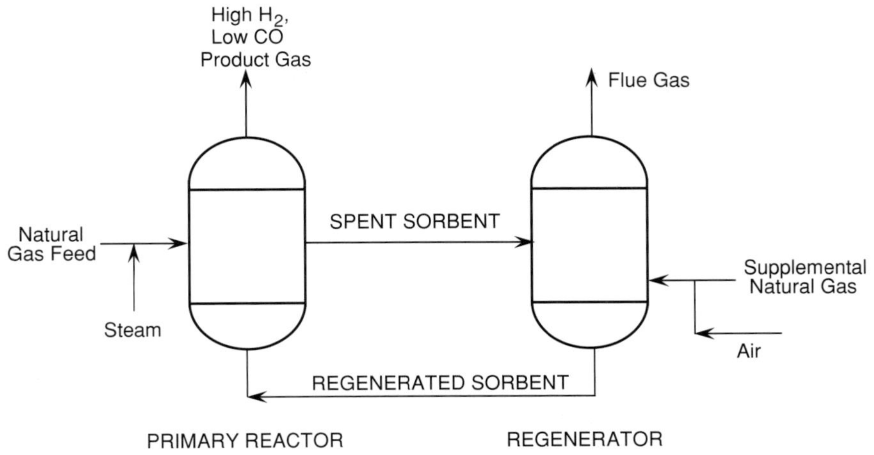

Figure 10.2 Schematic of a fluidized bed TSR system. (Figure adapted from Harrison and Peng (2003) with permission. © 2003 Berkeley Electronic Press.)

steam reforming of methane (Han and Harrison, 1997; Balasubramanian et al., 1999; Lopez Ortiz and Harrison, 2001; Harrison and Peng, 2003). Figure 10.2 provides a schematic of the fluidized bed concept. Fluidized beds are well mixed so the contacting shown in the figure is not advantageous for equilibrium-limited reactions. Plug flow in the production step of the cycle results in higher adsorption and reaction rates. The contacting pattern is not a disadvantage for regeneration.

An alternative to reactors with moving particles is to move the bed as a whole. The simplest way to move a bed is to rotate the bed around an axis through the centerline of the vessel. Figure 10.3 shows how such a rotating bed would operate on the Claus reaction. Rotating chromatographic reactors were considered in Chapter 6. As we pointed out in that chapter, either the bed may rotate or the feed and take-off ports at the ends of the bed may rotate. Both types of systems have been operated experimentally. Rotating bed heat exchangers enjoy fairly wide use, but rotating reactors are a rarity. Agar (2003) and Menge et al. (2003) discuss rotating temperature swing systems, but descriptions of laboratory units are not given. Of course, even if the operation of the reactors is continuous, the catalyst–adsorbent mixture experiences a periodically changing environment.

Segregation of catalyst and adsorbent is an alternative to a homogeneous mixture. Segregation has been studied for simulated countercurrent moving-bed chromatographic reactors (see Chapter 7), but only Agar (2003) and Menge et al. (2003) mention this possibility for TSRs. A possible arrangement is given in Figure 10.4. Here the reactants flow consecutively into a bed of catalyst and a bed of adsorbent. At least two catalyst beds are necessary to take advantage of product adsorption. Three catalyst beds are shown in the figure.

Segregation provides operating flexibility at the expense of capital outlay for reactors and piping. For example, with separate catalyst and adsorbent beds, the adsorbent or "capture" agent could be regenerated at temperatures that would deactivate the catalyst. Different temperatures or even catalysts might be used in successive catalyst beds.

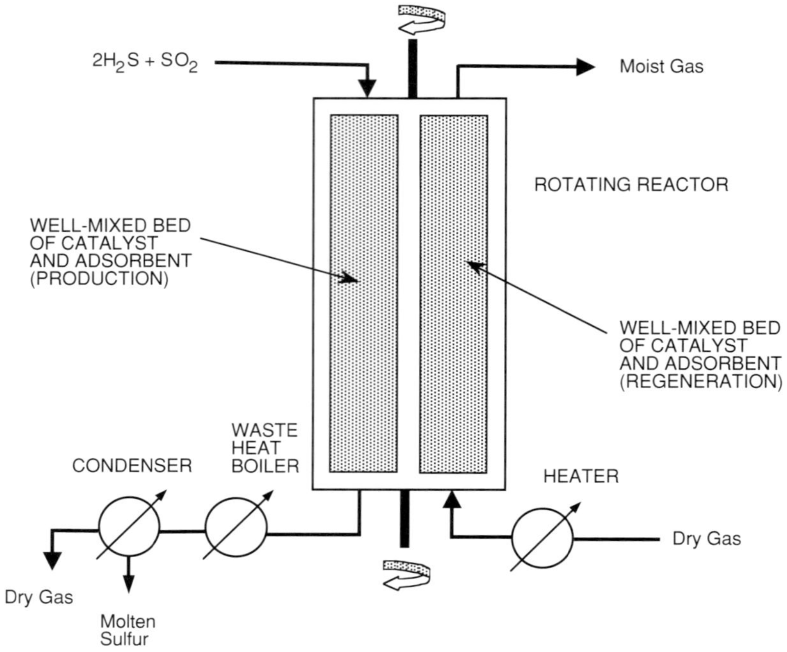

Figure 10.3 Schematic operation of a two-section, rotating bed TSR for the Claus reaction.

Researchers at the University of Dortmund (Germany) (Menge et al., 2003) have proposed a rotating reactor with separate catalyst and adsorbent compartments. A schematic of their design is shown in Figure 10.5. There are entrance and exit catalyst beds but just a single adsorbent bed. Hot carrier gas for regeneration passes only through the adsorbent bed. Although Figures 10.3 and 10.5 suggest two-step processes, any number of steps are possible by adding feed and take-off connections.

Moving-bed reactors and segregated catalyst and adsorbent systems have been introduced in this chapter rather than in the previous one because moving solids or a rotating bed are impractical, though not impossible, with large pressure changes that are encountered in pressure swing systems. Segregated systems are possible in pressure swing applications but they have not been discussed in the literature.

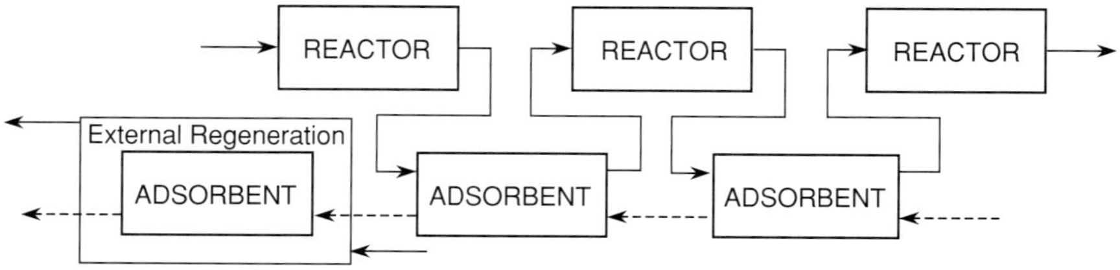

Figure 10.4 Schematic of a TSR with catalyst and adsorbent isolated in different beds. (Figure adapted from Menge et al. (2003) with the authors' permission.)

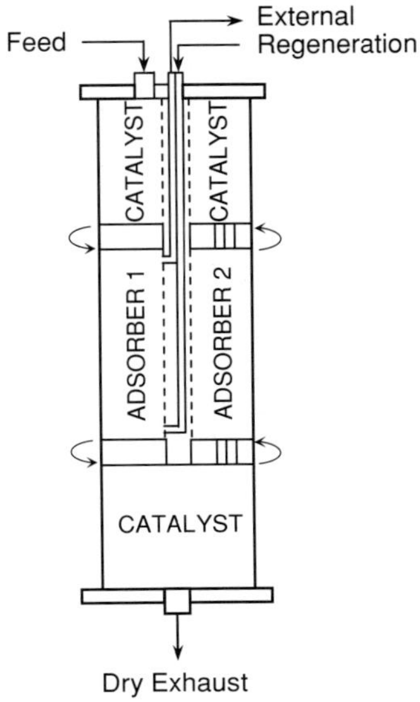

Figure 10.5 Schematic of a rotating bed TSR with catalyst and adsorbent isolated in different beds. (Figure adapted from Menge et al. (2003) with the authors' permission.)

10.2 Modeling

Most numerical studies of temperature swing separating reactors (TSR) in the literature assume a one-dimensional, pseudo-homogeneous model for an isotropic mixture of adsorbent and catalyst. Reactors are not isothermal so the energy balance must be invoked. Thus, the equations used are those given in Tables 3.2 and 3.4 of Chapters 3 and repeated in Chapter 9 as Equations 9.33–9.35 and 9.40. They are given also in Table 9.2 when partial pressure is used as the variable. For reactor design, the pseudo-homogeneous assumption may not be adequate so the heterogeneous model given in Table 3.3 may be preferable.

Initial conditions for the models are the continuity condition on the change from one step to the next that can be stated as

$$\begin{aligned} C_i(z, t_k)_k &= C_i(z, 0)_{k+1} \\ C(z, t_k)_k &= C(z, 0)_{k+1} \\ T(z, t_k)_k &= T(z, 0)_{k+1} \\ P(z, t_k)_k &= P(z, 0)_{k+1} \end{aligned} \tag{10.1}$$

k in the above condition is the step index for a cycle of K steps and t is time so the condition states that the properties at an axial point in the bed at the beginning of the next step, $k + 1$, of the cycle must be the same as the properties at the end of the previous step, namely at t_k. When $k = K$, the number of steps, $k + 1 = 1$. The continuity condition cannot be used for velocity as there is a discontinuity on reversal of flow direction.

As we have mentioned in the previous chapter, it is the initial conditions at each step and the inputs at each reactor port that distinguish each type of separating reactor. Generally,

Danckwerts boundary conditions (Froment and Bischoff, 1990; Westerterp et al., 1993) are used at the exit or entrance to the bed. At the flow entrance, either $z = 0$ or $z = L$ depending on flow direction

$$u(C_i)_{0^-} = u(C_i)_{0^+} - D_{z_i}\left(\frac{\partial C_i}{\partial z}\right)_{0^+}$$
$$u(C)_{0^-} = u(C)_{0^+} - D_z\left(\frac{\partial C}{\partial z}\right)_{0^+} \tag{10.2}$$
$$u\rho_g C_{p_g} T_{0^-} = u\rho_g C_{p_g} T_{0^+} - k_z\left(\frac{\partial T}{\partial z}\right)_{0^+}$$

At the flow exit, either $z = L$ or $z = 0$

$$\frac{\partial C_i}{\partial z} = \frac{\partial C}{\partial z} = \frac{\partial T}{\partial z} = 0 \tag{10.3}$$

Various boundary conditions are possible for pressure, P. It is often convenient to specify an inlet pressure, such as P_f. Alternatively, superficial or nozzle velocity can be specified. In this case, the inlet pressure can be calculated. If velocity is used in the model in place of pressure, the calculation is unnecessary unless the inlet pressure is specified. Only a single boundary condition is needed because a phenomenological relation (the Darcy or Ergun equation) is used in the models in place of a momentum balance. The values of C_i, C, T and P used in Equations 10.2 depend on the cycle chosen.

10.3 Simulations

The temperature swing system for the Claus reaction shown in Figure 10.1 has been studied numerically by Elsner et al. (2002a,b) assuming adiabatic and isobaric operation. The model discussed above was employed. A two-step cycle was used: (1) production and (2) regeneration. To simplify calculations, feed temperatures were constant in each step. Of course, these temperatures were different. Simulations were undertaken for a commercial alumina catalyst and a zeolite 3A adsorbent. Adsorption on the catalyst was neglected. Formation of carbonyl sulfide from the parasitic reaction of H_2S with CO_2 was not included in the reaction model.

Experiments were performed on the catalyst to measure the kinetics of the Claus reaction as well as to test kinetic models given in the literature. An empirical model was selected:

$$R = 5292e^{-\frac{6011.25}{T}} P_{H_2S}^{0.95} P_{SO_2}^{0.22} - 1.252 \times 10^6 e^{-\frac{10415.7}{T}} P_{H_2O}^{0.99} \tag{10.4}$$

Units of the reaction rate, R, are mmol/s kg catalyst. Pressures are in mbar and temperatures are in K (Menge et al., 2003). Separate experiments were performed on zeolite 3A to obtain isotherms for water at different temperatures. These data were employed to estimate parameters of a Freundlich isotherm model:

$$q_{H_2O}^* = 0.65e^{-\frac{30.5}{RT}} C_{H_2O}^{0.75} \tag{10.5}$$

Units of the adsorbate load are mmoles/kg adsorbent. R, the gas constant, is in kJ units. Rate of adsorption was closely represented by a linear driving force model, Equation 9.35,

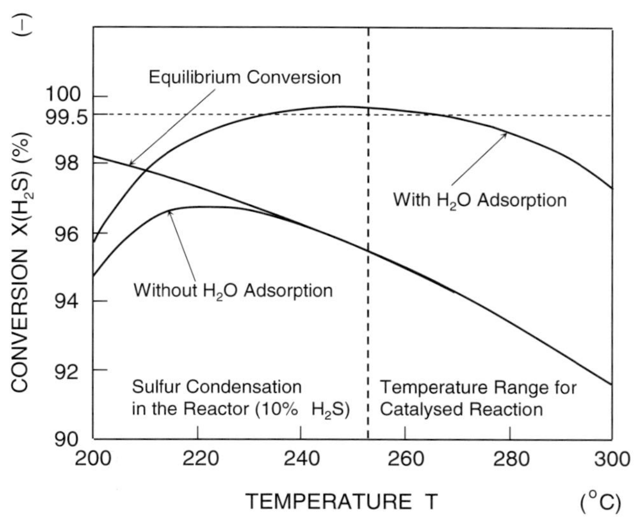

Figure 10.6 Actual and equilibrium conversion of H_2S for isothermal operation with and without adsorbent as a function of temperature assuming a feed space velocity of 100 Nm^3/m^3 catalyst / h and 10% H_2S and 5% SO_2 in the gas feed. (Figure adapted from Elsner et al. (2002a) with permission. © 2002 Elsevier Science Ltd.)

where the composite mass transfer coefficient, k_c, is given by

$$k_c = \frac{60}{d_p^2} \frac{D_{m_{H_2O}}}{1 + \dfrac{0.4875 e^{\frac{30.5}{RT}}}{\psi C_{H_2O}^{0.25}}} \tag{10.6}$$

Units of k_c are s^{-1}. $D_{m_{H_2O}}$ is the diffusivity of water in the gas phase. Menge et al. assign a value of 3.94×10^{-7} to this diffusivity.

Using just equilibrium data and assuming isothermal operation, the limiting performance of the two-bed system is given in Figure 10.6. It is evident that the 99.5% removal target can be met in the two-bed system illustrated in Figure 10.1.

Employing simulation, Elsner et al. (2002a,b) explored the effect of the adsorbent to catalyst volume ratio on H_2S breakthrough time in the first step of the cycle. They observed that a 3A zeolite at 43 vol% maximized the breakthrough time. With lower ratios, the reverse reaction mounts in importance leading to earlier breakthrough, while at higher ratios reactions rates are reduced because of the lower catalyst density in the bed.

Menge et al. (2003) have extended the simulation study to examine the effect of varying the volume fraction of adsorbent continuously as well as stepwise in the bed. Continuous variation from 0 to 100 % with a mean 45 vol% or a two-section bed of equal lengths with the first 20 vol% and then 80 vol% reduced the cycle period by about 20%. This implies higher throughput. On the other hand, these investigators found that reversing the slope of the variation of adsorbent volume percent or the order of the section with the high volume percent adsorbent greatly increased the cycle period needed to obtain H_2S removal equal to that obtained with a homogeneous adsorbent–catalyst mixture. They appear to have extended their work to particles formed from mixtures of adsorbent and catalyst rather than a mixture of particles of each material. Indeed, they have looked at layered particles with adsorbent centers or even catalyst centers (Menge et al., 2003).

The HCN synthesis from ammonia and carbon monoxide via the slightly endothermic gas-phase reaction,

$$2CO + NH_3 \leftrightarrow HCN + CO_2,$$

using a two-step temperature swing cycle has also been studied numerically by the University of Dortmund team (Elsner et al., 2002a,b; Menge et al., 2003). An isothermal, one-dimensional, pseudo-homogeneous model that incorporated axial dispersion was used. Kinetics for the catalyst, Fe_3O_4/Cr_2O_3, were measured and modeled, while the adsorption isotherm and the linear driving force rate expression developed by Ding and Alpay (2000b) was selected for CO_2 adsorption on the potassium-modified hydrotalcite. Most of the research effort on this synthesis was focused on experiments so we defer further discussion until the next section.

Although multibed TSR systems were suggested early in the development of this technology, Yongsunthon and Alpay (1999) seem to be the first investigators to attempt a comprehensive analysis of the advantages multiple beds might bring. They sought to evaluate improvements in conversion, product separation and purity possible through TSR operation by undertaking an optimization of TSR systems using a two-step cycle. The first step was pressurization, that is, introduction of reactant into a mixed bed of catalyst and adsorbent, with removal of a product stream at the discharge port. In the second step, feed and discharge ports were closed and a purge or regenerant was introduced, often at a higher temperature, to desorb product and sweep remaining product and reactant out of the reactor through the exhaust port. Figure 10.7a shows these operations for a single bed. Purge flow through each bed can be in the same direction as the feed or it can be in the reverse direction. Multiple catalyst–adsorbent beds are feasible. Serial operation of multiple beds is shown in the upper portion of Figure 10.7b. Purge can be added at the first bed or split between beds. In the bottom portion of (b) part of the product can be removed between stages in both the pressurization and purge steps.

For their investigation, the authors used the pseudo-homogeneous, one-dimensional, adiabatic model given in Equation 9.32, in Table 3.2 and repeated below in Equation 10.7.

$$\left(\varepsilon_t \rho_g C_{p_g} + \rho_b C_s\right) \frac{\partial T}{\partial t} = k_z \frac{\partial^2 T}{\partial z^2} - u \rho_g C_{p_g} \frac{\partial T}{\partial z} - \rho_b \sum_{i=1}^{N_C} \Delta H_{a_i} \frac{\partial q_i}{\partial t} - \varepsilon_t \sum_{i=1}^{N_C} \Delta H_{r_i} r_i \qquad (10.7)$$

They assumed isobaric conditions in the bed or beds and so neither the Darcy or Ergun equations were used. Axial dispersion and conductivity were considered, however. Boundary conditions are as stated in Table 9.4. However, in the staged or multiple bed arrangements, the feed or purge to beds other than the first are the discharges from the upstream beds when serial connections are used. With serial–parallel connections, a mixer is located at the entrance to the bed in the flow direction so the feed or purge composition will be the mixture of discharge from the previous bed and the fresh feed or purge. Of course, the splits of fresh feed or purge to the various beds and the splits for the discharge of each bed to the product line and the following bed must be known. The Yongsunthon and Alpay simulation used as a model reaction the dehydrogenation of methylcyclohexane (MCH) to

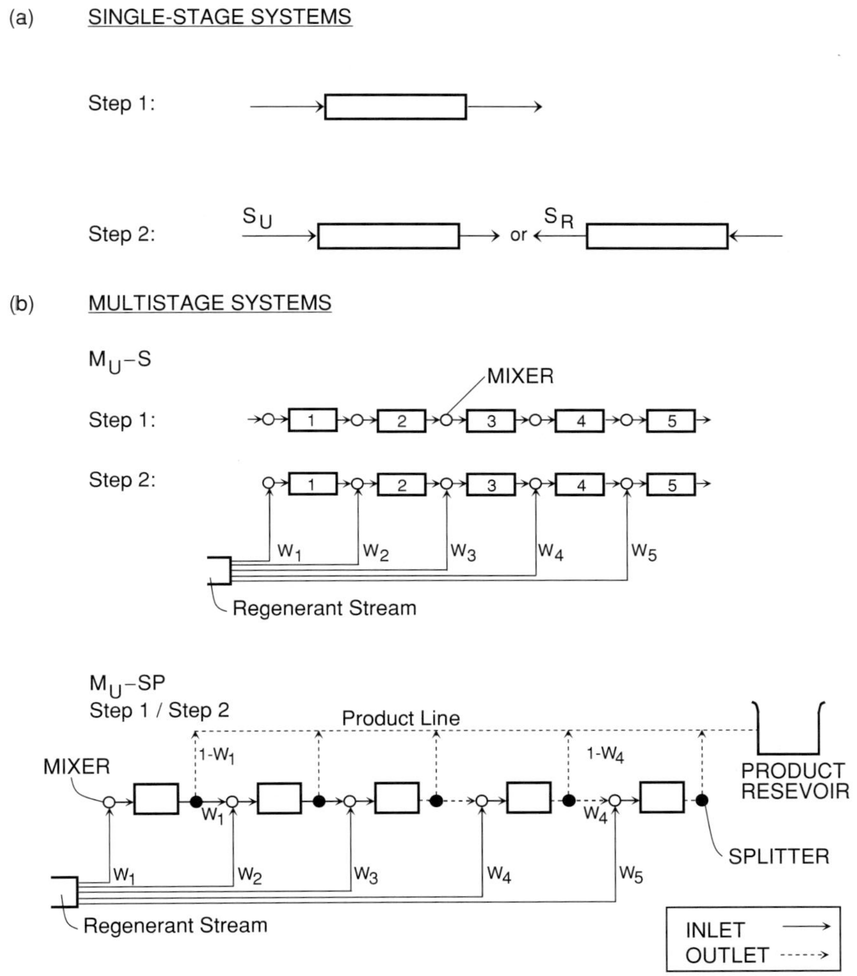

Figure 10.7 Mixed catalyst–adsorbent bed and flow arrangements considered in the optimization study of a TSR. M_U and M_R refer to the direction of feed and purge flow through each bed, U = unidirectional and R = reversed direction; S and P refer to the connection of the beds or stages, S = serial, P = parallel. (Figure adapted from Yongsunthon and Alpay (1999) with permission. © 1999 Elsevier Science Ltd.)

toluene over a Pt/Al_2O_3 catalyst with zeolite 5A as the adsorbent. This material preferentially adsorbs toluene. The MCH reactant is just weakly adsorbed. The kinetic model was taken from a paper discussed above (Alpay et al., 1994), while adsorption isotherms and heat of adsorption were drawn from experimental measurements (Alpay et al., 1996).

The search for the optimal TSR system used quadratic programming. First, the optimal adiabatic plug flow reactor performance, namely the feed rate of MCH for a specified toluene production rate, was determined in a variable space of feed temperatures between 298 and 623K, inlet velocities between 0.2 and 10 m/s and bed lengths between 1 and 5 m. Then, the search sought the minimal feed rate for each TSR arrangement for the specified toluene production rate. Constraints on the search were that the rate of energy input had to be less than that for the plug flow reactor, recovery of toluene had to exceed 70%, and purities of toluene in the exhaust stream or hydrogen in the discharge stream had to be at least 70%. In addition, the feed and purge duration was limited to10 000 s and total bed length had to be less than the length of the adiabatic plug flow reactor. The plug flow reactor contained the catalyst–adsorbent mixture, but at steady state the adsorbent acts as an inert. The feed rate included MCH in the fresh feed and any MCH entering with the purge. Production rate of toluene included the entire cycle and all toluene in the discharges. These refer to product eluted from the bed or beds during either the feed or purge (regeneration) steps. Toluene recovery as a percent was the toluene recovered from the exhaust during the purge divided by the total toluene eluted in the cycle. The simulation assumed that energy input came from heating the feed stream or the purge stream entering a single bed or the multiple beds of the staged system from 25°C to the temperature necessary to meet the heat duties in the pressurization and purge steps.

To limit the computing resources required for the optimization, arrangement of catalyst–adsorbent beds was specified. With multiple stages, only results for five stages were calculated. Table 10.1 summarizes the optimization results. The first column in the table gives the performance for the adiabatic PFR. Whatever the arrangement and number of stages or beds, it is evident from the table that the TSR system has substantially increased conversion to toluene for this equilibrium-limited dehydrogenation reaction. This can be seen by a large decrease in the MCH feed rate that is needed to produce 0.020 mol/m^{2}s of toluene. Multiple stages increase conversion significantly, but at a cost of slightly lower toluene recovery in the purge stream eluted from the beds as the exhaust. The purity of the toluene in all the TSR arrangements is greater than 80% showing that both good separation and an increase in

Table 10.1 Optimization results for various PFR arrangements and flow directions in a TSR system (Yongsunthon and Alpay, 1999)

	Adiabatic PFR	One-stage unidirectional flow	One-stage reverse flow	Five stages (serial)	Five stages (serial/parallel)
Operating policy					
Feed temperature (K)	531	623	623	623	623
Purge temperature (K)		474	352	474	487
Performance indices					
MCH feed rate for 0.020 mol/m^{2}s of toluene (mol/m^{2}s)	0.087	0.041	0.051	0.039	0 034
Energy input rate (10^4 J/m^{2}s)	6.43	6.33	5.61	6.27	6.27
Toluene recovery in purge (%)		76	76	72	72
Toluene purity in purge discharge (%)	11 in product	83	86	85	100
Hydrogen purity in product discharge (%)	33	72	68	74	79

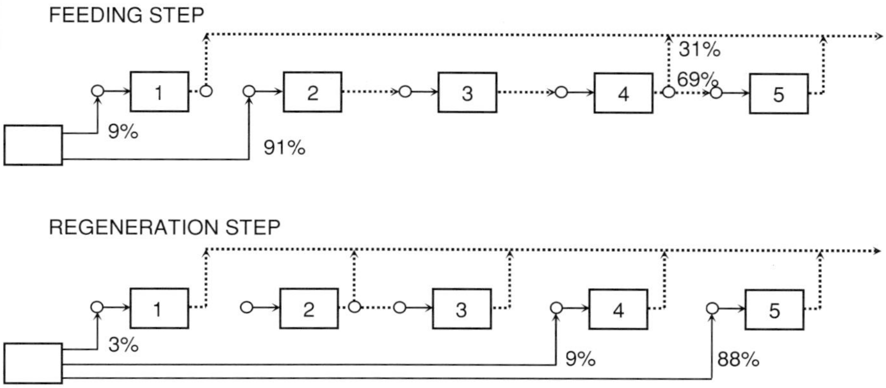

Figure 10.8 Feed and regeneration gas distribution for the optimal five-bed TSR with unidirectional flow. (Figure adapted from Yongsunthon and Alpay (1999) with permission. © 1999 Elsevier Science Ltd.)

conversion can be achieved in TSRs. The table also shows that the TSR systems offer a small energy saving. Reverse flow operation with a single bed provides about a 10% saving in the energy supply rate. If the product streams heat the feed stream, the total energy requirement is the same for all TSR arrangements and the adiabatic PFR as it depends solely on the toluene production rate. With respect to reactor design, the optimization gave the same reactor size for the PFR and all TSR arrangements. The gas loading or inlet velocity was also the same. With multiple beds, the optimization results gave some rather unusual connections for the series-parallel case. The optimal arrangement appears in Figure 10.8. A rather surprising result of the optimization exercise is that most of the energy input occurs through the feed. With an endothermic reaction that is understandable, but with concomitant exothermic adsorption of toluene and endothermic desorption in the purge step a more balanced input would have been anticipated.

In a follow-up paper (Yongsunthon and Alpay, 2000), flow in the reverse direction was permitted. Indeed, all flow arrangements were possible in what the authors call total connectivity. The network that arises is shown in Figure 10.9. Note that each bed is preceded by a mixer and followed by a splitter. Recycle about each unit is also possible. The mixer prior to any bed is connected to the feed, the regenerant reservoir and to the outlets of all beds. The splitter after each bed proportions flow to the product stream and to the mixers preceding each bed. The optimization approach pursued began with the performance of the optimal unidirectional series-parallel network, that is, the minimum feed rate subject to the constraints mentioned above, and attempted to find a total connectivity scheme that would decrease the feed rate.

The search required a huge expenditure of computing resources and, by nature, did not identify a global optimum. Nevertheless expanding the connection possibilities did provide a further, but small, reduction in the MCH feed rate. The network is quite surprising (see Fig. 10.10). Much of the feed flows to the second bed where reaction takes place, but very little adsorption occurs. Product from these beds goes mainly to the product line. Adsorption and finishing of the reaction to give high conversion occurs in the final two beds. Of course, the regenerant flows mainly to the final two beds. Recycle is absent. The authors note that

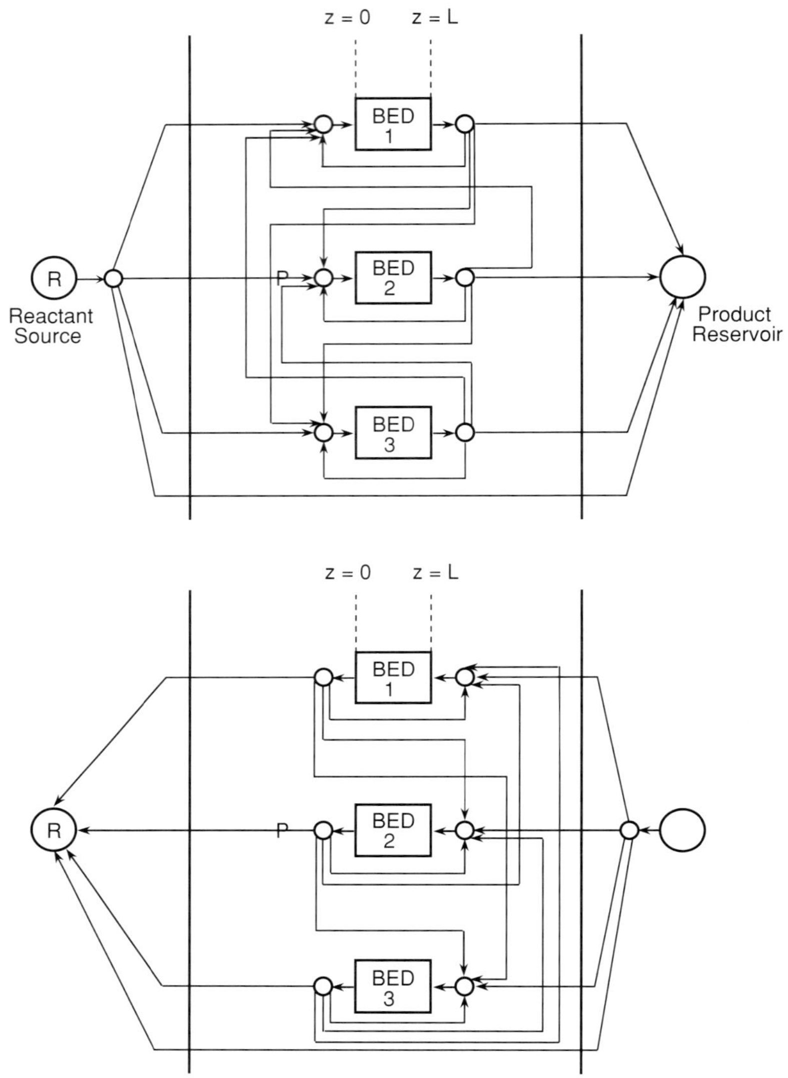

Figure 10.9 Total connectivity network explored for improved TSR performance. (Figure adapted from Yongsunthon and Alpay (2000) with permission. © 2000 Elsevier Science Ltd.)

the adsorbent appears to play a minor role in enhancing yield and conversion. The main contribution arises from the temperature distribution provided by the hot feed gas.

Yongsunthon and Alpay (2000) conclude further that the complexity introduced by the connections shown in Figure 10.10, which imply gas compression, are probably not justified by the small increase in performance. They suggest that the simpler unidirectional series-parallel network given in Figure 10.8 is preferable. The advantage of the total connectivity investigation is that it suggests the best performance attainable in a network and thus

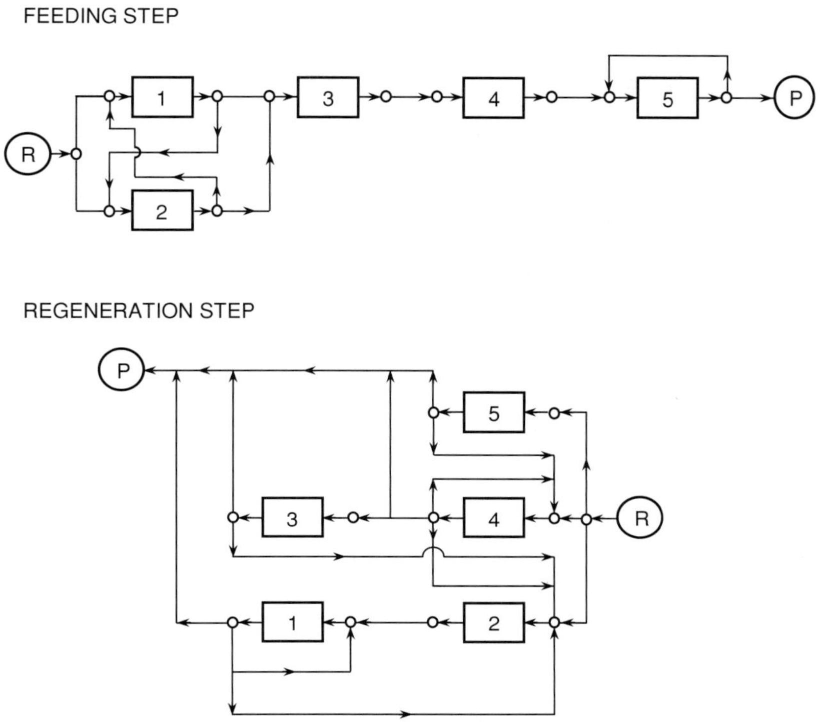

Figure 10.10 Optimal network for a five-bed TSR system when total connectivity is permitted. (Figure adapted from Yongsunthon and Alpay (2000) with permission. © 2000 Elsevier Science Ltd.)

provides a measure for comparing the performance achieved in simpler flow arrangements or even with a single bed.

10.4 Experimental

It is rather surprising that for temperature swing separating reactors experimental study preceded simulation. Han and Harrison (1994) investigated CO_2 uptake by calcined dolomite as an alternative to the water gas shift reaction in the production of hydrogen for industrial use. A packed-bed reactor, shown in Figure 10.11, was employed by their program. The experimental system feeds carefully controlled amounts of pure gases into a mixer and then into the reactor. The feed is heated in upflow around the bed. Reaction products are cooled to condense out water and pressure is reduced upstream from the analysis section. At temperatures above 823 K, the shift reaction is forced by CO_2 adsorption into the dolomite via the exothermic reaction: $CO_2 + CaO \leftrightarrow CaCO_3$.

Variables investigated were bed temperature, pressure, synthesis gas composition, space velocity and dolomite properties. With temperatures between 773 and 873 K and at 15 bar, the phases reached equilibrium at a space velocity as high as 3400 h^{-1}. This meant 99.5% conversion of CO as well as CO_2 removal. Carbon oxide concentration in the reactor off gas

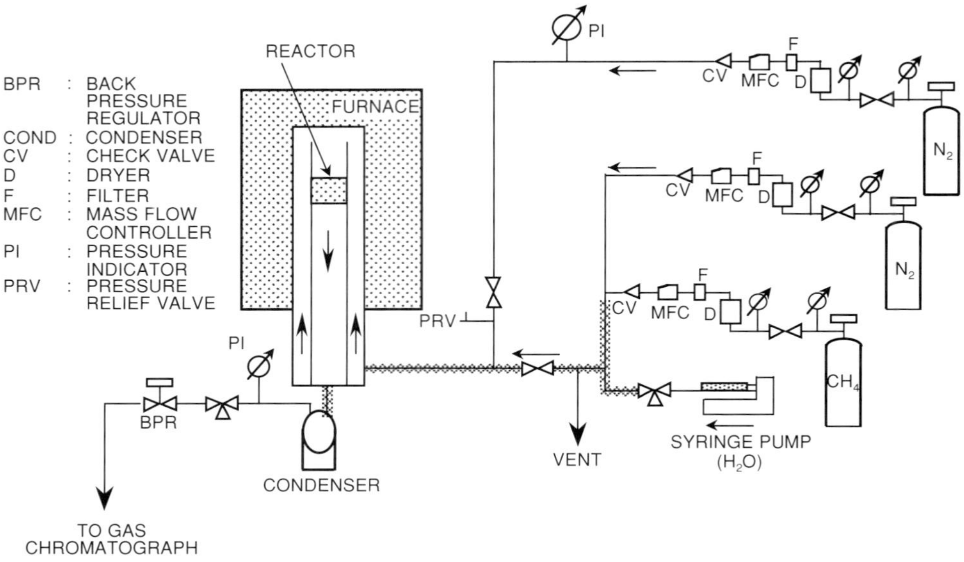

Figure 10.11 Experimental TSR for the study of a high temperature shift reaction and the steam reforming of methane. (Figure adapted from Balasubramanian et al. (1999) with permission. © 1999 Elsevier Science Ltd.)

was below 400 ppm by volume on a dry basis. CO conversions are higher than those from the high temperature catalytic water gas shift followed by scrubbing with an amine solution. Han and Harrison point out that at 823 K and 25 bar almost 100% conversion is possible.

Economic viability of a CO_2 acceptor process for converting CO to H_2 in the presence of steam depends critically on the stability of the dolomite acceptor over many cycles. This question was considered by Han and Harrison (1997). They used a commercial dolomite (54.5 wt% $CaCO_3$, 45 wt% $MgCO_3$) that was calcined at 1023 K in N_2. This material was loaded into the reactor shown above. Their dolomite bed, containing 10–15 g of solid, exhibited an abrupt breakthrough of both CO and CO_2. Time on stream up to breakthrough was used to characterize the durability of the dolomite acceptor. A dimensionless time on stream, θ^*, based on the throughput of carbon oxides was used to allow for variations of space velocity and feed composition:

$$\theta^* = \frac{n_g y_{CO+CO_2}}{n_{CaO}} t \tag{10.8}$$

In this expression, n_g is the molar gas flow rate and n_{CaO} is the moles of calcium oxide in the dolomite sample. Breakthrough is illustrated by Figure 10.12. It sets in at about $\theta^* = 0.7$. Prior to breakthrough, the CO_2 level is about 300 ppm (vol). When the adsorption front reaches the end of the bed, it rises rapidly to about 5 vol%.

Plotting dimensionless breakthrough time versus cycle number, Figure 10.13 demonstrates a steady deactivation of the dolomite acceptor. Each cycle represents exposure to a synthesis gas at 823 K and regeneration of the solid at 1023 K in N_2. Note that breakthrough is defined either by when the CO concentration reaches 100 ppm or when CO_2 reaches 500 ppm (vol). Up to breakthrough, the removal of carbon oxides is about 99.6% while the

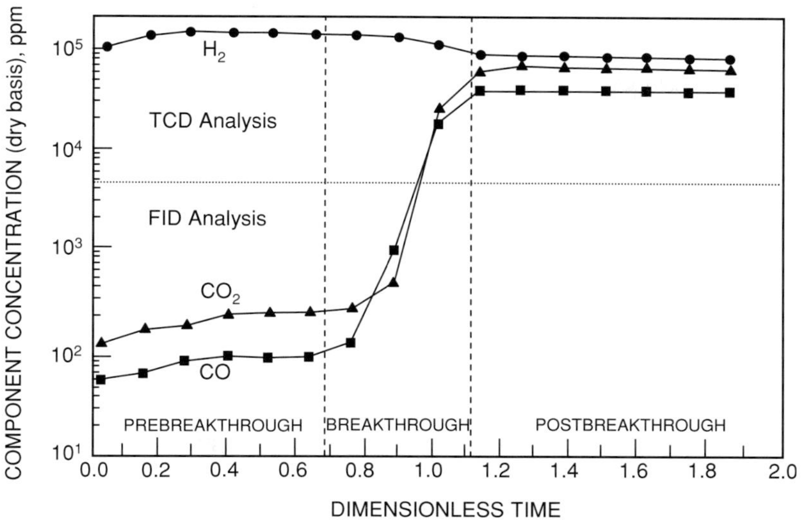

Figure 10.12 Composition of gas leaving a bed of calcined dolomite as a function of dimensionless time for $T = 823$ K, $P = 15$ bar, $SV = 1425\ h^{-1}$ with a feed stream of 5.5 vol% CO, 26.4 vol% H_2O, 3.1 vol% CO_2 and 4.5 vol% H_2 in N_2. (Figure adapted from Han and Harrison (1997) with permission. © 1997 Elsevier Science Ltd.)

CO level in the product gas is about 30 ppm. Deactivation was found to be independent of synthesis gas composition and space velocity, but it increased with temperature in both the production and regeneration steps. Addition of steam to the carrier gas in the regeneration step reduced the rate of deactivation.

The Harrison team has completed several studies on hydrogen production via the catalytic steam reforming of methane using calcined lime (Balasubramanian et al., 1999; Harrison

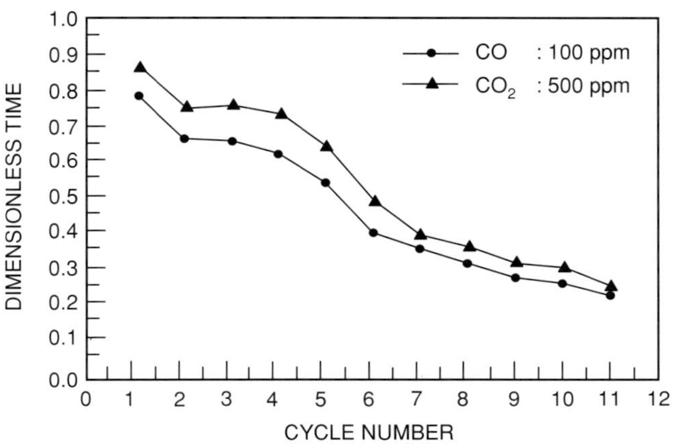

Figure 10.13 Change of the dimensionless breakthrough time with the number of successive regeneration cycles for a calcined dolomite exposed to a shift-carbonation feed gas (5.6 vol% CO, 20.0 vol% H_2O in N_2) at 823 K and 15 bar. Regeneration with dry N_2 at 1023 K. (Figure adapted from Han and Harrison (1997) with permission. © 1997 Elsevier Science Ltd.)

and Peng, 2003) or calcined dolomite (Lopez Ortiz and Harrison, 2001) as CO_2 acceptors or adsorbers. The shift in methane conversion that can be forced through the addition of CO_2 capture is discussed in Section 11.1 in the next chapter. The steam reforming process currently used commercially has been described in Section 9.7, where it was mentioned that the process incorporates three reaction steps, two separation ones as well as several heat exchange and compression steps to produce high purity hydrogen with CO at a low ppm level. A further reaction step is necessary to achieve CO levels compatible with hydrogen use in a PEM fuel cell. The endothermic reforming reaction runs at a relatively high temperature of between 1100 and 1150 K so that like a temperature swing process, heat must be supplied at the highest temperature in the system. Consequently, the high temperature penalty for regeneration of the CO_2 captures material in the two-step TSR for high-purity H_2 may not be an obstacle in this application.

All of the studies mentioned above used the experimental unit shown schematically in Figure 10.11. A methane feed replaced the synthesis gas source. A burnt lime, CaO, produced by calcining a high-purity $CaCO_3$, was used by Balasubramanian et al. (1999) and by Harrison and Peng (2003). A commercial nickel catalyst, NiO/Al_2O_3, was homogeneously mixed with the lime. Particle sizes of the materials differed somewhat as sieving was used, but both were less than 200 μm. Experiments were undertaken at 15 bar, temperatures between 723 and 1023 K, steam/CH_4 from 3 to 5 and two levels of catalyst percent by weight, 23 and 50%. Only the feed + production step was studied. Figure 10.14 illustrates the time behavior of the catalyst and CO_2 acceptor bed observed. For the feed rate used, the dry H_2 concentration in the product gas builds up over the first 70 min on stream to reach 94% and then declines slowly over the next 80 min as CH_4 begins to breakthrough. The acceptor

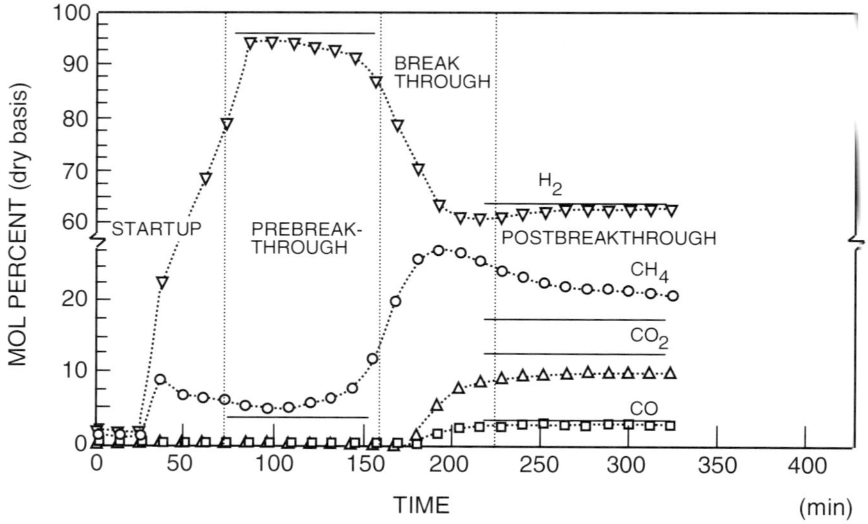

Figure 10.14 Effluent gas composition versus time for a mixed bed of a NiO catalyst and a CaO adsorbent at 923 K, 15 bar for a feed of CH_4 and steam only in a 1:4 ratio. Catalyst/adsorbent = 3:10 by weight. Solid lines and dotted lines express equilibrium levels and experimental results, respectively. (Figure adapted from Balasubramanian et al. (1999) with permission. © 1999 Elsevier Science Ltd.)

prevents carbon oxide breakthrough for a further 20 min. Prior to breakthrough, chemical equilibrium for the gas–solid system is closely approached. Similarly, after breakthrough, the gas phase equilibrium is nearly attained. This indicates that the catalytic reactions and the rate of uptake are relatively fast. Figure 10.14 suggests a series of concentration or reaction fronts that move through the packed bed of catalyst and adsorbent during the feed + production step. A temperature swing reactor (TSR) would exploit this behavior by switching the flow from feed to hot gas regeneration once CH_4 begins to breakthrough. Hot gas would be created by burning CH_4 in compressed air or in a compressed off gas recycle, enriched in oxygen, if CO_2 sequestration is to be undertaken. By-products along with 95% hydrogen with just ppm levels of carbon oxides would be power and waste heat or steam. Compared to conventional steam reforming, hydrogen purity, methane and steam consumption are lower and the by-product values and CO levels are similar (Balasubramanian et al., 1999).

The investigators found that increasing bed temperature up to about 970 K in the feed + production step raises the CH_4 conversion and H_2 purity slightly, but increasing the CH_4 content of the feed had a larger effect on purity. Increasing the steam/CH_4 ratio improved conversion and the H_2 production rate.

Harrison and Peng (2003) extended the experimental work, but used a different commercial nickel catalyst and changed the procedure for making the lime CO_2 acceptor. Their effort focused on CO content in the product gas. Again only the production step was considered. Pressure and temperature were lowered to 5 bar and 750 K, respectively. These investigators found that CO levels under 10 ppm (vol) can be achieved. H_2 purity just under 98% is possible. In the range of variables studied, the CO content was independent of the volumetric feed rate and vol% CH_4 in the feed. The content, however, was affected by temperature. About 15 ppm (vol) of CO appeared in the stream at 775 K.

Carbonation of lime and calcined dolomite and their regeneration in a hot gas stream were studied by using an electro balance (Silaban and Harrison, 1995; Silaban et al., 1996). CO_2 uptake was examined at 1–15 bar, 823–1023 K with a feed containing up to 15 vol% CO_2, while regeneration was studied in the same pressure and CO_2 concentration range, but at temperatures of 1023–1173 K. Understanding the loss of acceptor capacity through successive carbonation and regeneration cycles is crucial to the economics of a temperature swing-based process. Figure 10.15a and b shows the weight and pore volume changes over a single cycle beginning with fresh calcium carbonate for both uptake and regeneration at 1023 K. Calcination drops the solid by almost 50%, but this loss is not recovered in the subsequent carbonation. Indeed weight gain is initially rapid, but then abruptly slows to a very low rate. Pore volume expands enormously on calcination, but is largely lost in the next step. Only about half of that volume reappears at the end of the first regeneration. Despite the volume loss, the mean pore diameter increases suggesting the irreversible closure of the fine micropores in lime. This behavior has been observed by several investigators (Silaban and Harrison, 1995). The rate behavior suggests a hot zone moving inward from the particle surface that is retarded by the endothermic decarbonation reaction. The reverse step is believed to be rapid kinetically, but diffusion controlled. Initially, diffusion control is at the outer surface of the particle. A layer of carbonate forms there and at the abrupt change of rate seen in Figure 10.15b, diffusion through solid carbonate becomes controlling. This is a very slow process.

The phenomena illustrated in Figure 10.15 continue to occur in subsequent cycles, however, the loss in weight and pore volume on regeneration decreases. After many cycles, a cyclic stationary state will be attained, but the acceptor capacity for that state will be just

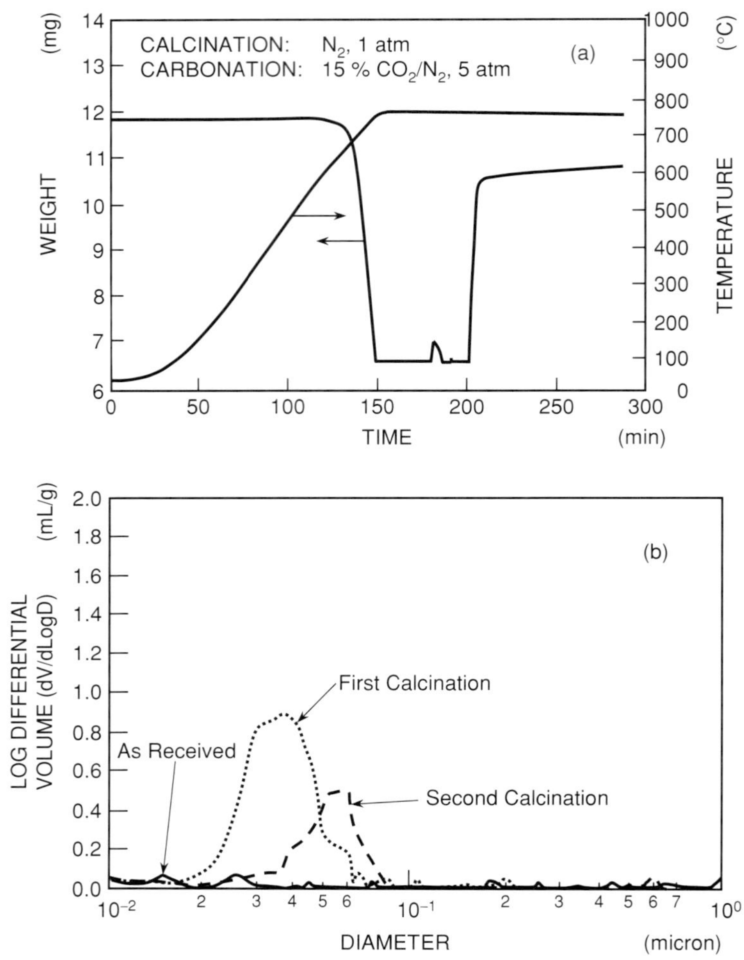

Figure 10.15 Particle weight changes as a function of time (a) and pore volume distribution in the raw material and at the end of two calcination and carbonation steps (b). Calcination takes place in N_2 and carbonation in 15 vol% CO_2 at 5 bar and 1023 K. (Figure adapted from Silaban and Harrison (1995) with permission. © 1995 Gordon and Breach Science Publishers SA.)

a small fraction of the capacity after the initial calcination. This implies continual replacement of the lime adsorbent in the TSR. Losses in weight over successive cycles for lime are presented in Figure 10.16.

Dolomite is a physical mixture of calcium and magnesium carbonate with less than 1 wt% of silicon, iron and aluminium oxides. Magnesium oxide, MgO, in calcined dolomite is not carbonated under the conditions expected in hydrogen production using a TSR process. Silaban et al. (1996) demonstrated that this is the case using the equipment and methods just described. It is also evident from Figure 10.17 that gives weight loss and gain in the first two calcinations–carbonation cycles for a commercial dolomite. Recarbonation of

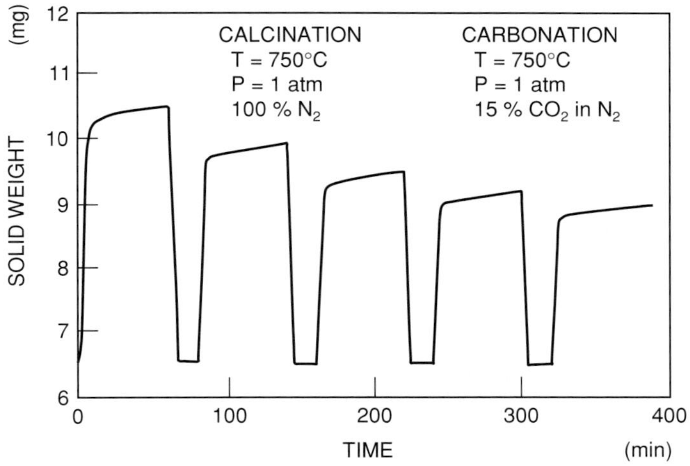

Figure 10.16 Decline in adsorbent weight in successive carbonation and regeneration cycles. Conditions as in Figure 10.15. (Figure adapted from Silaban and Harrison (1995) with permission. © 1995 Gordon and Breach Science Publishers SA.)

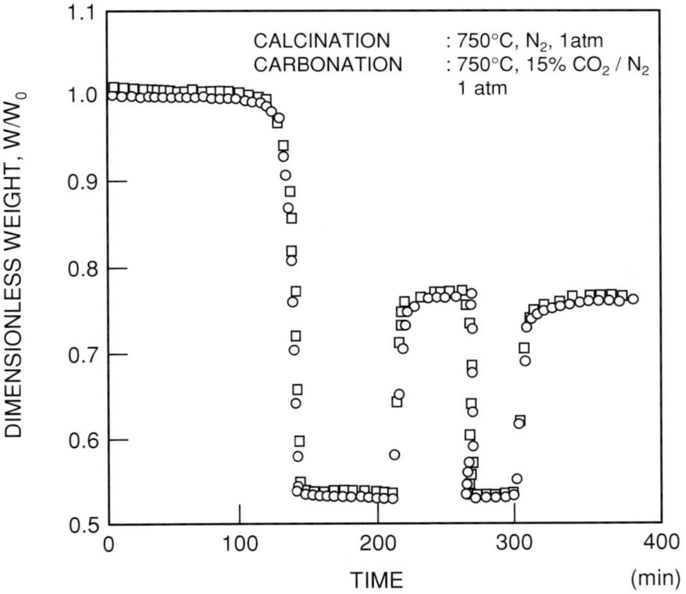

Figure 10.17 Change in adsorbent weight over two calcination–carbonation cycles starting with raw dolomite. Measurements made at 1 bar, 1023 K with only N_2 in the calcination and 15 vol% CO_2 in N_2 for the carbonation steps. (Figure adapted from Silaban et al. (1996) with permission. © 1996 Gordon and Breach Science Publishers SA.)

the calcined dolomite in the figure proceeds initially very rapidly and then abruptly changes and weight increases exceedingly slowly proving that carbonation continues. Weight loss on calcinations, even in the second cycle is very fast. Indeed, the behavior resembles closely that for lime shown in Figure 10.15a. Change in pore volume, though not presented above, is also similar to that for CaO. The microscopic processes of pore opening and closure due to expansion of the solid phase from carbonate formation are the same for lime and dolomite. Presence of MgO, however, prevents closure of some of the pore structures in dolomite. Even though dolomite has only 42% of the CO_2 capacity of lime because of MgO, the stability of this capacity over repeated cycles makes it a better adsorbent choice. Silaban et al. (1996) show that carbonation as a fraction of the ultimate CO_2 capacity of the solid reaches about 0.9 after 2 min exposure to 15% CO_2 in N_2, but reaches 97% after 5 min. It may be seen from Figure 10.18 that this does not change over 20 carbonation–calcination cycles. These authors report that 1023 K is about the lowest temperature that can be used for regeneration of the calcined dolomite. As expected, increasing carbonation temperature decreases CO_2 capacity. Other variables have only minor effects. The behavior shown in the figures does not change when a synthesis gas is used in place of CO_2. Adding steam during regeneration increases the CO_2 capacity of the dolomite.

Evaluation of other adsorbent properties besides fractional capacity indicates that there is a gradual decline in the uptake properties of dolomite. Thus, a temperature swing process must incorporate adsorbent replacement and probably some type of fluidized bed must be used to facilitate removal of spent material and the addition of makeup. Dolomite, nevertheless, emerges as the CO_2 acceptor of choice for hydrogen production.

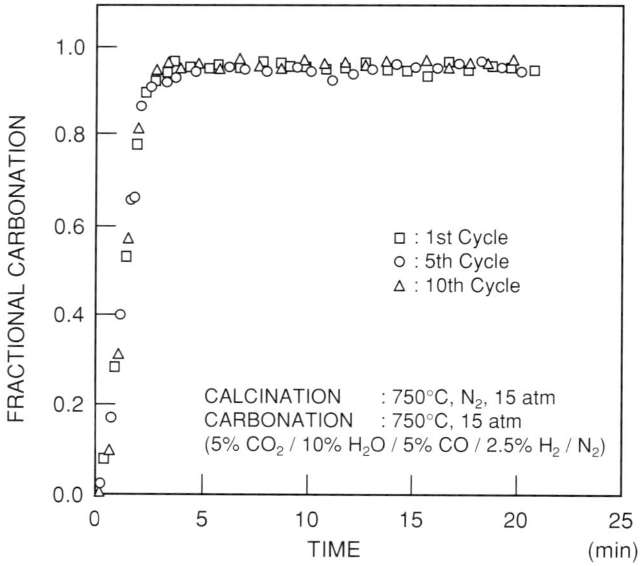

Figure 10.18 Fraction of the ultimate CO_2 capacity utilized as a function of time in carbonation using a gas feed containing 15 vol% CO_2 in N_2 at 1023K and 1 bar. (Figure adapted from Silaban et al. (1996) with permission. © 1996 Gordon and Breach Science Publishers SA.)

Table 10.2 Experimental conditions for the study of a TSR process for hydrogen production

Materials	
Catalyst	18 wt% NiO on an alumina support (United Catalyst C11-9-02)
Particle size	(1) 75–150 μm, (2) 300–425 μm
Sorbent	Calcined dolomite (53.78 wt% $CaCO_3$, 45.89 wt% $MgCO_3$)
Particle size	(1) 75–150 μm, (2) 300–425 μm
Operating conditions for the feed half cycle	
Temperature	923K
Pressure	15 bar
Feed composition	12 vol% CH_4, 48 vol% steam, 40 vol% N_2
Operating conditions for the regeneration half cycle	
Temperature	1073, 1123, 1173 and 1223 K
Pressure	1 bar
Carrier gas composition	(1) 100 vol% N_2, (2) 4 vol% O_2 in N_2, (3) 100 vol% CO_2

Multicycle performance of a temperature swing process for hydrogen based on dolomite has been explored by Lopez Ortiz and Harrison (2001). Table 10.2 gives the experimental conditions used by these researchers. Their research unit has been discussed in the previous paragraphs (see Fig. 10.11). Composition of the carrier gas in the regeneration step was examined because in the presence of oxygen and a sufficiently elevated temperature oxidation of nickel can occur. This would deactivate the catalyst, which is reduced by the presence of methane in the feed step.

The packed bed used in the Lopez Ortiz experiments exhibits a breakthrough behavior in the feed step that is illustrated in Figure 10.14. Methane appears in the outlet first, followed by carbon dioxide. Regeneration of the bed would begin, normally, once CO_2 breakthrough commences, but in these experiments, it was undertaken well after, when the adsorbent was fully carbonated. Most experiments were carried out over five cycles. Minimal loss of sorbent capacity was observed with a pure N_2 carrier gas at 1073 and 1123 K. Extending the run time to 25 cycles indicated sorbent deterioration primarily due to a loss in carbonation capacity (ca. 30%). Increasing the regeneration temperature to 1223 K accelerated the capacity loss but only after about 10 cycles. Time to breakthrough decreased as well and this was attributed to catalyst deactivation. When CO_2 was used in place of N_2 at 1223 K, there was an abrupt drop in capacity after the first cycle. Thereafter, the loss was about the same as that with a N_2 carrier gas at 1073 and 1123 K.

In a further investigation of sorbent durability, Lopez Ortiz and Harrison used an electro balance with cycling between 1073 and 1223 K in pure CO_2 at 1 bar. Results from those experiments for run lengths of 33 cycles and 144 cycles are shown in Figure 10.19. Data for a 25-cycle fixed-bed experiment with H_2 production at 923 K followed by regeneration in pure CO_2 at 1223 K are shown for comparison. Good agreement is evident so the electro balance experiments appear to represent adequately the fall in sorption capacity.

Figure 10.19 demonstrates that continuous replacement of sorbent will be necessary even with the relatively stable dolomite. Roughly a 5 wt% exchange of sorbent per cycle would maintain the sorption capacity at about 50% of its value for freshly calcined dolomite.

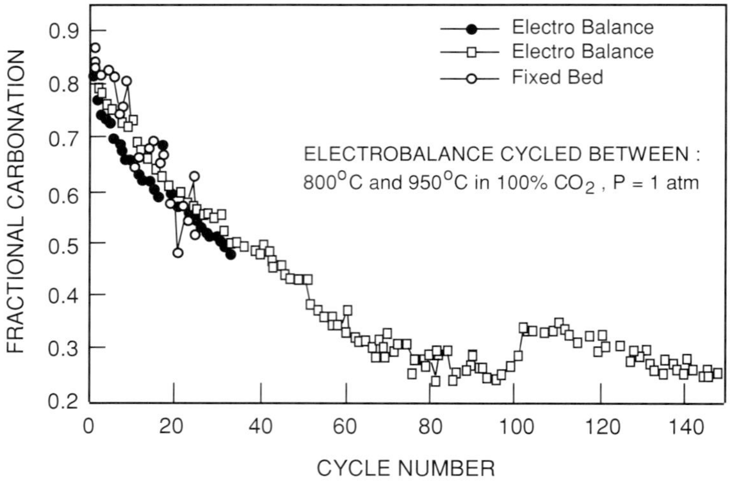

Figure 10.19 Loss of sample weight with successive cycles of carbonation at 1073 K in CO_2 and regeneration also in CO_2 at 1223 K. (Figure adapted from Lopez Ortiz and Harrison (2001) with permission. © 2001 American Chemical Society.)

Certainly replacement at this rate would mandate use of a fluidized bed or a transport reactor system. Loss of carbonation capacity seems to be associated with sintering. Addition of steam to the carrier gas may suppress sintering, but this requires evaluation. In addition to sorbent capacity loss, Ni crystallite growth was observed. This was relatively rapid in the first five cycles on stream, but slowed appreciably in the succeeding 25 cycles studied at 1073 K. Regeneration at 1223 K caused continued rapid crystallite growth. This growth would be manifested in the reactor as a loss in catalyst activity.

Development of sorbents for use in temperature swing reactors, particularly in fluidized beds or transport reactors, is discussed by Satrio et al. (2005). Compressive strength and abrasion resistance are essential for such particles. Satrio et al. propose to achieve this by forming an alumina shell around a dolomite core. They then impregnate the shell with a Ni salt solution so that on calcination a single particle contains both catalyst and sorbent. Strongest particles were formed when 5–10 wt% limestone was added to the alumina used to make the particle shell. The authors hypothesize that a calcium aluminate is formed that bonds both alumina and the alumina shell to the dolomite.

The impregnated particle was tested with steam reforming of propane using a 3:1 volumetric ratio of steam to propane at 923 K followed by regeneration at 1023 K in the reaction mixture. Weight and temperature change with time are shown in Figure 10.20 for this electro balance experiment. Clearly, the reaction proceeds with CO_2 sorption by dolomite. Heating by just 100 K releases CO_2. Some loss in capacity can be seen over the first cycle. Apparently this study is continuing.

Kato et al. (2003) discuss vehicle-based hydrogen production for a fuel cell employing steam reforming of methane in a TSR. The purpose of their study was to see if a TSR process could be used to generate fuel cell hydrogen. Experiments were performed. Reforming catalyst was a supported Ni material of industrial origin and the sorbent was lime, generated in situ from $CaCO_3$. These solids were intimately mixed and used only to collect data for

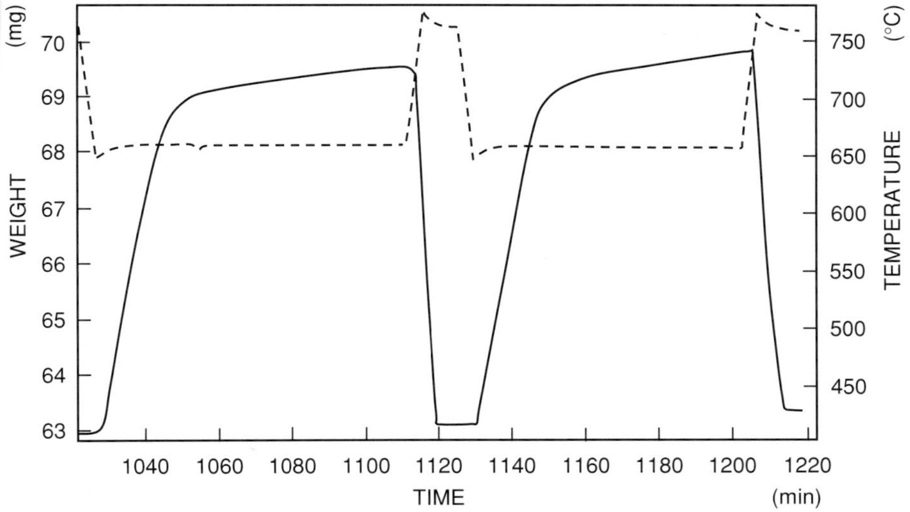

Figure 10.20 Weight (solid line) and temperature (broken line) change with time for a composite catalyst–dolomite particle exposed to a 3:1 steam/propane mixture in an electro balance. (Figure adapted from Satrio et al. (2005) with permission. © 2005 American Chemical Society.)

the feed step of a TSR cycle. Results were compared with those for a conventional PFR packed only with catalyst. Experiments were performed using a quartz tube, 15 mm i.d. by 1000 mm, packed with 2.7 g of NiO catalyst, 8.1 g of $CaCO_3$ (CaO precursor) and 21.9 g of alumina as a diluent. Prior to use, this was calcined at 1103 K in Ar for 4 h. The tube was mounted in a furnace to maintain isothermal conditions. Feed was a steam/CH_4 mixture with a ratio of about 4:1. Temperatures used ranged from 823 to 1023 K at 1 bar. Total flow rate was maintained at 150 mL/min.

Concentrations of H_2 leaving the reactors are compared in Figure 10.21a for different reactor temperatures. The delivery H_2 concentration is significantly higher for the reactor containing mixed adsorbent and catalyst. Equilibrium calculations are also shown in the figure. The upper curve is based on the free energy change for the coupled steam reforming and lime carbonation reactions, whereas the lower curve just employs the free energy change for the reaction alone. It is evident that coupling the two reactions alters the free energy and thus the equilibrium composition of the effluent gas. This can be viewed as an explanation of the increased hydrogen concentration with the mixed catalyst and adsorbent system. Measurements and equilibrium calculations for CO appear in Figure 10.21b. Now, the mixed solids substantially reduce the concentration of CO leaving the reactor. Equilibrium calculations, allowing for the coupling of reforming and carbonation, predict the much lower concentrations.

From the experimental results shown partially in the figure, Kato et al. estimate that about 7.5 kg of CaO would be required in a vehicle to generate sufficient H_2 (1 kg) to travel for 100 km. For this amount of hydrogen, 2.1 kg of CH_4 and 4.8 kg of water are necessary. The authors deem that these weights are feasible. The sorbent would be removed from the vehicle in a service station and regenerated there with collection of any effluent CO_2. Thus,

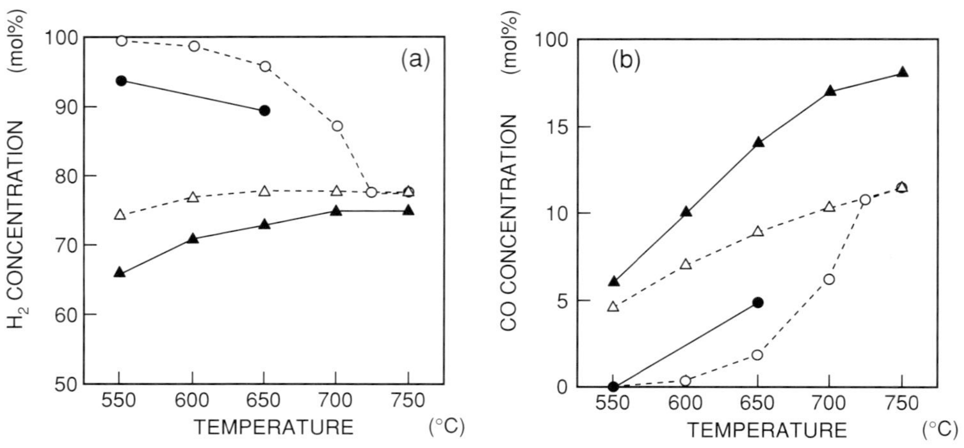

Figure 10.21 Comparison of experimental (solid lines and closed keys) and calculated equilibrium (broken lines and open keys) concentrations leaving identical reactors packed with only catalyst (triangle) and with a mixture of catalyst and adsorbent (circle): (a) hydrogen, (b) CO. Pressure is atmospheric and feed concentration and space time are identical for the experimental reactors. (Figure adapted from Kato et al. (2003) with permission. © 2003 Society of Chemical Engineers of Japan.)

there would be no carbon oxide emissions during travel. The concept is interesting because the parts of the TSR cycle are carried out in different locations.

One of the outcomes of the massive research effort on oxidative coupling of methane that was undertaken in the 1980s was the realization that the failure to develop an economically feasible process was the facile further oxidation of ethylene, the desired product, to CO_2 and water. This led to the application of an SCMCR to oxidative coupling by Carr and coworkers (Tonkovich et al., 1993; Tonkovich and Carr, 1994a,b). We examined this work in Sections 7.4 and 7.5 of Chapter 7. A temperature swing reactor is another alternative to removing ethylene from the reactor before it is further oxidized. This alternative was examined experimentally by Machocki (1996). Separate catalyst and adsorbent beds operating at widely different temperatures were used. The 10 mm i.d. catalyst bed, consisting of calcined $Na_2CO_3/CaCO_3$ particles, was maintained at 1073 K and accepted a methane/oxygen feed at 20:1. A high flow rate resulted in a short contact time of 0.4 s/g cm^3. An adsorbent bed of the same size was packed with a zeolite 5A. This bed was maintained at about 300 K. A blower recycled the effluent from the adsorber back to the front of the catalyst bed. Consumption of CH_4 and O_2 were monitored and make up added to maintain a constant flow rate through the reactor.

Machocki observed that neither methane nor ethane were trapped by the zeolite bed. He ran his experiment for 3 h. Ethylene ($C_2^=$) did not breakthrough the adsorber during this time. Ethylene was recovered by desorbing the zeolite bed at 573 K off-line. It is only the adsorbent bed that undergoes a temperature swing. Machocki did not discuss heat integration or how the 273 K temperature changes might be handled in a commercial scale plant. His results, however, were impressive. Methane conversion over the 3 h run was 94%, yield of C_{2+} hydrocarbons was 66%, while the $C_2^=$ yield was 58%. CH_4 conversion in the catalyst bed operating alone as conversion per pass was 8.2% with a C_{2+} hydrocarbon yield of 6.9%. This performance is certainly comparable to that reported by Tonkovich et al. (1993) for the same process in a SCMCR. Using a different catalyst, reactor and adsorbent temperatures and CH_4/O_2 ratio, these researchers report a C_1 conversion of about 65%, an 80% selectivity to C_2 with about 25% $C_2^{=}$ in the C_2 product (see Section 7.5). Use of temperature swing for oxidative coupling does not seem to have been pursued further.

The Machocki solution to the problem of different reaction and adsorption temperatures through separating the reaction and separation steps and using recycle represents, of course, the conventional approach of sequential operations. It is cyclical because all adsorption operations are cyclical viewed from the prospective of the adsorbent. The conventional approach must always be kept in mind when separating reactors are considered. What becomes important when performance of a conventional and a separating reactor system are about the same is the complexity and the opportunity for heat integration. Some insight into this comes from examining the flow sheet for the conventional approach with continuous feed. A process schematic is given in Figure 10.22.

It is evident that the conventional sequence is complicated. It becomes even more so if the adsorbent beds must be purged to heat or cool the adsorbent or flush out undesorbed product.

An adiabatic pilot scale reactor with an internal diameter of 60 mm and a 1-m length was employed by Menge et al. (2003), using an optimal mixture of the catalyst and 3A zeolite, to validate the model used by Elsner et al. (2002a,b) and by themselves in the Claus reaction simulations discussed in Section 10.2. Runs were also made with just catalyst and an inert in the reactor. The steel reactor was maintained nearly isothermal by circulating a heat transfer fluid through the reactor mantel. Adequacy of the model for the first step of the two-step cycle is clearly indicated by Figure 10.23. The investigators, however, have not published experimental results for the two-step cycle. They have found that the conversion of the two reactants, H_2S and SO_2, are not the same as they should be according to the Claus reaction stoichiometry. Lower conversion of SO_2 is explained by Menge et al. by differences in the adsorption of the reactants on the alumina surface, rather than by the parasitic formation of carbonyl sulfide, COS, from H_2S with traces of CO_2 in the reactor feed.

As mentioned in the previous section, experiments have also been reported by the University of Dortmund team on the HCN synthesis from ammonia and carbon monoxide via the gas phase reaction:

$$2CO + NH_3 \leftrightarrow HCN + CO_2 + H_2$$

The equilibrium conversion is only 17% for this reaction at 673 K and a stoichiometric ratio of reactants. Elsner et al. (2002a,b) discovered through measurements in a 10 mm i.d. by 100 mm reactor that the conventional high temperature water gas shift catalyst, Fe_3O_4/Cr_2O_3, also catalyzes the HCN reaction. With a reaction temperature greater than 523–573 K, chemisorption capacity is limited so the investigators turned to a gas–solid reaction for CO_2 capture. Their early choice, the exothermic formation of lithium carbonate from the zirconate,

$$Li_2ZrO_3 + CO_2 \leftrightarrow Li_2CO_3 + ZrO_2$$

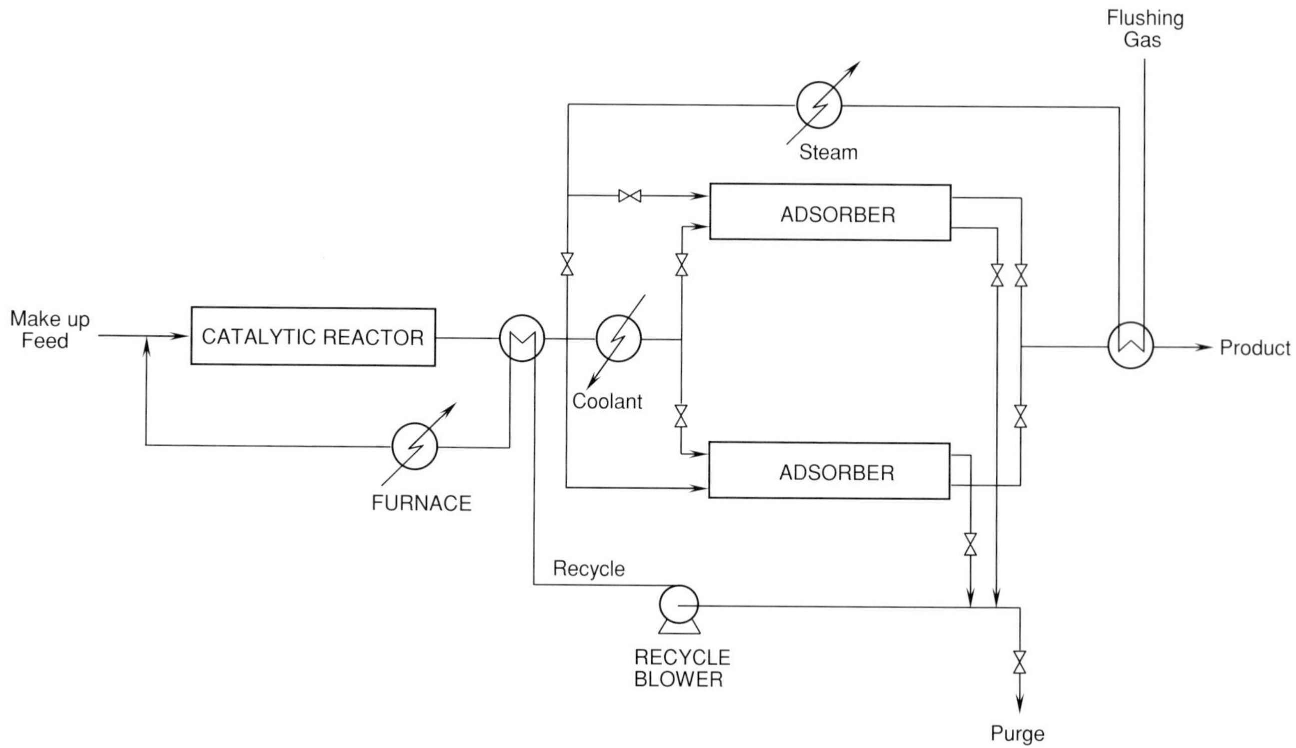

Figure 10.22 Schematic of a conventional sequential reactor separator arrangement with recycle of unreacted feed.

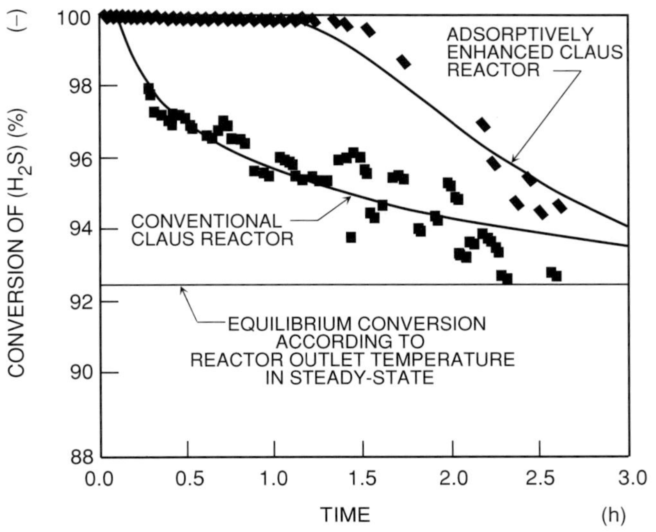

Figure 10.23 Comparison of experimental and predicted H_2S conversion for a 1-m-long reactor operating at 1 bar and 623 K and packed with either a uniform mixture of alumina catalyst and 3A zeolite adsorbent or catalyst and a silicon carbide inert. (Figure adapted from Elsner et al. (2003) with permission. © 2003 Elsevier Science B.V.)

was found to proceed too slowly in the range of operating conditions studied. Instead, a potassium form of hydrotalcite was employed. This "capture" material was studied for the steam reforming of methane using PSR systems and has been discussed in the previous chapter (Section 9.7).

Using a somewhat larger, but still isothermal, reactor, Elsner et al. explored just the feed and production step of a two-step temperature swing cycle over a wide range of operating conditions (temperature, pressure, adsorbent to catalyst, NH_3/CO and reactants to diluent ratios). They found that it is not possible to enhance the equilibrium conversion. Indeed, in the presence of a N_2 diluent, HCN was not detected at the reactor outlet. Furthermore, catalyst activity appeared to decline with time on stream. The author's analysis suggested that this unexpected failure was caused by the Boudouard reaction,

$$2CO \leftrightarrow C + CO_2$$

which also will be forced in the forward direction by CO_2 "capture". Carbon deposition probably explains catalyst deactivation as well. Coadsorption of HCN and CO_2 by the hydrotalcite adsorbent contributed to the absence of enhancement, although the hydrotalcite capacity for HCN is much lower than that for CO_2.

This negative outcome crystallizes the pitfalls that can occur in developing pressure or temperature swing separating reactors. For the HCN synthesis, these were (1) an important side reaction, (2) "capture" of both reaction products by the adsorbent, (3) fouling of the catalyst by the side reaction and, with respect to the proposed lithium zirconate adsorbent for CO_2, (4) low rates of CO_2 adsorption compared to the rate of reaction.

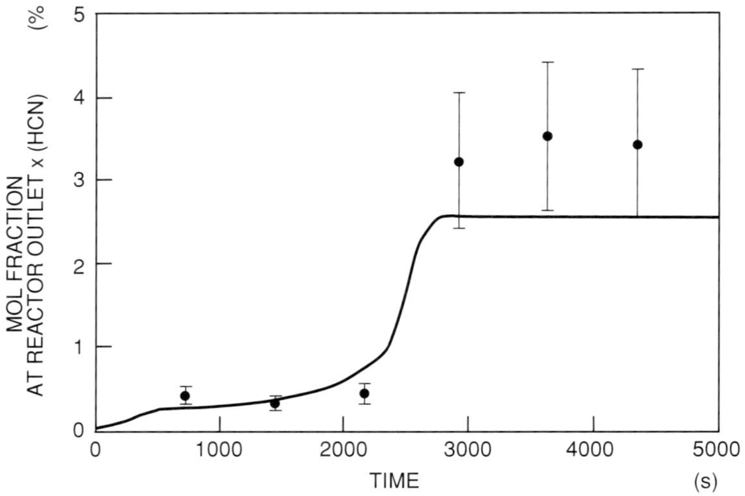

Figure 10.24 Comparison of measured and calculated concentrations of HCN in the outlet of a mixed bed of catalyst and adsorbent for an experiment using excess ammonia (NH_3:CO = 0.5). Other conditions are $T = 673$ K, $P = 300$ kPa, adsorbent:catalyst ratio = 2, space velocity = 537 h^{-1} (STP). (Figure adapted from Elsner et al. (2002a) with permission. © 2002 Elsevier Science Ltd.)

When terms for the reaction and coadsorption of HCN are introduced into the model mentioned in Section 10.2, Figure 10.24 shows that the model represents reasonably well HCN breakthrough in the feed and production step of the cycle. The figure shows the change of HCN in the reactor outlet with time after startup for the 160-mm bed used by Elsner et al. Vertical lines through the experimental points indicate uncertainty in the measurements.

Chapter 11

Combined Pressure and Temperature Swing Reactors

11.1 Concept

Regeneration of the adsorbent in a swing reactor can be more challenging than the reaction and adsorption step. So far, we have considered using depressurization and a temperature increase for regeneration. In the preceding section on chromatographic reactors, a third method, flushing with a carrier gas was also introduced. It is also possible to regenerate the adsorbent by reacting the adsorbate with a component in the regenerative gas to form a vapor phase species that leaves with the gas. Furthermore, it is possible to combine two or more regeneration methods. Recently, a combined regeneration method has been applied to achieve both high conversion to and high purity of a desired product. The requirement of high conversion and high purity is encountered in the production of hydrogen for use in fuel cells. Various hydrocarbons, such as methane, low molecular weight paraffins, methanol or ethanol, can be used as feed stocks in a steam reforming water gas shift process. The purpose is to produce a stream rich in hydrogen, but with a carbon monoxide content less than 30 ppm. To achieve good separation in a PSA system for removing CO from H_2, deep adsorbent beds generally must be used. Work undertaken in late 1990s indicates that deep beds will also be needed in a PSR system for high-purity hydrogen production. With deep beds, the problem of regeneration becomes particularly acute. A simple cycle with just depressurization is no longer adequate. A purge step must be added and a repressurization step is also necessary.

Production of hydrogen by steam reforming of natural gas is an important industrial process, not just a process for making fuel cell hydrogen. Two reactions are involved: (1) steam reforming:

$$CH_4 + H_2O \leftrightarrow CO + 3H_2$$

and (2) the water gas shift reaction:

$$CO + H_2O \leftrightarrow CO_2 + H_2.$$

The first reaction is endothermic, whereas the water–gas shift reaction is exothermic. Because of the endothermicity of reforming, relatively high temperatures are needed for complete conversion of methane. The reaction goes to completion under typical industrial operating conditions of 6:1 H_2O/CH_4 and 446 kPa at about 1070 K. If CO_2 can be removed from the equilibrium mixture, the temperature can be reduced substantially. Figure 11.1 shows the effect of removing CO_2 by an adsorbent on the dry equilibrium mixture under industrial

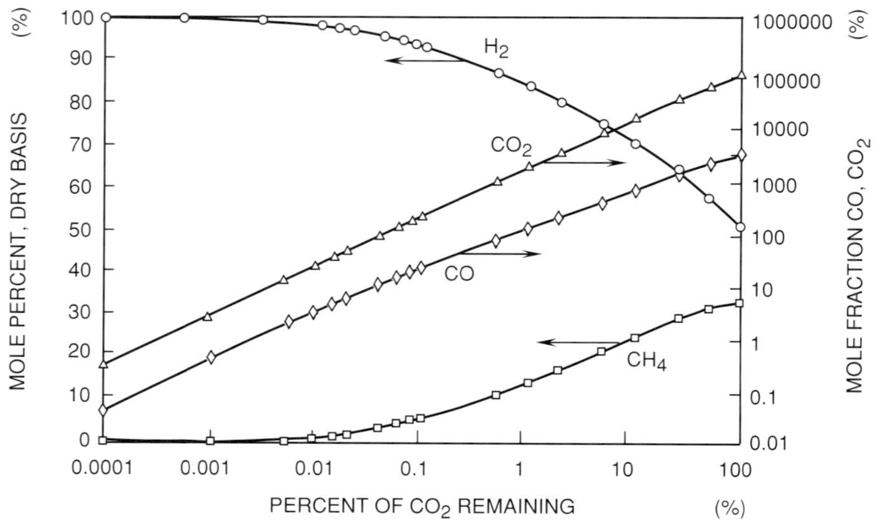

Figure 11.1 Dry composition of the equilibrium mixture as a function of the fraction of CO_2 in the original equilibrium gas remaining after selective removal of CO_2 by an adsorbent. Feed to the reformer used $H_2O/CH_4 = 6{:}1$. (Figure adapted from Hufton et al. (1999) with permission. © 1999 Air Products and Chemicals, Inc.)

operating conditions and a temperature of 1070 K. Ninety-nine percent removal of CO_2 reduces CO in the mixture to less than 10 ppm and CO_2 to about 30 ppm. Hydrogen purity approaches 100%. It is this consideration that has stimulated interest in applying swing reactors to hydrogen production.

Production of fuel cell grade H_2 by reforming natural gas in a swing reactor has been studied, as discussed in Chapters 9 and 10, experimentally by Sircar and coworkers (Hufton et al., 1999; Waldron et al., 2001) and through both simulation and experiment by Ding and Alpay (2000a,b). Rodrigues and coworkers (Xiu et al., 2002a,b, 2003a) have investigated the process using simulation. All of these studies, except that of Ding and Alpay (2000a), considered multistep PSR processes with a different regeneration technique. Hufton et al. (1999) used a five-step cycle, employing for regeneration of the adsorbent depressurization, a purge with methane, followed by a short purge with the hydrogen product to sweep most of the methane out of the reactor. This cycle was simulated by Xiu et al. (2002b), who explored the effect of bed design and operating conditions on CH_4 conversion, H_2 purity and production. We discussed this work in Chapter 9. Later Waldron et al. (2001) tested a four-step cycle that employed a sub-atmospheric purge using superheated steam. Xiu et al. (2002a, 2003a) also studied this cycle. Their analysis suggested that the CO_2 adsorbed could be removed by reducing its partial pressure in the gas phase by forcing the reverse reforming reaction, that is, by generating methane (Xiu et al., 2002b). They also observe that catalyst and adsorbent at the exit end of the reactor are poorly utilized in the pressure swing process. They suggest that methanation of the carbon oxides could be forced at that end by reducing the bed temperature locally. These suggestions of Rodrigues and his coworkers result in a combined pressure and temperature swing operation. Moreover, two heretofore untried regeneration processes, a carrier gas flush and reaction of the adsorbate, were introduced.

11.2 Simulation

The reactive regeneration process proposed by Xiu et al. (2002b) is shown as Case 2 in Figure 11.2. After the feed step the reactor is depressurized through the exhaust port. It is then purged at a lower temperature with a stream containing a portion of the hydrogen product in a nitrogen carrier gas. After this, the reactor is flushed with steam and repressurized with this reactant, but at a higher temperature. The reason for changing temperature is that the reforming reactions are strongly endothermic whereas the methanation reaction taking place during the purge is exothermic. The LHS of the figure, Case 1, depicts the four-step PSR process used by Waldron et al. (2001) and simulated by Xiu et al. (2002a,b, 2003a). The experimental study and the simulation were discussed in Chapter 9.

For the simulation, the model given in Table 3.2 of Chapter 3 was used, however the solids are stationary in the swing reactor so $U_s = 0$ in all equations. Equations 9.33–9.35 and 9.40 restate the model in the form used by Xiu et al. (2002a,b, 2003a,b). Table 9.2 gives the material balances in dimensionless form in terms of component partial pressures. Nevertheless, for convenience, Table 11.1 summarizes the model in a form appropriate for combined pressure and temperature swing operation. It should be borne in mind that the cycle employed is reflected in the boundary conditions rather than in the model.

In multistep cyclic processes there is also a set of continuity conditions,

$$\begin{aligned} C_i(z, t_k)_k &= C_i(z, 0)_{k+1} \\ C(z, t_k)_k &= C(z, 0)_{k+1} \\ T(z, t_k)_k &= T(z, 0)_{k+1} \\ P(z, t_k)_k &= P(z, 0)_{k+1} \end{aligned} \tag{10.1}$$

k in the above condition is the step index for a cycle of K steps and t is time so that the condition states that the properties at an axial point in the bed at the beginning of the next step, $k + 1$ of the cycle, must be the same as the properties at the end of the previous step, namely at t_k. When $k = K$, the number of next step, $k + 1 = 1$. The continuity condition cannot be used for velocity as there is a discontinuity on reversal of flow direction.

Any model is completed by boundary and initial or startup conditions and a set of model parameters. Initial conditions are not important except for the time required for convergence. Usually, a bed filled with a nonadsorbing product is assumed at the temperature and pressure of the first step in the cycle. Danckwerts boundary conditions are normally assumed at the flow entrance and exit of the bed. Walls are zero flux for mass, but are assumed to be at constant temperature for the energy equation. Boundary conditions are specific to the cycle studied as are the values of t_k in the continuity relation, Equation 10.1. Model parameters, rate models and the isotherms for evaluating q_i are determined by the chemical system and the adsorbent chosen.

For the study of the two cases in Figure 11.1, the boundary conditions are summarized in Table 11.2 where:

$$\delta_1 = \left(\frac{\partial C_i}{\partial z}\right)_{z=0}, \delta_2 = \left(\frac{\partial C_i}{\partial z}\right)_{z=L}, \delta_3 = u_{z=0} = \left(\frac{\partial u}{\partial z}\right)_{z=0}, \delta_4 = u_{z=L} = \left(\frac{\partial u}{\partial z}\right)_{z=L},$$

$$\delta_5 = P_{z=0} = \left(\frac{\partial P}{\partial z}\right)_{z=0}, \delta_6 = P_{z=L} = \left(\frac{\partial P}{\partial z}\right)_{z=L}, \delta_7 = \left(\frac{\partial T}{\partial z}\right)_{z=0}, \delta_8 = \left(\frac{\partial T}{\partial z}\right)_{z=L}$$

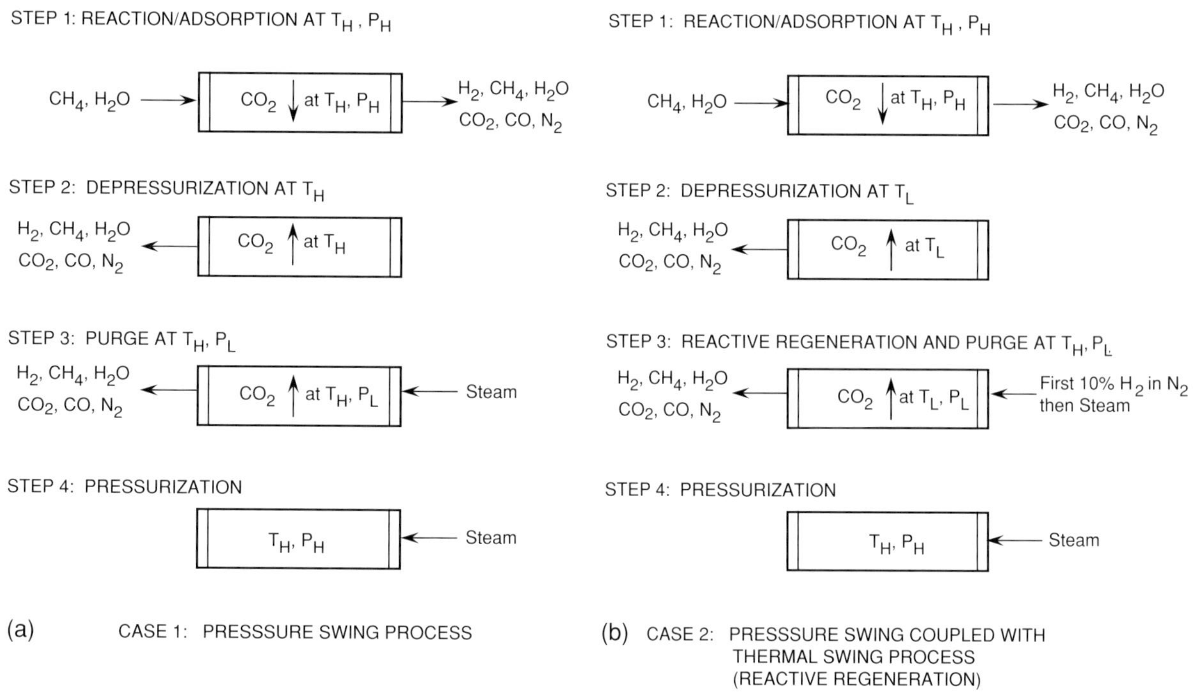

Figure 10.2 Four-step swing processes considered for producing high-purity hydrogen by steam reforming of methane. Case 1: Isothermal pressure swing, Case 2: Nonisothermal combined pressure swing and reactive regeneration of the adsorbent. (Figure adapted from Xiu et al. (2002a) with permission. © 2002 Elsevier Science Ltd.)

Table 11.1 Model equations for combined pressure and temperature swing cycles

Component mass balance	$\varepsilon_t \frac{\partial C_i}{\partial t} + (\rho_b)_{\text{ads}} \frac{\partial q_i}{\partial t} = \varepsilon_b \frac{\partial}{\partial z}\left(D_{ax_i} \frac{\partial C_i}{\partial z}\right) - \frac{\partial u C_i}{\partial z} - (\rho_b)_{\text{cat}} \sum_{j=l}^{N} \nu_{ij}\eta_j R_j$	(11.1)
Total mass balance	$\varepsilon_t \frac{\partial C}{\partial t} + (\rho_b)_{\text{ads}} \frac{\partial q}{\partial t} = -\frac{\partial u C}{\partial z} - (\rho_b)_{\text{cat}} \sum_{i=1}^{n} \sum_{j=l}^{N} \nu_{ij}\eta_j R_j$	(11.2)
Momentum balance as an empirical relation (Ergun equation)	$\frac{\partial \rho_g u}{\partial z} = \frac{dP}{dz} = -J_v u - J_k u^2$	(11.3)
Energy (thermal) balance	$\varepsilon_t C_{p_g} \frac{\partial(\rho_g T)}{\partial t} + ((\rho_b)_{\text{ads}} + (\rho_b)_{\text{cat}}) C_{p_s} \frac{\partial T}{\partial t} = k_z \frac{\partial^2 T}{\partial z^2} - C_{p_g} \frac{\partial(u \rho_g T)}{\partial z}$ $-(\rho_b)_{\text{ads}} \sum_{i=1}^{N_C} (-\Delta H_{a_i}) \frac{\partial q_i}{\partial t} - (\rho_b)_{\text{cat}} \sum_{i=1}^{N_C} \Delta H_j \eta_i R_j + \frac{4 h_W}{d_r}(T_W - T)$	(11.4)
Adsorption rate model (linear driving force assumption)	$\frac{\partial q_i}{\partial t} = k_i (q_{s_i} - q_i)$	(11.5)

$P^* = -(P_f - P_L)/t_2, (P_f - P_L)/t_4$ for steps 2 and 4, respectively, ** = wall temperature and depends on case. Note that velocity u is always a positive number and z is measured from the feed entrance.

For the kinetics of the two reforming reactions and the water gas shift reaction, Xiu et al. (2002b) used the model of Xu and Froment (1989) that was presented in Chapter 9 as Equations 9.41–9.43. As mentioned in that chapter, Ding and Alpay (2000a) have verified the adequacy of these kinetics. A constant effectiveness factor of 0.8 was assumed for all reactions. The importance of intraparticle diffusion on pressure swing systems was dealt with in Chapter 9. Adsorption of CO_2 on the hydrotalcite employed the isotherms devised by Ding and Alpay (2000b) from their experiments for the presence and absence of steam.

Table 11.2 Boundary conditions for a combined pressure and temperature swing reactor

Conditions at $z = 0$					
Step	δ_1	δ_3	δ_5	δ_7	
1	$-\frac{u_{1_0}(C_{i_f} - C_i)}{\varepsilon_b C D_z}$	u_{1_0}	P_f	$-\frac{u_{1_0}\rho_g C_{p_g}(T_f - T)}{k_z}$	
2	0	0	P^*	0	
3	0	0	0	0	
4	0	0	0	0	
Conditions at $z = L$					
Step	δ_2	δ_4	δ_6	δ_8	T_w
1	0	0	0	0	T_f
2	0	0	0	0	**
3	$\frac{u_{3_0}(C_{i_f} - C_i)}{\varepsilon_b C D_{z_i}}$	$-u_{3_0}$	P_L	$\frac{u_{3_0}\rho_g C_{p_g}(T_f - T)}{k_z}$	T_f
4	$\frac{u_{4_0}(C_{i_f} - C_i)}{\varepsilon_b C D_{z_i}}$	0	P^*	$\frac{u_{4_0}\rho_g C_{p_g}(T_f - T)}{k_z}$	T_f

Case 1 has been discussed in Chapter 9 and it was mentioned that the model was tested against the experimental results of Waldron et al. (2001) for a 6-m bed and reasonably good agreement with experiment was found, given uncertainties with adsorption data and the use by Waldron et al. of a steam purge at sub-atmospheric pressure in their experiments. For their simulation, Xiu et al. (2003a) assume all feed and wall temperatures, T_f and T_W, are held at 723 K. The feed pressure, P_f, is 445.7 kPa and the exhaust pressure, P_e, is 125.7 kPa. Duration of the steps, t_k, are 500, 150, 450 and 100 s, respectively. For case 2, the combined pressure and temperature swing operation, $T_f = T_W = 723$ K in steps 1 and 4, but $T_f = T_W = 673$ K in step 3. There is no flow into the reactor in step 2, however the temperature is held at 673 K. Duration of each step is the same, although the third step consists of a 400 s purge with H_2 in nitrogen, followed by a 50-s purge with low-pressure steam. High and low pressures are the same. Xiu et al. (2003a) assume that cooling of the bed takes place instantly when the wall temperature is lowered by 50°C. At the end of the purge period the wall temperature rises instantly by 50°C.

Reactive regeneration is compared with a steam purge at the feed temperature in Figure 11.3. The figure shows the adsorbate concentration on the hydrotalcite at the end of the first three steps for the 2-m-long bed studied. It can be seen that the first 0.5 m of the bed are saturated at the end of the first step. Depressurizing desorbs only a small amount of CO_2. Adsorbate concentration drops enormously in the purge step as the four bottom curves in the figure show. Nevertheless, the reactive purge is more effective than the steam purge at 723 K. Comparing the hydrogen purge curves at three different temperatures, there appears to be an optimal temperature for this purge of about 673 K.

Figure 11.4 compares rates of CO_2 desorption with the sum of the rates of the reverse water gas shift and methanation at 673 K. In the back end of the reactor, adsorbate concentrations are low so CO_2 desorption is slow. The figure shows, however, that by adding a reaction that consumes CO_2, desorption is accelerated and it equals the rate of consumption at the end

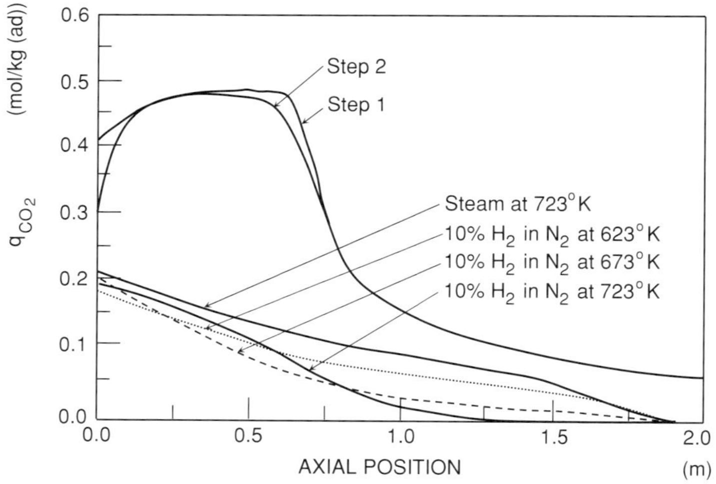

Figure 11.3 Adsorbate concentration on hydrotalcite at the end of cycle steps as a function of axial position for a 2-m-long reactor operating with $u_{1_0} = 0.08$ m·s^{-1}, $u_{3_0} = 0.3$ m·s^{-1} and cycle durations given in the text. (Figure adapted from Xiu et al. (2002a) with permission. © 2002 Elsevier Science Ltd.)

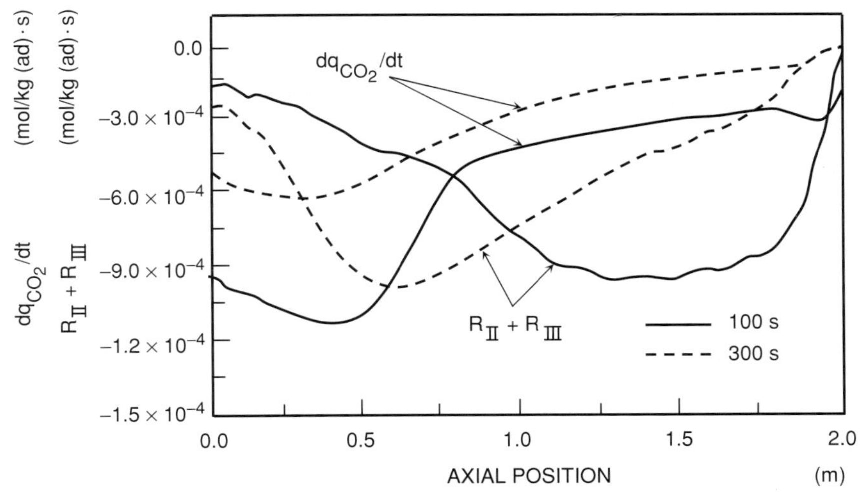

Figure 11.4 Rates of CO_2 desorption from hydrotalcite and consumption by reaction with hydrogen at two times in the purge step as a function of axial position for a 2-m long reactor operating at 673 K with other conditions as given in Figure 11.3. (Figure adapted from Xiu et al. (2002a) with permission. © 2002 Elsevier Science Ltd.)

of the reactor. Lower temperatures drive down desorption further within the reactor and drop reaction rates. However, high gas phase CO_2 concentration raises rates at the reactor entrance.

The Xiu simulation showed that a 2-m bed is too short to reduce the CO concentration down to the target level of 30 ppm for the combined cycle, even though the performance is better than Case 1 (discussed in Chapter 9). Methane conversion goes up from 47.5 to 51%, purity from 78 to 80.5% and CO in the product drops from 150 to 110 ppm. Because some of the product is used for the purge, production for the 2-m bed decreases from 0.56 to 0.41 mol/kg of solid/cycle. Carbon monoxide level in the product is much too high for direct use in a PEM fuel cell. The permissible level for such cells is about 30 ppm.

In Chapter 9, it was observed that increasing bed length increased H_2 purity as well as CH_4 conversion. Xiu et al. (2003a) tested the effect of bed length in their numerical study. Using the same step durations, feed velocities, temperatures and pressures as indicated in the above text and in Figures 11.3 and 11.4, a 6-m bed brought the CH_4 conversion to 62% and the H_2 purity to 86.5% for the combined pressure and temperature swing cycle. CO contamination of the product fell to about 50 ppm. H_2 production rate dropped drastically to 0.17 mol/kg of solid/cycle, unfortunately. Waldron et al. (2001) used a 6-m bed in their experimental study of a four-step PSR.

Increasing the length of the mixed catalyst and adsorbent bed while keeping feed rates and cycle period constant creates a large region that has quite low adsorbate levels at the product discharge end of the bed. This can be readily seen in Figures 11.5(a) and (b) showing the adsorbate concentration profiles at the end of the cycle steps. Profiles in both parts assume the same operating parameters, cycle periods and step durations. Only bed length differs. Although there is little difference in the adsorbed CO_2 levels at the end of the bed by the end of step 4, the much greater amount of adsorbent significantly reduces the adsorbate level

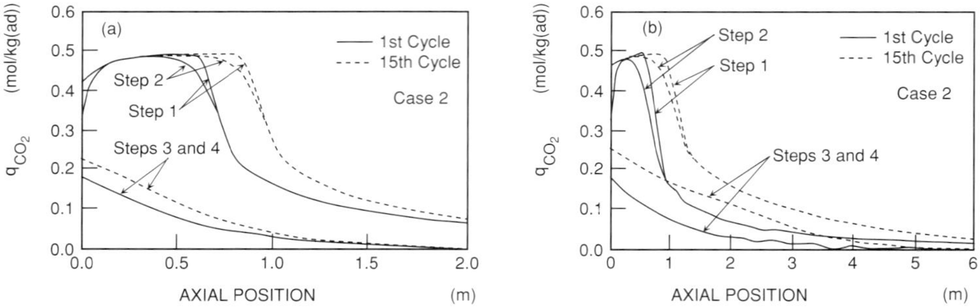

Figure 11.5 CO_2 adsorbate distributions along the bed at the end of cycle steps in the combined pressure and temperature swing cycle for the catalytic steam reforming of CH_4 in the presence of an adsorbent: (a) 2-m bed, (b) 6-m bed. (Figure adapted from Xiu et al. (2002a) with permission. © 2002 Elsevier Science Ltd.)

at the end of the step 1, the production step. Note from the figures that most of the CO_2 is held in the first meter of length for both beds at the end of step 1 and that the adsorbent is saturated in this region.

Reducing the duration of the feed step increases conversion, H_2 purity and lowers CO content of the product gas. A feed duration of 350 s in place of 500 s, gives a product gas that meets the CO content specification for fuel cell use. Production rate, however, falls further to 0.11 mol/kg solids/cycle. Cycle period for this operation is 1050 s. Duration of depressurization is 150 s while the purge step accounts for 450 s, but in the last 50 s, 673 K steam replaces 10% hydrogen in a nitrogen carrier gas used as the purge. Repressurization occurs over 100 s. The shorter duration results in a 10% increase in conversion and an improvement in H_2 purity from 86.5 to 89%. Even with the shorter feed duration, the four-step PSR cycle (Case 1) was not able to reach the CO level requirement.

In general, the simulation showed that reactive regeneration results in higher conversion and hydrogen purity, but lower productivity per kg catalyst than the four-step isothermal pressure swing cycle. Reaching a cyclic steady state required about eight cycles or about 2.5 h for both cycles considered. Xiu et al. (2003b) claim that the use of reactive regeneration considerably reduces the duration of the regeneration step for the same hydrogen purity and CO level in the product gas.

The problem of the combined cycle is the poor utilization of catalyst and adsorbent. Methane conversion occurs primarily in the front portion of the bed, close to the entrance. Removing all of the carbon oxides through adsorption means, also, that regeneration of the adsorbent becomes limiting. Xiu et al. (2003b) suggest that these problems can be improved by dividing the reactor into sections and operating with different temperatures and catalyst to adsorbent ratios in each section. The concept underlying breaking the bed into sections is that different processes predominate at different points. Catalytic reactions occur in the front section, while in the middle section adsorption becomes important and acts to hinder reverse reactions that consume product. At the end of the bed, adsorption, though slow, operates to reduce the concentrations of the undesired components in the product gas. Xiu et al. (2003b) propose to enhance these different roles. Thus, for the steam reforming of methane, they suggest increasing the catalyst to adsorbent ratio in the front section of the bed, section 1, and operating at the highest temperature to force the endothermic reforming reactions to completion. In the middle of the bed, section 2, the catalyst to adsorbent ratio is sharply decreased, but temperature is kept high to remove carbon oxides in order to stifle

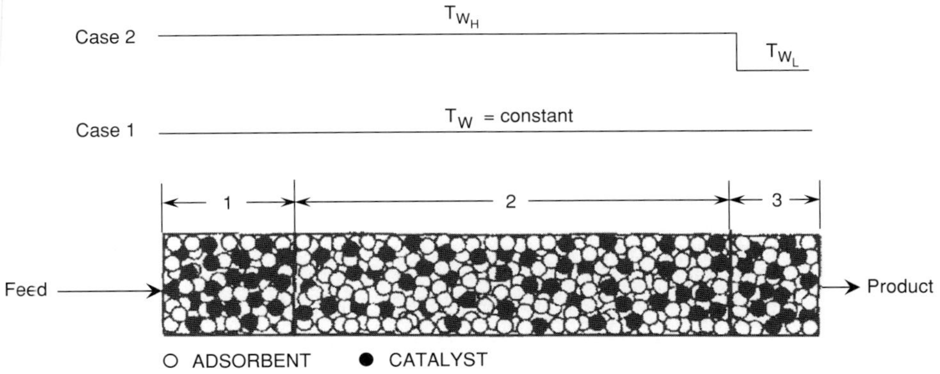

Figure 11.6 Schematic of the three-section, mixed catalyst and adsorbent bed investigated with the proposed wall temperature profiles for the two cases studied. (Figure adapted from Xiu et al. (2003b) with permission. © 2003 Elsevier Ltd.)

the back reactions such as methanation and the reverse water gas shift. At the rear of the bed, section 3, the temperature is lowered by withdrawing heat through the wall and the catalyst to adsorbent ratio is increased. Lowering the temperature shifts the equilibrium from reforming to methanation and forces CO toward CO_2 by the reverse water gas shift. This change reduces the CO level in the product gas without adding to the adsorption burden and serves to make the adsorbent regeneration step less critical. Figure 11.6 shows the division of the bed into sections as proposed by Xiu et al. (2003b) and the wall temperature profile for the two cases these authors studied numerically.

Xiu et al. (2003b) continued to explore the four-step swing reactor cycle considered in their earlier papers on hydrogen production. Their emphasis, as in the paper discussed above, was on a cycle that would reduce both CO_2 and CO content of the product gas. Their target level for CO was less than 30 ppm by volume. Hydrogen productivity, as mol/kg solid/h, is important. Methane conversion is less important because CH_4 is tolerated by PEM fuel cells and waste gas from the process is consumed to supply the heat demand. The four-step cycle, shown in Figure 11.7, is much like the cycle shown in Figure 11.2b except for the purge and repressurization steps. A steam purge is used at a higher temperature in Case 1 and, in Case 2, at a lower temperature. Steam repressurization is at the high or low temperature depending on the purge used.

For their simulation, the investigators retained from their previous work the 2:1 adsorbent to catalyst weight ratio ($(\rho_b)_{cat} = 233.3$, $(\rho_b)_{ads} = 466.6$ kg/m^3 reactor volume) in bed sections 1 and 3, but the ratio is doubled in section 2 so that $(\rho_b)_{cat} = 116.6$ and $(\rho_b)_{ads} = 583.3$ kg/m^3 reactor volume. Total bed depth was taken to be 6 m. Depths of sections 1, 2 and 3 were 0.1, 4.3 and 0.6 m, respectively. H_2O to CH_4 ratio at 6:1 and high and low pressures of 445.7 and 125.7 kPa were held the same as in the authors' earlier work. Temperatures, however, were different. T_H was mainly 763 K and T_L was 693 K. Effects of the higher pressure and the two temperature levels on performance were explored.

The reactor model used was discussed above. Xiu et al. (2003b) made use of the Xu and Froment (1989) kinetics for steam reforming and the Ding and Alpay (2000b) formulation for adsorption on hydrotalcite. Transport properties, such as diffusivities and heat transfer coefficients, were taken from standard references, a practice the authors followed in their

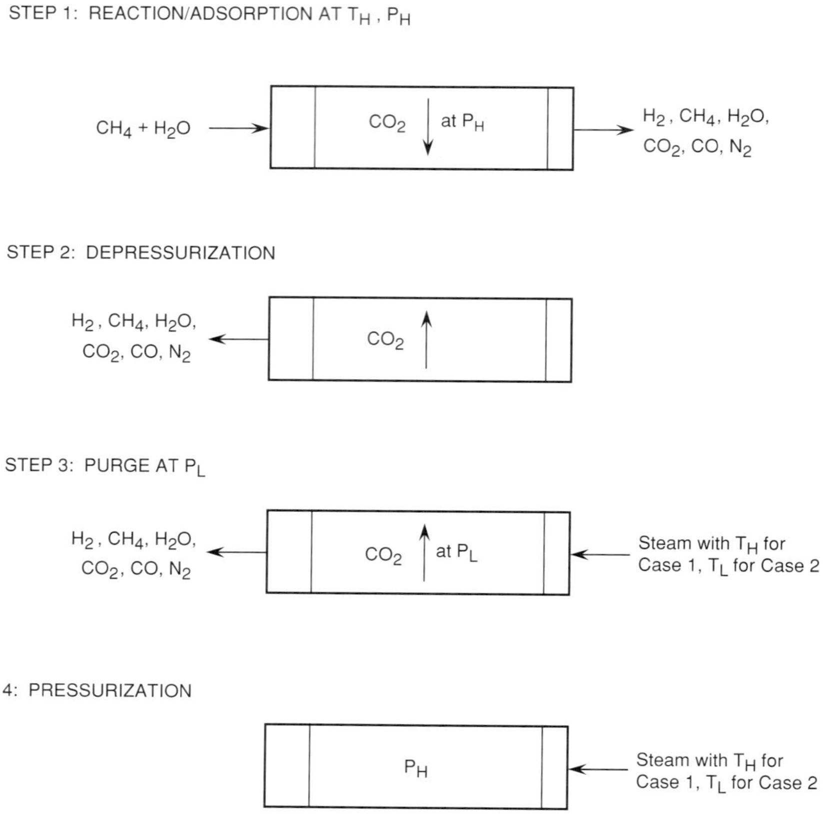

Figure 11.7 Four-step swing reactor process with a three-section mixed catalyst and adsorbent bed for producing high-purity hydrogen by methane steam reforming. Case 1: isothermal pressure swing (PSR), Case 2: nonisothermal combined pressure and temperature swing operation. (Figure adapted from Xiu et al. (2003b) with permission. © 2003 Elsevier Ltd.)

earlier papers. Performance criteria continued to be CH_4 conversion, composition of the product gas and H_2 productivity.

Operating variables explored were T_H, T_L, P_H and t_1, the duration of the production step. Lowering temperature raises H_2 production, but reduces CH_4 conversion. Composition of the product gas is not effected. Duration of the production step has a strong effect in the same direction as lower temperature and it also adversely affects product gas purity, in particular the CO content. It increases, however, the utilization of the catalyst and adsorbent as reflected by large increases in H_2 productivity. Reducing P_H by one half lowers production by about that amount, but methane conversion increases significantly. Product gas purity changes little. Xiu et al. (2003b) comment that acceptable results can be obtained at least for the critical CO level over a fairly wide range of operating conditions provided that temperature swing is used along with a pressure change.

Lowering the wall temperature in section 3 and purging at the lower temperature are critical if CO levels under 30 ppm are to be attained. This is illustrated by Figure 11.8, a

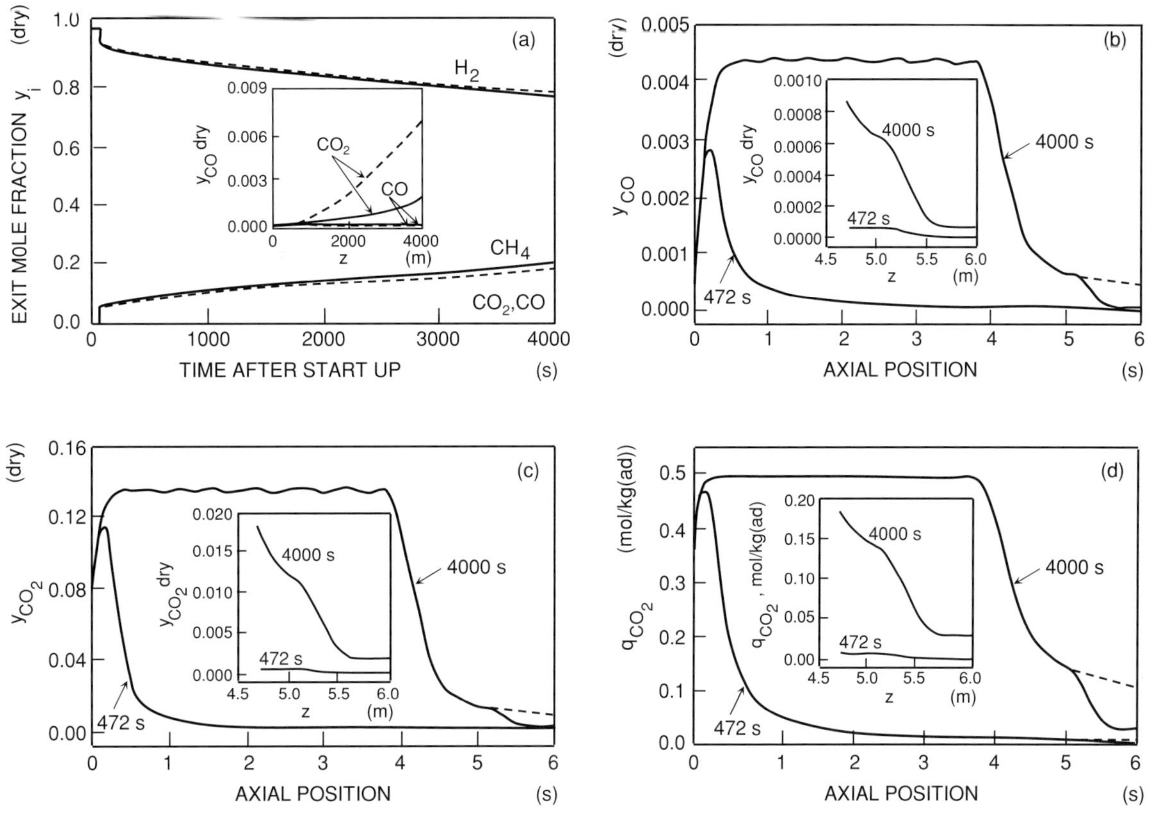

Figure 11.8 Product gas composition leaving the discharge port at the end of the reactor (a), the axial distribution of CO (b) and CO_2 (c) in the gas phase, and CO_2 adsorbate loading (d) at different times, 472 s and 4000 s in the production step for H_2 production in a combined temperature and pressure swing reactor for the two section wall and reverse flow temperature cases given in Figure 11.6. Solid lines = Case 2 while dashed lines = Case 1. (Figure adapted from Xiu et al. (2003b) with permission. © 2003 Elsevier Ltd.)

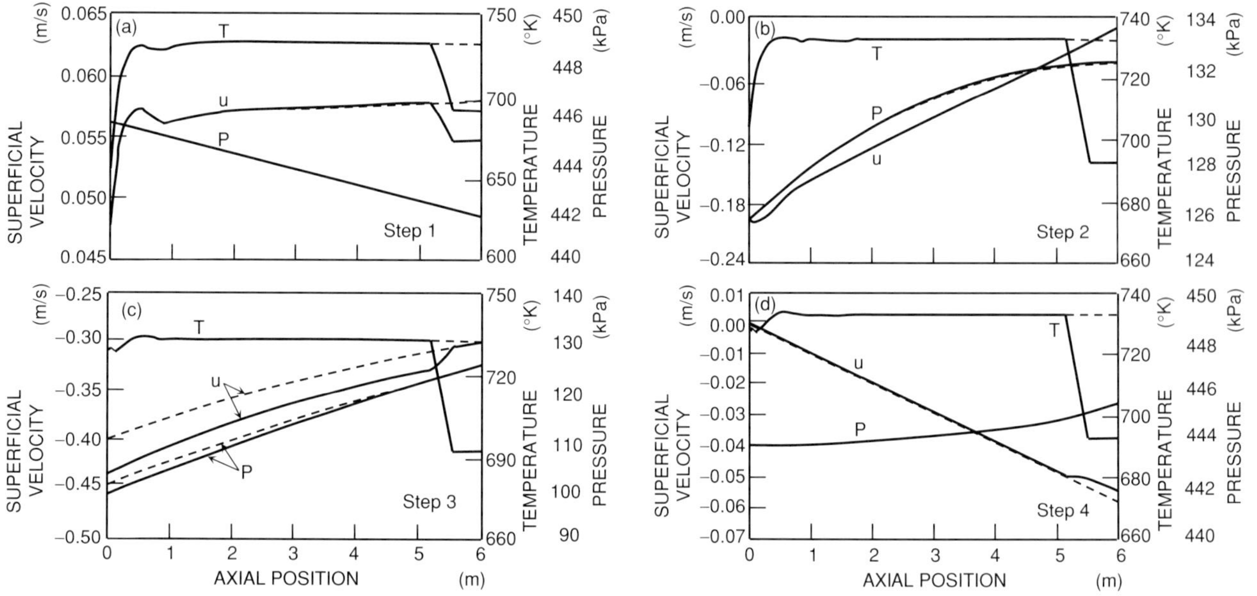

Figure 11.9 Temperature, pressure and superficial velocity profiles in the mixed bed reactor at the end of each of the four steps in the cycle for H_2 production using a combined pressure and temperature swing reactor for the two section wall and reverse flow temperature cases given in Figure 11.6. Solid lines = Case 2, dashed lines = Case 1. (Figure adapted from Xiu et al. (2003b) with permission. © 2003 Elsevier Ltd.)

comparison of simulations for Cases 1 and 2 with the same set of operating conditions, cycle period, bed and bed section dimensions. The only difference, shown in Figure 11.7, is that the wall and purge temperature in step 3 of Case 2 are at 693 K as are the repressurization and wall temperatures in step 4. Figure 11.8a shows the change in composition of the product gas with time in the production step, while the remaining parts of the figure show the axial profiles at the end of the production step. The inserts in Figure 11.8 blow up the profiles to show conditions at the end of the reactor, that is the time behavior at long feed times in (a).

It is easily seen from the above figure that the discharge composition is virtually the same in the two cases except in the last 2000 s of the feed step where reducing the wall temperature by 40°C in section 3 substantially reduces the level of the carbon oxides. The wall temperature effect can also be seen at the end of reactor, the last meter, in Figures 11.8b, c and d. The lack of difference in the profiles from 0 to 5 m or in the profile at $t = 472$ s seems to suggest that the temperature changes in the purge and repressurization steps do not have an important effect on behavior. This conclusion is reinforced by Figure 11.9 that shows the temperature, pressure and superficial velocity profiles at the end of each of the four steps in the swing reactor cycle (see Fig. 11.6) after a cyclic stationary state has been reached. With the exception of the purge step, step 3, that lasts for 600 s, the profiles through the bed are the same except in the final 0.6 m. The effect of the purge temperature is greatest on velocity because the boundary conditions are the same except for temperature. Figure 11.9 also shows how rapidly the wall temperature propagates into the 12.5 mm i.d. bed that Xiu et al. assumed for their study. The duration of repressurization is 100 s. Figure 11.9d shows the bed temperature has reached the wall temperature 1 m into the bed despite the 40 K lower feed temperature and wall temperature of the first 0.6 m. Xiu et al (2003b) find that about 10 cycles or about 3 h are needed to reach a cyclic stationary state for the dual temperature case.

The Xiu simulation showed that operating at constant wall and feed temperature (Case 1), a strictly PSR mode, a CO level of 30 ppm could not be reached for the range of operating variables considered.

Experimental confirmation of the Xiu et al. (2003b) simulation results has not appeared.

Chapter 12

Periodically Pulsed, Trapping and Extractive Reactors

12.1 Introduction

Three quite different cyclically operated reactors are examined in this chapter. The first of these, the periodically pulsed reactors (PPR), can be described mathematically and operated in the laboratory, but they are probably not practical as commercial devices because, like a simple chromatographic reactor, their use of catalyst and adsorbent is inefficient. This class of cyclic reactors uses mixed or segregated beds of catalyst and adsorbent and employs a carrier gas without reactant or product to regenerate the adsorbent in a two-step cycle. Regeneration by flushing with a carrier gas implies quite weak adsorption of reaction products and thus relatively poor separation. Strong adsorption would indicate a large heat demand for adsorbent regeneration that would be difficult to supply without heating the carrier gas and making the operation a temperature swing one.

Cyclically operated trapping reactors (PTR) are the inverse of a separating reactor. An adsorbent is used, but it serves to remove a reactant from a fluid stream, concentrate it and introduce that reactant into the reaction system. We consider such operations in this monograph because the reactor–adsorbent systems resemble separating reactors in design and they employ similar mathematical models.

Use of a liquid phase instead of a solid one to remove one or more products from a reacting system introduces a separation method that has not been considered in this monograph except for a brief mention in Chapter 1. Cyclic operation of such separating reactors may not be necessary because the liquid phase is readily withdrawn from the reactor and can be conveniently regenerated externally. We have chosen to discuss cyclically operated extractive reactors (CER) in this chapter because when a solid is used, either as a distributor or as a catalyst, extractive processes are frequently mass transfer controlled. Reactors containing a solid with mobile gas and liquid phases constitute a three-phase system. A well-known example of such systems is the trickle bed reactor widely used in the refining industry. There is a large body of data demonstrating that the performance of three-phase reactors can be improved by periodic flow modulation (Silveston and Hanika, 2004).

12.2 Periodically pulsed reactors

In this section we are concerned with isothermal regeneration using flushing by a carrier gas or an eluent. We will refer to a separating reactor functioning with periodic flushing as

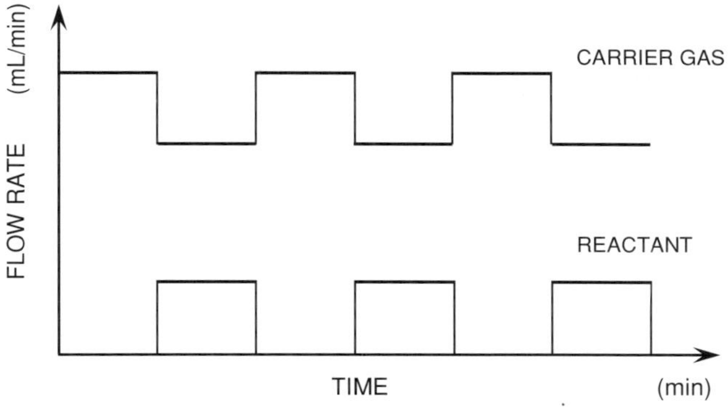

Figure 12.1 Flow rate variation with time at the feed port of a separating reactor operating with periodic pulsing.

a periodically pulsed reactor (PPR). Unlike desorption of product or possibly reactant from an adsorbent by reducing pressure, a PPR can be applied to systems with either gas or liquid mobile phases. Of course, flushing can also be part of a combined operation using depressurization and heating as well. Indeed, flushing is always used in temperature swing systems as may be seen in Chapter 11. Some PSR cycles use a low-pressure purge (see Section 9.2).

Flushing with a nonadsorbing fluid changes the adsorbate partial pressure in the mobile phase. This disturbs the gas phase-adsorbate equilibria causing desorption. The partial pressure change may be brought about by either removing reactants and products from the gas flow or by increasing that flow. The former is usually employed.

When changing reactant concentrations in the carrier gas or eluent, PPRs closely resemble reactors operating under composition modulation as can be seen from a schematic of their function in Figure 12.1. The schematic shows a constant volumetric flow rate with time with a composition change, but it is also possible to impose reactant flow on a time-invariant eluent flow so that both feed composition and flow rate vary periodically. Symmetrical square-wave pulses are shown in the schematic. The fraction of the cycle, or cycle split (s), devoted to reactant feed can be greater or smaller than 0.5. The split, like the cycle period, depends on reaction and adsorption kinetics. A very small split, typically $s < 0.1$, represents a chromatographic reactor. Many periodically pulsed reactors operate with $s > 0.5$. Experimental and theoretical studies of this type of reactor have been underway since the 1980s. The application of composition modulation to the catalytic dehydration of alcohols to produce olefins is a good example of this technique. The work is thoroughly discussed by Silveston (1998) in his book on composition modulation. In alcohol dehydration reactions, both reactant and product are adsorbed, but in most applications of cyclically operated separating reactors discussed so far adsorption of just one of the products is sought.

Periodically removing reactant from the reactor feed halts product formation. This reduces product concentration in the fluid phase allowing product desorption and thus adsorbent regeneration. The alternative, as discussed in Chapters 5 and 6, is to move adsorbent countercurrent to the fluid phase. This replaces a periodic operation by a continuous one with respect to feed and product flow. If the catalyst is fixed in place, it remains at steady state at any point in the reactor, but the moving adsorbent is not at steady state.

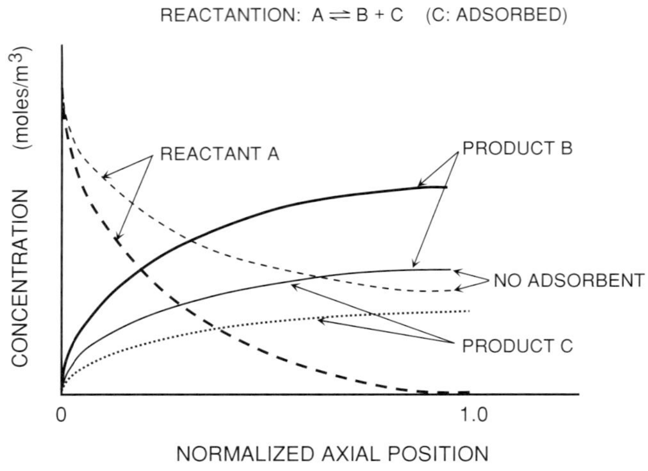

Figure 12.2 Time-averaged concentration profiles when A is present in the reactor feed in a periodically pulsed reactor with catalyst + adsorbent for product C (heavy lines) and with catalyst and a nonadsorbing diluent (light lines) for an equilibrium-limited reaction.

The consequence of periodic pulsing or countercurrent adsorbent flow is to effectively reduce the amount of adsorbate held by the absorbent at any point in the separating reactor. As can be appreciated from Figure 12.1, the adsorbed product is separated from the nonadsorbed one. Considering an equilibrium-limited reaction, $A \leftrightarrow B + C$, in a PPR, the effect of separation is illustrated in Figure 12.2. In the figure, the time average behavior, while A is flowing, is compared for two reactors of equal size operating at equal space times where one reactor, the PPR, contains a catalyst and an adsorbent for only product C while the other reactor contains just catalyst and a nonadsorbing solid that replaces the adsorbent volumetrically. The PPR is subjected to the pulsing shown in Figure 12.1 (reactant + carrier gas, carrier gas only). Note that with periodic pulsing, the product B leaves the reactor during the portion of the cycle when A is present in the feed, while C exits mainly when the feed contains just the carrier gas. Both products are diluted by carrier gas, but C more so than B. In this illustration, complete conversion of A has been assumed as well as equilibrium adsorption of C when adsorbent is present. Concentrations of both products B and C in the PPR exit vary with time between high and low values. When reactor effluent flows are time averaged over the part of a cycle that A enters the reactor, the profiles given in Figure 12.2 are obtained. The conversion of A and recovery of product B in the PPR effluent increase over those when just catalyst and a nonadsorbing solid are used. C is adsorbed and leaves the PPR with the carrier gas so it emerges at a lower concentration. Of course, the amount of C recovered must be the same as the amount of B recovered. Product C leaves the reactor, however, when A is in the feed and when it is absent. The same amount of carrier gas flows through the reactor with inert in place of adsorbent so product B concentrations are lower. How much the product B concentrations or reactant conversion exceed those for the continuous PFR packed with just catalyst, as shown in Figure 12.2, depends on the cycle parameters, operating conditions, adsorption equilibria and the reaction rate.

If a countercurrent moving bed of adsorbent is employed in place of periodic pulsing, C emerges with the adsorbent, and its profile in Figure 12.2 will be inverted. Profiles for the other components do not change significantly and the effect of adding an adsorbent to the reactor charge on reactor performance is the same.

Equipment-wise, a PPR is the same as a CR or a two-port PSR. They consist of a packed bed of intimately mixed catalyst and adsorbent with feed and exit ports at either end. There have been some theoretical studies and experimentation with bed segregated into alternating bands of catalyst and adsorbent. Segregated beds can be advantageous when catalyst and/or adsorbent must be periodically replaced.

Utilization of catalyst and adsorbent in a PPR can be improved by using a moving bed or, as discussed in Chapter 3, breaking up the reactor into a cascade of packed beds and switching the feed, purge and take-off locations periodically to create a simulated, countercurrent moving bed of catalyst and adsorbent. In this variant, the PPR becomes essentially an SCMCR. Such a reactor is discussed by Mensah and Carta (1999) for esterification using an immobilized enzyme and an organic liquid carrier. Products of the reaction are an ester, which is flushed away by the carrier, and water, which is trapped by the adsorbent. In this situation, a reactor consisting of just two beds can be used, although better use of the adsorbent can be attained by using multiple beds where just one of the beds undergoes regeneration as shown in Figure 12.3. The multiple beds allow a reactant wave to propagate downward through the cascade. Before breakthrough, the top bed is shifted to the eluent stream and the regenerated bed is added to the bottom of the cascade. Generally, the eluent will be a solvent for the product and will be different from the carrier, unlike a chromatographic system.

12.2.1 Simulations

Periodically pulsed reactors using fully mixed beds of catalyst and adsorbent are generally modeled as pseudo-homogeneous by the partial differential equations given in Chapters 9 and 10 or in Table 3.2. If mass transfer of reactant or a product in the reaction network is slow enough to influence the overall reaction rate, a heterogeneous model, as given in Table 3.3, must be used. Usually the model equations are normalized to yield the equations presented in Tables 3.4 and 3.5. Boundary conditions are identical to those chosen for fixed bed chromatographic reactors. As for these reactors, initial conditions depend on the startup mode.

Mensah et al. (1998a,b) and Mensah and Carta (1999) employed the pseudo-homogeneous packed-bed model in Table 3.2 to describe a cyclically operated adsorber–reactor that they used for the esterification of propionic acid with isoamyl alcohol over a lipase immobilized on a macroporous anion-exchange resin. Carrier for the reactants was hexane and the eluent or desorbent (Fig. 12.3) was isoamyl alcohol. A cationic ion exchange resin was also used to remove water. Product adsorbs on both the enzyme catalyst and on the adsorbent. Thus, Equation 3.1 in Table 3.2 becomes

$$\varepsilon_t \frac{\partial C_i}{\partial t} + (\rho_b)_{\text{cat}} \frac{\partial q_i^{\text{cat}}}{\partial t} + (\rho_b)_{\text{ads}} \frac{\partial q_i^{\text{ads}}}{\partial t} = D_x \frac{\partial^2 C_i}{\partial x^2} - \frac{1}{A} \frac{\partial n_i}{\partial x} - (\rho_b)_{\text{cat}} \nu_i r \qquad (12.1)$$

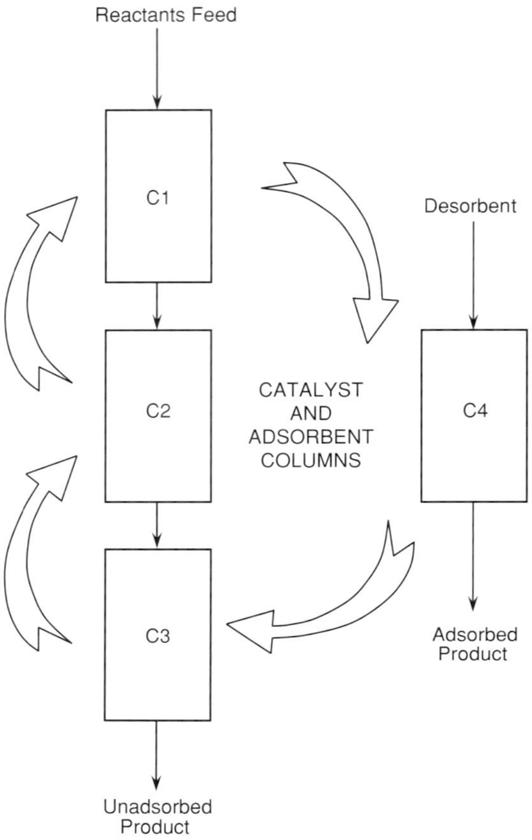

Figure 12.3 Operation of a cyclic adsorptive reactor in which a catalyst is deactivated by a reaction product. (Figure adapted from Mensah and Carta (1999) with permission. © 1999 John Wiley & Sons, Inc.)

where q_i^{cat} and q_i^{ads} represent adsorbate concentrations on the catalyst and adsorbent respectively, with units of mol/g of solid, and ε_t is the total porosity (bed + particle). Mensah and Carta further assume that adsorption is mass transfer controlled, while esterification is kinetically controlled. This assumption changes Equation 12.1. The time derivative of the adsorbate must be replaced by a mass transfer term. An adsorbate material balance then relates the mass transfer of component i to the adsorbent to the rate of change of the adsorbate density. Mensah and Carta employ a linear driving force model for the mass transfer controlled adsorption rate. The lipase catalyst loses activity in the presence of adsorbed water and is entirely deactivated when free water is present, which explains the use of a pulse reactor system for the esterification. The reaction is also reactant-inhibited so the rate model is complicated.

$$r = \frac{k_m C_{\text{acid}} C_{\text{alcohol}}}{C_{\text{acid}} C_{\text{alcohol}} + K_{\text{alcohol}} C_{\text{acid}} \left(1 + \frac{C_{\text{acid}}}{K_{\text{acid}}}\right) + K_{\text{acid}} C_{\text{alcohol}} \left(1 + \frac{C_{\text{alcohol}}}{K_{\text{alcohol}}}\right)} \quad (12.2)$$

In this expression, the rate constant depends on adsorbed water as

$$k_m = 3.6 \times 10^{-3}\left(1 + 0.028q_{\text{water}}^{\text{cat}} - 0.016\left(q_{\text{water}}^{\text{cat}}\right)^2 + 1.11 \times 10^{-3}\left(q_{\text{water}}^{\text{cat}}\right)^3 - 2.3 \times 10^{-5}\left(q_{\text{water}}^{\text{cat}}\right)^4\right) \quad (12.3)$$

provided the adsorbed water is below 0.022 mol/g catalyst. Adsorption equilibria on the immobilized lipase and the cation-exchange resin were measured by the researchers in terms of activity and fitted to quadratic relations so

$$\begin{aligned} q_{\text{water}}^{\text{ads}} \times 10^3 &= 56.4a_{\text{water}} - 106a_{\text{water}}^2 + 105a_{\text{water}}^3 \\ q_{\text{water}}^{\text{cat}} \times 10^3 &= 21.5a_{\text{water}} - 57.8a_{\text{water}}^2 + 61.4a_{\text{water}}^3 \\ q_{\text{acid}}^{\text{ads}} \times 10^3 &= 57.2a_{\text{acid}}/(1 + 19.7a_{\text{acid}}) \end{aligned} \quad (12.4)$$

Activities were calculated using the UNIFAC model. Boundary conditions for the model are given in Table 3.7.

The above model was used by Mensah and Carta to design their laboratory unit. They chose a total bed length to provide the desired conversion and then manipulated the switching time, τ_s, to ensure that the water activity did not exceed 0.9. When $a_{\text{water}} = 1$, irreversible deactivation of the lipozyme occurs. Details are given in Mensah et al. (1998b).

Experiments were undertaken to validate the model. These and model predictions are discussed in Section 12.3.

Segregated beds of catalyst and adsorbent can be built up from a shallow layer of catalyst followed by a layer of adsorbent. An interesting alternative has been suggested by Štěpánek et al. (1999). They propose using a square channel monolithic structure and coating the walls with separate bands of catalyst and adsorbent. This should offer a compact, low-pressure drop reactor, one that could be miniaturized to a microscale. The structure they propose is illustrated in Figure 12.4. A microreactor built from a monolith should be attractive for

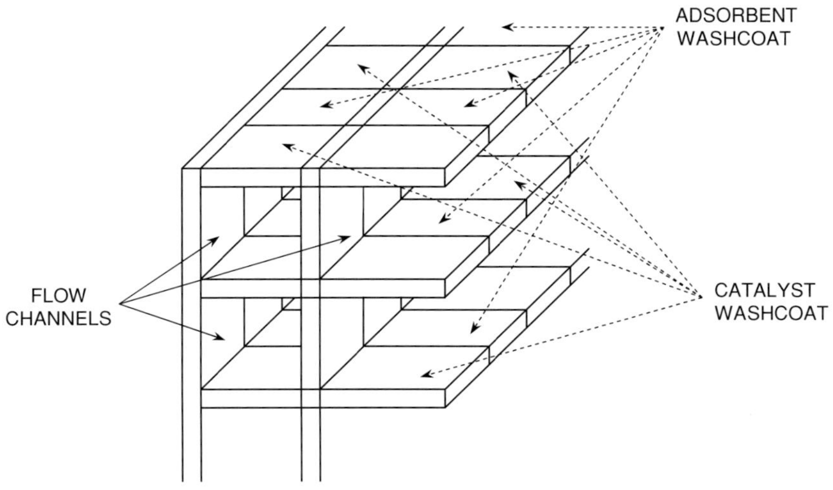

Figure 12.4 Square channel, monolith microreactor with separate bands of catalyst and adsorbent along the flow path. (Figure adapted from Štěpánek et al. (1999) with permission. © 1999 Elsevier Science Ltd.)

a biochemical reaction with a low product demand. Reaction would be catalyzed by an immobilized enzyme, while an ion-exchange resin would carry out the separation function. Construction of such a monolith microreactor is not discussed by the authors; only a reference is given to a monograph on structured catalytic reactors (Cybulski and Moulijn, 1997).

The objective of the Štěpánek study was to demonstrate that a large yield enhancement can be obtained through a cyclically operated separating reactor. A further objective was to examine reactor optimization in a situation in which both yield and separation must be considered. As an example, Štěpánek et al. (1999) chose production of nicotinamide by the hydration of cyanopyridine employing an immobilized nitrile hydratase.

Because of the distinct structure, heterogeneous models are generally employed for the segregated or banded system illustrated in the above figure. The set of equations given in Table 3.3 of Chapter 3 apply, but they require some modification because the active materials are coated on walls and are not distributed on particles intercepting fluid flow. Because monoliths provide good heat conduction and low-pressure drop, they are often assumed to be isothermal and isobaric. These assumptions apply particularly to microreactors. Consequently, energy balances, Equation 3.3, and momentum balances, Equation 3.4, are routinely neglected.

For the reactor shown in Figure 12.4, Štěpánek et al. (1999) write

$$\frac{\partial C_i}{\partial t} = -u\frac{\partial C_i}{\partial x} + a f_{\text{cat}}\frac{1-\varepsilon}{\varepsilon}(\rho_b)_{\text{cat}}\nu_i r_{\text{cat}} - a(1 - f_{\text{cat}})\frac{1-\varepsilon}{\varepsilon}(\rho_b)_{\text{ads}} r_{\text{ads}_i} \quad (12.5)$$

For $f_{\text{cat}} = 0$,

$$\frac{\partial q_i}{\partial t} = r_{\text{ads}_i} \quad (12.6)$$

They assume that adsorption is rate or mass transfer controlled and neglect axial dispersion. $f_{\text{cat}} \subset (0,1)$ where f_{cat} is the fraction of the wall surface in each band that is catalytically active so that where the monolith wall has a catalytic washcoat, the rate of adsorption term in Equation 12.5 vanishes. The rate term, in turn, drops out when the band has an adsorbent washcoat. Also, a is the surface area of the wall per unit volume, ε is as usual the void fraction in the reactor cross section, but u is the area mean fluid velocity in the channel. Rate terms are measured per unit weight of active material and ν_i is the stoichiometric coefficient.

The band structure with $f_{\text{cat}} \subset (0,1)$ implies that at $x = kz, C_i\ (kz-) = C_i(kz+)$ where z is the width of a washcoat band (assuming all bands are of equal width) and k is a band index, 1, 2, 3 … N. N is the number of bands in the reactor. Furthermore, at a cyclic stationary state

$$C_i(x, t) = C_i(x, t+\tau), \quad q_i(x, t) = q_i(x, t+\tau) \quad (12.7)$$

where n is a large integer. Boundary conditions at the reactor entrance are simpler than those given in Table 3.7 for the chromatographic reactor because of the absence of a dispersion term in Equation 12.5, namely,

$$\text{For } (n-1)\tau \le t \le sn\tau. \quad C_i(x=0, t) = C_i^{\text{feed}},\ u = u^{\text{feed}} \quad (12.8)$$

$$\text{For } sn\tau \le t \le (n+1)\tau. \quad C_i(x=0, t) = 0,\ u = u^{\text{carrier}} \quad (12.9)$$

These conditions allow a velocity change between parts of a cycle.

The investigators simplified their analysis by assuming the cyanopyridine feed contained a pH buffer solution that promotes substrate adsorption, while the carrier was a second solution buffered at a higher pH to favor product desorption. Because hydrogen ions are neither produced nor consumed, a balance for this species was not needed. Kinetics for the enzyme-catalyzed reaction are complex and indicate substantial product and substrate inhibition:

$$r_{\text{cat}} = \frac{k_{\text{cat}} C_c}{C_c + K_{\text{m}}\left(1 + \frac{C_n}{K_n} + \frac{C_c^2}{K_c}\right)} \tag{12.10}$$

The rate constant, k_{cat}, in the kinetic expression depends on pH. K_{m}, K_c, and K_n are the Michaelis–Menten constant and constants for substrate and product inhibition, respectively. The rate of adsorption of the nicotinamide on the ion-exchange resin is given by

$$r_{\text{ads}} = k_{\text{ads}}\left(q_n^* - q_n\right) \tag{12.11}$$

$q_n{}^*$ is the equilibrium adsorption of nicotinamide on the exchange resin at its concentration in the fluid phase. Štěpánek et al. assume a linear adsorption isotherm with the adsorption constant dependent on pH. The relationships with respect to pH are

$$\begin{aligned} k_{\text{cat}} &= k_{\text{cat}}^{\max} e^{-\left[\frac{pH-\mu_r}{\sigma_r}\right]} \\ K_{\text{ads}} &= K_{\text{ads}}^{\max} e^{-\left[\frac{pH-\mu_{\text{ads}}}{\sigma_{\text{ads}}}\right]} \end{aligned} \tag{12.12}$$

where $\mu_r, \sigma_r, \mu_{\text{ads}}$, and σ_{ads} are constants. The μ_r and μ_{ads} represent optimal pH values. K_{ads} is a distribution or Henry's law constant. Since, however, the authors assume pH undergoes a step change on switching from feed to carrier and back, the rate constant k_{cat} and adsorption constant take on just two values during a cycle.

Using values of model constants taken from the literature and assuming a back-mixed reactor, Štěpánek et al. demonstrate that the kinetics of the enzymatic reaction, in particular the strong product and substrate inhibition, can result in two steady states depending on cyanopyridine and nicotinamide concentrations in the reactor and on the space time. High levels of adsorbed product suppress multiplicity. Trajectories in the cyanopyridine–nicotinamide plane during a cycle depend strongly on flow rate through the reactor.

Given the kinetic complexities, Štěpánek et al. undertook simulation of the monolith microreactor performance for a 20 cm-long unit with 20 bands of equal width and a channel cross-sectional area of 10^{-5} m^2. An attempt was made to find a cycle period, τ, a cycle split, s, and a velocity, u, that would simultaneously optimize the throughput of the nicotinamide product and its separation from the substrate subject to practical constraints on u ($u^{\text{feed}} = u^{\text{carrier}}$) and on τ. The model PDEs were solved by the method of lines used together with Merson's method for integration. Simulation results appear in Figure 12.5 for two of the cases run.

In Figure 12.5a, the cycle period is short and the cycle split is large so that the higher concentration of nicotinamide in the reactor effluent means that the production rate is higher than in (b). Figure 12.5b shows results for a longer cycle period and symmetrical pulsing. In (a), the effluent contains the cyanopyridine substrate for 80% of the time. Thus, separation is poorer than in the second case in which the effluent contains none or very little substrate for 50% of the time. The simulations illustrate the typical trade-off of production

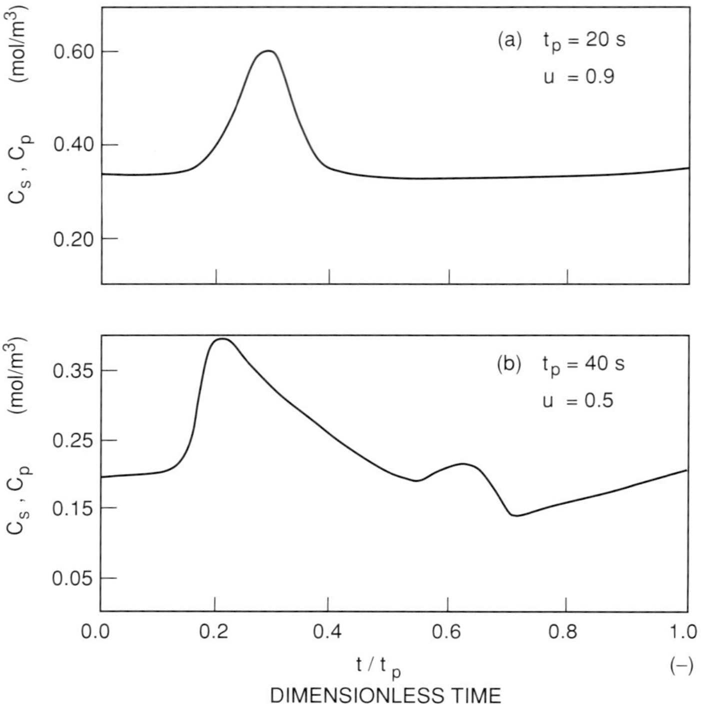

Figure 12.5 Outlet concentrations of nicotinamide product and cyanopyridine substrate as a function of normalized time within a cycle for a 20-cm monolith microreactor with 20 alternating bands of an immobilized enzyme and an ion-exchange resin. Results assume $u = 7.5$ mm/s and catalyst/adsorbent bands of equal width. (Figure adapted from Štěpánek et al. (1999) with permission. © 1999 Elsevier Science Ltd.)

rate and product purity encountered with periodically pulsed reactors. They too demonstrate that a PPR can be an effective operation when the reaction exhibits strong product inhibition.

The trade-off between production rate and product purity suggests that optimization of a monolith microreactor should employ two objective functions, F_x, the production rate per monolith channel and F_y, the separation required downstream of the reactor. Clearly F_x must be maximized and F_y must be minimized. In the optimization, channel length and cross section are fixed; variables are distribution of enzyme and resin, flow rates in the two parts of the cycle, cycle split and period. Because of the trade-off, the optimum must be stated in terms of Pareto sets. Figure 12.6 shows the sets as functions of the split, s, defined as duration of desorption divided by cycle period, and the cycle period. The solid lines are for a low flow rate, while the dashed lines are for a flow rate about 60% higher. Slopes are positive so choices of variables that increase F_x decrease F_y. Pareto sets were discussed in Chapter 7 for an SCMCR. They typically arise in separating reactor optimization.

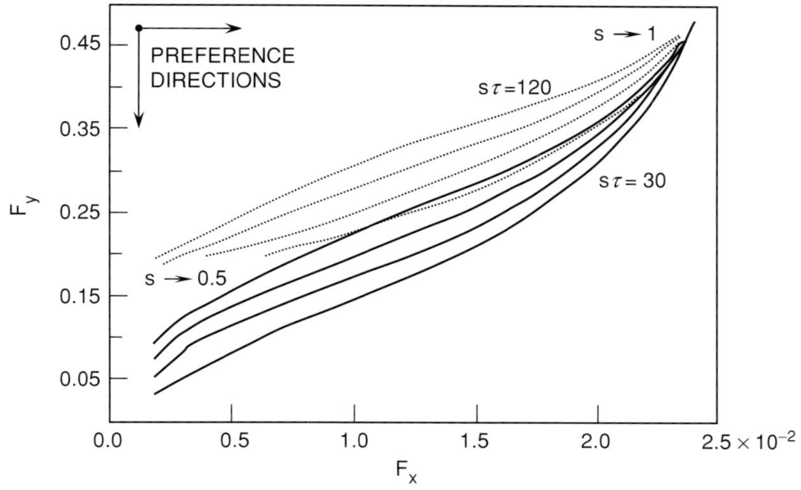

Figure 12.6 Variation of the objective functions F_x, production rate per channel, and F_y, downstream separation requirement, as a function of flow rate per channel, cycle split and cycle period for equal enzyme and resin band widths. Dashed lines are low flow rates while solid lines are high flow rates. (Figure adapted from Štěpánek et al. (1999) with permission. © 1999 Elsevier Science Ltd.)

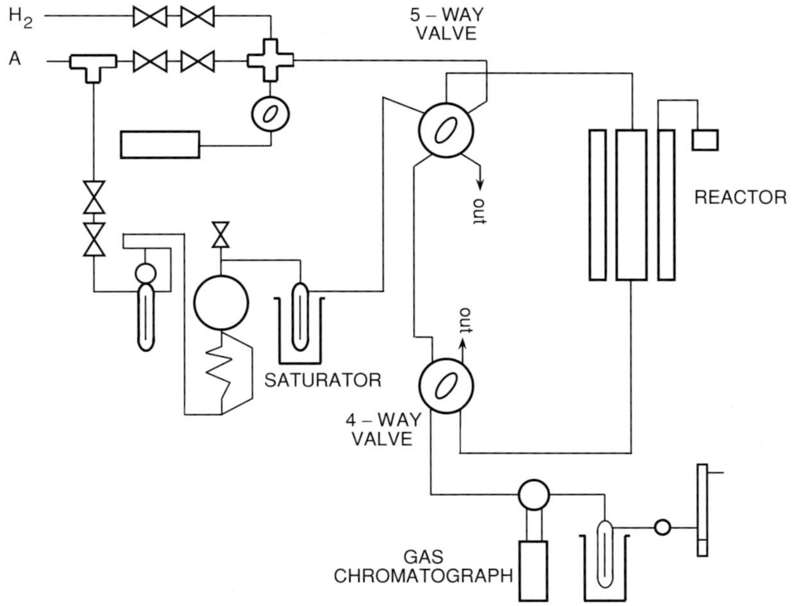

Figure 12.7 Schematic of the laboratory single pulse and periodically pulsed reactor used by Goto and coworkers. (Figure adapted from Goto et al. (1995) with permission. © 1995 Japan Petroleum Institute.)

12.2.2 Experimental studies

Goto and coworkers at Nagoya University explored the application of PPRs to catalytic dehydrogenation and dehydroaromatization using metals and metal alloys as hydrogen acceptors. Fully mixed catalyst and adsorbent as well as segregated solids were investigated. The experimental unit used in all studies is given in Figure 12.7. In the Goto experiments, helium or argon served as the carrier and purge gas. The hydrogen feed, shown in the diagram, was used for conditioning the catalyst and hydrogen storage metals as well as for measurement of metal or alloy sorption capacity. As shown, the paraffin reactant was introduced into the carrier stream via a saturator. A four-way solenoid valve affected the switch from feed to purge gas.

Catalytic dehydrogenation of cyclohexane to benzene at ambient pressure was undertaken by Goto et al. (1993) using a Pt/Al_2O_3 catalyst and a calcium–nickel alloy, $CaNi_5$, in particulate form as a highly selective adsorbent for H_2. The unit shown in Figure 12.7 was used. Helium served as carrier gas and purge and the temperature range of the experiments was between 423 and 463 K. Under the conditions used, cyclohexane conversion is equilibrium limited. Most of the Goto experiments used a single pulse and they explored the influence of space time, temperature and the duration of the helium flow on cyclohexane conversion. A shallow bed was used with an alloy to catalyst weight ratio of 5.

Feed to the reactor was cyclohexane in helium. Discharge from the reactor during the short feed step was benzene in the helium carrier. Using a shallow bed, alloy regeneration by reducing total pressure to near vacuum and by flushing the mixed catalyst–adsorbent bed with carrier gas alone was compared for a single cycle operation at 190°C. For the flushing test, Goto et al. used helium with about a 20:1 ratio of purge to feed flow rate. Thus, a very dilute H_2 stream was removed through the exhaust port. With a regeneration duration of 20 min, flushing with helium realized a C_6H_{12} conversion of 42.9%, while drawing a vacuum gave a conversion of 39.1%. For a 60-min duration, the respective conversions were 58.2% and 54.9%. Clearly, under comparable conditions, flushing is at least equivalent, if not superior, to creating a low total pressure.

Increasing bed depth by a factor of 2.5, a PPR was investigated by the researchers with the results shown in Figure 12.8. Conversion at the conditions used reached 100% for about

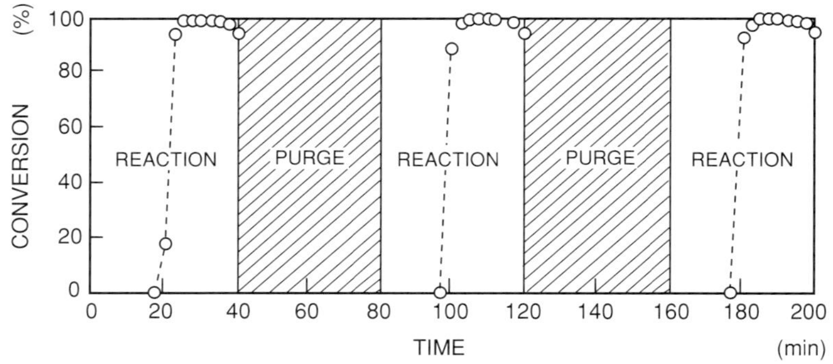

Figure 12.8 Variation of cyclohexane conversion with time in a PPR using a 40-min feed pulse in an 80-min cycle: $T = 190°C$, $(C_{CH})_0 = 2.6$ mol/m^3, weight ratio of alloy to catalyst = 5. (Figure adapted from Goto et al. (1993) with permission. © 1993 Society of Chemical Engineers of Japan.)

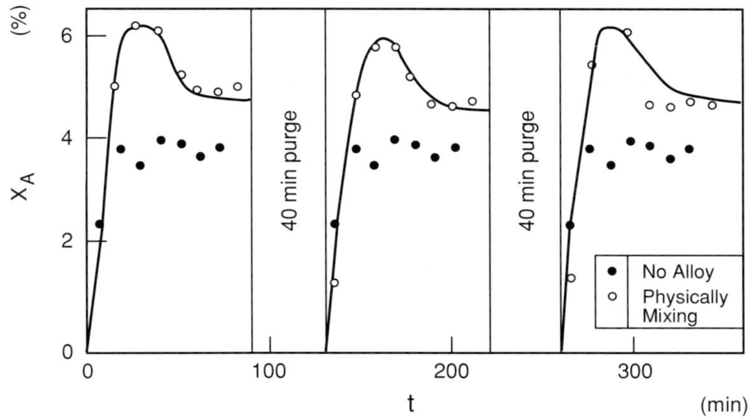

Figure 12.9 Comparison of conversion of *n*-hexane to benzene as a function of time in a PPR consisting of a Zn/H ZSM-5 catalyst and a $Mg_{51}Zn_{20}$ hydrogen adsorbent with conversion in the same sized reactor consisting of Zn/H ZSM-5 admixed with quartz: $(C_{C6})_0 = 3.2$ mol/m^3, $T = 573$ K, weight ratio of alloy to catalyst = 13. (Figure adapted from Goto et al. (1995) with permission. © 1995 Japan Petroleum Institute.)

20 min of the 80-min cycle. This means a relatively pure benzene product in helium is obtained in this half cycle, while H_2 is removed as a dilute product when purge gas flows through the reactor. The results nicely illustrate the benefits and drawbacks of a PPR for an equilibrium-limited reaction: complete conversion can be obtained without recycle and reactants can be well separated reducing recovery cost, but reactor throughput is well below that of a same size PFR operating at identical feed conditions.

The dehydroaromatization of paraffins over ion-exchanged ZSM-5 with metallic hydrogen adsorbents was also explored by Goto et al. (1995, 1996). These were primarily screening studies and as such they will be examined again in Chapter 14. The 1995 study examined the dehydroaromatization of *n*-hexane to benzene over a Zn-exchanged hydrogen ZSM-5 zeolite. Various magnesium-based hydrogen storage alloys were studied as well as the arrangement of the packed bed. Fully mixed catalyst and adsorbent were compared with catalyst diluted with an equal volume of a nonadsorbing solid, and with adsorbent and catalyst segregated into layers. Only the fully mixed arrangement was tested in a PPR. Conversion to benzene was significantly improved in the pulsed operation, although conversion was quite low because of the catalytic system and operating conditions examined. Figure 12.9 illustrates the Goto results. This was obtained with a $Mg_{51}Zn_{20}$ alloy, which is about a 50:50 mixture of Mg and Zn by weight, fully mixed with the ZSM-5 catalyst. Cycle duration was 130 min in which *n*-hexane was fed for 90 min. Cycle structure is clearly not optimal.

Several single pulse experiments were carried out to compare fully mixed ZSM-5 catalyst and $Mg_{51}Zn_{20}$ with the catalyst and alloy separated into three bands of catalyst and two of alloy. These showed that the highest conversions were achieved with the fully mixed catalyst and alloy. However, even the segregated catalyst and alloy significantly increased conversion over an experiment in which catalyst was mixed with an inert diluent so as to maintain a constant spacetime.

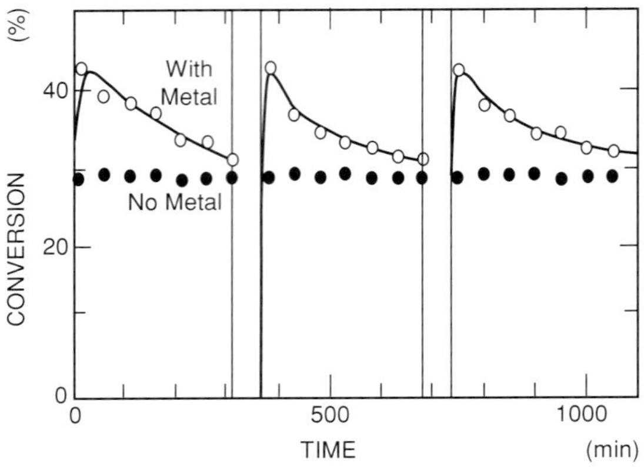

Figure 12.10 Comparison of propane conversion at 793 K using a mixed bed of Zn/H ZSM-5 catalyst and finely crushed titanium metal adsorbent at a metal to catalyst weight ratio = 6.5 with conversion for the same amount of catalyst mixed with a quartz diluent at the same weight ratio for a propane feed rate of 2.53×10^{-5} mol/s. (Figure adapted from Goto et al. (1996) with permission. © 1996 Japan Petroleum Institute.)

Dehydrogenation of propane to a mixture of paraffins, cyclohexane and aromatics over a partially exchanged Zn/H ZSM-5 was investigated in a second study (Goto et al., 1996). This equilibrium-limited reaction sequence proceeds to a greater extent over the zeolite catalyst than dehydroaromatization of *n*-hexane due to the higher temperature employed. Addition of a hydrogen adsorbent is also effective. Once again, the investigators primarily screened different metal adsorbents, adsorbent particle size and bed arrangement. We consider this work in Chapter 14. When run in a packed bed of mixed catalyst and adsorbent functioning as a PPR, a substantial increase in propane conversion was observed as Figure 12.10 illustrates. Selectivity to aromatic products, however, was unchanged. Note the remarkably high cycle split, $s = 0.86$, used. The PPR operation increased the time-average propane conversion by about 20%. Aromatics production also increased.

The enzymatic propionic acid–isoamyl alcohol esterification in a hexane solvent investigated by Mensah and Carta (1999) using a simulated moving-bed version of a periodically pulsed reactor (PPR) typifies reactions under study for the preparation of fine chemicals. Water formed in the reaction interferes with the enzyme and must be continuously removed. Because adsorbed water is strongly held by both the enzyme support and the ion-exchange resin, regeneration is difficult so the bed undergoing regeneration is isolated from the cascade and flushed with either solvent at a high flow rate or a special eluent. Isolation is illustrated by the schematic of the cyclic system shown in Figure 12.3. Valve arrangement required in the operation on a laboratory scale is presented in Figure 12.11. The acid and alcohol reactants in a hexane solvent are directed by valving consecutively through three beds packed with immobilized lipozyme and the ion-exchange resin that serves as a trap for water. The fourth bed is isolated from the sequence and is flushed with eluent, in this case, the reactant isoamyl alcohol.

Experiments undertaken by the investigators were intended to validate the model as well as to compare operation of the mixed bed of enzyme and adsorbent with just the immobilized lipozyme. The effect on performance of the number of beds, keeping the

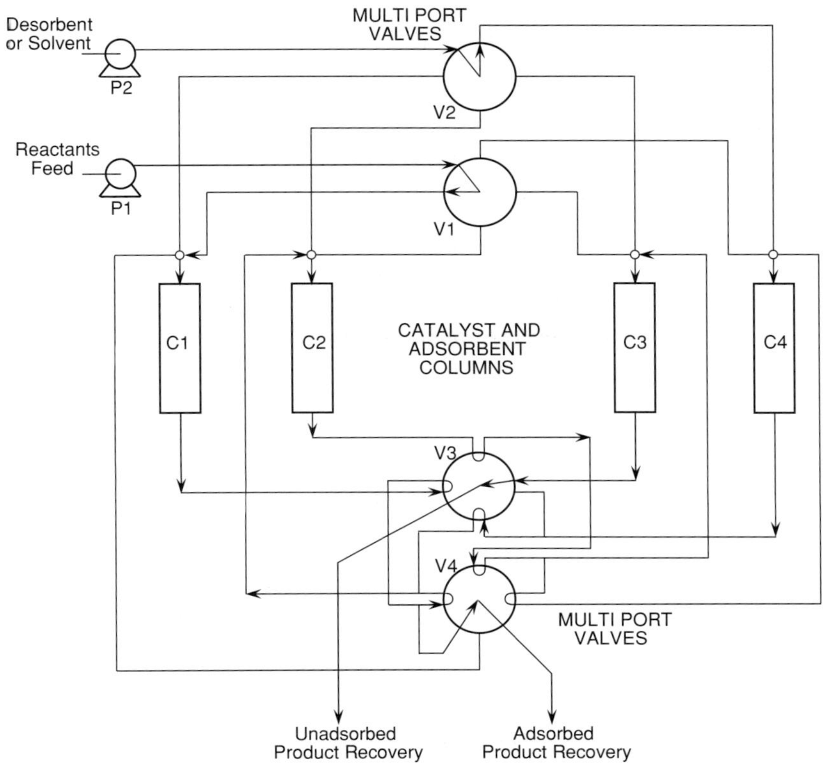

Figure 12.11 Schematic of a laboratory-scale simulated moving-bed adsorptive reactor showing the valve switching and flow sequence. (Figure adapted from Mensah and Carta (1999) with permission. © 1999 John Wiley & Sons, Inc.)

amount of catalyst constant, of varying the amount of enzyme and of switching time was also explored. Figure 12.12 gives composition of the effluent from the reaction–adsorption cascade through six cycles from startup. To obtain these data, just two beds were placed in the reaction–adsorption sequence instead of three shown in Figure 12.11. The switching time used, 4 h, was the largest that would maintain the water activity $a_{\text{water}} < 0.9$ at any point in the consecutive beds according to a simulation based on the model discussed in Section 12.3. The bed undergoing switching was briefly flushed with the hexane solvent to remove interstitial isoamyl alcohol. It can be seen from Figure 12.12a that the effluent from the reaction–adsorption sequence is primarily the ester. A small amount of isoamyl alcohol also leaves, as it is not adsorbed. There is no propionic acid leaving the last bed. Figure 12.12b is a simulation of the concentration profiles in the two consecutive beds at the end of a cycle. Figure 12.12c shows the spatial variation of the activity of water held in the ion-exchange resin at the start and end of a cycle. There is virtually no water in the second bed that has just been switched into the sequence.

Figure 12.12a indicates that the model given in Section 12.3 for this simulated moving-bed PPR reproduces the shape of effluent variation properly, but does not give the peak values accurately. Nevertheless, it seems to be reliable for design and analysis purposes.

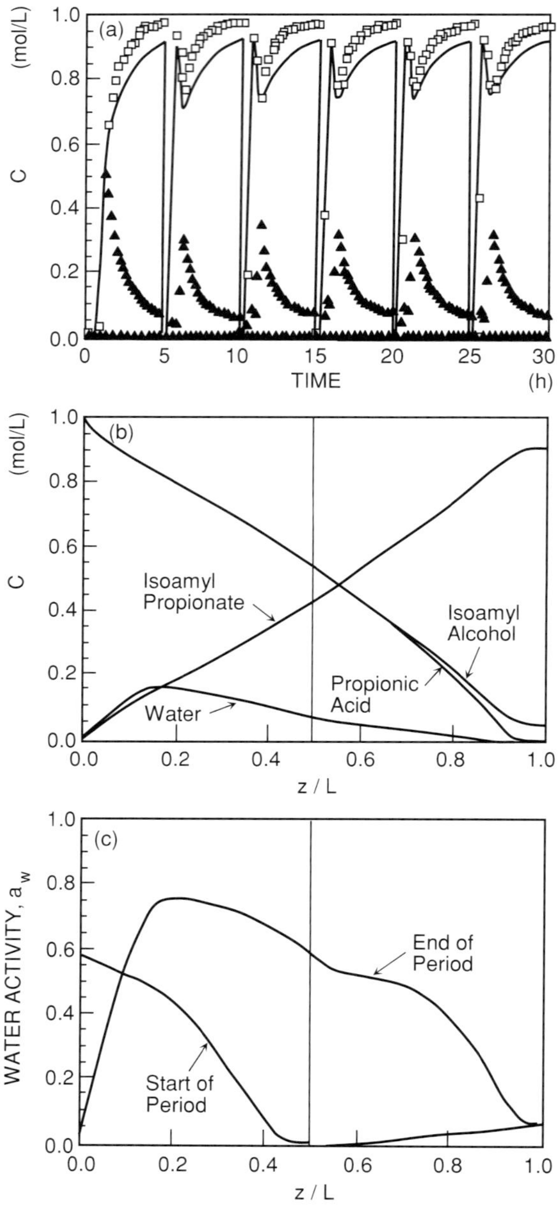

Figure 12.12 Experimental and predicted time and position profiles for the esterification of propionic acid and isoamyl alcohol in a cyclically operated adsorptive reactor using an immobilized lipozyme as a catalyst and a strong-acid cationic exchange resin as adsorbent. Lipozyme to adsorbent ratio was 2:1 by weight. In (a), □ = ester, ▲ = isoamyl alcohol and solid lines = model predictions; (b) spatial concentration profiles in the reaction–adsorption double bed at the end of a cycle; (c) variation of water activity in the resin phase. (Figure adapted from Mensah and Carta (1999) with permission. © 1999 John Wiley & Sons, Inc.)

Interestingly, when the switching time is increased so that a_{water} reaches 1.0 within the bed according to the simulation, the experimentally measured production rate per unit of enzyme drops significantly and the model fails to predict the effluent concentrations. It is known that when the water activity approaches unity, the enzyme is irreversibly deactivated. Simulations predict that ester production per unit of enzyme increases with switching time as long as a_{water} remains below 1.0 at any point in the bed.

The Mensah and Carta experiments demonstrate higher productivity using a 2:1 mixture of lipozyme and adsorbent compared to the use of the immobilized enzyme alone. Increasing the number of beds in the reaction–adsorption sequence from one to two while holding the amount of enzyme constant almost doubles productivity, but a further division of the beds in the sequence increases productivity by a much smaller amount.

Aida et al. (1999) and Na-Ranong et al. (2002) have shown that bang-bang periodic switching between reactants in the reduction of NO by CO over a Pt/Al_2O_3 catalyst increases NO reduction. The reaction is important in a three-way catalytic muffler. The mechanism for the switching effect seems to be that strong adsorption of CO by the catalyst inhibits NO adsorption and thus reduces the rate of the reduction reaction. Bang-bang cycling allows NO adsorption during the NO portion of the cycle. At the beginning of the CO portion, the adsorbed NO reacts with adsorbing CO. Eventually, CO displaces NO and the reaction halts. In deeper catalyst beds, desorption of unreacted NO deforms the NO pulse resulting in a NO peak moving at the speed of the CO pulse and overlapping the front of that pulse. Reaction rate is reduced in deeper beds because the NO and CO pulses, 180° out of phase at the entrance, move into phase. To prevent this phase shift, Aida et al. (2002) proposed using a mixture of catalyst and 5A molecular sieve to maintain separation of the pulses by slowing the movement of CO through the bed. Their calculations indicated an increase of NO reduction for longer cycle periods if adsorbent is added to the reactor charge. When the authors tested their proposal experimentally using either a mixed bed of catalyst and adsorbent or catalyst and adsorbent separated into sequential layers, little improvement was observed over bang-bang cycling using just catalyst for short cycle periods and a spacetime of 180 s. When a longer spacetime was used, 450 s, corresponding to a deeper bed, a significant increase in NO reduction was observed for both long and short cycle periods. Examination of their results led Aida et al. to conclude that the source of the improvement was trapping of NO by the adsorbent. In the absence of adsorbent at the space times used in the experiments, NO breaks through the bed before switching to CO. Adding molecular sieve traps NO in the bed thereby increasing reduction.

12.3 The periodically operated trapping reactor

The concept of trapping a gas or liquid phase contaminant in low concentration by adsorption, raising the temperature of the adsorbent and flushing with an eluent to augment considerably the contaminant concentration in the eluent has been practiced for many years. If the concentration in the carrier can be raised sufficiently, the contaminant can be recovered if it has value. More often, the adsorption cycle is used to raise the concentration high enough so that the contaminant can be destroyed in a subsequent reactor. Agar and Ruppel

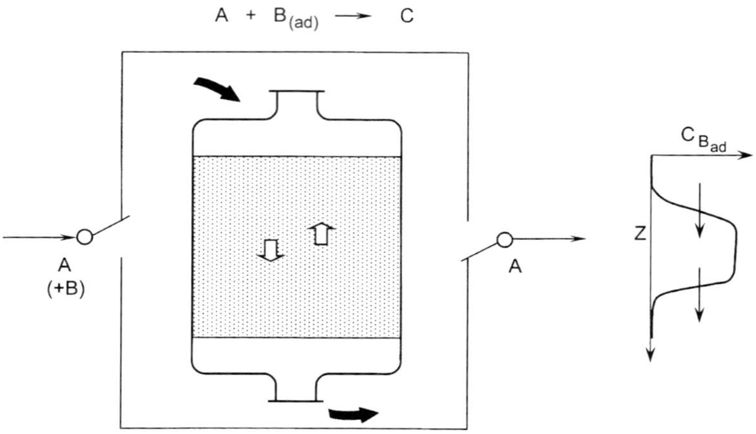

Figure 12.13 Schematic of a PTR for trapping a reactant B using periodic switching of the flow direction and co-feed of reactant B with A only in the first part of each half cycle. (Figure adapted from Agar and Ruppel (1988a) with permission. © 1988 VCH Verlagsgesellschaft mbH.)

(1988a,b) demonstrated that the trapping and reaction functions could be carried out in a single vessel. It is only this type of periodic trapping reactor (PTR) that we will consider.

Agar and Ruppel were concerned with ammonia slip in the catalytic reduction of nitrogen oxides, NO_x, over a noble metal catalyst capable of adsorbing ammonia. In order to ensure complete NO_x conversion with time fluctuating concentrations, a slight excess of NH_3 is used. This leads to a periodic loss of NH_3 in the effluent. This loss is referred to as ammonia slip. Ammonia itself is a contaminant so slip must be avoided. Agar and Ruppel demonstrated that NH_3 consumption could be reduced and slip largely avoided by mixing NH_3 with NO_x at the entrance of the bed for a portion of the cycle or by injecting NH_3 into the center of the catalyst bed. In either case, the flow direction of the NO_x contaminated gas through a packed bed de-NO_x reactor is periodically switched. Because of NH_3 adsorption by the catalyst or its support, NH_3 is trapped within the reactor. Once the flow direction is switched, the entering NO_x strips off the adsorbed NH_3 by reaction. Unreacted NH_3 is also stripped off and carried inward where it is again trapped by regenerated catalyst. If the catalyst does not adsorb the slip reactant, a mixture of catalyst and a suitable sorbent could be employed. Alternatively, the catalyst could be sandwiched between beds of sorbent. Figure 12.13 shows schematically the reverse flow operation investigated by Agar and Ruppel using mixing of ammonia at the bed inlet during the first half of each cycle.

12.3.1 PTR simulation

Agar and Ruppel (1988a) simulated the trapping operation shown in Figure 12.13 for a square channel monolith with a catalyst washcoat by a 1-D, plug flow, pseudo-homogeneous model, which assumed that both adsorption and reaction were kinetically controlled. At the start of the downflow step, their results showed adsorbed NH_3 at the top of the reactor from the upflow portion of the previous cycle. Addition of NH_3 greatly increases gas phase and adsorbed NH_3 in this portion of the bed during the first part of the downflow portion. Just before the NH_3 supply is interrupted half way through the downflow portion of the cycle, the NH_3 front has just about reached the end of the bed. Gas phase NO is present only at the inlet. There is no NH_3 entering the bed. With just the NO containing waste gas flowing downward, an NO front moves through the bed. At the front, NO_x reduction takes place consuming adsorbed NH_3. Before NO_x breakthrough, the cycle ends leaving a small amount of adsorbed NH_3 at the end of the bed. Thus, neither NO_x nor NH_3 leaves the reactor.

The first investigations of a periodic trapping reactor (PTR) employed 1-D, pseudo-homogeneous models (Agar and Ruppel, 1988a,b; Falle et al., 1995; Snyder and Subramanian, 1998). These models fail to capture the full spectrum of dynamic behavior that may arise so that recent researchers have used heterogeneous models (Salden and Eigenberger, 2001; Jeong and Luss, 2003). Models have been given in the earlier chapters. Most are versions of those in Table 3.2 when the pseudo-homogeneous assumption is adopted or in Table 3.3 for the 1D heterogeneous version.

Jeong and Luss (2003) employed simulation to examine the PTR performance as a function of adsorbent capacity and the cycle parameters. They were interested as well in the optimal level of the reactant that is trapped and potential benefits of periodic introduction of that reagent. A reaction of the type $A + B \rightarrow C$ was considered where A would represent NO_x and B the NH_3 educt. The reaction occurs under dilute conditions so isothermal operation was assumed, but dynamic behavior was important so the researchers used a two-phase heterogeneous model. Following previous investigators (Kallrath et al., 1994; Falle et al., 1995), Jeong and Luss assumed the educt was injected into the midpoint of the reactor. This requires a minor modification of the gas phase material balance for B. If mass transfer controls, it now becomes,

$$\varepsilon_t \frac{\partial C_B}{\partial t} = -fu\frac{\partial C_B}{\partial z} + \varepsilon_t D_{ax}\frac{\partial^2 C_B}{\partial z^2} - a_V k_m (C_B - C_B^*) + J_B \delta\left(z - \frac{L}{2}\right) \qquad (12.13)$$

The Dirac delta function, $\delta(x)$, assumes a non-zero value only at the midpoint, namely at $x = L/2$ where L is the reactor length. Because flow direction switches periodically, the f modifying the convection term changes sign from + to – or vice versa on the switch. $C_B{}^*$ represents the concentration of the trapped species in the pore space of the adsorbent. On normalizing position and rendering time dimensionless, model parameters become N_{Pe}, N_{Da}, a modified Sherwood number and D a measure of the educt deficiency, namely, $D = 1 - J_B/u(C_A)_0$. The concentrations are in the feed, while J_B is the molar flux of the educt. More stringent restrictions on reactant discharge can be handled by a penalty term, $p \geq 1$. Thus, effluent concentration to be minimized in any optimization of the trapping reactor would be $C_P = (C_A)_{exit} + p(C_B)_{exit}$.

Jeong and Luss compare the time average C_P of a PTR with that of a packed-bed reactor containing catalyst–adsorbent. With midpoint injection of the educt, performance of a packed bed is also improved. As might be anticipated, C_P depends strongly on p and D. It also depends on N_{Da} and for the PTR on the dimensionless cycle period τ. Figure 12.14 compares the variation of C_P with these variables for the two reactor types. The dimensionless half-cycle period is 7000 times the residence time of A in the bed so this represents slow cycling that is conceivable only for very dilute systems.

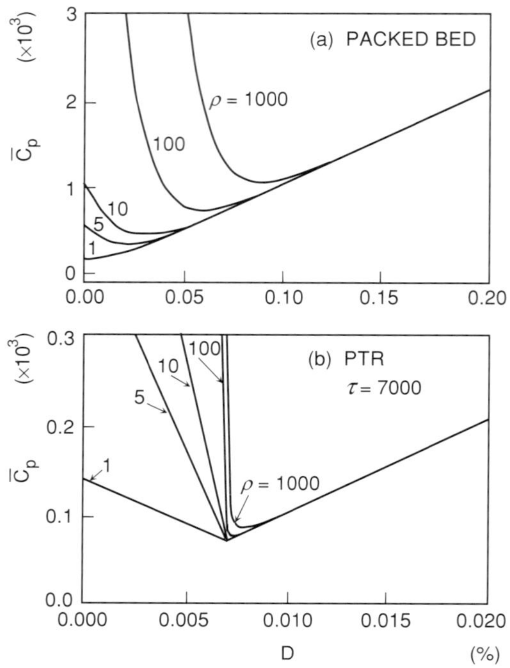

Figure 12.14 Comparison of the variation of the effective pollutant concentration in the reactor effluent with the educt penalty ρ and deficiency D for midpoint injection of B in a unidirectional packed-bed reactor and in a PTR. (Figure adapted from Jeong and Luss (2003) with permission. © 2003 Elsevier Science Ltd.)

A measure of the effectiveness of a PTR for this simple, dilute phase reaction can be had by considering the ratio of the minimum C_P for any value of p and D for the packed-bed reactor with the time average C_P for the PTR at the same values. This measure then depends on τ. Jeong and Luss observe that there is an optimum τ at any D and p. At large values of τ, the reagent B is found only in one half of the reactor, just as in the unidirectional packed bed, so that enhancement due to trapping by flow reversal is small.

Because of trapping through flow reversal, educt does not have to be added continuously. It can be added, for example, in just one half of the cycle. Jeong and Luss consider this case and conclude that a better performance is obtained with continuous addition. However, the difference in these modes is not very large at small and large deficiencies, D. Increasing the dimensionless rate constant, N_{Da}, increases the differences.

A problem with trapping organic materials is degradation during heat up prior to catalytic oxidation that can lead to cracking and tar formation. Salden and Eigenberger (2001) have devised a clever but complicated PTR design for such situations illustrated in Figure 12.15.

Operation of the system shown in the figure consists of at least two steps: the adsorbent strips organic material from the gas stream flowing through the reactor in a long first step. In the second step, the flow is switched to air and the heater toward the end of the bed is switched on. After a short time, for the adsorbate in the neighborhood of the heater to reach ignition temperature, incineration of the adsorbate begins. Careful adjustment of air flow allows a combustion front to move slowly upstream. The countercurrent movement of this front avoids large temperature excursion that may occur when a front moves in the same direction as the flow. In this way, the incineration temperature can be maintained at a modest increment above the ignition temperature so that degradation may be avoided. Because of the relatively low incineration temperature and the controlled air flow rate, carbon monoxide can be present in the combustion products. This residual is oxidized catalytically at the end of the reactor following the introduction of secondary air.

A model for the system was proposed by Salden and Eigenberger that supported the operation described above. Assuming dilute styrene that polymerizes on a zeolite adsorbent at about 150°C, the model was used to describe the behavior of the system. Experiments in a 55 mm i.d. × 800 mm, well-instrumented reactor confirmed the model predictions.

12.3.2 Experimental studies

In many of the previous chapters, we have mentioned that there is a moving-bed system that corresponds to a periodic operation or vice versa. This is the case for the problem discussed by Salden and Eigenberger. Arsenijevic et al. (1999) examined the catalytic oxidation of gaseous discharges containing low levels of ethylene oxide (ETO). They observed strong adsorption of ethylene oxide on Pt/Al_2O_3 at temperatures below the ignition temperature of the waste stream as well as residue formation on the catalyst surface. To exploit these observations, Arsenijevic et al. proposed a moving-bed trapping reactor for treating the ETO containing discharge. Circulation of the catalyst–adsorbent utilizes a spouted bed. In principle, the spouted bed operates continuously. The catalyst–adsorbent in the bed, however, undergoes periodic exposure to several different gas environments just as in other moving-bed reactors we have considered, e.g. a CMCR.

The spouted bed used by Arsenijevic et al. consisted of a riser and an annular bed. The riser was fed with air and carried the low temperature catalyst–adsorbent to the top of the bed where the particles settled through a hot section of an annular, dense fluidized bed. This section was fed with air and its cross-sectional area was smaller than the lower section of the bed but enough to provide sufficient residence time for burn-off of residues on the particles. If the ETO concentration in the waste gas was high enough, the heat released in combustion met the thermal demand of the gas treatment system. Cross section of the

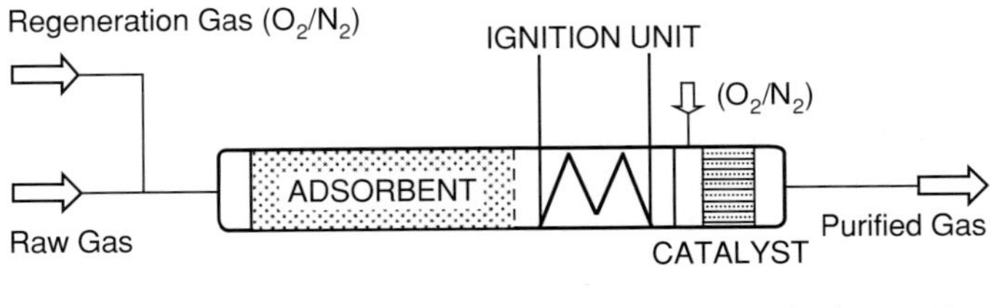

Figure 12.15 PTR combining noncatalytic incineration of an organic adsorbent with a catalytic afterburner. (Figure taken from Salden and Eigenberger (2001) with permission. © 2001 Elsevier Science Ltd.)

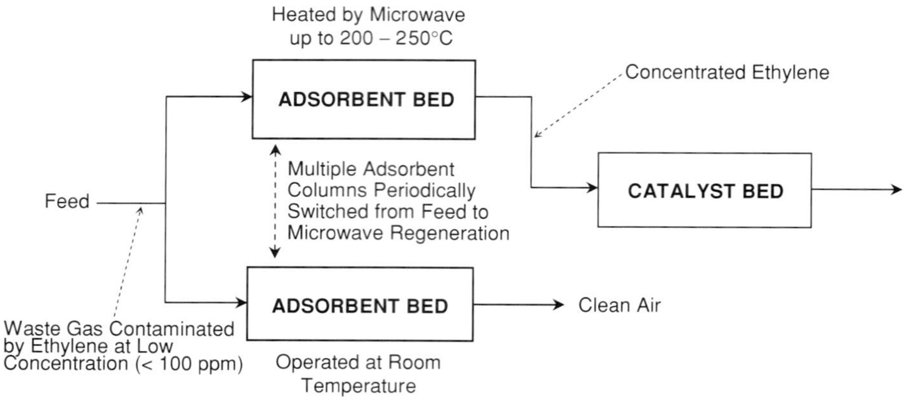

Figure 12.16 Schematic of a trapping reactor for incineration of a VOC containing industrial effluent.

lower portion of the annular bed was larger, permitting faster settling of the particles. This section was fed by the cold waste gas discharge. Thus, the section functioned as a preheater as well as an adsorber. Hot gas containing ETO that was not adsorbed flowed upward into the upper reactive section of the bed. Adsorbed ETO was carried out of the bottom section by the countercurrent flow of catalyst and then carried to the top of the annular bed in the spout section or riser. Temperature of the top reactive section of the annular bed was about 475 K, hot enough for combustion of adsorbed or gas phase ETO and tar-like residues. The temperature in the lower section was 100°C lower, but adequate for ETO adsorption from the waste gas.

Curiously, Arsenijevic et al. (1999) found it was difficult to control residence time and the desired temperature distribution with continuous operation and achieved a better thermal economy, in terms of auxiliary heating, when they operated the riser in a pulse mode at about 4–5 s on and 55 s with no upward flow in a 60-s cycle. This resulted in a temperature oscillation of 50°C amplitude in the middle of the upper, reactive section of the annular bed. Cyclic variations of other temperatures in the bed were negligible or at most just several degrees.

An interesting variant of a trapping reactor has been investigated by Kim et al. (2005) in which beds of adsorbent and catalyst were employed to remove VOC components like ethylene from a waste gas. A schematic of the Kim system is given in Figure 12.16. In the figure, the catalyst and adsorbent bed are separate, but a mixed bed could be used. Waste gas containing VOCs flows to one adsorber where a synthetic zeolite removes the VOC as well as water vapor in the gas stream. Adsorption of water is essential because in the regeneration step adsorbed water absorbs microwave energy to heat the adsorbent and desorb both water and the trapped VOC. For regeneration, the waste gas continues to flow through the bed. The bed is now irradiated by microwaves of suitable wavelength for absorption by water. As shown in Figure 12.16, the off gas from regeneration flows to a catalyst bed where the concentrated and hot VOC is oxidized. In principle, a single mixed bed of catalyst and zeolite adsorbent could be used.

Kim et al. (2005) studied several synthetic zeolite (mordenites) and catalysts (Pt/Al_2O_3, Co_3O_4) in their experiments. They observed that the concentration of ethylene in the exhaust after microwave irradiation exceeded considerably its original concentration in the feed. The reaction order was found to be 0.5 with respect to ethylene. The Kim experiments showed much higher throughputs for the cycled adsorbent system compared to a single continuously operated catalyst bed.

12.4 Cyclic extractive reactor

Development of this class of cyclically operated reactors is much less advanced than the adsorptive reactors considered heretofore. Two-phase, liquid–liquid extractive reactors, of course, exist and have been discussed in the past (Piret et al., 1960; Trambouze and Piret, 1960; Goto and Matsubara, 1977; Schügerl et al., 1988; Minotti et al., 1998). They operate continuously and therefore will not be considered in this section. Most applications have been to biochemical systems, particularly fermentation, and the literature is large (e.g. Reschke and Schügerl, 1984a,b; Schügerl et al., 1988; Yabannavar and Wang, 1991; Freeman et al., 1993; Pai et al., 2002).

A surprising example of an extractive reactor is the well-studied oxidation of sulfur dioxide with air or oxygen over activated carbon (Hartman and Coughlin, 1972; Mata and Smith, 1981; Haure et al., 1989). Several activated carbons are superb catalysts for this reaction, but unfortunately, the sulfur trioxide reaction product is strongly adsorbed by the carbon thus inhibiting the reaction. Flushing the carbon surface with water extracts the trioxide as sulfuric acid allowing the reaction to proceed. Although oxidation and extraction can take place continuously, Canadian researchers demonstrated that the SO_2 oxidation rate is significantly enhanced by carrying out flushing intermittently (Haure et al., 1989; Metzinger et al., 1994; Lee et al., 1996). In place of cocurrent downflow of the gas and water phases, they maintained steady gas flow but cyclically switched the liquid flow on and off. Duration of liquid flow in their experiments varied fractionally from 0.01 to 0.5 of the cycle period. The investigators referred to the fraction of liquid flow as the cycle split, s. Experimental variation of SO_2 removal by conversion to acid with cycle period and split is illustrated in Figure 12.17.

Experiments undertaken in the Canadian program employed various industrial activated carbons packed as a granular material in a bed as well as structured packings with fine carbon particles cemented on to steel mesh in different forms. Time average rates of SO_2 removal and conversion with intermittent water flow were consistently 15–30 % higher than those with continuous cocurrent water downflow.

Some of the improvement in oxidation rate results from reduction in liquid hold up in the bed when liquid flow is halted. Lower hold up reduces wetting of the carbon external surface or at least the depth of the liquid film on this surface. As a consequence, mass transfer of SO_2 and O_2 from the gas phase to the external surface of the carbon increases. Both reactants enter the trickle bed in the gas phase and have limited solubility in water or dilute sulfuric acid. Reactions in trickle beds are usually mass transfer controlled. Explanation of the shape of Figure 12.17 is that at short durations of water flow (a small cycle period or a small split), sulfuric acid collected in the static hold up in the bed during the zero flow portion of the cycle is not washed out. SO_2 and O_2 solubility decreases sharply as acidity increases. Absorption rates depend on solubility. Thus, increasing cycle period at a constant split or split at a constant period enhance the removal of acid from the bed so that SO_2 removal rates rise. However, in the figure it is not shown that as split increases above 0.5, bed hold up mounts and greater mass transfer resistance causes rates to fall gradually.

Two other phenomena also contribute to the improvement of this periodic extractive reactor (PER). Periodic interruption of liquid flow reduces heat removal from the trickle bed for this exothermic oxidation reaction. Lower heat removal raises the time average temperature of the trickle bed and thereby also the SO_2 removal rate. Temperature variations in the trickle bed with intermittent water flow are pronounced despite the low partial pressure of SO_2 in the gas flow to the bed. Variations are shown in Figure 12.18 for an 80°C water flush. Continuous gas feed to the trickle bed at 25°C was 1.3 vol% SO_2 in air. As we discussed in earlier chapters, temperature variation must be expected in cyclically operated reactors. Large variations shown in Figure 12.18 are unusual and arise from evaporation of liquid hold up and the heat of reaction as well as the difference in gas and water feed temperatures. Data points are shown only for a thermocouple located just below the top of the carbon bed. Solid and dashed curves in the figure are drawn through data from thermocouples located at the axial middle and at the bottom of the bed. These curves show very little variation between cycles. The large variation in readings from the top thermocouple results from unavoidable variation in flow distribution of water to the top of the bed for the low flows used.

Higher acidity in the liquid phase reduces SO_2 and O_2 solubility. Thus high water flow rates for short durations wash out acid and provide periods of low acidity static hold up, whereas a steady flow at a uniform much lower rate maintains a constant acidity level. The contribution of these abrupt reductions in acidity to improved performance is uncertain. It is, of course, a specific characteristic of the SO_2–O_2–H_2O system. A more complete exposition of the Canadian work is given by Silveston and Hanika (2004).

Yamada and Goto (1997) use switching between two different liquids in a trickle-bed reactor to affect an extractive reaction. This is an unusual four-phase system. An intermediate

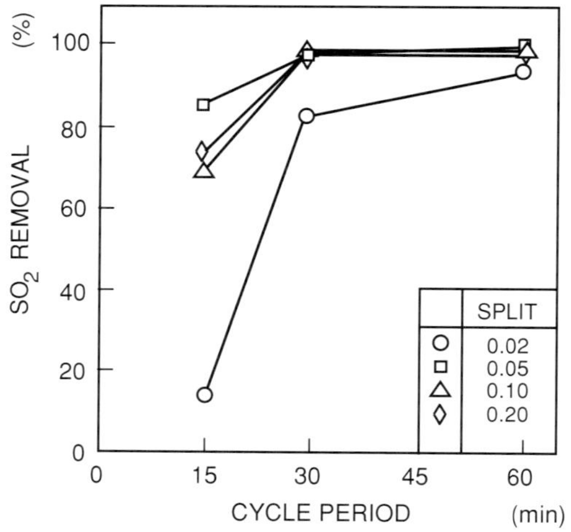

Figure 12.17 Variation of SO_2 removal from a dilute, oxygen containing gas stream and its conversion to acid in a trickle bed reactor packed with grains of Centaur^{TM} activated carbon with on–off water flow. (Figure adapted from Lee et al. (1996) with permission. © Canadian Society of Chemical Engineering.)

in the production of phenylalanine is carbobenzoxyphenylalanine. The final product is obtained from this intermediate by hydrogenolysis over a carbon-supported palladium catalyst. Toluene is a by-product. Carbobenzoxyphenylalanine and toluene are soluble in the organic media used, dichloroethane, while phenylalanine is insoluble. Thus, during the reaction the hydrogenolysis product crystallizes on the catalysis surface. These solid deposits block active sites and inhibit the reaction. Noting that phenylalanine is water soluble, Yamada and Goto fed an aqueous stream along with the organic solution to their experimental reactor, that is, they resorted to an extractive reaction. Because the streams segregated while flowing downward through the catalyst bed, this continuous extractive strategy failed to prevent catalyst deactivation. However, these investigators were successful when they changed to a periodic extractive operation by periodically throttling the flow of the organic solution. Switching alternately between the organic solution and an aqueous wash also maintained activity. Their experimental results are shown in Figure 12.19. Conversions measured while interrupting the flow of the organic solution containing the reactant are shown in Figure 12.19a. Conversions for flow switching appear in Figure 12.19b.

We began this section by mentioning the application of continuous reactive extraction to biochemical processes. One of these is the recovery of penicillin G from a fermentation broth. The process is important industrially and has been well studied (Reschke and Schügerl, 1984a,b,c; Lee et al., 1993; Yang and Cussler, 2000; Pai et al., 2002). A related process is the enzyme-catalyzed hydrolysis of penicillin G to 6-aminopenicillanic acid (APA) and a by-product phenylacetic acid (PAA). In the present industrial process several pH adjustments are needed so it is undertaken in stages and inorganic salts are generated. These salts must

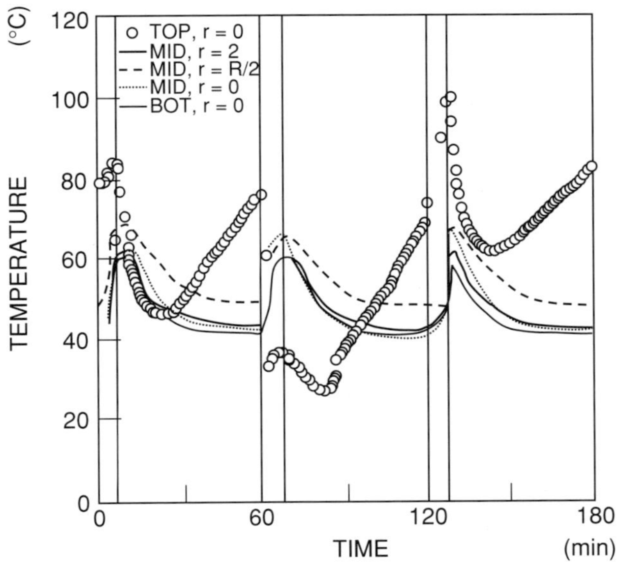

Figure 12.18 Measured temperature variations in a 300-mm-deep laboratory trickle bed reactor running SO_2 oxidation in air over a bed of granular BPL activated carbon. Gas flow at 25°C is continuous while liquid flow at 80°C is for just 5 min in every 60-min cycle. (Figure adapted from Lee et al. (1995) with permission. © 1995 Elsevier Science Ltd.)

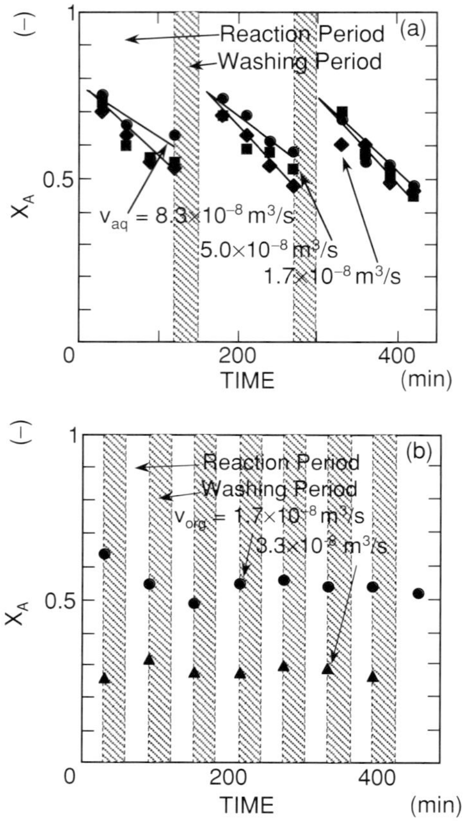

Figure 12.19 Conversions for the hydrogenation of carbobenzoxy phenylalanine over a 0.5 wt% Pd/C catalyst at 323 K in a four-phase trickle bed with (a) periodic interruption of 1, 2-dichloroethane at various flow rates and with (b) periodic switching between dichloroethane and a water wash at 0.17 mL/s. Cycle period = 150 min. (Figure adapted from Yamada and Goto (1997) with permission. © Society of Chemical Engineers of Japan.)

be separated from the fluid stream and disposed of. These problems can be avoided through extraction. den Hollander et al. (2002) demonstrated a continuous process using butyl acetate as the extractant with an immobilized penicillin acylase in a cascade of mixer-settlers. An aqueous solution of the potassium salt of penicillin G at pH = 2.6 served as the feed. This, together with butyl acetate, was introduced into a well-stirred mixer containing the immobilized enzyme. Thereafter, the mixed phases went to a settler where they were separated. A three-stage cascade of mixer-settlers significantly increased conversion of penicillin G and the APA yield over what can be achieved in the batch operation that is used industrially. Source of the improvement is the separation of APA and the by-product PAA that eliminates the reverse reaction. The point of this discussion is that in place of a cascade of mixer-settlers the hydrolysis could be carried in a cyclically operated, downflow, submerged bed employing switching between the aqueous feed and the butyl acetate extractant.

Chapter 13

Swing Reactors: Overview, Assessment, Challenges and Possibilities

13.1 Overview and assessment

There are important differences between swing reactors and chromatographic ones. The first of these is, of course, the regeneration of the adsorbent. Because the latter class of separating reactors use a flushing or carrier gas or an eluent, products and, indeed, the feed is diluted by this gas or liquid. Separation of the products from the flushing gas or from the eluent becomes necessary. This adds a further processing step. In swing reactors, the product withdrawn from the product or discharge port, the product most weakly adsorbed, can be obtained in high purity. On the other hand, the strongly adsorbed product, withdrawn from the exhaust port, will be a mixture of products and reactants. A CRAC or SCMCR usually offers a better separation of products and can be operated for complete conversion. Apart from the presence of carrier gas or eluent, high purity for both products is possible.

The feed and production step in the swing reactors can run on pure feed, whereas, in an SCMCR, reaction occurs in zones or sections 2 and 3 where the feed is diluted by the carrier gas or eluent. Higher concentrations result in higher rates and larger temperature effects. Thus, swing reactors may offer higher productivity than their chromatographic reactor counterparts. Of course, temperature effects should be more important in the swing reactors.

Research reviewed in this section demonstrates that swing reactors, pressure, temperature or their combination, are capable of significantly affecting conversion in equilibrium-limited single reactions and selectivity when multiple reactions occur. Perhaps the main advantage these reactors offer is to provide reactant conversion along with separation of reactants and products in the same vessel. Experimental studies, although limited, show some but hardly remarkable conversion increases; they do support theoretical work demonstrating that substantial separation can be accomplished.

Evidence in the research literature suggests that allowing a reaction to proceed in a swing adsorber does not dramatically affect the adsorber performance in terms of separation achieved. Indeed, if the reaction proceeds rapidly and is effectively irreversible, separation performance is changed just minimally provided the swing reactor operates isothermally in the feed and production step of a cycle. There is little difference in carrying out the reaction within the adsorber or in a separate reactor prior to the adsorber. However, reactions that proceed slowly because of their intrinsic kinetics or because of the rate of the reverse reaction will affect purity of products obtained at the delivery and exhaust ports. Larger effects on separation are anticipated with multiple reactions, but such systems have not been explored.

Far fewer reaction systems have been studied in swing reactors compared to those examined in chromatographic reactor systems (Table 8.1). Systems examined experimentally are listed in Table 13.1.

13.1.1 Pressure swing reactors

Two PSR versions are under study: one with an empty recycle volume at the discharge end of the reactor and a short cycle period (RPSR) and another using cycle periods an order of magnitude greater. Not enough work has been done to choose one version over the other. Perhaps the choice depends on the reaction system considered. Nevertheless, until further study, it is our opinion that the longer cycle period PSR should receive more attention. This version offers greater flexibility because it is not tied to a reflux volume so this alternative can accommodate back-fill cycles or those employing back flushing with streams other than the product stream. The advantage of the RPSR version is that it provides a nearly constant product delivery rate throughout a cycle. The longer period PSR has varying product delivery rates or there is no delivery for several steps in a cycle.

Pressure swing reactors, we believe, have commercial promise. There are several reasons for this conclusion. The first is that pressure swing adsorbers have found wide application in the chemical industry. An example is the use of this technology for hydrogen production. Consequently there is a firm foundation of practical design know-how. Furthermore, catalyst–adsorbent in PSRs are in packed beds. Design, therefore, of a PSR and scale-up of experimental units relies on the fluid dynamics of packed beds. These are well understood making the engineering of isothermal units, at least, straightforward. Perhaps the only uncertainty in the design and operation of PSRs is provision for large heat effect that will arise for systems with reactant concentrations of commercial interest. Just a start has been made on this potentially important problem by Yongsunthon and Alpay (1999).

13.1.2 Temperature swing reactors

Temperature swing and pressure swing reactors probably will not compete for an application because of differences in energy demand and response times. Pressure change propagates at the speed of sound and wave intensity is damped mainly by particle porosity and adsorption. Temperature change, on the other hand, propagates much more slowly and amplitude is diminished as the change penetrates a bed of solids by the heat capacity of those solids. Endothermic processes, like desorption, also contribute. As a consequence, TSR cycle periods are much longer compared to cycle periods in a PSR. Comparisons of the two swing classes have not been made so criteria for the choice of one class over the other in an application do not exist.

For this class of swing reactors, experimental studies predominate as may be seen in Table 9.1. Both improved conversion and separation of products has been demonstrated. Temperature swing appears to be feasible using fluidized bed reactors, as demonstrated by the work of Harrison and coworkers (Han and Harrison, 1994, 1997; Balasubramanian et al.,

1999; Lopez Ortiz and Harrison, 2001; Harrison and Peng, 2003), or employing packed beds as has been done by Elsner et al. (2003) and Menge et al. (2003). Packed beds operate as PFRs and offer advantages of higher productivity per mass of catalyst compared to a fluidized bed that tend to be well mixed. Fluidized beds offer the possibility of continuous addition of fresh adsorbent or catalyst to overcome loss of adsorbent capacity or deactivation. On the other hand, attrition of solids in the bed with the accompanying problem of fines removal may affect process economics. Heat supply from an external source is usually simpler in fluidized beds. Consequently, choice of reactor must depend on the reaction system and operating conditions. Yongsunthon and Alpay (1999, 2000) observed that a significant improvement in yield, productivity and energy demand can be obtained by staging TSRs. This is likely to be the case with PSRs, but this possibility has not yet been considered in the research literature.

As was the case for PSRs, TSRs are a rather straightforward extension of adsorption technology. The large literature and engineering know-how on adsorptive separation with thermal regeneration of the adsorbent is thus available for modeling TSR operation, for design and for scale-up. Consequently prospects for commercial application are promising.

13.1.3 Combined pressure and temperature swing reactors

Only one study of a combined swing operation has been published (Xiu et al., 2002a). This study, employing simulation, used just a small temperature change in a flushing step but demonstrated a significant improvement in product recovery. In our opinion, the combined swing reactor will be the best choice for many reaction systems. This is because desorption is always an endothermic step so that heat must be added. It is simpler and usually less costly to add heat through the fluid phase. This means adjusting the temperature of this phase and thus providing a temperature swing even if a pressure reduction is also introduced.

13.2 Modeling

Comments on modeling of chromatographic reactors in Chapter 8 apply generally to the modeling of swing reactors. If isothermality can be realistically assumed, as it can be perhaps for bench-scale reactors, one-dimensional pseudo-homogeneous models are adequate provided mass transfer is not rate limiting. Heterogeneous models find application when both reaction and adsorption rates are very fast. Nonisothermal reactors are also well represented by one-dimensional models. Two-dimensional models have not been proposed. Unlike modeling of chromatographic reactors, isobaric operation cannot be assumed because in many designs both the delivery rate and the exhaust rate are functions of the pressure drop.

In practice, temperature effects mean that swing reactors cannot be assumed to be isothermal. An energy balance must be introduced. Unfortunately, this balance has been neglected in many of the early simulations of pressure swing reactors given in the literature. The error introduced in this way has not been determined because model validation was not then a concern. More recent studies, such as Ding and Alpay (2000a,b), have not assumed

Table 13.1 Chemical and biochemical reaction systems investigated

Reaction system	Swing reactor type
Catalytic oxidation of CO	Pressure swing
Water gas shift	Temperature swing
Reverse water gas shift	Pressure swing
Butene dehydrogenation	Pressure swing
Propene disproportionation	Pressure swing
Steam reforming of methane	Pressure and temperature swing
Claus reaction	Temperature swing
HCN synthesis	Temperature swing

isothermality. These researchers also tackled model validation. They demonstrated a one-dimensional pseudo-homogeneous model adequately represented experimental data. Xiu et al. (2002a,b; 2003a,b) also introduced an energy balance but these investigators employed a heterogeneous model for the adsorbent. Validation was indirect. They showed experimental data in the literature was well predicted.

Validation of isothermal models would be useful or, at least, a comparison between the predictions of nonisothermal and isothermal models for identical reactors and operations. This would allow researchers to know under what conditions the much simpler isothermal representation of a swing reactor can be used. Experimental work on pressure swing reactors has lagged behind theoretical studies and certainly this is why the adequacy of an isothermal model has not been tested. Of the several experimental investigations of PSR systems given in Table 13.1, only one (Vaporciyan and Kadlec, 1989) compared simulation and experiment. This comparison suggested that the model or rather the simplifications employed are not fully satisfactory (see Fig. 9.20). Part of the model inadequacy may be the isothermal assumption even though the feed to the experimental PSR was dilute. Vaporciyan and Kadlec blame the use of a linear isotherm for the poor agreement.

Models employed by all investigators assumed that processes on the particle scale are very fast so that a pseudo-steady-state can be assumed. This needs investigation. Just as in chromatographic reactors, catalyst and adsorbent particles experience a constantly changing environment. Diffusion effects tend to be larger in transient operation than under steady state so these could contribute to simulation errors. Furthermore, investigators have used steady state kinetic models and assumed a linear driving force for mass transfer to the particle surface. An investigation is needed to verify under what conditions such assumptions apply.

Yongsunthon and Alpay (1999) studied the energy flows for steps in TSR cycles. When energy flow was considered, a reactor with separate vessels serving as stages rather than a single vessel was advantageous with respect to throughput and product purity. Also a serial-parallel arrangement of the stages can be better than a simple cascade. It seems likely that staging will be beneficial for pressure swing reactors, but this has not yet been undertaken. A theoretical study of the performance of a cascade or an array of PSR vessels by simulation seems warranted.

13.3 Design

Systematic procedures for the sizing and dimensioning of swing reactors have not been discussed in the technical literature. Nonetheless, some conclusions can be drawn on the basis of the much more extensive treatment of simulated countercurrent moving bed chromatographic reactors (SCMCRs). For this reactor, design procedures follow those for separators with an additional step of checking the residence times to ensure they are long enough to attain the conversion, selectivity or yield required. For fast, essentially irreversible reactions, design for separation is usually satisfactory without changing the residence time to allow for reaction. With intrinsically slow reactions or when equilibrium results in only a small conversion, residence times must be increased causing a reduction in productivity. It is likely that this will be the situation for swing reactors. Nevertheless, this supposition needs to be checked. Large heats of reaction relative to heats of adsorption and desorption could affect how design is undertaken.

The work of Yongsunthon and Alpay (1999) has shown that staging of swing reactors improves performance and heat economy. Thus, design of a swing reactor system must consider this possibility. Their second study (Yongsunthon and Alpay, 2000) suggested that rather simple serial-parallel networks are adequate. Nonetheless, design of a swing reactor system must begin with simulation to determine whether a single reactor or a network of reactors should be used. This simulation would also give the network structure if staging is advantageous.

13.4 Research needs

Much work is needed to realize the commercial potential of swing reactors. Indeed, needs are so large that opportunities for innovation are difficult to identify.

Table 13.1 lists just nine reaction systems and suggests that there is a pressing need to study experimentally more systems in order to understand what properties and what operating conditions result in successful swing reactors. Candidate systems are those gas phase reactions that have been investigated with chromatographic reactors such as hydrogenation of mesitylene, methanol synthesis and aromatics alkylation. All of these reactions have been carried out in SCMCRs similar enough to swing reactors so that suitable adsorbents and operating conditions can be identified. Temperature swing should be applicable to liquid phase reaction systems as well as gas phase ones. Some exploratory study of this application would be rewarding. However, biochemical reactions are probably not good candidates for a liquid phase TSR because the enzymes used may be deactivated by rising temperature. Probably a better study choice would be an ester hydrolysis, such as methyl formate hydrolysis, a reaction that has been investigated by Falk and Seidel-Morgenstern (2002) using an SCMCR. An activated carbon served as the adsorbent. Formic acid is more strongly adsorbed by the carbon than methanol.

Only single reactions in PSRs have been studied experimentally. Although systems such as steam reforming or butene dehydrogenation may contain multiple reactions, these systems have been looked at under conditions such that just a single reaction is important. Certainly expansion to systems with multiple reactions is needed. Separating reactors may substantially alter selectivity. This was a need identified in Chapter 8 for chromatographic reactors as well.

We observed in Section 13.1.3 that a combined pressure and temperature swing should be an attractive mode of operation. There are no comprehensive experiments reported in the literature on the combination. Simulation of the combined operation would indicate what magnitude of performance improvement in terms of yield and productivity may be achieved compared with a PSR. Indeed, Xiu et al. (2003a) considered a temperature change in just one step of a five-step PSR cycle and found a recovery improvement. Nevertheless, experimental results are essential to validate simulations and to guide further development. A possible reaction for such a study is propene disproportionation over a Rh catalyst using a zeolite adsorbent. This reaction was the subject of a PSR investigation by Gomes and Yee (2002). An alternative could be the reverse water gas shift studied by Carvill et al. (1996). Other pressure swing systems mentioned in Chapter 9 suffer from loss of adsorbent capacity or strong adsorption of a reactant.

Yongsunthon and Alpay (1998a) compared nonisothermal and isothermal models through simulation for a well-mixed bed of catalyst and adsorbent and showed that heat effects are important. Chromatographic reactor studies considered in Part II indicate that laboratory-scale reactors with dilute feeds can be treated as isothermal systems with little error. It remains uncertain under what conditions the simpler isothermal representation is satisfactory for swing reactors. An extension of Yongsunthon and Alpay (1998a) simulation would supply this information.

Loss of adsorbent capacity has been observed experimentally as reported in Chapter 11. Catalyst deactivation can be expected as well. Study of these time-on-stream changes is needed to determine their effect on reactor performance. It is likely that some performance loss can be overcome by reducing throughput or through manipulating cycle or temperature–pressure conditions. Any study of time-on-stream changes needs to explore what can be done to mitigate problems introduced by adsorption capacity loss or catalyst deactivation.

With respect to modeling and simulation needs, further testing and perhaps development of the "shooting" methods for predicting reactor performance at the cyclic stationary state limit are needed. These methods were discussed in Section 9.4 of Chapter 9. Although most of the literature concludes that these methods reduce computing times, several papers have suggested this is not always true. In some cases, integration of the model equation in the time dimension from the initial conditions until there is no further change in a cycle has been observed to be faster because of time needed to repeatedly estimate the model eigenvalues. Large CPU requirements are a serious problem in simulating the periodic operation of both swing reactors and simulated countercurrent moving-bed chromatographic reactors.

A further need in the realm of simulation is an examination of the benefits of staging a PSR system. This was mentioned above in Section 13.3 and arises because of the work of Yongsunthon and Alpay (1999, 2000) showing improved performance when TSRs are staged. There are many choices for the reaction system to be used.

Systematic design of swing reactors should be examined. Can these units be designed as separators with an adjustment only in flow rate or bed length to insure conversion or yield requirements are met? Or, must design build from the reaction and undergo adjustment to meet separation targets.

We have not discussed periodically pulsed, trapping or exploratory reactors in this chapter. Periodically pulsed reactors, discussed in Chapter 12, fall between swing and chromatographic reactors. In terms of research needs, they are like multiple pulse chromatographic reactors. Indeed, they are best considered to be an extreme case of such reactors (no chromatographic separation takes place). Research needs for the development of this reactor type will be the same probably as those for CRs.

Trapping reactors were mentioned in Chapter 12 because they exploit adsorption. They are not separating reactors so their research needs will not be discussed.

Exploratory reactors, a CR or a well-mixed bed of catalyst and adsorbent, will be examined in the next chapter. It is appropriate, however, to consider research needs at this point. First of all some theoretical work is needed to define the characteristics of reaction–separation systems suitable for PSR applications. The use of a well-mixed reactor to explore candidate systems, choose adsorbents and suggest operating conditions has been suggested by Sheikh et al. (1998). Operation of well-mixed reactors with both catalyst and adsorbent has been demonstrated and used to show that adsorption or reactively trapping a product can increase conversion. However, application of the single, well-mixed reactor to screening has not been carried out. This application deserves experimental study. It is potentially an alternative to a single pulse chromatographic reactor as a technique for choosing systems suitable for combined reaction and separation.

13.5 Research Opportunities

Development of swing reactors is not sufficiently advanced for us to speculate on variations that could improve performance and thus represent a research opportunity. However, there are substantial research opportunities for swing reactors. These can be resolved into three areas: (1) discovery of suitable high temperature adsorbents, (2) identification of reaction systems whose overall performance and separation efficiency could be improved by using swing reactors and (3) new applications.

With regard to adsorbents, it appears that the primary application of swing reactors will be for solid-catalyzed gas phase reactions. Such reactions are usually carried out at temperatures greater than 500 K. With physical adsorption and chemisorption, capacity of the adsorbent decreases with rising temperature so that there are few conventional adsorbents, like activated carbon or silica gels, that can be used above 500 K. The research challenge as well as the research opportunity is to develop high temperature adsorbents or reactive solids. Hydrotalcite is a probably a reactive solid rather than an adsorbent for CO_2. It has properties of a base and a carbonate appears to form. Unfortunately, the chemical composition of the hydrotalcite or its production needs further work in order to eliminate loss of capacity with repeated cycling. Synthetic zeolites operating on size exclusion may be attractive, however, synthetic zeolites collapse at higher temperatures. The challenge is to find high capacity, size selective materials that are stable at temperatures well above 500 K.

Because of the commercial promise of PSRs and TSRs, it is our view that the development of the technology should make identifying suitable reaction–separation systems for swing reactor application a top priority. This must be largely an experimental activity. The literature reviewed suggests that cycle operation and perhaps the advantage of staging will differ depending on the application. Consequently, any experimental investigation of a PSR application would have to have a modeling component directed at developing a suitable cycle and exploring staging.

Despite our disclaimer that innovations will be difficult to identify until the development of swing reactor technology progresses further, there are several opportunities that are worth mentioning. One of these is the use of nonisotropic adsorbent–catalyst beds. Only homogeneous beds have been examined. One type of nonisotropic bed that should be investigated is one with variation of catalyst to adsorbent ratio. At the front of the bed where the reactant concentrations are high and product concentrations are low during the feed step, a high catalyst to adsorbent ratio might be suitable because the adsorbent has little effect on the reverse reaction. At the other end of the bed where reactant concentration would be low but product concentrations high, the ratio could be reversed to diminish the rate of the reverse reaction. Adsorption–reaction systems can be expected to be nonlinear. It is precisely in this situation that variations in the catalyst adsorbent ratio should exert an effect. Xiu et al. (2003b) simulated the performance of a multizone bed with a different ratio in the final zone and showed performance improved. The strategy needs experimental study. Another arrangement might be to change the adsorbent used part way through the bed. With multiple reactions occurring, one adsorbent might accelerate an initial equilibrium-limited reaction while a second adsorbent could affect the products obtained. This might also be accomplished by changing the type of catalyst mixed with adsorbent in the bed. The effect of a two-catalyst policy was examined by Aida et al. (1999) under periodic operation. Because reaction kinetics are frequently nonlinear, synergistic effects could be realized with a two-catalyst policy. Variation of catalyst activity through the bed is another possibility. This could be accomplished through temperature variation or by the loading of the active catalytic material on its inert support. This latter suggestion draws on the work of Cote et al. (1999). There is also precedent in the literature for using layers of different catalyst to improve selectivity for multiple reactions (Shi et al., 1996). The two adsorbents or two-catalyst arrangements would be easier to implement in separate beds. Use of multiple beds or stages is feasible in both SCMCR and PSR systems.

Furthermore, some of the novel operating modes mentioned in Chapter 8 for SCMCRs, such as changing solvent composition within a step or modulating feed flow or composition, might be applicable to swing reactors.

One new application, the use of a liquid feed near its boiling point or use of a gas–liquid mixture, could be worthwhile and challenging to explore. In the regeneration step, partial or complete vaporization would take place producing rapid mass transfer. Candidates for investigation would be systems used in catalytic distillation, such as ester synthesis. Several different ion-exchange resins could be used to adsorb the water product. An SCMCR, in principle, competes with catalytic distillation for esterification and ether formation. The swing reactors may also be competitors.

Part IV
System Screening and Development

Chapter 14
Screening Reactors

14.1 Requirements for separating reactors

Properties of reaction systems that would benefit from combined reaction and separation may be readily enumerated: (1) single reactions that are equilibrium limited at temperatures and pressures that provide reasonable reaction rates, (2) single reactions that are product inhibited, or even reaction systems in which the feed contains a nonreacting inhibitor, (3) parallel reactions with just one desired product, often reactions in which there is an equilibrium step, and (4) sequential reactions in which an intermediate product is desired. Parasitic reactions that consume the sought-after product could represent this class. Of course, identifying such systems is the starting point for any investigation of separating reactors. The exercise is not necessarily trivial. The extent of the equilibrium limitation or the selectivity problem must be considered. If the equilibrium conversion or the selectivity to the desired product is as high as 95% or as low as 1%, is the use of a separating reactor worth considering? An answer can be found only by looking at other alternatives and through undertaking an economic analysis. In an in-between range, of say 25–75%, this is rarely a question. An economic analysis is important, but it would be undertaken only after a separating reactor had been designed. We return to reaction system properties after considering adsorbents.

Identifying suitable adsorbents is a further requirement. Properties required are (1) adequate adsorption capacity for one of the products in the temperature range of the reaction, (2) significant differences in adsorption capacity for the different products, (3) low adsorption capacity for reactants, (4) low or moderate heat of adsorption so desorption by reducing pressure or purging can be affected without a large increase in temperature and (5) a rate of adsorption that matches the rate of reaction as equilibrium is approached. The desirable situation for (4) is illustrated in Figure 14.1 that plots bed lengths to affect adsorption and to drive the reaction to completion or equilibrium as a function of temperature. The adsorber length increases with temperature because temperature reduces adsorbate capacity.

The figure indicates that at a temperature of about 250°C adsorption rates and reaction rates are of comparable magnitude. Under this condition product adsorption will have the largest effect on reaction rate and thus on conversion. If reaction rates greatly exceed adsorption rates, adsorption cannot influence reaction. If this is reversed in the neighborhood of equilibrium, the reactor may be oversized or the temperature may be too high. Figure 14.1 reflects the Claus reaction: $SO_2 + 2H_2S \leftrightarrow (3/8)S_8 + 2H_2O$.

Other considerations for the adsorbent are that it should be catalytically inactive (unless it is both adsorbent and catalyst) and mechanically stable at the reaction temperature or over the range of temperature employed if a temperature swing cycle is contemplated. It is

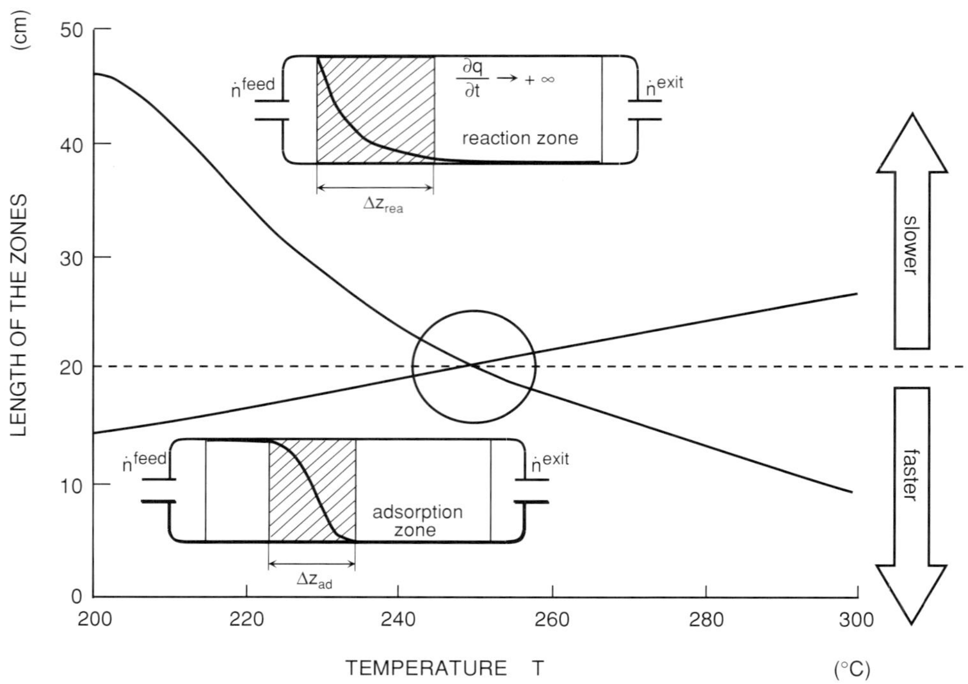

Figure 14.1 Equal bed length condition for separating reactors. Temperatures and bed length correspond to the Claus reaction. (Figure reproduced from Elsner et al. (2002a) with permission. © 2002 Elsevier Science Ltd.)

hardly surprising that discovering or developing a suitable adsorbent is a major obstacle to creating a separating reactor system.

Further requirements for the reaction system, presumably a catalytic system, are that products should be just weakly adsorbed, side reactions and deactivation should be very slow if not absent altogether. Catalyst particles must have mechanical stability. They should be of millimeter scale to avoid diffusional limitations.

It would be a mistake to assume that an adsorbent can be selected a priori. What is possible is that a group of materials can be chosen on the basis of past research. From this group, the best candidate must be identified, but this demands experimentation. Fortunately, a chromatographic or pressure swing system is not necessary for the experiments. Relatively simple screening reactors may be used.

14.2 Screening reactors

14.2.1 Well-mixed systems

Sheikh et al. (1998) analyze an isothermal, isobaric well-mixed reactor continuously fed with a solid adsorbent or a liquid absorption medium as well as reactant. If the reaction requires a catalyst, it could be fixed within the reactor or it could be fed continuously along with the adsorbent. We consider the Sheikh study in this review because this relatively simple separating reactor performs in part like a PSR, a PPR or a CMCR. The difference is that usually the adsorbent is not regenerated, at least not within the reactor. Physically, the Sheikh reactor could be realized by a circulating fluidized bed with continuous feed and removal of catalyst and adsorbent. A moving bed of adsorbent through a fixed bed of catalyst with external recycle of the fluid phase would also serve. A fluidized bed of catalyst with a very fine adsorbent blown in with the fluid phase and removed overhead, or with large adsorbent particles sinking through the fluidized bed could also be used. Indeed, the latter system was studied experimentally by Brun-Tsekhovoi et al. (1986).

Other options that are perhaps easier to construct and operate would place catalyst and test adsorbent in a basket in the form of a flat bladed impeller mounted on a rotating shaft. The fluid entering and leaving the vessel is continuously stirred by the impeller creating a well-mixed fluid phase environment. A design of this type, known as a Carberry reactor, is used for measuring the kinetics of catalytic reactions. Sealing the rotating shaft sets an upper pressure limit for rotating basket reactors. Fixing the basket within the reactor and recirculating the fluid phase at a high rate relative to the feed flow avoids the pressure limitation and provides much better temperature control so that isothermal operation becomes possible. When a blower is located within the vessel, two designs are commercially available. The Berty reactor places the blower beneath the catalyst bed and draws fluid downward through the bed. If the blower is located above the bed, the system goes under the name of an Amoco reactor. Externally located blowers are also widely used in these fixed-bed recirculating systems.

Of course, with adsorbent in a fixed bed, these screening reactors must be operated periodically. Usually, with short cycle periods of just several minutes, adsorbate build-up is small so that fluid samples withdrawn in either portion of the cycle will not change composition with time. These samples then show the performance of the adsorptive reactor with fresh adsorbent. Increasing the cycle period to tens of minutes or to about 1 h will show how the system performance changes as the adsorbent becomes saturated. Longer cycle periods require the collection of multiple samples. Usually this is done employing multiloop samplers. Recirculating reactors use shallow beds so that conversion per pass is differential. Ideally, conversions should be about 1% per pass. Recirculation rates are often greater than tenfold the feed rate.

If the conditions chosen slow the reaction rate sufficiently, periodic switching between feed and a flushing gas can be avoided. Change in reaction rate as the adsorbent becomes saturated can be followed by repetitive sampling. Indeed, in this case a batch reactor could be used.

Sheikh et al. (1998) suggest that their well-mixed reactor could serve as a screening system to identify reaction systems suited to a PSR or, indeed, an SCMCR process. They suggest further that their reactor can be used to select adsorbents to use with a chosen catalyst or to select operating pressures and temperatures for the process.

The theoretical analysis in the Sheikh et al. paper does not reveal new insight into PSR or SCMCR performance; it confirms the modeling results discussed under isothermal operation in Chapter 9. A general reversible reaction: $aA + bB + \ldots \Leftrightarrow rR + sS + \ldots$ and a consecutive reaction: $A \rightarrow B \rightarrow C$ are considered. The appropriate model is given by Equation 3.73 where the time derivatives drop out at steady state. For the reversible reaction, conversion is related to a modified Damköhler number, $(N_{\mathrm{Da}})'$, written as

$$(N_{\mathrm{Da}})_f' = \frac{(N_{\mathrm{Da}})_f}{\prod_k (1+\varphi_k)^{\nu_k}} \tag{14.1}$$

for the forward reaction where k is the reactant index and ν_k is the stoichiometric coefficient for the kth reactant. φ_k is the adsorbent capacity for reactant k. For the reverse reaction,

$$(N_{\mathrm{Da}})_r' = \frac{(N_{\mathrm{Da}})_r}{\prod_i (1+\varphi_i)^{\nu_i}} \tag{14.2}$$

i is the index for the products. An analytical expression for conversion can be derived only for simple reactions. For example, for A $\Leftrightarrow$ B + C,

$$X_{\mathrm{A}} = \left(\frac{(N_{\mathrm{Da}})_f'}{(N_{\mathrm{Da}})_r'} + \left[\frac{(N_{\mathrm{Da}})_f'+1}{2(N_{\mathrm{Da}})_r'}\right]^2\right)^{1/2} - \frac{(N_{\mathrm{Da}})_f'+1}{2(N_{\mathrm{Da}})_r'} \tag{14.3}$$

and the modified Damköhler numbers in Equations 14.1 and 14.2 for the simple reaction are $(N_{\mathrm{Da}})_f' = \dfrac{(N_{\mathrm{Da}})_f}{1+\varphi_{\mathrm{A}}}$ and $(N_{\mathrm{Da}})_r' = \dfrac{(N_{\mathrm{Da}})_r}{(1+\varphi_{\mathrm{B}})(1+\varphi_{\mathrm{C}})}$. The conversion expression shows that increasing the adsorption of reactant A reduces conversion while increasing the adsorption of one or both of the products increases conversion. Sheikh et al. derive an expression for the enhancement of conversion by introducing an adsorbent and demonstrate how the adsorption capacities and the Damköhler numbers influence the enhancement.

The modified conversion expression for a PSR permits a numerical assessment of the advantage of introducing an adsorbent into the reactor. An enhancement factor can be defined as

$$\eta_X = \frac{X_{\text{with adsorbent}}}{X_{\text{without adsorbent}}} \tag{14.4}$$

If rate constants and adsorption capacities are known, the enhancement can be predicted from the expressions given above. For a reaction of stoichiometry A $\Leftrightarrow$ B

$$\eta_X = \frac{(N_{\mathrm{Da}})_f'}{(N_{\mathrm{Da}})_f}\,\frac{1+(N_{\mathrm{Da}})_f-(N_{\mathrm{Da}})_r}{1+(N_{\mathrm{Da}})_f'-(N_{\mathrm{Da}})_r'} \tag{14.5}$$

For a reaction with stoichiometry A $\Leftrightarrow$ B + C

$$\eta_X = \frac{(N_{\mathrm{Da}})_r}{(N_{\mathrm{Da}})_r'}\,\frac{\left(\left[1+(N_{\mathrm{Da}})_f'\right]^2+4(N_{\mathrm{Da}})_f'(N_{\mathrm{Da}})_r'\right)^{1/2}-\left(1+(N_{\mathrm{Da}})_f'\right)}{\left(\left[1+(N_{\mathrm{Da}})_f\right]^2+4(N_{\mathrm{Da}})_f(N_{\mathrm{Da}})_r\right)^{1/2}-\left(1+(N_{\mathrm{Da}})_f\right)} \tag{14.6}$$

The influence of these parameters on selectivity or yield for the consecutive reaction is easily derived. Selectivity is the difference of the extent of reaction 1 (A $\rightarrow$ B) and the extent of reaction 2 (B $\rightarrow$ C) divided by the extent of reaction 1. The extent of reaction 1 for unit stoichiometry is

$$\xi_1 = \frac{n_{\mathrm{A}_0}(N_{\mathrm{Da}})_1}{1+(N_{\mathrm{Da}})_1+\varphi_{\mathrm{A}}} \tag{14.7}$$

Consequently, the selectivity for B is

$$\Delta\xi_1/\xi_1 = \frac{1+\varphi_B}{1+\varphi_A}\frac{n_{A_0}(N_{Da})_1}{1+(N_{Da})_2+\varphi_B}\left(1-\frac{(N_{Da})_1}{1+(N_{Da})_1+\varphi_A}\right) \tag{14.8}$$

Yield is the product of the extent of reaction 1 and selectivity. It can be seen that increasing the adsorption capacity for the reactant decreases extent of reaction 1 or reactant conversion. Yield of B is also decreased, but the effect on selectivity is smaller. Raising the adsorption capacity for the intermediate B generally raises selectivity to B and the yield of B unless the second reaction step is slow so that $(N_{Da})_2$ is small. As expected, selectivity and yield are sensitive to $(N_{Da})_1/(N_{Da})_2$.

The analysis just summarized indicates how screening experiments should be done. An initial set of experiments should be undertaken without the use of an adsorbent. This provides a baseline to see what enhancement is produced in either conversion or selectivity when different adsorbents are introduced.

Although isothermal experiments are easier to perform on a laboratory scale, they will not indicate the full advantage of combining separation and reaction. The screening reactor does not indicate the separation possible. Because they are isothermal, they will also not indicate rate and selectivity enhancement through modification of the conversion-temperature trajectory. This shortcoming can be overcome by carrying out screening experiments in an adiabatically operated reactor.

Equation 14.4 may be used to quantify the performance improvement for the adiabatically operated reactor, but the other analytical expression given above will not apply.

14.2.2 Tubular reactor systems

Behavior of a reaction system in the presence of catalyst and adsorbent can also be assessed by a breakthrough experiment. This is a rather familiar experiment. A bed of well-mixed adsorbent and catalyst is packed into a tube of small diameter, say 10–15 mm i.d. × 100–150 mm, held in a constant temperature bath. Carrier gas or, in the case of a liquid system, solvent or eluent circulates through the system to bring the column to an initial steady state. The experiment is initiated by introducing the feed mixture and the effluent of the column is continuously monitored. Breakthrough composition, as the reaction products begin to emerge from the bed, or pre-breakthrough composition can be compared to the effluent composition after a long time on stream. The adsorbent, then, is saturated and no longer provides separation. Without a model it is not possible to quantify conversion or yield improvement, but a qualitative assessment is possible through comparing product concentrations prior to breakthrough with concentration after an extended time on stream. Higher concentration of the first product to emerge, the raffinate stream, means the adsorbent is shifting reaction extent.

Another version of this experiment uses a single feed pulse in place of a step change in feed composition. This is, of course, a chromatographic experiment. These have been discussed in Chapter 4 for multiple, well-separated pulses. For assessment, interpretation is simple. Use of an adsorbent is attractive if the effluent profiles of the products concentrations are well separated and the reactant peaks are small. This is illustrated by Figure 14.2 for an

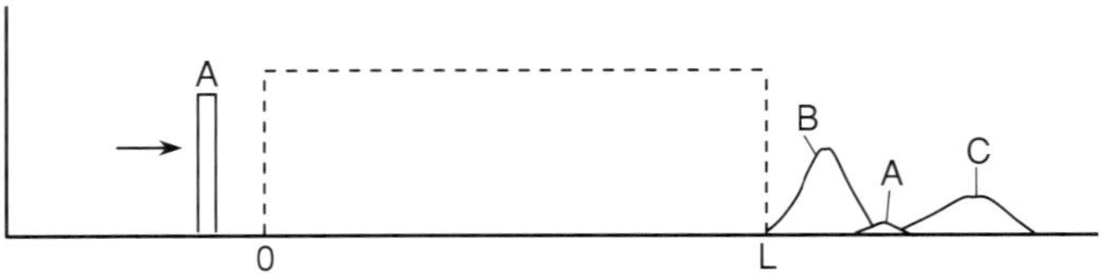

Figure 14.2 Qualitative interpretation of a packed-bed pulse experiment for an A → B + C reaction.

A → B + C reaction. Requirements for single or multiple pulse chromatographic experiments are discussed in the chapter just mentioned.

Goto and coworkers (Goto et al., 1993, 1995, 1996) used step change or multiple pulse experiments in a packed-bed tubular reactor to explore the use of metals and metal alloys for hydrogen capture. Their experiments are reviewed at the end of the next section.

14.3 Experimental studies

14.3.1 Well-mixed screening reactors

Use of a fluidized bed of catalyst as a well-mixed reactor to test the addition of adsorbents to increase conversion was explored experimentally by Brun-Tsekhovoi et al. (1986) well before the theoretical study of Sheikh et al. (1998). Brun-Tsekhovoi et al. were investigating steam reforming of methane at 900 K and pressure up to 100 bar to produce H_2. Their catalyst was microspheres of Ni/Al_2O_3 while their adsorbent for the CO_2 produced in reforming was a CaO–MgO dolomite. Adsorbent particles were very much larger than the catalyst and thus descended through the fluidized catalyst and were collected at the bottom. They were calcined in a separate vessel and recycled to the reactor. The stream leaving the fluidized bed contained 92–96 vol% H_2 and small amounts of CH_4 and CO_2. Conventional steam reforming produces a H_2–CO_2 mixture with roughly the same levels of CO and unconverted methane found in the Brun-Tsekhovoi work. The reactor utilized by Brun-Tsekhovoi et al. was not, of course, a PSR, but resembled one by using an adsorbent or more accurately a solid reactant to augment conversion. The process proposed by the investigators is interesting, but does not appear to have been pursued further.

Dolomite was also used as a CO_2 trap in the water gas shift reaction by Han and Harrison (1994) to augment conversion in this equilibrium-limited reaction. The reaction was run in a fixed bed mixture of shift catalyst and dolomite at 823 K and 15 bar. The authors observed a H_2 content in the off gas well above that dictated by thermodynamic equilibria assuming no CO_2 removal by the dolomite. The system considered is another example of a screening reactor. However, in the Han and Harrison study, the reactor served to demonstrate the role of an adsorbent or trap in raising conversion.

Batch experiments were undertaken by Zhu et al. (1999) on ethylene oxide formation by bubbling a mixture of ethylene, oxygen and water through a well-mixed slurry of catalyst suspended in silicone oil. The authors claimed selectivity to ethylene oxide of over 80%. They also demonstrated near isothermal operation. Experimental results were compared to a model that considered only the reactants. This is tantamount to assuming irreversible absorption of ethylene oxide in silicon oil and no further reaction of the oxide product with oxygen. The authors report a small increase in CO_2 in the gaseous effluent so the assumption of negligible parasitic reactions was clearly incorrect. Nevertheless, the results indicate that a continuous operation in which absorbent flows through the reactor would substantially improve the synthesis performance. The Zhu et al. contribution is a further example of a screening reactor for examining candidate systems to which periodically operated separating reactors could be applied.

14.3.2 Tubular screening reactor

The second approach to screening employs a tubular reactor packed with the catalyst or adsorbent to be tested. Usually the solids are well mixed, but segregated beds formed from alternating layers of the solids can also be used. By keeping the bed shallow, the reactor is easily maintained in a furnace or constant temperature bath to assure isothermal operation. In operation the catalyst after it and the adsorbent are loaded into the reactor would be conditioned. Then the reactor would be thoroughly purged at the desired reactor temperature usually by the carrier gas to be used in the screening experiment. Reactant is then added to the carrier gas feed as a step change up in concentration. Duration of the step up is often an experimental variable. Composition of the effluent is measured continuously. The flow, finally, is switched back to the carrier gas and the effluent composition continues to be measured. Alternatively, if a PSR operation is under study, a vacuum would be drawn on the bed and the gas withdrawn would be subject to analysis. It was this type of system and procedure that Imai et al. (1985) and Goto et al. (1993, 1995, 1996) employed.

The intent of the Goto study and earlier work by Imai et al. (1985) was to explore the feasibility of using metal hydrogen adsorbers (acceptors) to force catalytic dehydrogenation reactions toward completion. A number of metals and alloys, such as titanium, tantalum, Ni alloys with magnesium or calcium, magnesium alloys of copper and iron alloys of titanium, were tested. Test reactions were dehydroaromatization of hexane and dehydrogenation of propane or cyclohexane. Catalysts for these reactions were also studied. Since the work of Goto and coworkers at Nagoya University also explored the application of PPRs to catalytic dehydrogenation reactions, it has been partially discussed in Chapter 12. We will deal here only with the screening work.

The screening reactor used by Imai et al. and the Goto team is illustrated in Figure 14.3. A schematic with more detail is given in Figure 12.7. In the Goto experiments, helium or

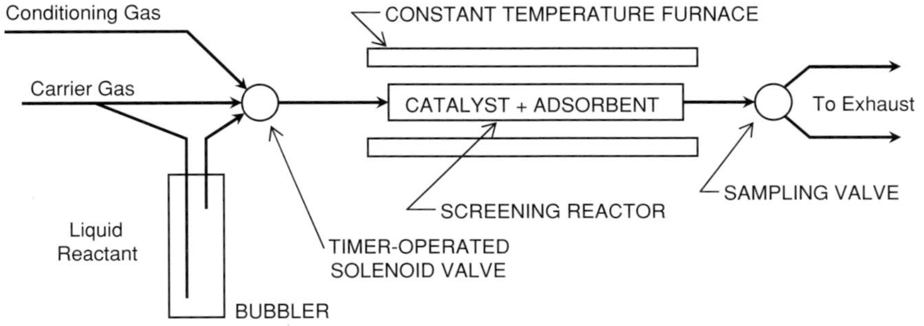

Figure 14.3 Packed-bed screening reactor for pulse or step change experiments.

Table 14.1 Hydrogen storage alloys tested by Imai et al. (1985)

Alloy	Surface area in m^2/g
FeTi	0.047
$Fe_{0.9}Mn_{0.1}Ti$	0.034
$CaNi_5$	0.064
$LaNi_{4.7}Al_{0.3}$	0.020
Mg_2Ni	0.534
Mg_2Cu	2.07

argon served as the carrier and purge gas. As shown, the paraffin reactant was introduced into the carrier stream via a bubbler. A four-way solenoid valve affected the switch from feed to purge gas.

Catalytic dehydrogenation of cyclohexane to benzene at ambient pressure was undertaken by Imai et al. (1985) using a Pt/Al_2O_3 catalyst and the alloys shown in Table 14.1. Their experiments, carried out at 150°C and 1 bar with an alloy to catalyst weight ratio of about 10:1, showed that the low surface area adsorbents were somewhat better than the higher surface ones, but all roughly doubled the cyclohexane conversion compared to catalyst mixed with inert support. Despite its low surface area, $CaNi_5$ gave significantly better results than the other alloys in Table 4.1. This observation led to experiments by Goto et al. (1993) using a step change from an argon carrier gas to a feed consisting of the carrier saturated with cyclohexane. Production of benzene immediately after the change greatly exceeded the equilibrium limit, but then slowly declined to this limit as the adsorbent became saturated with hydrogen. Multiple pulse experiments were also undertaken. These are discussed in more detail in Section 12.3.

Goto et al. (1995) screened several magnesium-based hydrogen storage alloys using the dehydroaromatization of *n*-hexane to benzene over a Zn-exchanged hydrogen ZSM-5 catalyst as the test reaction. Table 14.2 gives the composition of the alloys tested and their H_2 capacity. The Imai experiments showed that surface area of the alloys was not an important factor in their use as an H_2 acceptor.

Step change experiments were used. In these, fully mixed catalyst and adsorbent were compared with catalyst diluted with an equal volume of a nonadsorbing solid, and with adsorbent and catalyst segregated into layers. Goto et al. observed that conversion to benzene

Table 14.2 Hydrogen storage alloys tested by Goto et al. (1995)

Alloy	Weight % Mg	H_2 capacity in 10^{-8} g H_2/g alloy*
Mg_2Ni	49.1	2.65
Mg_2Cu	54.7	2.65
$CeMg_{12}$	79.8	1.30
$CaMg_{1.8}Ni_{0.5}$	45.2	5.55
$Mg_{51}Zn_{20}$	49.1	4.39
$Mg_{2.2}La_{0.8}Ni$	23.1	3.62

*Measured at 0.1 MPa and 573 K.

was significantly improved by the presence of an H_2 acceptor. The best alloys in terms of the initial increase in conversion were the Mg_2Ni, Mg_2Cu or $Mg_{51}Zn_{20}$ alloys, which are about a 50:50 mixture of Mg and the second metal by weight. Adsorbent fully mixed with the ZSM-5 catalyst gave the highest conversion of *n*-hexane.

Only the fully mixed arrangement was tested in a PPR. We have already examined the results with a 50:50 magnesium zinc adsorbent in Chapter 12 where Figure 12.9 illustrates what was observed. The figure compares the response to a step change from carrier gas to carrier gas saturated with *n*-hexane when a $Mg_{51}Zn_{20}$ alloy is present to the response when only catalyst and an inert solid is used. The former shows a large hexane conversion overshoot on the switch followed by a gradual decay to the equilibrium-limited conversion, while the latter just goes to the equilibrium limit.

Dehydrogenation of propane to a mixture of paraffins, cyclohexane and aromatics over a partially exchanged Zn/H ZSM-5 was investigated in a second study (Goto et al., 1996). This equilibrium-limited reaction sequence proceeds to a greater extent over the zeolite catalyst than dehydroaromatization of *n*-hexane. Addition of a hydrogen adsorbent is also effective. In this study the investigators tested titanium and tantalum, examining the effect of adsorbent particle size and bed arrangement. Screening was done using a step change procedure. Finely divided titanium powder was the most effective adsorbent. With this powder, the initial propane conversion after the step change was about 80%, more than threefold the conversion seen when an inert is used in place of titanium.

When run as a PPR experiment with an unusually high cycle split ($s = 0.86$), the time-average propane conversion increased by about 20%. Aromatics production also increased. Selectivity to aromatic products, however, was unchanged. These results are shown in Figure 12.10 and discussed in Section 12.3.

Silva and Rodrigues (2002) explored the use of a packed-bed adsorptive reactor for the synthesis of diethylacetal or acetal employing a strong acid, cationic exchange resin as both adsorbent and catalyst. They used a step input. The objective was to see if the production of this important industrial chemical could be carried out in an SCMCR, that is, can the conversion be driven past the equilibrium limit and can the products be separated. For this purpose, Silva and Rodrigues used a 2.2 cm i.d. × 28 cm bed packed with 46 g of Amberlyst 18 ion-exchange resin ($d_p = 0.5$ mm, $\varepsilon_p = 0.39$). Because they also wanted to determine the feasibility of an SCMCR for the acetal synthesis by constructing a model, experiments were extended to obtain parameters for the model.

The stoichiometric relation for acetal formation is acetaldehyde + 2ethanol ↔ diethylacetal + water. The resin adsorbs water in preference to acetaldehyde or ethanol, followed by acetal. Ethanol was proposed as the eluent. The step change experiment was performed by introducing a downflowing, nearly stoichiometric mixture of ethanol and acetaldehyde into the resin bed saturated with the ethanol eluent. Breakthrough behavior is shown in Figure 14.4.

The figure shows acetal breakthrough beginning at 5 min. Water and acetaldehyde appear at about the same time, about 12 min after feed introduction. Thus complete conversion of acetaldehyde and separation of the products is feasible. Interestingly, just before water breakthrough, the acetal concentration is higher than what would be achieved at steady state in a packed-bed reactor.

Breakthrough experiments were also carried out for nonreacting binary pairs, e.g. acetal in ethanol, acetaldehyde in water. Data were used to fit a competitive Langmuir isotherm model used to represent adsorption equilibrium. Breakthrough data, such as given in Figure

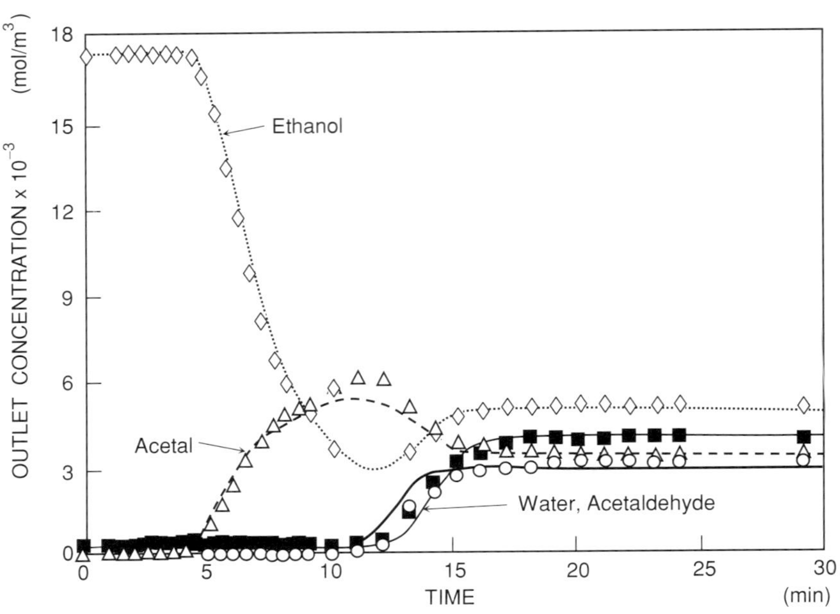

Figure 14.4 Breakthrough of products and reactants for the synthesis of diethylacetal from acetaldehyde and ethanol over a strong acid ion-exchange resin for a step introduction of a near stoichiometric feed into a bed saturated with ethanol. (Figure adapted from Silva and Rodrigues (2002) with permission. © 2002 AIChE.)

14.4, were exploited too for estimates of mass transfer coefficients for the heterogeneous model Silva and Rodrigues proposed for an SCMCR. Resin particle size was varied to demonstrate that diffusion in the resin pore structure was rate controlling. The kinetic model and rate constant were determined in separate batch experiments. Tracer experiments on the bed indicated Peclet numbers of about 270 at the flow rates used for their step change experiments.

14.4 Assessment

The choice of screening reactor is a matter of convenience. It depends on what is available in the research laboratory. If a reactor must be constructed, the simplest is certainly a tubular packed-bed reactor. All that is required is a length of tubing, appropriate fittings, a constant temperature bath or a tube furnace and temperature controller, and a four- or six-port timer operated solenoid valve. Several simple, three-port solenoids can be used in place of multiport valves. A suitable sampling valve, preferably a multiloop sampling valve, is necessary at the effluent end if chromatography is used for concentration measurements. The fixed bed recirculating screening reactor is about as simple to build. It requires, however, a high speed, positive displacement blower. The various name reactors, e.g. the Berty reactor, are off-the-shelf items from several laboratory supply companies.

Systems involving flow through adsorbent or catalyst and adsorbent require development and are much more difficult to build and operate because of the necessity of continuously feeding and removing solids. They are also limited to operating near atmospheric pressure.

All screening reactors require accurate flow control, temperature monitors and a means of precisely measuring fluid composition. Continuous monitors, such as an IR would be ideal except the spectra of products and reactants must be suitable. Most often sampling systems will be used, either a GC or an HPLC or a GC–MS system. Pulse or periodically operated screening reactors need a timer. Now, this function can be carried out through a computer in conjunction with monitoring and data collection.

Screening reactors employing both flowing fluids and solids operate at steady state. This simplifies sampling. The fixed-bed-circulating fluid systems can be operated either continuously with reactant or periodically by switching between reactant and an eluent. In the former case, when adsorbent is present, the reactor effluent changes with time. In periodic operation with adsorbent, a cyclic stationary state is attained eventually. By using a brief cycle period to avoid high adsorbate levels in the reactor, concentration in the effluent remains almost constant during each half cycle. This also simplifies sampling. For screening, steady-state operation has the disadvantage that two experiments must be conducted to determine the advantage of using an adsorbent: one experiment with the adsorbent and another with an inert replacing the adsorbent. If a number of adsorbents are being screened, this is just a minor disadvantage.

Step change, packed-bed experiments and the fixed solids-circulating fluid systems when operated with long cycle periods will show overshoot with a gradual decline to a steady-state characteristic of the catalyst without adsorbent. Thus, the advantage of an adsorbent is obvious immediately. Different adsorbents can be compared semi-quantitatively by the height of the overshoot and the time it takes to reach a steady state.

For the chromatographic screening reactor, peak heights, or areas, of the products and peak separation indicate the preferred adsorbent and offer a basis for comparing adsorbents.

Screening reactors of all types can be used to obtain parameters for models of all the potentially commercially interesting adsorptive reactors, whether they be CRACs, SCMCRs or PSRs. Parameters usually derived from screening reactors are effectiveness factors for the catalyst, and mass transfer coefficients as linear driving force (LDF) models are widely used. Coefficients from fluid-particle mass transfer correlations do not apply when the LDF assumption is made. Sometimes dispersion coefficients, effective axial conductivities and bed to wall heat transfer coefficients will be estimated when more detailed models are used. Although kinetic rate constants and adsorption isotherm constants can be estimated using screening reactors, these parameters are normally taken from separate experiments. Thermodynamic properties, such as activity coefficients, are now universally drawn from correlations.

When employed for parameter estimation, care must be taken that the screening reactor is operated with catalyst and adsorbent properties and particle sizes that represent those to be used in the pilot or full-scale adsorptive reactor. Of course, temperature range, pressure and fluid velocities used in the screening reactor must be close to those planned for the adsorptive reactor. When parameter estimation is the objective, the screening reactors operating at steady state have the advantage that a steady-state model will be applied. A steady-state model is easier to use and often requires a smaller computation effort for parameter extraction than when data is collected employing a pulse or periodically operated screening reactor. The latter uses partial differential equations that must be integrated in time as well as space. Laplace or Fourier transforms to remove the time variable can rarely be used because of nonlinearity in the models. Determination of parameters for separating reactor models is discussed in Chapter 15.

Chapter 15
Development of a Cyclic Separating Reactor

15.1 Developing the cyclically operated separating reactor system

Identifying reaction systems that would benefit from combined reaction and separation is the starting point for development of any type of separating reactor. Such reaction systems will be found in one of four groups: (1) single reactions that are equilibrium limited at temperatures and pressures that provide reasonable reaction rates; (2) single reactions that are product inhibited, or even reaction systems in which the feed contains a nonreacting inhibitor; (3) parallel reactions with just one desired product, often reactions in which there is an equilibrium step; and (4) sequential reactions in which an intermediate product is desired. Parasitic reactions that consume a sought-after product are part of this group. The decisive task in the string that leads to development of a commercial process is, however, the identification of an adsorbent with sufficient capacity at reaction conditions to significantly alter conversion or selectivity. As mentioned in the previous chapter, identification is necessarily an experimental activity.

Further development follows the conventional route for reactors of all types. Experiments provide the reaction kinetics and adsorption equilibria, or occasionally adsorption kinetics, needed for a separating reactor model. Once a model has been formulated, experiments are necessary to validate the model and, if the model is not satisfactory, to identify model inadequacies. All these experiments can be carried out in laboratory-scale reactors, that is, reactors sized to run in a research laboratory.

It is often advantageous to use specialized reactors to determine reaction kinetics. These specialized units are discussed in many textbooks (e.g. Froment and Bischoff, 1990; Fogler, 1992). Two such specialized reactors are the isothermally operated packed-bed tubular reactor or the CSTR, that is, a reactor with a well-mixed fluid phase. Both types were discussed in Chapter 14 as types of screening reactors. They can also be used to measure kinetics. Indeed, adsorption isotherms can be determined with these reactors. Reaction kinetics will be obtained using only a catalyst in the packed bed, while the isotherm will be measured with just adsorbent present. Screening reactors, of course, are laboratory reactors.

As has been made clear in the previous chapters, the material and energy balances for the different adsorptive reactors are the same. Only boundary conditions and perhaps initial conditions differ. Consequently, a screening reactor can serve for model validation, even a specialized reactor with either well-mixed or plug flow characteristics. The latter reactors, however, must be operated periodically or with a flowing adsorbent bed or adsorbent–catalyst bed. Nevertheless, investigators in industry and their employers usually will have more confidence if model validation is conducted in the type of reactor tentatively selected for the design, that is, in an SCMCR or a PSR. This reactor may then be considered as a demonstration reactor. A demonstration reactor can be built on a laboratory scale. Caution is needed. Validation should be done on an isothermal model with isothermal operation of the separating reactor because of large heat loss at the laboratory scale. Heavy insulation is not an option because the heat capacity of the insulation affects the response in cyclically operated reactors. Furthermore reactor feeds should be dilute to reduce the magnitude of heat effects. Running a reactor with a dilute feed and a diameter of about 1 cm isothermally is generally simpler than attempting adiabatic operation.

Once a validated model is available the full-scale reactor can be designed. In this exercise the model is used to test the effect of operating conditions and cycle parameters on the reactor design and its performance. With the capabilities of the current generation of computers, this may be done in the context of optimization. The problem of heat effects make a full-scale design from a model validated on a laboratory scale risky. It is common practice to go from a laboratory scale to an intermediate pilot scale and than to a semi-commercial scale. The problem of heat loss is much less serious for a larger unit, with a diameter of the order of 10 cm, so that higher reactant concentrations can be used. Any model for a larger unit should include a heat balance. Often more detailed kinetic and adsorption equilibria models will be introduced. Mass and heat transfer coefficients should be taken from measurements. With this increase in model sophistication, validation must be undertaken once again. Of course, this validation must use the pilot unit.

Decision on the use of a separating reactor depends on an economic evaluation. This can be made using the full-scale design based on the validated model even if this model is isothermal. Certainly, the economic evaluation will have to be repeated once the pilot-scale unit has been operated and its performance determined. A more detailed, nonisothermal model can be formulated then and the economic study based on that model.

Development of a periodically operated separating reactor has been discussed in the literature. In Section 15.4, we illustrate the process discussed above using the work of Mazzotti et al. (1996, 1997a) on an SCMCR for the catalytic esterification of ethanol with acetic acid.

15.2 Models

Reactor models have been thoroughly discussed in Chapters 3, 5, 7, 9 and 10. These models are applicable to both industrial and laboratory-scale reactors. Sophistication of the model must match the application. The simplest forms often can be used for preliminary design, that is, choosing the type of separating reactor and probably identifying suitable operating temperature, pressure and space velocities. They are often useful for preliminary selection of cycle parameters. When actually built, the reactor is referred to as a demonstration reactor. Generally one-dimensional pseudo-homogeneous models are appropriate for cyclic separating reactors. They would be certainly selected for preliminary investigations and sizing a demonstration reactor. Dispersion is in most cases a minor transport process so a plug flow model should be considered first. The mass balance for this model has various forms. In Chapter 3 the material balance was given as

$$\varepsilon_t \frac{\partial C_i}{\partial t} + (\rho_b)_{\text{ads}} \frac{\partial q_i}{\partial t} = D_x \frac{\partial^2 C_i}{\partial x^2} - \frac{1}{A} \frac{\partial n_i}{\partial x} + U_s (\rho_b)_{\text{ads}} \frac{\partial q_i}{\partial x} + (\rho_b)_{\text{cat}} \nu_i r \tag{15.1}$$

for a chemical species i taking part in the reaction. The total porosity term, ε_t, can be expanded to $\varepsilon_b + \varepsilon_{cat}(\varepsilon_p)_{cat} + \varepsilon_{ads}(\varepsilon_p)_{ads}$. Sometimes the dispersion term, D_x, is written as $\varepsilon_t D_x$. The choice depends on whether or not D_x is defined in terms of the fluid or in terms of the reactor. The dispersion term drops out if plug flow is assumed. The solids density term, ρ_b, can be replaced by $\varepsilon_{cat}(\rho_p)_{cat}$ for example and the product, $\nu_i r$, is the disappearance rate, $-r_i$, if i indicates a reactant. If the adsorbent and catalyst are fixed in a bed, $U_s = 0$.

Usually an energy balance would be ignored and the system assumed to operate isothermally in the first round of modeling. If the energy balance is used, its one-dimensional pseudo-homogeneous form is

$$\left(\varepsilon_t C_{p_g} C + \rho_b C_{p_s}\right)\frac{\partial T}{\partial t} = k_x \frac{\partial^2 T}{\partial x^2} - C_{p_g}\frac{n}{A}\frac{\partial T}{\partial x} + U_s \rho_b C_{p_s}\frac{\partial T}{\partial x} - \frac{P}{A}\frac{\partial}{\partial x}\left(\frac{n}{C}\right)$$
$$- \rho_{b_{ads}} \sum_{i=1}^{N_C} H_{a_i}\frac{\partial q_i}{\partial t} + \rho_{b_{cat}} \Delta H_R r - \frac{4h_0}{d_c}(T - T_a) \qquad (15.2)$$

Justification of a pseudo-homogeneous model for the energy balance is the observation that particles in packed beds are locally at a uniform internal temperature and that fluid–particle temperature differences are negligible. The balance in Equation 15.2 assumes that the catalyst and adsorbent have the same heat capacity and that both flow through the reactor when a CMCR is being modeled. If the heat capacities are quite different, the $(\rho_b)_s(C_p)_s$ term must be replaced by $(\rho_b)_{ads}(C_p)_{ads} + (\rho_b)_{cat}(C_p)_{cat}$. The boundary work term, fourth on the RHS, is very rarely significant. Sometimes, the effective conductivity term incorporating k_x can be neglected. If the reactor operates adiabatically, $h_0 = 0$.

Chromatographic reactors are assumed to operate isobarically, but usually this assumption cannot be used for PSR or TSR reactors. The Ergun equation, from Table 3.2, relates fluid velocity and the rate of pressure drop,

$$\frac{dP}{dz} = -J_v u - J_k u^2 \quad \text{with } J_v = \alpha \frac{\mu_g\left[\lambda_s(1-\varepsilon_b)\right]^2}{d_p^2 \varepsilon_b^3} \text{ and } J_k = \frac{\lambda_s(1-\varepsilon_b)\rho_g}{d_p \varepsilon_b^3} \qquad (15.3)$$

Note that in the simplest form, plug flow, isothermal and isobaric operation, the model contains only adsorption and reaction kinetics parameters. A total mass balance is not necessary. If these assumptions cannot be made, there are six parameters that must be evaluated experimentally. Three of these from the Ergun equation are easily obtained.

More sophisticated material balances will be used in the latter stages of design. The pseudo-homogeneous model may still be used if mass transfer and adsorption are rapid relative to the rate of reaction. However, the dispersion contribution in Equation 15.1 normally will be considered. Velocity will change with pressure drop and heat effects so velocity must be adjusted. Alternatively, this can be done using a total mass balance to calculate C:

$$\varepsilon_t \frac{\partial C}{\partial t} + (\rho_b)_{ads}\sum_{i=1}^{N_C}\frac{\partial q_i}{\partial t} = D_x \frac{\partial^2 C}{\partial x^2} - \frac{1}{A}\frac{\partial n}{\partial x} + U_s(\rho_b)_{ads}\sum_{i=1}^{N_C}\frac{\partial q_i}{\partial x} + (\rho_b)_{cat}\sum_{i=1}^{N_C}\nu_i r \qquad (15.4)$$

If mass transfer to or from the adsorbent is a rate-controlling step, a one-dimensional heterogeneous model must be used with an additional material balance for the adsorbent. Such a model is given below. Note that in Equation 15.5 mass transfer to the catalyst is assumed to be fast. If mass transfer interference is significant, an effectiveness can be introduced into the rate term r.

$$\varepsilon_t \frac{\partial C_i}{\partial t} = D_x \frac{\partial^2 C_i}{\partial x^2} - \frac{1}{A}\frac{\partial n_i}{\partial x} - k_m a_m (\rho_b)_{\text{ads}} \left(C_i - \frac{q_i}{K_i} \right) + (\rho_b)_{\text{cat}} \nu_i r \tag{15.5}$$

$$\frac{\partial q_i}{\partial t} = U_s \frac{\partial q_i}{\partial x} + k_m a_m \left(C_i - \frac{q_i}{K_i} \right) \tag{15.6}$$

A special case occurs when the same solid serves as both catalyst and adsorbent. Our discussion in Section 5.6 considers this case. Then,

$$\varepsilon_t \frac{\partial C_i}{\partial t} = D_x \frac{\partial^2 C_i}{\partial x^2} - \frac{1}{A}\frac{\partial n_i}{\partial x} - k_m a_m \rho_b \left(C_i - \frac{q_i}{K_i} \right) \tag{15.7}$$

$$\rho_b \frac{\partial q_i}{\partial t} = U_s \rho_b \frac{\partial q_i}{\partial x} + k_m a_m \rho_b \left(C_i - \frac{q_i}{K_i} \right) + \rho_b \nu_i r \tag{15.8}$$

As usual, $U_s = 0$ for cyclically operated separating reactors.

Models are completed by initial and boundary conditions. These are specific for each type of adsorptive reactor. For brevity, they will not be given here. They are given in Table 3.7 for the chromatographic reactors, in Table 9.4 for pressure swing reactors and in Section 10.2 for temperature swing reactors. However, the boundary conditions depend on the modeling assumptions made, such as the assumption of plug flow.

All reactor models require submodels for the rate of reaction and, if mass transfer or adsorption are slow steps, submodels for these steps. However, if adsorption is fast, a submodel for adsorption equilibrium will be needed. Nonisothermal models will need submodels for heat transfer. When dispersion is important, a sub-model will be needed for that process.

15.2.1 Dispersion models

Dispersion is a term used to describe a group of local mixing processes, such as spatially varying local velocity, turbulent mixing induced by wakes around particles or by fluid acceleration, diffusion in porous solids or into stagnant regions in a packed bed. It is assumed that this local mixing, which adds a transport mode when a concentration gradient is present, follows Fick's law. Many measurements of dispersion have been made and various correlations have been published. A correlation widely used is

$$N_{\text{Pe}} = \frac{u d_p}{D_{ax}} = 0.20 + 0.011 (N_{\text{Re}})_p^{0.48} \tag{15.9}$$

This correlation was developed for one-dimensional heterogeneous models, but it is used as well, but improperly, for one-dimensional pseudo-homogeneous models. Dispersion coefficients are empirical and they incorporate model approximations. Thus, the pseudo-homogeneous model neglects mass transfer between phases. Any error this introduces into estimation of the coefficient will make the coefficients different for these two models. Fortunately, dispersion is a minor transport process in many separating reactors so that errors in the dispersion coefficient have little effect on prediction of performance.

15.2.2 Kinetic models

Catalytic reactions proceed through adsorption on a catalyst surface site where an intermediate adatom species is formed. This may be strongly bound and rendered unreactive by the site, it may be destabilized by a neighboring site or adsorbed species, or it may be so weakly bound that the adatom can migrate on the surface to another site. If destabilized, the adatom may be stretched or even split by the neighboring site to form two different adatoms. Or, reaction with a neighboring adatom could result in the formation of a new adatom on one of the sites. It is probable that the chemical sequence from reactants to products proceeds through a great many rapid interactions through many adatom intermediates. Eventually, an adatom desorbs from a surface site to yield a reaction product. It is thought that the sequence is not unidirectional so that for every 100 forward steps there might be 99 reverse steps. It is believed that the reverse reaction, from product back to reactant, proceeds through the same series of adsorption, reaction or migration steps with the same intermediates. Adatoms that interact with a neighboring site or a neighboring adatom can be thought of as unstable to differentiate them from inert or unreactive adatoms, although "unstable" is a relative term as it depends on the neighboring sites, whether or not these sites are occupied, or the type of adatom adsorbed on the sites.

Individual bond rupture and bond creation are very fast, perhaps picosecond events. Net changes, the difference between forward and reverse events, are many magnitudes slower. Net change too depends upon populations of the adatom species. Net change we refer to as overall rates of reaction. Most often, one of the sequence of rate steps on the surface or perhaps an adsorption or desorption event is intrinsically slower than all other steps. This is called the rate-controlling step. If the remaining steps are intrinsically faster they can be considered to be equilibrated so that their adatom densities on the catalyst surface are exactly related. By successive substitution, the densities of the adatoms involved in the rate-controlling steps can be replaced by measurable reactant or product concentrations in the gas or liquid phases. This leads to fairly simple models for reaction rates unless competitive adsorption or inhibition events occur in the surface reaction sequence.

The simplest of these is a first order rate model,

$$r = k_1 C_A \quad \text{or} \quad = k_1' p_A \tag{15.10}$$

that might represent the reaction: A $\rightarrow$ B. Strictly, we should use the activity of A in the fluid phase, namely a_A, in place of concentration. Often there is little difference in their numerical values. For a gas phase reaction, there is the alternative of representing activity by the fugacity or, when the ideal gas conditions apply, by the partial pressure, p_A, as shown in Equation 15.10. The subscript on the rate constant means it is the rate constant for a first order reaction. This index is useful because it indicates the units of the rate constant. Nevertheless, common practice is to not use an index. Equation 15.10 assumes the reaction is irreversible, that is, the rate of the reverse reaction is negligible. Otherwise, the first order rate model for the reaction, A $\leftrightarrow$ B, should be written as

$$r = k_f C_A - k_r C_B \tag{15.11}$$

Now the subscripts on k indicate the forward and reverse reaction, respectively. The equation could have been written in terms of partial pressure. Frequently, Equation 15.11 is written as

$$r = k_1 \left(C_{\mathrm{A}} - \frac{C_{\mathrm{B}}}{K_{\mathrm{AB}}} \right) \tag{15.12}$$

This formulation is exact if the reaction is elementary or if there is just one sequence of reaction and adsorption steps from reactant to product and the reverse reaction follows this sequence. Furthermore, the rate-controlling step must be an order of magnitude slower than all other steps so that equilibrium may be assumed for all other steps. In practice, the reaction subscript on the equilibrium constant is dropped and K is used.

If there are two reactants so that $\mathrm{A} + \mathrm{B} \rightarrow \mathrm{C} + \mathrm{D}$, Equation 15.10 can still be used provided one of the reactants is limiting, that is, $C_B \gg C_A$. In this situation, the concentration of B changes little during the course of the reaction. It can be treated as a constant and incorporated into the rate constant. In general, however, rate would be expressed as,

$$r = k_2 C_{\mathrm{A}} C_{\mathrm{B}} \tag{15.13}$$

This is a statement of the law of mass action that states the rate of reaction between two unlike molecules is proportional to the product of the concentration of those molecules. The law of mass action applies to gases or liquids. It is an empirical law.

If the reverse reaction is significant,

$$r = k_2 \left(C_{\mathrm{A}} C_{\mathrm{B}} - \frac{C_{\mathrm{C}} C_{\mathrm{D}}}{K_{\mathrm{ABCD}}} \right) \tag{15.14}$$

Equation 15.13 is also referred to as a power law kinetic model. This group of empirical models is based on the law of mass action. Over rather narrow ranges of concentration, Equation 15.13 can be generalized to

$$r = k C_{\mathrm{A}}^m C_{\mathrm{B}}^n \tag{15.15}$$

where m and n are the orders of the reaction for each reactant. $(m + n)$ is the overall order of reaction. Orders can be integers or fractions and may be negative. Usually, $-1 < m, n < 2$. Often there is a mechanistic basis for the order.

Industrially important reactions frequently have side reactions or reactor feeds may be contaminated by impurities. These may inhibit the reaction and concentrations of side products or impurities may be included in a rate expression. Also a product can be so strongly adsorbed that enough surface sites are occupied by the product to affect the reaction rate. This is also inhibition. To allow for these two sources of inhibition, a rate expression for the reaction $\mathrm{A} + \mathrm{B} \rightarrow \mathrm{C} + \mathrm{D}$, where both reactant A, product D and an impurity E in the feed stream are strongly adsorbed, could be,

$$r = k_2 \frac{C_{\mathrm{A}} C_{\mathrm{B}}}{1 + K_{\mathrm{A}} C_{\mathrm{A}} + K_{\mathrm{D}} C_{\mathrm{D}} + K_{\mathrm{E}} C_{\mathrm{E}}} \tag{15.16}$$

Constants in the denominator are often assumed to be adsorption equilibrium constants, but this assumption is not necessary. Inhibition models, such as Equation 15.16, are empirical models.

An alternative to empirical models of the type just described are those based on a reaction mechanism. In order for these models to be mathematically tractable in structure, the reaction mechanism usually must be grossly oversimplified. For example, chemisorption is assumed, even though adsorption likely takes place through a sequence of physisorption, surface migration and eventual capture on to an active site.

Consider a reaction, $A + B \rightarrow C$, involving adsorption from, say, a gas phase, a surface reaction and desorption of the C product into the same phase. Assume that the surface reaction proceeds through the adsorbed adatoms and that this step is rate controlling. Chemisorption of A and B to form reactive adatoms A and B are rapid and so chemisorption can be represented as at equilibrium. Desorption of the product C is also rapid so that equilibrium can be assumed as well. If we represent the density or surface concentration of a chemisorbed species i as $n_{\text{total}}\theta_i$ where n_{total} is the number of sites/g of catalyst or sites/m^2 of catalyst surface and θ_i is the fraction of the total sites occupied by the i adatoms. Then,

$$\theta = \sum_{i=1}^{N} \theta_i \tag{15.17}$$

where N is the number of species present in the system. The fraction of sites free for chemisorption is then $1-\theta$. Treating chemisorption as a gas–solid phase reaction or $A + (1-\theta) \rightarrow \theta_A$, the chemisorption equilibrium becomes

$$K_A = \frac{\theta_A}{p_A(1-\theta)} \tag{15.18}$$

The site density, n_{total}, cancels out. Similar expressions arise for the other two chemisorption equilibria.

Assuming the law of mass action is applicable to surface reaction between adatoms,

$$r = k_s \theta_A \theta_B \tag{15.19}$$

k_s is the rate constant for the surface reaction between adatoms A and B. It incorporates n_{total}. Replacing the fractional occupancy of sites using the equilibrium relations (Equation 15.18) for reactants and products and introducing Equation 15.12 to eliminate the $(1-\theta)$ term,

$$r = \frac{k_s K_A K_B p_A p_B}{(1 + K_A p_A + K_B p_B + K_C p_C)^2} \tag{15.20}$$

where K_A, K_B and K_C are adsorption equilibrium constant incorporating the density of active sites, n_{total}. It must be noted that these equilibrium constants cannot be calculated from standard free energy and enthalpy tables.

On the other hand, if the adsorption of reactant A is rate controlling, then

$$r = k_{\text{ads}} p_A (1-\theta) \tag{15.21}$$

After substitution of the equilibrium relations for θ,

$$r = \frac{k_{\text{ads}} p_A (K K_B p_B - K_C p_C)}{K K_B p_B (1 + K_B p_B + K_C p_C)} \tag{15.22}$$

Or, if desorption of C from the surface controls,

$$r = k_{\text{des}} \theta_C \tag{15.23}$$

Substitution for θ_C results in a quadratic relation in $(1-\theta)$ so that a simple algebraic relation for r cannot be written. The rate, however, will be bounded by

$$r = \frac{k_{des} K K_A K_B p_A p_B}{(1 + K_A p_A + K_B p_B + K_C p_C)^2}, = \frac{k_{des} K K_A K_B p_A p_B}{(1 + K_A p_A + K_B p_B)^2} \quad (15.24)$$

where K is a reaction equilibrium constant written for the adatoms.

The algebraic form of the rate model obtained from these simplified mechanistic model depends, consequentially, on the stoichiometry of the reaction, the adsorption model and the rate-controlling step. When the rate-controlling step is not established, experiments can be undertaken using models such as those given by Equations 15.20, 15.22 and 15.24 to identify the step, or at least narrow down the possibilities. Models based on simplified mechanisms are referred to as Langmuir–Hinshelwood–Hougen–Watson models, honoring the scientists who proposed and then promoted the approach. Of course, these mechanism-based models reduce to mass action law models if the adsorption equilibrium constants are small or if the reactants and products are dilute. On the other hand, if one of the equilibrium constants is large relative to others, or if a reactant or product exists in large excess, these models can be simplified to power law models with fractional or even negative exponents.

All of the reaction rate models discussed in the preceding paragraphs are steady-state models and are strictly applicable to stationary systems where inputs such as flow, concentration or temperature are not changing with time. Nevertheless, these models are widely employed for dynamic systems as we have seen in Chapters 4–7 and 9–12. Doing so enormously simplifies the reactor model, even in the rare cases where heterogeneous models are employed. The simplification is justified in almost all cases because net surface reaction rates and adsorption–desorption from surface sites are often orders of magnitude faster than the rates of change in concentration or temperature due to time-varying conditions in the reactor. When reaction or adsorption rates are comparable to system rates of change, steady-state models should not be used. Instead, knowledge of the reaction mechanism is required and it is the rate-controlling step that must be modeled. Then, material balances must be written for all the species participating in the rate-controlling step. Usually, in this case, a pseudo-homogeneous reactor model cannot be employed. It is very unusual to be able to estimate rate and equilibrium constants used in steady-state models from theory, such a collision theory, or from thermodynamic data. When such models are used in dynamic systems, this is doubly true. Constants in rate equations must be estimated from experimental data.

There are a number of considerations that lighten the task of selecting or developing reaction models. First of all, first order models, such as given in Equations 15.10–15.12, often represent observed rates well over a range of reactant concentration. The range that first order can be used may be expanded by segmenting the relation and using a different constant for each section. Thus, if concentration goes out of a permissible range in a reactor, the rate constant can be changed. This tactic speeds up a simulation appreciably. If curvature over the concentration range is too great to justify a first order model, a power law can be adopted. Some caution is advisable in using these simplifications. In dynamic systems where surface concentrations are varying, albeit slowly, it is a good practice to employ an inhibition model incorporating concentrations of the time-varying surface concentrations.

Despite many similarities between catalytic and enzymatic reactions, particularly when the enzyme is immobilized on a support, biochemical reactions have different reaction rate models. The fundamental reason for this difference is that the cell or enzyme activity depends on the local environment, such as pH or ion concentration, as well as on past history, that is, how the enzyme was immobilized and activated or how the cell developed. Effect of temperature is dramatic. Often, above a critical temperature, the enzyme or cell is completely deactivated. Thus, the Arrhenius relation may not represent the temperature behavior well. If the reaction is mediated by living cells, such as a yeast, age may be important.

Often, rate is best represented by assuming a living cell or an enzyme has a maximum activity that is reduced by the presence of the reactant or a product, that is, inhibition occurs. This assumption is embodied in the various forms of the Michaelis–Menten relation: on a volumetric basis,

$$r = \frac{V_{\max} C_S}{K_m + C_S} \tag{15.25}$$

In the relation, $V_{\max}$ is the maximum rate of reaction per unit concentration of the substrate S. K_m is the Michaelis–Menten constant. Symbols in the relation are those traditionally used. The relation applies to a fluid phase. Another form allows for greater inhibition by the substrate (Sarmidi and Barker, 1993a,b):

$$r = \frac{V_{\max} C_S}{K_m + C_S + \frac{C_S^2}{K_i}} \tag{15.26}$$

K_i in the relation is an inhibition constant. When the enzymatic reaction is reversible, another form arises (Lee et al., 1976):

$$r = \frac{V_{\max}(C_S - C_S^{\text{equil}})}{K_m' + (C_S - C_S^{\text{equil}})} \tag{15.27}$$

where

$$K_m' = \left(\frac{1}{(K_m)_r - (K_m)_f}\right)[((K_m)_r + (K_m)_f K)C_S^{\text{equil}} + (K_m)_r (K_m)_f] \tag{15.28}$$

and $(K_m)_r$ and $(K_m)_f$ are the Michaelis–Menten constants associated with the reverse and forward reactions, K is the equilibrium constant for the reversible reaction and C_S^{equil} is the substrate concentration at equilibrium.

When there are two substrates and inhibition also must be taken into account, the models become more complicated. Enzymatic esterification is described by a bi-bi-ping-pong mechanism (Mensah et al., 1998a,b) that leads to

$$r = \frac{V_{\max} C_{\text{alcohol}} C_{\text{acid}}}{C_{\text{alcohol}} C_{\text{acid}} + K_{m_{\text{alcohol}}} C_{\text{acid}}(1 + \frac{C_{\text{acid}}}{K_{i_{\text{acid}}}}) + K_{m_{\text{acid}}} C_{\text{alcohol}}(1 + \frac{C_{\text{alcohol}}}{K_{i_{\text{alcohol}}}})} \tag{15.29}$$

In this relation, $V_{\max}$ is a strong function of water, an esterification product, adsorbed onto or dissolved in the enzyme phase.

The enzymatic deacylation of penicillin occurs through a uni-bi-ping-pong mechanism (van der Wielen et al., 1996) to yield two products, P and Q, and a hydrogen ion (H^+). S represents the substrate. The rate on a volume of immobilized enzyme basis is

$$r = \frac{k_{\text{enz}} f_{\text{enz}}^0 \left(C_S - \frac{C_Q C_P C_{H+}}{K} \right)}{K_{\text{m}_S} \left(1 + \frac{C_S}{K_{\text{m}_S}} + \frac{C_Q}{K_{i_Q}} \right)\left(1 + \frac{C_P}{K_{i_P}} \right)} \quad \text{with } f_{\text{enz}}^0 = \frac{1}{1 + \frac{C_{H+}}{K_{a\text{enz}}}} \tag{15.30}$$

The f_{enz}^0 = allows for the activity dependence on pH.

The various rate terms given above are rooted in mechanisms for the enzymatic or cell-mediated reactions. These are much like those used to develop the Langmuir–Hinshelwood–Hougen–Watson models and are given in standard bioengineering textbooks, such as Bailey and Ollis (1986).

15.2.3 Adsorption equilibria and rate models

In the earlier chapters, we have seen that quite a variety of materials have been used as adsorbents: alumina, activated carbon, hydroxy bicarbonates, synthetic zeolites, ion-exchange resins, and even metals. It will not be surprising, then, that these materials function through different mechanisms. Indeed, for several of these materials more than a single mechanism is at work. Ion-exchange resins capture an adsorbate by an exchange of an ion in the regenerated resin with either an ion from the liquid phase or by ionizing a solute in that phase. Resins swell, changing pore size, and that leads to entrapment of the solute. With activated carbon and alumina, for example, physisorption takes place along with chemisorption. Physisorption involves van der Waals forces between the solid surface and the adsorbate. There is just a small heat release when this weak bonding occurs so the sorption is easily reversed. Chemisorption, conversely, involves some degree of electron donation between orbitals of surface atoms and those in the adsorbate. Heat effects are much larger. Adatoms formed can be strongly held and a significant energy input must be brought in to affect desorption. Further, in materials exhibiting a fine pore structure, such as most activated carbons, capillary condensation can take place when the fluid phase is gaseous so that the adsorbate exists as a condensed phase within the solid. Synthetic zeolites entrap an "adsorbate" through slow configural diffusion into the solid phase. Metals adsorb through physisorption on the external surface, followed by diffusion into the bulk. Absorption rather than adsorption might be a more apt term.

The different sorption phenomena playing out on different materials offer a modeling challenge. Three approaches are possible. Adsorption can be modeled as an equilibrium distribution between phases in contact, just as in the familiar unit operation of extraction. It can be treated as a physisorption process that is modeled by an isotherm: the change in equilibrium uptake with changing fluid concentration of the adsorbate at constant temperature. Finally, it can be modeled as a chemical reaction between an adsorbate molecule and a surface site to create an adatom. Different approaches can be applied to the same material. Mazzotti et al. (1996) and Lode et al. (2003a,b) used the same ion-exchange resin for different esterification reactions. The former researchers used an equilibrium distribution between liquid and resin phases, while Lode et al. employed a competitive Langmuir isotherm to represent adsorption equilibrium.

The phase equilibrium model finds applications when an ion-exchange resin functions as a combined catalyst and adsorbent or when the resin serves as adsorbent alone. The two phases identified in the model are a liquid phase, free of resin, and a resin phase that is assumed to be a gel. The simplest model relates concentrations in the two phases by a distribution constant

$$C_i^R = K_i C_i^L \quad \text{or} \quad q_i = K_i C_i \tag{15.31}$$

It is a linear relation analogous to Henry's law. The distribution constant must be obtained experimentally.

More sophisticated models are certainly available. By definition of equilibrium, for each species in the liquid–resin phases:

$$a_i^L = a_i^R \tag{15.32}$$

The liquid phase activity is related to concentration through the activity coefficient as $a_i^L = \gamma_i C_i$. Also, each species must conform to a material balance

$$n_i^0 = n_i^R + n_i^L \tag{15.33}$$

where n_i^0 is the moles of i entering with the feed or introduced into a batch reactor. Distribution of each species is given conveniently as a volume ratio

$$v_i = \frac{n_i^P V_i}{V_P^0 + \sum_{j=1}^{N} n_j^P V_j} \tag{15.34}$$

where V_P^0 is the dry volume of the resin in the system.

Liquid phase activities in Equation 15.32 come from $a_i^L = \gamma_i^L C_i$ where the activity coefficients, γ_i^L, are estimated by the UNIFAC group contribution method (Fredeslund et al., 1977; Reid et al., 1986). In this method,

$$\ln \gamma_i^L = \ln \gamma_i^C + \ln \gamma_i^R \tag{15.35}$$

where the superscripts C and R denote combinatorial and residual contributions. The combinatorial contribution is given by

$$\ln \gamma_i^C = 1 - V_i + \ln V_i - 5q_i(1 - \frac{V_i}{F_i} + \ln(\frac{V_i}{F_i})) \tag{15.36}$$

while the residual contribution is given by

$$\ln \gamma_i^R = \sum_k v_k^i(\ln \Gamma_k - \ln \Gamma_k^i) \tag{15.37}$$

The terms in these expressions are

$$V_i = \frac{r_i}{\sum_j x_j r_j}, \quad F = \frac{q_i}{\sum_j x_j q_j}, \quad \ln \Gamma_k = Q_k(1 - \ln \sum_m \theta_m \Psi_{mk}) - \sum_m \frac{\theta_m \Psi_{km}}{\sum_n \theta_n \Psi_{nm}}$$

and the parameters of these relations are calculated from group contributions that are tabulated (Fredeslund et al., 1977; Reid et al., 1986). Experimental data are not needed.

Of course, the correlations at the heart of the UNIFAC estimating procedure are based on experimental vapor–liquid equilibrium data.

Activity of each species in the resin phase is estimated from the extended Flory–Huggins model. For the variables used, the model can be written as

$$\ln a_i^P = 1 + \ln v_i - \sum_{j=1}^{N+1} m_{ij} v_j + \sum_{j=1}^{N+1} \chi_{ij} v_j - \sum_{j=1}^{N+1} \sum_{k=1}^{j-1} m_{ik} v_j v_k \chi_{kj} + \eta V_i (\frac{5}{3} v_P^{1/3} - \frac{7}{6} v_P) \tag{15.38}$$

In this expression, χ_{ij} is the binary interaction parameter from the classical Flory theory ($\chi_{ii} = 0$ and $\chi_{ij} = \chi_{ji} m_{ij}$), η is an elasticity parameter and $m_{ij} = V_i / V_j$, the ratio of molar volumes of i and j, while v_i is the volume fraction (Equation 15.34). The model is completed by the swelling ratio, q,

$$q = 1 + \frac{\sum n_i^P V_i}{V_P^0} \tag{15.39}$$

where n_i^R is the moles of i in the resin phase and $n_i^0 = n_i^R + n_i^L$ must hold.

The parameters in Equation 15.38, χ_{ij}, η must be estimated from distribution and swelling data for the nonreacting binary pairs. Interaction parameters for the reacting pairs are set equal to zero. n_i^R and n_i^L are estimated from distribution data, y versus x, and swelling data along with the material balances given by Equation 15.33. Molar volumes are available in the literature. The resin volume, V_R^0, must be measured. These relations are substituted into Equation 15.33 to obtain the distribution of each component between the liquid and solid or rather gel phases.

Adsorption equilibrium can be represented also by an isotherm. The simplest of these, a linear isotherm is given by Equation 15.31. Isotherms are generally reserved for physisorption on surfaces, but in separating reactors it is usually chemisorption that is involved because separation of physisorbed species requires large beds of adsorbents. Nevertheless, isotherms are used when chemisorption is the process happening. The simple Langmuir isotherm is used when just one species is adsorbed

$$q_i = \frac{q_i^{\text{sat}} K_i^l C_i}{1 + K_i^l C_i} \tag{15.40}$$

In this statement of the isotherm, q_i^{sat}, is the moles or g of i adsorbed when the surface is saturated. K_i^l is the Langmuir constant. As C_i increases, q_i approaches q_i^{sat}. The Langmuir isotherm assumes an ideal surface and low occupancy so that adsorbate–adsorbate interactions are absent. At higher coverage of the surface, adsorbate interactions arise and a more appropriate isotherm is

$$C_i = \frac{1}{K_i} \left(\frac{q_i}{1 - q_i} \right) \exp \left(\frac{q_i}{1 - q_i} \right) \tag{15.41}$$

This isotherm is difficult to use and has not been employed for adsorptive reactors. When there is low occupancy, but more than one adsorbing species, a modified Langmuir isotherm may be used. It has the form,

$$q_i = \frac{q_i^{\mathrm{sat}} K_i^l C_i}{1 + \sum_j K_j^l C_j} \tag{15.42}$$

The Langmuir isotherm will be used for gaseous systems, where p_A replaces C_A in Equations 15.40–15.42, but it can be applied when a liquid is present. For gases an extension of the Langmuir isotherm, popularly called the BET isotherm, is

$$\frac{1}{q_i} = \frac{1}{K_i^{\mathrm{BET}} q_i^{\mathrm{sat}}} \frac{(p_i^0 - p_i)}{p_i} + \left(\frac{K_i^{\mathrm{BET}} - 1}{K_i^{\mathrm{BET}} q_i^{\mathrm{sat}}} \right) \left(\frac{p_i^0 - p_i}{p_i^0} \right) \tag{15.43}$$

The isotherm assumes noncompetitive adsorption and allows for multilayer adsorption so that it is a more realistic representation of adsorption at moderate to high partial pressures of the adsorbate than the Langmuir isotherm. Despite its greater generality, the isotherm has not been used for the modeling of separating reactors. The variable q_i^{sat} refers to monolayer coverage of the adsorbent surface.

The Freundlich isotherm is often considered to be an isotherm for adsorption from the liquid phase, particularly aqueous ones. It is, however, applicable to adsorption from gases as well. Unlike the simple Langmuir isotherm, the isotherm has two adjustable parameters:

$$q_i = K_i^F C_i^{1/n} \tag{15.44}$$

An expression resembling the Freundlich isotherm results when adsorption leads to condensation in the micropores of the adsorbent. Figure 15.1 reveals that the isotherm fits aqueous solution data very well.

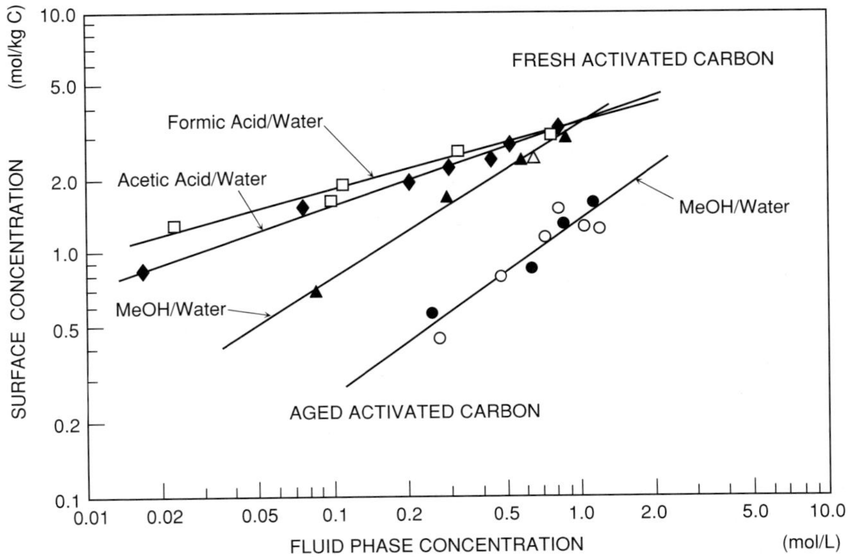

Figure 15.1 Application of the Freundlich isotherm to adsorption on activated carbon from aqueous binary mixtures. (Figure adapted from Cho et al. (1980a) with permission. © 1980 Pergamon Press Ltd.)

A simplistic mechanism for chemisorption treats the process as a gas–solid reaction. If A is the adsorbate, the stoichiometric representation would be A + site $\leftrightarrow$ A – site. Assuming, the process to be elementary, then,

$$q_i = K_i C_i \left(q^{\text{sat}} - \sum_j q_j \right) \tag{15.45}$$

where q^{sat} is the capacity of adsorbent when all adsorption sites are occupied as kmol/kg adsorbent. If only a single component of fluid phase adsorbs, then Equation 15.45 rearranges and becomes Equation 15.40, the Langmuir isotherm. Another statement of chemisorption equilibria is given by Equation 15.18.

Physisorption appears to be very fast. Mass transfer is probably rate limiting as it likely is as well for adsorption into gels. Chemisorption, however, may be a slow step. In such a case, equilibrium cannot be assumed and a rate model for chemisorption would have to be added to a model for an adsorptive reactor. Because chemisorption is viewed as a fluid–solid reaction,

$$r_{\text{ads}_i} = \frac{dq_i}{dt} = k_{\text{ads}_i} p_i \left(q_i^{\text{sat}} - q_i \right) \tag{15.46}$$

In the same way, if the desorption step is slow, its rate is given by

$$r_{des} = -\frac{dq_i}{dt} = k_{des_i} q_i \tag{15.47}$$

15.2.4 Mass transfer rate models

Mass transfer must occur whenever different phases are present in a system. Thus, it must be a factor in separating reactors. Mass transfer also arises when porous solids participate in the reaction system. Both catalysts and adsorbents used in adsorptive reactors are porous because large surface areas are needed for high catalytic reaction rates or large adsorption capacities. In many systems, mass transfer is the slowest rate step, but fortunately, it may be fast enough compared to system changes in cyclically operated systems, so that it can be neglected, or rather the mass transfer step can be rolled into an overall reaction rate. It is likely that this is what is being done when a pseudo-homogeneous model is used for the dynamic operation of a reactor. As mentioned earlier in this chapter, pseudo-homogeneous models dominate the modeling of cyclically operated separating reactors. If mass transfer between phases is to be explicitly recognized in a reactor model, a heterogeneous model must be used.

If mass transfer interferes with reaction rate within a porous solid in a dynamic system, a new partial differential equation should be introduced into the reactor–adsorber model. This makes solution of the model very much more difficult. In steady-state systems, solution of an additional PDE is avoided by introducing an effectiveness factor that estimates the overall effect of the mass transfer interference on the reaction rate. Jain et al. (1985) have shown that the same assumption may be used for transient operation. Indeed, the steady-state effectiveness factor turns out to be a good approximation for the effectiveness factor that represents transient operation. In this treatment, mass transfer within a porous solid is lumped into a reaction rate model.

With adsorptive reactors we are concerned with modeling the rate of mass transfer between phases when mass transfer controls the adsorption process. Mass transfer between phases takes place by diffusion and micro-scale convection of turbulent eddies in a mobile fluid. Both the diffusion pathway and turbulence intensity varies spatially as well as temporally. Common practice is to simplify the mass transfer model by postulating that transfer goes on across a surface bordered by a stagnant film of uniform depth. Diffusion and convection are lumped together in a mass transfer coefficient so that the rate is given by

$$r_i = k_i^{\text{mass}} a_V \left(C_i - C_i^{\text{surface}} \right) \tag{15.48}$$

In this film model, k_i^{mass} is an empirical mass transfer coefficient, sometimes written as $(k_m)_i$, and a_V is the interfacial surface area per unit of volume, usually the volume of the adsorbent or catalyst phase. Product of the two is the volumetric mass transfer coefficient. The driving force term is the concentration difference across the hypothetical film.

The mass transfer or volumetric mass transfer coefficient may be measured on a suitable model system or it can be extracted from one of the many mass transfer correlations in the Engineering literature. One of these correlations, given by Ruthven (1984), is

$$N_{\text{Sh}} = \frac{1.09}{\varepsilon_b} (N_{\text{Re}_p} N_{\text{Sc}})^{1/3} = \frac{k_{m_i} d_p}{D_{i_m}} = \frac{1.09}{\varepsilon_b} \left(\frac{u d_p}{D_{i_m}} \right)^{1/3} \tag{15.49}$$

Here, $(D_i)_m$ is the diffusivity of the transported species in the fluid mixture. Transport between a fluid and a particle is assumed so d_p is the equivalent spherical diameter of the particle. The variable u is the superficial fluid velocity.

With adsorbents, like catalysts, diffusion within the porous adsorbent may generate a concentration gradient within the solid that reduces adsorption or desorption rates. It is possible to handle rate estimation rigorously, but this requires adding further PDEs to an adsorber model. Glueckauf (1955) and later Kim (1989) show that a linear driving force (LDF) model that treats the concentration within the porous solid as uniform can be derived that reproduces the rate behavior with mass transfer interference quite accurately,

$$k_i^{\text{internal}} = \frac{10 \varepsilon_p^2 D_{i_m}}{d_p} \tag{15.50}$$

Kim's extension gives

$$k_i^{\text{internal}} = \frac{60 \varepsilon_p D_{i_{\text{eff}}}}{d_p^2 (\varepsilon_p + (1 - \varepsilon_p) K_i)} \tag{15.51}$$

It is probably a more accurate estimate. In Equation 15.51, $(D_i)_{\text{eff}}$ can be approximated as $\varepsilon_p^2 (D_i)_m$.

An overall mass transfer coefficient can be formulated from a resistance-in-series viewpoint,

$$\frac{1}{k_i^{\text{overall}}} = \frac{1}{k_{m_i}} + \frac{1}{\varepsilon_p k_i^{\text{internal}}} \tag{15.52}$$

Sometimes, $(k_m)_i$ or k_i^{overall} are obtained experimentally in the same way as reaction rate constants or adsorption distribution coefficients.

15.2.5 Heat transfer rate models

Many studies of cyclically operated separating reactors have assumed that these reactors operate isothermally. Although isothermality can be realized on a laboratory scale, it is unlikely for a commercial scale unit. In the latter operation, heat will probably be introduced and removed through the mobile or fluid phase. This suggests temperature fronts proceeding through packed beds. It also implies heat loss or possibly gain through the reactor wall.

The energy balance given as Equation 15.2 above, is a one-dimensional model that does not allow for radial variations. To use this model, a wall heat transfer coefficient, h_0, and an effective axial thermal conductivity, k_x, are needed. Although relations for estimating h_0 are available in the literature (see, for example, Dixson, 1985), they are unreliable because of heat generation or consumption accompanying adsorption and reaction. Overall coefficients should be obtained from measurements made with the separating reactor under consideration, probably at a pilot plant scale.

The above comment can be made for the effective axial thermal conductivity, but for this parameter an estimate drawn from a correlation may be adequate. Convective heat transfer through the mobile fluid phase is always present in the axial direction. With a liquid system, convective transport will outweigh conduction so that an error in the estimate of k_x will not engender a large error in calculating axial heat flow. On the other hand, for gaseous systems, heat transfer by conduction is about the same as transfer through the flowing gas. Consequently a better estimate of k_x would be desirable. There are several relations in the literature for effective axial thermal conductivity. One that has been used for many years stems from Yagi and Kunii (1964),

$$\frac{k_x}{k_g} = \frac{k_x^0}{k_g} + N_{\mathrm{Pr}} N_{\mathrm{Re}} \tag{15.53}$$

where the zero flow contribution is $\frac{k_x^0}{k_g} = \varepsilon_b + \frac{1-\varepsilon_b}{2.5 + \frac{2k_g}{3k_s}}$. In this relation, the subscript s and g indicate the solid and gas phases respectively. The model assumes the same conductivity for the catalyst and the adsorbent. The variable ε_b is the bed void fraction.

Allowing for heat loss on a commercial scale may require use of two-dimensional models. For these models, effective radial thermal conductivities and heat transfer coefficients are needed. The Engineering literature contains many models for effective radial thermal conductivity. Gianetto and Silveston (1986) suggest,

$$\lambda_r = \lambda_r^0 + (\lambda_r)_{\mathrm{fluid}}$$

where

$$\frac{\lambda_r^0}{\lambda_{\mathrm{fluid}}} = \varepsilon_b + \frac{1-\varepsilon_b}{\psi_r + \frac{2\lambda_{\mathrm{fluid}}}{3\lambda_{\mathrm{solid}}}} \quad \text{and} \quad \psi_r = 0.22\varepsilon_b^2 \tag{15.54}$$

$$\frac{u\rho_{\mathrm{fluid}}^2 d_p}{(\lambda_r)_{\mathrm{fluid}}} = 8.65\left[1 + 19.4\left(\frac{d_p}{d_{\mathrm{reactor}}}\right)^2\right]$$

In the above relations, d_p is an equivalent spherical particle diameter. With a two-dimensional model, boundary conditions require a wall heat transfer coefficient. This is given by

$$h_W = h_W^0 + (h_W)_{\text{fluid}}$$

where
$$\frac{h_W^0 d_p}{\lambda_{\text{fluid}}} = 2\varepsilon_b + \frac{1-\varepsilon_b}{\psi_W + \frac{\lambda_{\text{fluid}}}{3\lambda_{\text{solid}}}}$$

and
$$\psi_W = 0.0024(d_{\text{reactor}}/d_p)^{1.58} \tag{15.55}$$

$$\frac{(h_W)_{\text{fluid}} d_p}{\lambda_{\text{fluid}}} = 0.0835(N_{\text{Re}_p})^{0.91}(N_{\text{Pr}})^{1/3} \textit{ for } 10 < N_{\text{Re}_p} < 1200$$
$$= 1.23(N_{\text{Re}_p})^{0.53}(N_{\text{Pr}})^{1/3} \textit{ for } 1200 < N_{\text{Re}_p} < 10000$$

Two-dimensional models are difficult to use and should be avoided when possible. The value of the relations given in Equations 15.54 and 15.55 is that they can be used outside of the energy balance to estimate radial heat flux. Estimates of this flux may be used to justify avoiding the adoption of a two-dimensional model.

15.3 Parameter estimation

The kinetic and adsorption sub-models discussed in the previous section will always contain one or more model parameters that must be evaluated to complete the sub-model. An elementary methodology of parameter estimation is discussed in Reaction Engineering textbooks, such as Fogler (1992) or Froment and Bischoff (1990), but better sources of methods are textbooks on the subject itself, such as Beck and Arnold (1977). Many books on experimental design also describe parameter estimation. The concept behind all estimation methods is to optimize an objective function $F(\boldsymbol{p}, \varphi)$ where $\boldsymbol{p}$ is a vector of parameter in a model M that predicts a set of outcomes or observations, while φ is a vector of experimental outcomes or observations. For reaction kinetics model, elements of the vector φ are often conversion or an effluent concentration of a reaction product measured at different space velocities for a steady-state experiment. For a transient experiment, elements of the vector might be an effluent concentration at different times after start-up. Parameters of adsorption isotherms are normally measured in a transient experiment.

A widely used function F is the sum of the squares of the difference between the predicted and measured outcome or

$$F(\boldsymbol{p}, \varphi) = \sum_q (C_i(L) - C_i^*(\boldsymbol{p}, L))_q{}^2 \tag{15.56}$$

where $C_i(L)$ is the concentration of species i measured leaving a reactor of length L, while $C_i^*(\boldsymbol{p}, L)$ is the concentration predicted by model M for the vector of parameters. The set of model parameters is now obtained by regressing on the vector $\boldsymbol{p}$ to minimize the objective function $F(\boldsymbol{p}, \varphi)$. The $C_i^*(\boldsymbol{p}, L)$ term in Equation 15.56 will involve solving an algebraic equation if a CSTR is used for parameter evaluation, namely,

$$\frac{Q}{V_R}(C_i^0 - C_i^*(\boldsymbol{p}, L)) = \rho_{\text{cat}} \nu_i r = \rho_{\text{cat}} \nu_i r(\boldsymbol{p}, C_i(L), C_j(L)) \tag{15.57}$$

In this CSTR material balance, $r(\mathbf{p}, C_i(L), C_j(L))$ is the rate model for the reaction assuming it encompasses reactants i and j. If, indeed, the model includes concentrations for two reactants, a set of algebraic equations must be solved to calculate $C_i^*(\mathbf{p}, L)$.

If a tubular reactor is used for evaluation, an ordinary differential equation or possibly a set of simultaneous ODEs must be integrated to obtain $C_i^*(\mathbf{p}, L)$.

A least squares estimation method has been outlined above. A Bayesian method is also used with a different objective function.

Several commercial computer packages to carry out the regression analysis are available. Among these are "Simulink Parameter Estimation" from the Simulink group and the "Athena Visual Workbench" from Stewart and Associates Engineering Software Inc.

15.4 Demonstration unit and performance testing

The purpose of a demonstration unit is to determine if the reactor separator design can meet the design objectives or can be made to meet them by adjusting operating conditions. It is not an exercise in model validation. As mentioned at the beginning of this chapter, it can be undertaken in a laboratory-scale unit. On scale-up, the operation of the larger unit can be considered also as a demonstration so demonstration may occur several times en route to the full-scale commercial plant.

An important use of the laboratory-scale demonstration unit is to explore experimentally how the operating conditions or the cycle parameters for a periodically operated reactor affect performance. In this way the region of optimal operation can be identified. This can be considered as the testing phase of the reactor–separator development. An outcome of the testing phase is usually sufficient technical information to undertake an approximate economic evaluation of the design. Development beyond a laboratory-scale unit becomes expensive so an early economic evaluation is important. Often a rough estimate is adequate for that purpose.

Many laboratory-scale demonstration units will be model based. Thus, the operation may suggest model inadequacies and the testing phase may produce improved model parameters for the reaction kinetics or the adsorption isotherms. A laboratory unit, however, will generally operate isothermally so demonstration at that scale will not be the source of a new reactor–separator model.

15.5 Scale-up and economic evaluation

In development of a novel design, scale-up can be expected several times. Trade-offs are usually involved, such as vessel diameter versus vessel length, so scale-up takes place stepwise and iteratively. The earlier chapters have shown that for the periodically operated reactors residence or space–time, cycle period and cycle subdivisions or the flow ratio in different sections are important. The cycle parameters have no scale-up rules as yet so in the first iteration they will be held constant. Given in any scale-up is the throughput for the larger unit, while throughput in the existing unit is known. With packed-bed reactors, increased throughput while maintaining a constant space velocity can be handled by either increasing vessel diameter or by increasing reactor length. Either alternative distorts the vessel aspect ratio, but increasing length in scale-up also raises the superficial velocity and the pressure drop. Aspect ratio can affect heat transfer or heat loss and, perhaps, flow distribution. Large diameter reactors are more costly. Often, the trade-off dilemma is solved in the first iteration by maintaining the aspect ratio of the existing plant in calculating the diameter and length of the larger unit.

With sizes, throughputs and cycle parameters now set, the reactor model can be introduced and the performance determined. For an SCMCR, the second iteration will attempt to improve the performance by applying the design procedures discussed in Section 7.2 involving the Damköhler and modified Aris numbers. A sequential design procedure has yet to be proposed for swing reactors, so the effect of the cycle period and duration of the exhaust and purge steps on performance would be explored using the reactor model to improve the variable choice. Again, once a new set of sizes and cycle parameters or apparent flow ratios has been selected, the model will be invoked to recheck the predicted performance.

The scale-up outlined in the two preceding paragraphs would be applied to the design of a pilot scale reactor–separator as well as the design of a full-scale unit. However, at the pilot scale, throughput and size are often such that isothermal operation can be no longer assumed. Then a nonisothermal model, including an energy balance, must be introduced. Operation of the pilot-scale unit would be the source of data for the nonisothermal model and would be applied to estimate parameters of the model as has been outlined in Section 15.3. Scale-up to a commercially sized reactor–separator would employ the nonisothermal model.

The ultimate economic analysis of a proposed reactor separator should employ a model intended for the commercial scale design, that is, a model based on at least pilot scale operation. This model provides sizes for estimating capital costs and also the cycling parameters and heat flows that will generate the unit's operating cost.

15.6 A development example

There are several recent papers in the research literature that offer what could be viewed as a general development plan for a simulated countercurrent moving bed chromatographic reactor. One of these is a contribution from the Politechnico di Milano (Mazzotti et al., 1996, 1997a). The system considered in their papers is the equilibrium-limited esterification of acetic acid by ethanol. The reaction is catalyzed by Amberlyst 15, a strong acid, cationic exchange resin based on cross-linked polystyrene and divinylbenzene. Amberlyst 15 also serves as an adsorbent for the water product of esterification. The ester is rejected by the resin. A complication of the system is that the resin swells strongly in contact with ethanol that is used as the eluent for the adsorbent. Not only are esterification reactions important commercially, they are representative for the production of fuel oxygenates such as ETBE and MTBE that are equilibrium limited as well as catalyzed by ion-exchange resins. This example illustrates the procedure discussed at the beginning of this chapter.

Once a catalytic reaction system has been selected that might be carried out in a cyclically operated separating reactor and a potential adsorbent has been identified, the development steps that must be undertaken according to Mazzotti et al. are

1. selection of operating temperature and pressure
2. formulation of a CR model

3. determination of the reaction kinetics (rate model) and rate constant, or constants if more than a single reaction is involved
4. selection of an adsorption equibrium model and determination of equilibrium data for single components and binary mixtures
5. pulse or step change (breakthrough) experiments on a packed bed of catalyst and adsorbent
6. model validation using the CR data
7. reworking of the CR model into the appropriate model for a separating reactor (SCMCR)
8. design of a laboratory-scale SCMCR using the reactor model from Step 7
9. parametric study on reactor performance employing the laboratory-scale reactor
10. economic evaluation.

Reviewing the published literature suggests that the bottleneck for the development process just outlined is the reaction selection and adsorbent identification. Most research seems to be academically generated. It should be generated by industry because it is in that environment that candidate systems are most easily identified. Selection of operating conditions, temperature and pressure is normally trivial as they are suggested by the system. Thus, for example, the esterification system to be considered in the following example from Mazzotti et al. occurs in the liquid phase and pressure is of minor importance. Temperatures are limited by boiling points of the components. Ethanol has the lowest boiling point. Consequently, temperature will be in the 25–100°C range.

Modeling discussions in Chapters 8 and 13 indicate that one-dimensional, pseudo-homogeneous models are adequate for most separating reactors. For a chromatographic reactor (CR) the solid phase is fixed so that with a combined catalyst and adsorbent, Equation15.1 becomes,

$$\varepsilon_b \frac{\partial C_i}{\partial t} + \rho_b \frac{\partial q_i}{\partial t} = D_x \frac{\partial^2 C_i}{\partial x^2} - \frac{1}{A}\frac{\partial n_i}{\partial x} + \rho_b \nu_i r \tag{15.58}$$

CRs, indeed most cyclic separating reactors, have large aspect ratios so that $L \gg d_R$. For this condition, the contribution of dispersion is negligible. Since the reactor is assumed to be isothermal, the fluid velocity will be constant. Thus, introducing a resin phase concentration of component i in place of an adsorbed amount and letting z be the position variable, Equation 15.58 can be written as

$$\varepsilon_b \frac{\partial C_i}{\partial t} + (1-\varepsilon_b)\frac{\partial C_i^P}{\partial t} + u\frac{\partial C_i}{\partial z} = \nu_i(1-\varepsilon_b) r \tag{15.59}$$

The rate of reaction is now expressed in units of resin volume. C_i^P is the concentration in the resin phase. Boundary condition at $z = 0$ is $C_i = (C_i)_{\text{feed}}$. The index i = ethanol, acetic acid, ethyl acetate and water. A total concentration balance is not necessary. No energy balance is required as the system is liquid phase and heat effects are modest. The system is also isobaric.

Esterification kinetics were determined by Mazzotti et al. using a well-stirred batch reactor. Mass transfer interference was estimated by varying stirrer speed. Assuming equilibrium between the resin and liquid phases, kinetics were found to be

$$r = kC_{\text{alcohol}}^P C_{\text{acid}}^P \left(1 - \frac{1}{K}\frac{a_{\text{ester}}^P a_{\text{water}}^P}{a_{\text{alcohol}}^P a_{\text{acid}}^P}\right) \tag{15.60}$$

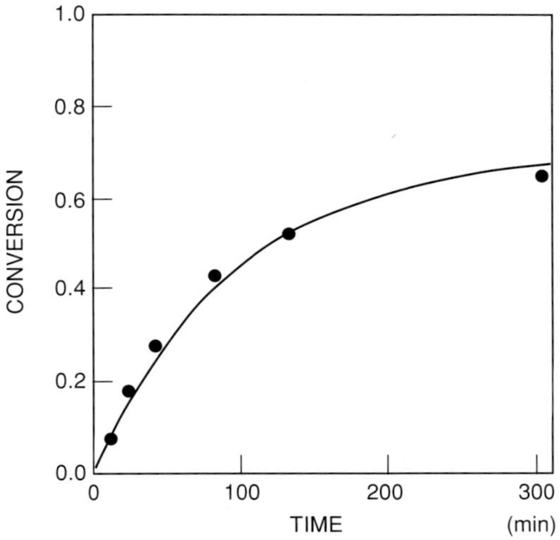

Figure 15.2 Comparison of the kinetic model with experimental data at 40°C for an equimolar initial mixture of ethanol and acetic acid. The curve gives the prediction of the kinetic model. (Figure adapted from Mazzotti et al. (1996) with permission. © 1996 Elsevier Science Ltd.)

In this expression, the equilibrium constant K can be calculated from thermodynamic data. Figure 15.2 shows that this model with the rate constant evaluated from the initial slope of the X versus t plot represents the batch reactor data well.

Equilibrium is usually represented by an isotherm. This is possible for the resin–ethanol/acetic acid system considered in this example. However, for swelling resins, as discussed above, equilibrium can be stated as a distribution between a liquid phase, free of resin, and a resin phase, free of liquid. Thus, $a_i^R = a_i^L$. Activity in the resin phase is expressed by Equation 15.38, which we repeat for convenience:

$$\ln a_i^P = 1 + \ln v_i - \sum_{j=1}^{N+1} m_{ij} v_j + \sum_{j=1}^{N+1} \chi_{ij} v_j - \sum_{j=1}^{N+1} \sum_{k=1}^{j-1} m_{ik} v_j v_k \chi_{kj} + \eta V_i (\frac{5}{3} v_P^{1/3} - \frac{7}{6} v_P) \tag{15.38}$$

The binary interaction parameter, χ_{ij}, is calculated from the classical Flory theory ($\chi_{ii} = 0$ and $\chi_{ij} = \chi_{ji} m_{ij}$), η is an elasticity parameter and m_{ij} is the ratio of molar volumes of i and j, while v_i is the volume fraction,

$$v_i = \frac{n_i^P V_i}{V_P^0 + \sum_{j=1}^{N} n_j^P V_j} \tag{15.34}$$

The swelling ratio, q, is

$$q = 1 + \frac{\sum n_i^P V_i}{V_P^0} \tag{15.39}$$

n_i^P is the moles of i in the resin phase where $n_i^0 = n_i^P + n_i^L$ must hold. It is a readily measured quantity that can be used along with distribution data to calculate the parameters

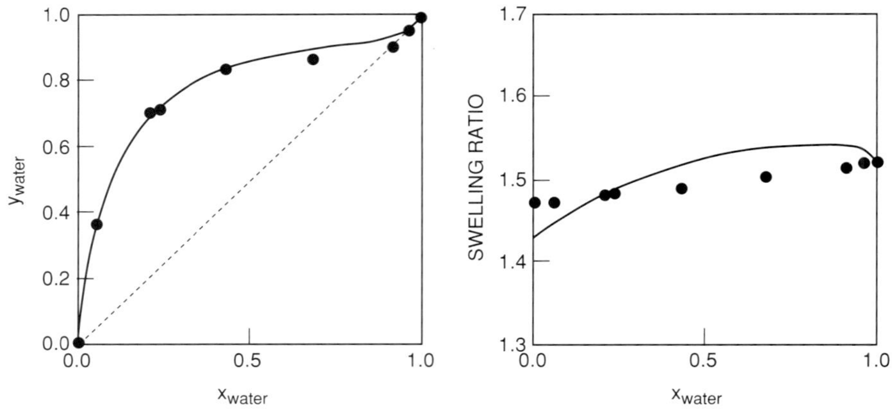

Figure 15.3 Sorption equilibria for the ethanol–water binary mixture in the presence of Amberlyst 15 resin at 25°C: *x* and *y* are mole fractions in the resin and liquid phases respectively. (Figure adapted from Mazzotti et al. (1996) with permission. © 1996 Elsevier Science Ltd.)

in Equation 15.38 employing Equations 15.34 and 15.39. Of course, only distribution and swelling data for the nonreacting binary pairs can be used. The interaction parameters for the reacting pairs, such as ethanol–acetic acid, and water–ethyl acetate, are set equal to zero. The activity in Equation 15.38 is the activity in the liquid phase. This is reliably predicted by the UNIFAC procedure mentioned earlier (Equation 15.35).

Figure 15.3 shows a set of the experimental binary data as an example. Equilibrium data were collected by Mazzotti et al. for the ethyl acetate–ethanol, the ethyl acetate–acetic acid and the acetic acid–water binary pairs. Binary experiments showed the component resin affinity to be in the sequence water > ethanol > acetic acid > ethyl acetate.

Step 5 in the development sequence given above is to validate the model using a single pulse CR or a step-change breakthrough experiment. Mazzotti et al. illustrate the development with the latter experiment. This was undertaken in a 38-cm i.d. tube containing 48.5 g of dry resin. At time zero an equimolar mixture of acetic acid and ethanol was introduced into the bed and the time-varying effluent composition was measured. These measurements are shown in Figure 15.4 for a downflow experiment. The experiment was repeated in upflow to see if density-driven mixing occurs. At the end of the breakthrough experiment, the eluent, ethanol, was fed to the bed in a second experiment to see if the resin could be completely regenerated.

The objective of the Mazzotti study was to develop an SCMCR and test its performance. Converting the CR model to an SCMCR model requires just altering the boundary conditions to represent the various steps in an SCMCR cycle. The SCMCR unit available to Mazzotti et al. had eight identical columns, each with a capacity for 3.1 g of resin and 7 cm in depth. The breakthrough experiment suggested regeneration of the resin is facile, but complete separation of water and ethyl acetate was a more difficult undertaking. Thus, a three-zone SCMCR was proposed with two columns or beds in zone 1 and five in zone 3. This arrangement could have been studied using the SCMCR model. Operating variables, namely the ethanol to feed ratio, the equimolar mixture feed rate and the switching time

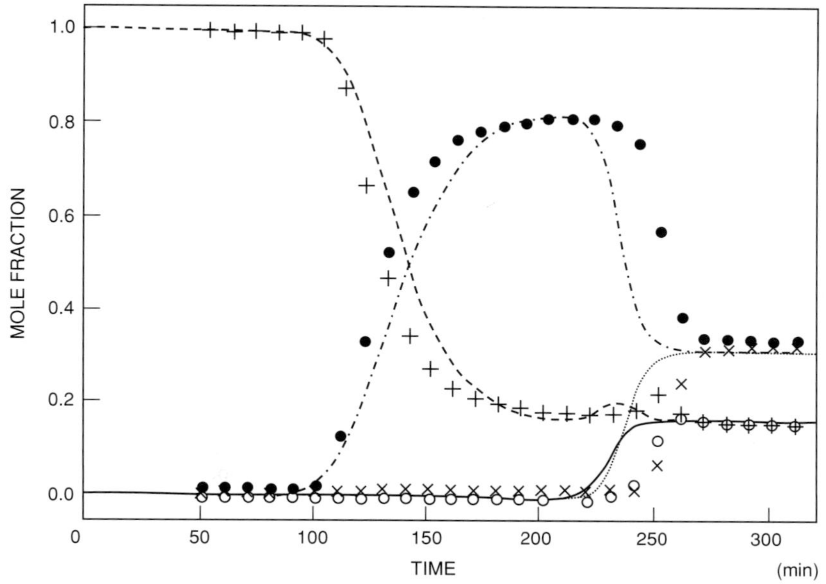

Figure 15.4 Time-varying effluent composition from a resin packed bed filled initially with ethanol after introduction at $t = 0$ of an equimolar mixture of ethanol and acetic acid. Curves in the figure are model predictions of concentration change. (Figure adapted from Mazzotti et al. (1996) with permission. © 1996 Elsevier Science Ltd.)

for the SCMCR, were set with help of the model. This is the eighth step in the development program.

The ninth step is a demonstration and a parametric study. For the demonstration, Mazzotti et al. (1996) report that with a 6 min switching time, an acetic acid feed rate of 15 g/h and an eluent to acetic acid flow rate ratio of 20, complete conversion and separation of the water and ester products were achieved. The tenth step, economic evaluation, was not undertaken.

Development of SCMCR systems for the esterification of acetic acid with methanol is illustrated by Lode et al. (2001) and by Gelosa et al. (2003) for the production of triacetine from acetic acid and glycerol. In the latter development, adsorption equilibrium was represented by the Langmuir isotherm even though the catalyst–adsorbent is an ion-exchange resin.

15.7 Reactor + separator alternative

In developing a separating reactor system, the alternative of a separate reactor followed by an adsorptive separation step should be kept in mind. Operating conditions for the two steps may be so different that their combination is not an efficient use of resources. Consider, for example, methane coupling discussed in Chapter 7 and hydrogen production by steam reforming that we examined in Chapter 13. It is not at all sure that a separating reactor is superior to a more conventional reactor followed by separation and recycle of unconverted educt. Machocki (1996) reports experiments on the oxidative coupling of methane using a periodically operated conventional reactor–separator sequence. A packed bed of catalyst, Na_2CO_3 impregnated onto a calcium carbonate carrier, operating at 1073 K was followed by a bed of Zeolite 5A that was cooled so its temperature was 300 K lower. The $C_2^=$ product was adsorbed by the adsorbent. The off gas was recycled to the reactor. A make-up stream was adjusted to maintain a stream containing 4.75% O_2 by volume with a CH_4:O_2 ratio = 20:1. In this way, together with a short contact time, a high selectivity to the C_2 product was achieved. A two-part cycle was used. In the second part, flow through the reactor was interrupted, the adsorbent bed temperature rose to 573 K and an inert, noncondensable gas circulated through this bed. Ethylene at high purity was condensed from the stripping gas.

A batch experiment lasting 3 h indicated a methane conversion of 94% with selectivity to ethylene and yield at 63 and 52 wt%, respectively. The SCMCR methane coupling study of Tonkovich et al. (1993) discussed in Chapter 7 showed a 65% conversion with a selectivity to C_2 of 80% but a $C_2^=/C_2$ ratio of 0.33. The conventional reactor–separator system results in a performance similar to that of the SCMCR, but in terms of heat transfer it is much simpler. Clearly, a separating reactor may not be the best solution for each application. Nevertheless, the experiments required to investigate a conventional solution will provide useful information for the development of a separating reactor.

Part V
Overview

Chapter 16

Periodically Operated Separating Reactors: Quo Vadis?

16.1 The role of separating reactors in reactor engineering

We introduced periodically operated separating reactors in Chapter 1 as a manifestation of a new process design paradigm that combines two unit operations rather than arranging them sequentially. A reactor in such a combination becomes multifunctional. The objective of this change in approach is process intensification, namely improved yield or conversion with equal or lower catalyst requirement, or, in other words, lower total cost. The literature surveyed in Chapters 4–7 and 9–12 demonstrates clearly that multifunctional reactors, combining reaction and adsorption, outperform a simple reactor with respect to product yield or reactant conversion, with often the same amount of catalyst, under identical feed conditions and throughput. The literature suggests equipment needs are reduced and the operation become simpler in many cases. It should be mentioned that adding adsorption to a reactor improves reactor performance, but may not improve the performance of the separation system at all. The inverse is also observed. Adding reaction to an absorber improves separation, but does not increase conversion or improve selectivity. There appears to be a truism here: combining two operations improves just the performance of one of the operations.

Process intensification by combining operations has its drawbacks. Compromises are necessary when operations are combined so it is seldom that either operation runs under optimal conditions. Flexibility to respond to changing feed rates or feed condition is also reduced. Increased feed rate for a stand-alone reactor could be handled, perhaps, by increasing reactor temperature. For a reactor–adsorber, this response, a temperature increase, will decrease adsorption capacity and so may adversely affect performance.

A further consideration is that a sequential arrangement of reactor and adsorber with recycle of reactant is always an option. Indeed, with an equilibrium-limited reaction or a consecutive reaction in which the intermediate product is valuable, the performance of the separate units with recycle can be made to match that of the reactor–adsorber. The decision between the alternatives would have to be made through an economic analysis.

Given these considerations, it is our opinion that periodically operated separating reactors will play a "niche" role in reactor engineering. They will not be universally applicable. Perhaps, this viewpoint extends to other types of separating reactors. Nonetheless, we believe that periodically operated separating reactors will be the best choice for many reaction systems. Likely, these systems will be constituted of multiple reactions or reactions with product inhibition where a sequential arrangement of reactor and adsorber cannot match the performance of a reactor in which catalyst and adsorbent are intimately mixed.

16.2 Current development status of different types of cyclic separating reactors

In our assessments of the state-of-the-art for chromatographic reactors in Chapter 8 and for swing reactors in Chapter 13, we classified different types of separating reactors into those that are now commercial or are at the threshold of commercial use, those that need further experiments, those that require substantial development and those unsuited for commercial exploitation. In the first group are the simulated countercurrent moving-bed chromatographic reactors (SCMCRs) and the continuous rotating annular-bed chromatographic reactors (CRACs). For the former, applications are mentioned in the literature. Laboratory-scale, off-the-shelf SCMCR units can be purchased. A design procedure in reasonable detail has been discussed (see Chapter 15). SCMCR applications are to aqueous phase reactions employing immobilized or mobile enzymes as catalyst and ion-exchange resins as adsorbents or ion-exchange resins serving both as catalyst and adsorbent. Reactions have been equilibrium-limited or inhibited. Nevertheless, complete conversion has been demonstrated along with very good separation of reaction products. Expanded use of SCMCRs will depend on how heating or cooling can be conducted as well as the importance of long-term loss of adsorption capacity.

The CRAC reactors are on the threshold of commercialization for the same classes of reactions that have been investigated for SCMCRs. Laboratory-scale units are not yet in the market, nor has a design procedure been published. Performance comparison of CRACs and SCMCRs under the same feed condition and flow rate still must be undertaken. The outcome of such a comparison will be critical for commercialization, as both reactors seem applicable to aqueous phase, enzymatic reactions. The SCMCR is probably a less costly option for large production rates, while the CRAC offers the possibility of isolating and collecting more than two products.

Pressure and temperature swing reactors, or their combination, are not ready for commercialization. More reaction systems need to be studied, particularly those with multiple reactions. Performance when large heat effects are present and allowing for heat loss or heat exchange through reactor walls should be examined, probably through simulation. Even though swing reactors appear to be a straightforward adaptation of swing separators, a systematic design or scale-up procedure has not yet been proposed. We anticipate that swing reactors will be used for solid-catalyzed gas phase reactions. However a SCMCR or even a CRAC reactor could be used as well. It is likely that a comparative study of a PSR–TSR and a SCMCR will be a key consideration in commercialization. Probably an economic analysis based on the experimental results will have to be carried out. In systems where heat must be added for an endothermic reaction or desorption, a TSR would be simpler to build or operate than a SCMCR.

Raining solids or staged fluidized beds CMCR designs are in only the earliest stage of development. A large research effort will be necessary if they are to find industrial application. There is, however, an important incentive for this development. The SCMCR, CRAC and PSR/TSR alternatives become impractical if adsorption capacity or catalyst activity diminishes with time-on-stream. They are fixed-bed reactor–adsorbers. Problems with the CMCR designs are discussed in Chapter 8. For the raining solids alternative, these are poor radial distribution of the moving solids, feeding and removal devices as well as attrition of the catalyst–adsorbent particles. Nevertheless, all of these problems have been solved in other moving-bed systems. With regard to the staged fluidized bed alternative, scale-up and perhaps solids removal seem to be the challenge.

Single or multiple pulse chromatographic reactors (CRs) are laboratory tools. There is no prospective of a commercial scale production reactor.

16.3 Predictions for the future

What can be expected in this field over the next 10 years? We are confident that the use of SCMCRs for enzyme-catalyzed reactions will become widespread. Use of this reactor for small or moderate volume production of fine chemicals will also develop. Our basis for this prediction is the current use of simulated countercurrent moving beds of adsorbent for large-scale product purification. Also, many of the papers discussed in Chapter 7 show total reactant conversion and very good separation of products. This is certain to attract industry's attention. The large number of publications since the start of the twenty-first century (12) show that a vigorous research effort is still underway.

Application of CRACs should start, but their commercial penetration will depend on how they compare performance-wise with SCMCRs. Because of their simplicity and versatility they will probably emerge as a popular reactor for small production runs.

Diminishing the rate of capacity loss for dolomites and synthetic adsorbents such hydrotalcite will be the key to commercialization of pressure and temperature swing separating reactors as these systems seem destined for solids-catalyzed gas phase reactions. We are optimistic that the deterioration of these materials can be stopped or at least greatly reduced. Thus, we expect to see semi-commercial scale if not full-scale units in use for synthesis gas and hydrogen production from the steam reforming of natural gas and biomass. We predict that those swing reactor–adsorbers will employ both pressure and temperature change for adsorbent regeneration. Use of swing reactors for lower temperature gaseous reactions will depend on how well they compare with SCMCRs. This must be determined by experimentation and economic analysis.

We believe development work will continue on CMCRs, but this will be a much less popular topic than work on SCMCRs or on swing technology. We do not expect this development to terminate in our 10-year horizon.

We expect a shift in focus from equilibrium-limited reactions to reactions in which selectivity is the main concern. It is our contention that separating reactors will have a unique effect on selectivity that cannot be achieved with a sequential reactor–adsorber arrangement and the use of recycle.

Development of adsorptive separating reactors will spur the development of adsorbents or reactive solids capable of sequestering reaction products so we anticipate over the next 10 years the discovery of new materials with high adsorption capacity, good stability and mechanical strength. These new adsorbents or reactive solids, in turn, will promote the further adoption of periodically operated separating reactors.

16.4 Expected direction of research on cyclic separating reactors

We devoted several pages in each of Chapters 8 and 13 to identifying the research needs for chromatographic and swing reactors and to speculating on novel research topics. Research needs will be obvious to any patient reader of the research literature on periodic separating reactors. Perhaps the research opportunities we discussed in the two chapters will also be evident. Consequently, we anticipate that these needs and challenges will frame the direction of research on periodic separating reactors in the next decade. Indeed, most of the research directions we predict are a continuation of those pursued now.

Identification of reaction systems that might benefit from combined separation and reaction and the closely allied subject of discovering or developing new adsorbents or reactive solids will form one important direction for future research. Work will have to be done through experimentation. Ideally this would be undertaken by the chemical industry, but given the low level of exploratory research in industry, research and development will probably fall to government or university laboratories.

A second direction for experimental work will be innovative operating schemes or bed designs. These have been mentioned in Chapters 8 and 13, but several are worth repeating here. Use of gradient methods, such as altering the eluent composition or changing eluents in different steps of a SCMCR, or at different angular positions in a CRAC, are one innovative operating scheme. Another is modulation of the feed flow or composition. Operating innovations appear to be applicable not only to SCMCRs and CRACs, but to PSR–TSRs as well. Bed design innovations are probably restricted to swing reactors and to CRACs. These are varying the adsorbent to catalyst ratio axially or using different adsorbents in the front and back end of beds. This should be viewed as a new research direction.

A third direction for experimental research will be in support of investigations undertaken through simulation. Their purpose will be both model validation and verification of predicted behavior or performance. This type of experiment, or its inverse, the modeling of experimental observations, has become a standard part of doctoral research in universities.

We expect a continued, but a rather small experimental effort on developing raining solids and fluidized bed systems for CMCRs. This is a fourth direction and is also a continuation of past experimentation.

The now low and still dropping cost of computation means that theoretical studies of periodically operated separating reactors through simulation can be undertaken by researchers in even the poorest nations. It is a certainty that such studies will flourish in the next decade.

One direction this computational research will follow is the investigation of increasingly complex reaction systems. These will likely be systems with multiple reactions, some of which will be equilibrium-limited and others subject to product or reactant inhibition. More concentrated feed streams will be considered along with adiabatic operation so that heats of reaction and adsorption will have to be dealt with. Adsorption isotherms will have to allow for competitive adsorption. Simple, power law kinetic models will no longer be adequate. As we have reported in Chapters 7, 10 and 11, work on more complex systems has already begun.

A second direction for computational studies will focus on the loss of adsorption capacity with time, the effect of loss on reactor performance and operational methods of containing this loss. Eventually this research direction will merge with studies of complex reaction systems, but initially kinetic complexity and loss of adsorption capacity or catalyst deactivation will be examined separately.

A third research direction will examine strategies for supplying energy to periodically operated separating reactors. Initially heat loss and heat addition through the walls as well as through the mobile phase will be investigated. Heat recapture and the energy efficiency of separating reactors with time will also be included. Eventually, like the second research direction, this type of study will become a part of the computational investigation of complex systems.

Innovative design or operating strategies, like axial variation of the adsorbent–catalyst ratio in a bed or modulation of reactant concentration in the feed, invite theoretical work. Studies of strategies represent a fourth research direction and one that is still not underway. Some effort might be directed at devising design procedures for CRAC and PSR–TSR reactors.

With the relatively large number of underemployed computational experts available, we can expect a continued attack on devising computational schemes that accelerate convergence to the cyclic stationary state and which reduce the computational effort. This is an extension of ongoing research.

A research need mentioned in Chapter 8 is a critical examination of the use of steady state submodels, such as kinetic or LDF mass transfer models, for dynamic systems. We do not expect investigators of periodic separating reactors to become deeply involved in such a fundamental topic. The adequacy question arises for other dynamic systems so we anticipate only a modest theoretical effort directed at cyclic separating reactors. It is a further research direction. Another research direction is economic analysis of separating reactors. This is a virgin topic.

16.5 Final word

Perhaps the most important conclusion from the overview in this chapter and, indeed, the entire book is that the field of periodically operated separating reactors offers investigators many exciting research opportunities. We anticipate a flood of research publications in the coming decade.

List of Symbols

A = reactant identifier
A = bed cross sectional area [m^2]
A_x = bed cross sectional area [m^2]
$a, b. c,$ = stoichiometric coefficients [−]
a = activity
= surface area usually per unit volume
a_i = activity of component i in solution [kgmol/m^3]
a_i^L = activity of component i in the liquid phase [kgmol/m^3]
a_i^P = activity of component i in polymer phase [kgmol/m^3]
a_i^R = activity of component i in resin phase [kgmol/m^3]
a_m = adsorbent surface area based on adsorbent mass [m^2/g]
= surface area of solid phase based on solid mass [m^2/g]
a_V = surface area of solid phase based on solid volume [m^2/m^3]
a_{water} = activity of water [kgmol/m^3]
B = reactant or product identifier
C = reactant or product identifier
C = total concentration in fluid phase [kgmol/m^3]
C_c = substrate (cyanopyridine) concentration [kgmol/m^3]
C_i = concentration of species i [kgmol/m^3]
C_n = product (nicotinamide) concentration [kgmol/m^3]
C_p = specific heat [kJ/kgmol K]
= pollutant concentration [kgmol/m^3]
C_S = substrate concentration [kgmol/m^3]
C_S^{equil} = substrate concentration at equilibrium in a reversible reaction [kgmol/m^3]
C_s = concentration in the solid phase [kgmol/m^3]
= specific heat of solid [kJ/kgmol K]
$\mathbf{C}_i$ = dimensionless concentration of species i
C_0 = initial or feed concentration [kgmol/m^3]
C_v = valve discharge coefficient [−]
C_i^L = concentration of species i in the liquid phase [kgmol/m^3]
C_i^P = concentration of species i in the polymer phase [kgmol/m^3]
C_i^R = concentration of species i in the resin phase [kgmol/m^3]

C_i^{surface}	= concentration of species i on the external surface of a particle [kgmol/m^3]
C^*	= surface concentration [kgmol/m^3]
C_i^*	= predicted concentration of component i [kgmol/m^3]
c_p, C_p	= specific heat [kJ/kgmol K]
CIS	= cells in series
CR	= chromatographic reactor
CRAC	= continuous rotating annular-bed chromatographic reactor
CMCR	= counter-current moving-bed chromatographic reactor
CPU	= central processing unit
D	= reactant or product identifier
D, d	= diameter [m]
D_x, D_z	= axial diffusivity or dispersion [m/s]
D_{ax}	= axial diffusivity or dispersion [m/s]
$(D_i)_m$	= diffusivity of component i in a mixture [m/s]
$(D_i)_{\text{eff}}$	= effective diffusivity of component i in a porous solid [m/s]
DOCFE	= double collocation on finite elements
d_c	= reactor diameter [m]
d_p	= particle diameter [m or mm]
d_R	= reactor diameter [m]
E	= activation energy [kgJ/kgmol]
ETO	= ethylene oxide
F	= objective function in an optimization exercise
F_0	= volumetric feed rate [m^3/s]
$(F_R)_i$	= fraction of group area contribution from group i
F_{seg}	= segregation function
$\mathrm{f}(y_i)$	= state equation or model equation in vector form evaluated at an assumed fixed point y_i
FTIR	= Fourier transform infrared spectroscopy
f	= function
	= $\pm$ function
f_{cat}	= fraction of band surface with catalyst washcoat
f_R	= reflux volume ratio for PSR with a reflux volume
f_f	= fraction of feed duration in PSR cycle [see Table 9.3]
f_d	= fraction of discharge duration in PSR cycle
f_e	= fraction of exhaust duration in PSR cycle
f_p	= pulsing frequency
f_{enz}^0	= fraction of immobilized enzyme that are active [−]
$f()$	= kinetic model for the forward rate term [various units]
GC	= gas chromatograph
GC-MS	= gas chromatograph with a mass spectrometer analyzer
$g()$	= kinetic model for the reverse rate term [various units]
H^+	= hydrogen ion
HPLC	= high pressure liquid chromatograph
$(H_a)_i$	= heat of adsorption of the ith species [kJ/kgmol]
$(H_r)_i$	= heat of reaction of the ith species [kJ/kgmol]

ΔH_r = heat of reaction [kJ/kgmol]
ΔH_j = heat of reaction for the jth reaction [kJ/kgmol]
$(\Delta H_r)_f$ = effective heat of reaction for flowing adsorbent (see Equations 9.37 and 9.38) [kJ/kgmol]
ΔH_a = heat of adsorption [kJ/kgmol]
h = fluid particle heat transfer coefficient [kJ/m^{2}s K]
h_0 = wall heat transfer coefficient [kJ/m^{2}s K]
h_w = wall heat transfer coefficient [kJ/m^{2}s K]
h_w^0 = wall heat transfer coefficient with no fluid motion [kJ/m^{2}s K]
$(h_{\mathrm{w}})_{\mathrm{fluid}}$ = fluid-wall heat transfer coefficient [kJ/m^{2}s K]
$\mathbf{I}$ = Identity matrix
i = species or component index
IR = infra red spectrophotometer
i.d. = inner diameter (of pipe or tube)
$\mathbf{J}(y_i)$ = Jacobean matrix evaluated at an assumed fixed point y_i
J_i = molar flux of i [mol/m^{2}s]
J_k = permeability term in the Ergun Eqn
J_v = inertial term in the Ergun Eqn
j = species or port index
K = reaction equilibrium constant [various units]
= number of reactions
K_i = adsorption equilibrium constant for species i, often a constant for a linear isotherm [various units]
= inhibition constant in forms of the Michaelis-Menten equation [kmol/m^3]
K_i^{BET} = adsorption equilibrium constant for species i in BET isotherm
K_i^F = adsorption equilibrium constant for species i in Freundlich isotherm
K_{ij} = equilibrium constant for the reaction $i \leftrightarrow j$
K_{ijkl} = equilibrium constant for the reaction $i + j \leftrightarrow k + l$
K_{eq} = equilibrium constant for the jth reaction [various units]
K_{ad} = adsorption equilibrium constant, often a constant for a linear isotherm [various units]
K_{m} = Michaelis–Menten constant [mol/cm^3]
$(K_{\mathrm{m}})_f$ = Michaelis–Menten constant for the forward reaction [mol/cm^3]
$(K_{\mathrm{m}})_r$ = Michaelis–Menten constant for the reverse reaction [mol/cm^3]
k = kilo [10^3]
k = permeability
= port index
= species identifier or index
= reaction index
= band index
= thermal conductivity [J/m s K]
k_a = adsorption rate constant [various units]
k_{ads} = adsorption rate constant [various units]
k_{des} = desorption rate constant [various units]
k_f = mass transfer coefficient [various units]
= rate constant of the forward reaction [various units]

k_i	= adsorption rate constant for the ith component
	= pre-exponential term for the ith component in the Arrhenius equation [various units]
k_i	= mass transfer coefficient for the ith component
$k_i^{external}$	= fluid-particle mass transfer coefficient for the ith component in a LDF model [various units]
$k_i^{internal}$	= intra particle mass transfer coefficient for the ith component in a LDF model [various units]
k_i^{mass}	= mass transfer coefficient for the ith component [various units]
$k_i^{overall}$	= overall mass transfer coefficient for the ith component [various units]
k_j	= rate constant for the jth reaction [various units]
k_k	= rate constant of the forward reaction where k indicates the rate in terms of the key component in the jth reaction [various units]
k_{-k}	= rate constant of the reverse reaction where k indicates the rate in terms of the key component in the jth reaction [various units]
k_m	= mass transfer coefficient [various units]
k_r	= rate constant of the reverse reaction [various units]
k_s	= rate constant of a surface reaction [kmol/kg s]
k_x, k_z	= effective axial thermal conductivity [kJ/s K]
k_1	= rate constant for a first order reaction [1/s]
k_2	= rate constant for a second order reaction [m^3/kmol s]
k'	= rate constant in partial pressure units [various units]
L	= bed length [m]
	= end of bed indicator
LDF	= linear driving force model for mass transfer or adsorption
LHS	= left hand side (of equation)
M	= mega [10^6]
M_i	= molecular weight of species i [kg/kmol]
MS	= mass spectrometer
m	= time step index
	= order of reaction by component [−]
m_i	= linear adsorption constant for component i [mol/kPa]
m_{ij}	= ratio of molar volumes of components i and j
m_{ad}	= mass flow rate of adsorbent [kg/s]
m	= mili [10^{-3}]
N	= total number of reacting or adsorbing components
	= number of adsorption sites [usually mol/m^2]
	= total number of stages or segments
	= total number of bands on surface banded with catalyst and adsorbent
N_{Ar}	= Aris number or α = ratio of solid and fluid velocities (see Table 3.6)
N_{Bo}	= Bodenstein number for mass
$N_{Bo,h}$	= Bodenstein number for heat
N_{col}	= number of columns or bed in a SCMCR
N_{Da_i}	= Damköhler number for species i (see Table 3.6)
N'_{Da}	= Temperature independent Damköhler number
N_{Da_f}	= Damköhler number for the forward reaction

N_{Da_b}	= Damköhler number for the reverse reaction
$(N_{Da})'$	= modified Damköhler number
N_d	= Bodenstein number for mass (see Table 3.6)
$(N_h)_g$	= Bodenstein number for heat (gas phase) (see Table 3.6)
$(N_h)_s$	= Bodenstein number for heat (solid phase) (see Table 3.6)
N_h	= Ratio of fluid-solid heat transfer to convective axial heat flux
N_{Ka}	= Kadlec number [see Table 9.3]
N_m	= Ratio of fluid-solid volumetric mass transfer coefficient to space velocity (see Table 3.6)
N_s	= Number of stages or sections
N_{Pe}	= Peclet number
N_{Pr}	= Prandtl number
N_R	= Number of reactions occurring
N_{Re}	= Reynolds number
$(N_{Re})_p$	= particle Reynolds number
N_{Ru}	= Ruthven number [see Table 9.3]
N_{Sh}	= Sherwood number
N_{Sc}	= Schmidt number
n	= total molar flow rate [mol/s]
	= moles
	= order of reaction by component [−]
	= integer
n^0	= moles originally present or molar flow rate entering
n_i	= molar flow rate of species i [mol/s]
	= moles of species i
n_i^L	= moles of component i in liquid phase
n_i^R	= moles of component i in resin phase
n_d	= molar flow rate of discharge [mol/s]
n_{CO2}	= adsorption capacity for CO_2 [mol/g adsorbent]
n_i	= Richardson and Zaki indices of fluidization for particles i
n_{total}	= density of sites on catalyst surface [kgmol/kg]
OC	= orthogonal collocation
OCFE	= orthogonal collocation on finite elements
OCM	= oxidative coupling of methane
ODE	= ordinary differential equation
o.d.	= outer diameter (of pipe or tube)
P	= product identifier
P	= total pressure [kPa]
P_d	= discharge pressure
P_e, P_v	= exhaust pressure
P_f	= feed pressure
P_i	= partial pressure of species i
p	= penalty function
$\mathbf{p}$	= model parameter vector
p_i	= partial pressure of component i [kPa]
PDE	= partial differential equation

PFR	= plug flow reactor, usually a fixed or packed bed
PPR	= periodically pulsed reactor
PSA	= pressure swing adsorber
PSR	= pressure swing reactor
Q	= volumetric flow rate [m^3/s]
	= product identifier
Q_{ads}	= adsorbent volumetric flow rate [m^3/s]
Q_k	= UNIQUAC group contribution
Q_L	= fluid (liquid) volumetric flow rate [m^3/s]
q	= swelling ratio [−]
	= observation index
	= surface area per mass of adsorbent [m^2/kg adsorbent]
q_i	= amount of component i adsorbed per mass of adsorbent [mol/kg adsorbent]
	= sum of group area contributions in UNIQUAC
q_i^{sat}	= capacity of the adsorbent for species i [mol/kg adsorbent]
q_s	= adsorbate capacity per mass of adsorbent
q_{si}	= capacity of the adsorbent for species i
$q_i^{\#}$	= amount of component i adsorbed per mass of adsorbent at equilibrium fraction of adsorbate i [−]
q_i^*	= equilibrium adsorbed phase concentration
R	= reactant or product identifier
R	= universal gas constant [various units]
	= rate of reaction [various units, mmol/s kg catalyst]
R_j	= rate of the jth reaction [various units]
RHS	= right hand side of equation
RPSR	= rapid cycle pressure swing reactor
r	= rate of reaction [various units, kgmol/m^3 s]
r_{ads}	= rate of adsorption [various units, kgmol/m^3 s]
r_{cat}	= rate of catalytic reaction
r_{des}	= rate of desorption
r_i	= rate of consumption or formation of component i
	= sum of group volume contributions in UNIQUAC
r_p	= particle radius [m]
r^{het}	= rate of a heterogeneous reaction [various units kg mol/m^3 s]
r^{hom}	= rate of a homogeneous reaction
S	= product identifier
SCMCR	= simulated counter-current moving-bed chromatographic reactor
ST	= space or residence time
s	= cycle split [−]
T	= temperature [K]
	= product identifier
T_a	= ambiant temperature [K]
T_g	= fluid temperature [K]
T_s	= solid temperature [K]
T_W	= wall temperature [K]

TSR	= temperature swing reactor
t	= time [s]
t_s	= switching period [s]
t^*	= contact time ($= V/Q$) [s]
U	= velocity, usually superficial velocity of fluid phase [m/s]
U_s	= solids velocity [m/s]
u	= velocity [m/s]
u_f	= superficial velocity of the fluid phase [m/s]
u_g	= superficial velocity of the gas [m/s]
u_L	= superficial velocity of the liquid [m/s]
u_s	= superficial velocity of the solid or adsorbent [m/s]
V	= volume [m^3]
V_{bed}	= volume of a bed or packed column [m^3]
V_i	= volume of component i in system [m^3]
	= fraction of group volume contributions in UNIQUAC
V_j	= volume of a section of a chromatographic reactor [m^3]
V_{max}	= maximum rate of an enzymatic reaction (constant in the Michaelis–Menten relation) [units of reaction rate]
V_p	= volume of reverse flow chamber in the pulsing system of a pulsed multistage separating reactor
	= volume of polymer phase [m^3]
V_R	= reactor volume [m^3]
v_{aq}	= volumetric flow rate of water [m^3/min]
v_i	= volume fraction or ratio [–]
v_{org}	= volumetric flow rate of organic phase [m^3/min]
v_P	= polymer volume fraction [–]
v_{slip}	= slip velocity [m/s]
v_x	= front or discontinuity velocity [m/s]
W	= weight of sample, usually catalyst [g]
W_k	= fraction of stream flowing from bed k to bed $k+1$
$\mathbf{W}_i$	= trajectory in the a phase plane fulfilling a material balance condition
w_k	= fraction of purge or regenerant entering bed k
X	= conversion [–]
X_i	= conversion based on component i [–]
$\mathbf{X}$	= dimensionless position variable in a phase plane representation
x	= axial position [m]
x_i	= mole fraction of component i usually in liquid phase [–]
Y_i	= yield of product i in a multiple reaction system
Y_i^*	= yield of product i in a multiple reaction system at the raffinate port
$\mathbf{Y}$	= dimensionless time variable in a phase plane representation
y_i	= mole fraction of the ith component usually in gas phase [–]
Z	= dimensionless total pressure [see Table 9.3]
Z_i	= dimensionless partial pressure [see Table 9.3]
Z_{R_i}	= dimensionless partial pressure in reflux volume [see Table 9.2]
z	= position variable, usually axial position [m]
z	= width of band on a surface with bands of washcoated catalyst and adsorbent

Greek

α = ratio of solid to fluid phase mass velocities (see Table 3.6)
= coefficient in Ergun equation
α_i = adsorption constant for component i [mol/kPa]
= exponent of the ith component in the rate term for the reverse reaction [−]
α'_i = adsorption constant for component i [mol/cm^3 solid/mol/cm^3 solution]
β = Prater number (see Table 3.6)
β_i = exponent of the ith component in the rate term for the forward reaction [−]
χ = capacity [see Table 9.3]
χ_{ij} = classical group interaction parameter
Δ = difference or increment
δx = spacing between inlet and outlet ports in a SCMCR [m]
$\delta \nu$ = sum of the stoichiometric coefficients in a reaction [−]
σ = ratio of solid and fluid velocities in a CMCR or the Aris Number, N_{Ar}
σ_i = ratio of adsorbate and fluid mass flow rates for species i
σ^j = ratio of velocities in section or bed j of a CMCR
ε = void volume in reactor bed [−]
ε_b = void volume in reactor bed or bed porosity [−]
ε_p = void volume in catalyst or adsorbent particle [−]
ε_{ads} = adsorbent fractional volume [−]
ε_{cat} = catalyst fractional volume [−]
ε_{inert} = fractional volume of inert solid [−]
ε_L = fluid (liquid) fractional volume
ε_t = total void volume in reactor bed (inter and intra particle) [−]
η = dimensionless fluid velocity (see Table 3.4)
η_i = effectiveness factor for the ith reaction or component [−]
Γ_i = purity of i as molar concentration ratio at a withdrawal port
Γ_k = group residual activity coefficient in UNIQUAC
γ_i = dimensionless capacity term for the ith component (see Table 3.6)
= dimensionless concentration of the ith component in the fluid phase (see Table 3.6)
= activity coefficient of component i [−]
γ_i^L = activity coefficient of component i in liquid phase [−]
γ_i^R = activity coefficient of component i in resin phase [−]
γ_b = effective thermal capacity for catalyst-adsorbent bed [kJ/kg K]
γ_f = effective thermal capacity for reactor with a moving bed of adsorbent and or catalyst [kJ/kg K]
η = effectiveness factor [−]
= enhancement factor [−]
= elasticity parameter
η_X = enhancement factor for conversion [−]
κ = relative adsorptivity (ratio of adsorption equilibrium constants [−]
κ_{ai} = dimensionless heat of adsorption for species i (see Table 3.6)
κ_{ri} = dimensionless heat of reaction for species i (see Table 3.6)
Φ_i = dimensionless adsorption capacity for species i

φ = vector of observations
φ_i, $(\varphi_i)_b$ = volumetric capacity of the adsorbent for species i per unit fluid volume in reactor = $\rho_a m_i RT/\varepsilon$
φ_i = dimensionless concentration of adsorbate i (see Table 3.6)
$(\varphi_i)_f$ = volumetric capacity of the adsorbent for species i per unit of feed volume in reactor = $m_a m_i RT/Q_0$
ϕ = dimensionless temperature departure (see Table 3.6)
= modified Thiele modulus
ϕ_i = volume fraction of species i
ϑ_i = productivity as the moles of key reactant i fed to the reactor per unit time and reactor volume [kg mol/h m^3]
λ = dimensionless axial position in reactor or PSR bed (see Table 3.6, 9.3)
= ratio of adsorption and forward reaction rate constants = Nk_a/k_f
= thermal conductivity [kJ/m s K]
= shape factor in Ergun equation
λ_r = effective thermal conductivity in the radial direction [kJ/m s K]
λ_x = effective thermal conductivity in the axial direction
λ^0 = effective thermal conductivity in the absence of fluid flow
λ^{fluid} = fluid flow contribution to effective thermal conductivity
μ = viscosity
= micro [10^{-6}]
ν_i = stoichiometric coefficient for component i
= dimensionless flow rate of species i (see Table 3.6)
= dimensionless absorbate concentration for species i
= volume fraction of component i
ρ_a, ρ_{ad} = bulk density of adsorbent in the bed [kg/m^3]
ρ_B = bulk density of bed
ρ_b = bulk density of bed
$(\rho_b)_{\text{cat}}$ = bulk density of catalyst
$(\rho_b)_{\text{ads}}$ = bulk density of adsorbent
ρ_{ad} = particle density of adsorbent
ρ_c = density of catalyst bed
ρ_g = density of gas phase
ρ_i = density of phase i
ρ_L = fluid (liquid) density
θ = dimensionless time = t/τ [−]
= dimensionless time (see Table 3.6)
= fraction of total surface that are occupied [−]
θ_i = fraction of surface sites occupied by component i [−]
= fraction of group area contribution in UNIQUAC
τ = cycle period [s]
τ_s = switching time in a SCMCR [s]
ζ = switching velocity in a SCMCR or the hypothetical solids rate of movement [m/s]
= any dependent variable
= denominator term in a rate eqn.

ζ_v = dimensionless viscous term in the Ergun equation (see Table 3.6)
ζ_k = dimensionless permeation term in the Ergun equation (see Table 3.6)
ξ = dimensionless carrier gas feed rate
= ratio of heat loss through wall to convective heat transport (see Table 3.5)
ξ_j = extent of the jth reaction [moles]
Ξ_i = selectivity to product i among various products j, k, l
Ξ_i^* = selectivity to product i among various products collected at a raffinate port
ω = ratio of solid and fluid heat capacities (see Table 3.6)
Ψ_{mk} = group interaction parameter
ψ = ratio of pressure and thermal energy (see Table 3.6)

Superscripts

' = modified
* = surface
∞ = at infinite dilution
0 = entrance, initial, start up, absence of fluid motion
app = apparent
BET = BET adsorption isotherm
c = combinatorial
cat = catalyst
equil = equilibrium
F = Freundlich adsorption isotherm
het = heterogeneous
hom = homogeneous
max = maximum
L = liquid
P = polymer or resin
R = resin
r = residual
sat = saturated (adsorbent surface)

Subscripts

A, B, etc = species indices
Ar = Aris
a = adsorption
= ambient
ad = adsorbent, adsorption
ads = adsorbent, adsorption
B = bed, bulk
Bo = Bodenstein
b = batch, bed, bulk
C = column, reactor bed
CH = cyclohexane
c = catalyst
= substrate (cyanopyridine)
cat = catalyst

col	= columns (beds)
D	= discharge
Da	= Damköhler
d	= discharge
	= dispersion
des	= desorption
e	= exhaust
eff	= effective
eq	= equilibrium
equil	= equilibrium
enz	= enzymatic
f	= flow, fluid,
	= forward reaction
	= feed
g	= gas or fluid
H^+	= hydrogen ion
h	= heat
i	= component or species index
	= inhibition
j	= component index
	= reaction index when multiple reactions occur
Ka	= Kadlec
k	= reaction index
	= kinetic
L	= fluid or liquid
m	= mass
	= mixture
	= Michaelis–Menten index
max	= maximum
mix	= mixed or mean
n	= product (nicotinamide)
opt	= optimum
Pe	= Peclet
Pr	= Prandtl
p	= particle
	= pulse
	= purge
R	= reflux
Re	= Reynolds
Ru	= Ruthven
r	= reaction,
	= reverse or back reaction
S	= substrate
	= superficial
Sc	= Schmidt
Sh	= Sherwood

s	= solid
	= switching
	= surface
seg	= segregated or segregation
V	= volume
v	= constant volume
	= exhaust
	= valve
vg	= constant volume of gas or fluid (heat capacity)
X	= conversion
x	= cross section
	= axial
z	= axial
0	= entrance or initial condition
1, 2	= reaction indices

References

Abdul-Kareem, H.K., Hudgins, R.R. and Silveston, P.L. (1980) Forced cycling of the catalytic oxidation of CO over a V_2O_5 catalyst II. Temperature cycling. *Chem. Eng. Sci.*, **35**(10), 2085–2088.

Agar, D.W. (1999) Multifunctional reactors: old preconceptions and new dimensions. *Chem. Eng. Sci.*, **54**, 1299–1305.

Agar, D.W. (2003) Dos and don'ts of adsorptive reactors. *4th International Conference on Unsteady-State Processes in Catalysis*. (USPC4), October 26–29, 2003, Montreal, Canada.

Agar, D.W. and Ruppel, W. (1988a) Multifunktionale Reaktoren für die heterogene Katalyse. *Chemie Ing.-Tech.*, **43**, 731–741.

Agar, D.W. and Ruppel, W. (1988b) Extended reactor concept for dynamic DeNOx design. *Chem. Eng. Sci.*, **43**(8), 2073–2078.

Agreda, V.H., Partin, L.R. and Heise, W.H. (1990) High-purity methyl acetate via reactive distillation. *Chem. Eng. Progr.*, **86**(2), 40–46.

Aida, T., Kobayashi, R. and Niiyama, H. (1997) Periodic operation of NO–CO reaction over Pt/Al_2O_3: effect of intraparticle diffusion. *Kagaku Kogaku Ronbunshu* (Japanese), **23**, 962–968.

Aida, T., Na-Ranong, D., Kobayashi, R. and Niiyama, H. (1999) Effect of diffusion and adsorption–desorption on periodic operation performance of NO–CO reaction over supported noble metal catalysts. *Chem. Eng. Sci.*, **54**, 4449–4457.

Aida, T., Sekine, T., Na-Ranong, D. and Niiyama, H. (2002) Combined system of catalyst and adsorbent for NO–CO reaction over Pt/Al_2O_3 under periodic operation. *17th International Symposium on Chemical Reaction Engineering*, August 25–28, 2002, Hong Kong, China (Preprint).

Akintoye, A., Ganetsos, G. and Barker, P.E. (1990) Preparative scale chromatographic systems as combined biochemical reactor-separators. In *Advances in Separation Processes, I. Chem. E. Symp. Ser.*, **118**, 21–28.

Akintoye, A., Ganetsos, G. and Barker, P.E. (1991) The inversion of sucrose in a semi-continuous countercurrent chromatographic bioreactor-separator. *Trans. I. Chem. E*, **69**C, 35–44.

Al-Juhani, A.A. and Loughlin, K.F. (2003) Simulation of a combined isomerization reactor & pressure swing adsorption unit. *Adsorption*, **9**, 251–264.

Alpay, E., Kenney, C.N. and Scott, D.M. (1993) Simulation of rapid pressure swing adsorption and reaction processes. *Chem. Eng. Sci.*, **48**(18), 3173–3186.

Alpay, E., Chatsiriwech, D., Kershenbaum, L.S., Hull, C.P. and Kirkby, N.F. (1994) Combined reaction and separation in pressure swing processes. *Chem. Eng. Sci.*, **49**(24B), 5845–5864.

Alpay, E., Haq, N., Kershenbaum, L.S. and Kirkby, N.F. (1996) Adsorption parameters for strongly adsorbed hydrocarbon vapours on some commercial adsorbents. *Gas. Sep. Purif.*, **10**(1), 25–33.

Altshuller, D. (1983) Design equations and transient behaviour of the counter-current moving bed chromatographic reactor. *Chem. Eng. Commun.*, **19**, 363–375.

Antonucci, F., Giordano, N. and Bart, J.C.J. (1978) Chemical reactions in chromatographic columns. Dehydrogenation of ethane over cadmium-exchanged zeolite 4A. *J. Chromatogr.*, **150**, 309–317.

Antos, D. and Seidel-Morgenstern, A. (2001) Application of gradients in the simulated moving bed process. *Chem. Eng. Sci.*, **56**, 6667–6682.

Arsenijevic, Z.L.J., Grbic, B.V., Grbavcic, Z.B., Radic, N.D. and Terlecki-Baricevic, A.V. (1999) Ethylene oxide removal in combined sorbent–catalyst system. *Chem. Eng. Sci.*, **54**, 1519–1524.

Bailey, J.E. (1977) Periodic phenomena. In L. Lapidus and N.R. Amundson (eds.), *Chemical Reactor Theory. A Review.* Prentice-Hall, Engelwood Cliffs, NJ, pp. 758–813.

Bailey, J.E. and Ollis, D.F. (1986) *Biochemical Engineering Fundamentals*, 2nd edn, McGraw-Hill, New York.

Balasubramanian, B., Lopez Ortiz, A., Kaytakoglu, S. and Harrison, D.P. (1999) Hydrogen from methane in a single-step process. *Chem. Eng. Sci.*, **54**, 3543–3552.

Barker, P.E. and Deeble, R.E. (1975) Sequential chromatographic equipment for separation of a wide range of organic mixtures. *Chromatogr. Rev.*, **8**(2), 67–79.

Barker, P.E. and Ching, C.B. (1980) Continuous liquid chromatographic process for separation of fructose–glucose mixtures. *Eur. Fed. Chem. Eng. Kemtek*, **5** (Copenhagen).

Barker, P.E. and Ganetsos, G. (1988) Chemical and biochemical separations using preparative and large scale batch and continuous chromatography. *Sep. Purif. Methods*, **17**(1), 1–65.

Barker, P.E., England, K. and Vlachogiannis, G. (1983) Mathematical model for the fractionation of dextran on a semi-continuous counter-current simulated moving bed chromatograph. *Trans. I. Chem. E. Chem. Eng. Res. Des.*, **61**, 241–247.

Barker, P.E., Ganetsos, G. and Thawait, S. (1986) Development of a link between batch and semi-continuous liquid chromatographic system. *Chem. Eng. Sci.*, **41**, 2695–2604.

Barker, P.E., Zafar, I. and Alsop, R.M. (1987a) A novel method for the production of dextrans and fructose. In G.W. Moody and P.B. Baker (eds.), *Bioreactors and Biotransformations*, Elsevier Science Publishers, Amsterdam, pp. 141–157.

Barker, P.E., Zafar, I. and Alsop, R.M. (1987b) Production of dextran and fructose in a chromatographic reactor—separator. In M.S. Verral and M.J. Hudson (eds.), *Separations for Biotechnology.* Ellis-Horwood, London, Chapter 9, pp. 127-151.

Barker, P.E., Ganetsos, G., Ajongwen, A. and Akintoye, A. (1992a) Bioreaction–separation in continuous chromatographic systems. *Chem. Eng. J. Biochem. Eng. J.*, **50**, B23–B28.

Barker, P.E., Ganetsos, G., Ajongwen, A. and Akintoye, A. (1992b)Continuous chromatographic bioreaction-separation. In D. Pyle (ed.), *Separations for Biotechnology*, Elsevier Science Publishers, London, pp. 549–557.

Barker, P.E., Ajongwen, N.J., Shieh, M.T. and Ganetsos, G. (1992c) Simulated counter-current chromatographic bioreactor-separators. In *Fundamentals of Adsorption, Proceedings of the Fourth International Conference on Fundamental Adsorption*, May 17–22, Kyoto, Japan, pp. 35–44.

Bassett, D.W. and Habgood, H.W. (1960) A gas chromatographic study of the catalytic isomerization of cyclopropane. *J. Phys. Chem.*, **64**, 769–773.

Beck, J.V. and Arnold, K.J. (1977) *Parameter Estimation in Engineering and Science.* John Wiley & Sons, New York.

Bennett, B.A.F., Cloete, F.L.D. and Streat, M. (1984) A systematic analysis of the performance of a new continuous ion-exchange technique. *Ion Exchange Proc. Ind.*, 133–143.

Bessling, B., Schembecker, G. and Simmrock, K.H. (1997) Design of procedure with reactive distillation line diagrams. *Ind. Eng. Chem. Res.*, **36**, 3032–3042.

Bjorklund, M.C. and Carr, R.W. (1995) The simulated counter-current moving bed chromatographic reactor: a catalytic and separative reactor. *Catal. Today*, **25**, 159–168.

Bjorklund, M.C. and Carr, R.W. (2002) Enhanced methanol yields from the direct partial oxidation of methane in a simulated countercurrent moving bed chromatographic reactor. *Ind. Eng. Chem. Res.*, **41**, 6528–6536.

Boreskov, G.K. and Matros, Yu.Sh. (1983) Unsteady-state performance of heterogeneous catalytic reactions. *Catal. Rev.-Sci. Eng.*, **25**(4), 551–590.

Boreskov, G.K., Matros, Yu.Sh., Kiselov, O.V. and Bunimovich, G.A. (1977) Realization of heterogeneous catalytic processes under unsteady-state conditions. *Dokl. Acad. Nauk USSR*, **237**, 160–163 (in Russian).

Boreskov, G.K., Matros, Yu.Sh., Lugovskoy, V.I., Bunimovich, G.A. and Puzhilova, V.I. (1984) Unsteady-state process of complete oxidation in reactor with fixed catalyst bed. *Teor. Osn. Khim. Technol.*, **18**, 328–334 (in Russian).

Brinkmann, M., Barresi, A.A., Vanni, M. and Baldi, G. (1999) Unsteady state treatment of very lean waste gases in a network of catalytic burners. *Catal. Today*, **47**(1–4), 263–277.

Broughton, D.B. (1961) US Pat. 2 985 589.

Broughton, D.B. (1968) Molex: case history of a process. *Chem. Eng. Progr.*, **64**(8), 60–65.

Broughton, D.B., Neuzil, R.W., Pharis, J.M. and Brearley, C.S. (1970) The Parex process for recovering paraxylene. *Chem. Eng. Progr.*, **66**, 70–75.

Brun-Tsekhovoi, A.R., Zadorin, A.N., Katsobashvili, Ya.R. and Kourdyumov, S.S. (1986) The process of catalytic steam-reforming of hydrocarbons in the presence of carbon dioxide acceptor. In *Hydrogen Energy Process VII, Proceedings of the World Hydrogen Energy Conference, Vol.* 2, Pergamon Press, New York, pp. 885–896.

Bryson, Jr., A. and Ho, Y. (1975) *Applied Optimal Control.* Hemisphere Press, Washington, DC.

Carta, G. (1991) Simultaneous reaction and chromatography. In Costa, C.A. and J.S Cabral (eds), *Chromatographic and Membrane Processes in Biotechnology.* Kluwer Academic Publishers, the Netherlands, 429–442.

Carvill, B.T., Hufton, J.R., Anand, M. and Sircar, S. (1996) Sorption-enhanced reaction process. *AIChE J.*, **42**(10), 2765–2772.

Champagnie, A.M., Tsotsis, T.T., Minet, R.G. and Webster, I.A. (1990) A high temperature catalytic membrane reactor for ethane dehydrogenation. *Chem. Eng. Sci.*, **36**, 2423–2429.

Chatsiriwech, D., Kershenbaum, L.S. and Kirkby, N.F. (1992) The enhancement of catalytic reactor performance by pressure swing operation. Internal Report, Imperial College, London.

Chatsiriwech, D., Alpay, E., Kershenbaum, L.S., Hull, C.P. and Kirkby, N.F. (1994) Enhancement of catalytic reaction by pressure swing adsorption. *Catal. Today*, **20**, 351–366.

Cheng, Y.S., Alpay, E. and Kershenbaum, L.S. (1998) Simulation and optimization of a rapid pressure swing reactor. *Comput Chem. Eng.*, **22**, S45–S52.

Ching, C.B. and Lu, Z.P. (1997) Simulated moving bed reactor: application in bioreaction and separation. *Ind. Eng. Chem. Res.*, **36**, 152–159.

Ching, C.B. and Ruthven, D.M. (1988) A liquid phase chromatographic study of sorption and diffusion in NaX and KX zeolite crystals. *Zeolites*, **8**, 68–73.

Cho, B.K. (1983) Dynamic behavior of a single catalyst pellet. I. Symmetric concentration cycling during CO oxidation over Pt/Al_2O_3. *Ind. Eng. Chem., Fundam.*, **22**, 410–420.

Cho, B.K. and West, L.A. (1986) Cyclic operation of Pt/Al_2O_3 catalysts for CO oxidation. *Ind. Eng. Chem., Fundam.*, **25**, 158–164.

Cho, B.K., Carr, R.W. and Aris, R. (1980a) A new continuous flow reactor for simultaneous reaction and separation. *Sep. Sci. Technol.*, **15**, 679–696.

Cho, B.K., Carr, R.W. and Aris, R. (1980b) A continuous chromatographic reactor. *Chem. Eng. Sci.*, **35**, 74–81.

Cho, B.K., Aris, R. and Carr, R.W. (1982) The mathematical theory of a countercurrent catalytic reactor. *Proc. R. Soc. Lond. A*, **383**, 147–189.

Chu, C. and Tsang, L.C. (1971) Behavior of a chromatographic reactor. *Ind. Eng. Chem. Process Des. Dev.*, **10**(1), 47–53.

Çinar, A., Deng, J., Meerkov, S.M. and Shu, X. (1987a) Vibrational control of an exothermic reaction in a CSTR: theory and experiments. *AIChE J.*, **33**, 353–365.

Çinar, A., Rigopoulos, K., Shu, X. and Meerkov, S.M. (1987b) Vibrational control of chemical reactors: stabilization and conversion improvement in an exothermic CSTR. *Chem. Eng. Commun.*, **59**, 299–308.

Clausen, B.S. and Nielsen, P.E.H. (ed.) (2000) Catalysis in membrane reactors, *Catal. Today*, **56**.

Cloete, F.L.D. and Streat, M. (1963) A new continuous solid–fluid contacting technique. *Nature*, **200**(4912), 1199–1200.

Cote, A.S., Delgass, W.N. and Ramkrishna, D. (1999) Investigation of spatially patterned catalytic reactors. *Chem. Eng. Sci.*, **54**, 2627–2635.

Croft, D.T. and Levan, M.G. (1994) Periodic states of adsorption cycles 1. Direct determination and stability. *Chem. Eng. Sci.*, **49**(11), 1821–1829.

Cybulski, A. and Moulijn, J.A. (1997) *Structured Catalysts and Reactors*. Marcel Dekker, New York.

Dalmon, J.A. (1997) Catalytic membrane reactors. In Ertl *et al.* (eds.), *Handbook of Heterogeneous Catalysis*. VCH, Weihheim, Germany, pp. 1387–1398.

Dautzenberg, F.M. and Mukherjee, M. (2001) Process intensification using multifunctional reactors. *Chem. Eng. Sci.*, **56**, 251–267.

de Deugd, R.M., Kapteijn, F. and Moulijn, J.A. (2003) Using monolithic catalysts for highly selective Fischer–Tropsch synthesis. *Catal. Today*, **79**(1–4), 495–550.

de Garmo, J.L., Parulekar, V.N. and Pinjala, V. (1992) Consider reactive distillation. *Chem. Eng. Progr.*, **3**, 43–50.

de Rosset, A.J., Neuzil, R.W. and Korous, D.J. (1976) Liquid column chromatography as a predictive tool for continuous countercurrent adsorptive separation. *Ind. Eng. Chem. Proc. Des. Dev.*, **15**, 261–266.

den Hollander, J.L., Zomerdijk, M., Straathof, A.J. and van der Wielen, L.A.M. (2002) Continuous enzymatic penicillin G hydrolysis in countercurrent water–butyl acetic biphasic systems. *Chem. Eng. Sci.*, **57**, 1591–1598.

Ding, Y. and Alpay, E. (2000a) Adsorption-enhanced steam-methane reforming. *Chem. Eng. Sci.*, **55**, 3929–3940.

Ding, Y. and Alpay, E. (2000b) Equilibria and kinetics of CO_2 adsorption on hydrotalcite adsorbent. *Chem. Eng. Sci.*, **55**, 3461–3474.

Dinwiddie, J.A. and Morgan, W. (1961) US Pat. 2 976 132.

Dittmeyer, R., Höllein, V., Quicker, P., Emig, G., Hausinger, G. and Schmidt, F. (1999) Factors controlling the performance of catalytic dehydrogenation of ethylbenzene in palladium composite membrane reactors. *Chem. Eng. Sci.*, **54**, 1431–1439.

Dixson, A.G. (1985) Thermal resistance models of packed-bed effective heat transfer parameters. *AIChE J.*, **31**(5), 826–834.

Dodds, R., Hudson, P.L., Kershenbaum, L. and Streat, M. (1973) The operation and modeling of a periodic countercurrent solid–liquid reactor. *Chem. Eng. Sci.*, **28**, 1233–1248.

Dünnebier, G., Fricke, J. and Klatt, K.-U. (2000) Optimal design and operation of simulated moving bed chromatographic reactors. *Ind. Eng. Chem. Res.*, **39**, 2290–2304.

Elsner, M.P., Dittrich, C. and Agar, D.W. (2002a) Adsorptive reactors for enhancing equilibrium gas-phase reactions two case studies. *Chem. Eng. Sci.*, **57**, 1607–1619.

Elsner, M.P., Menge, M. and Agar, D.W. (2002b) Assessment of adsorption enhanced reaction exemplified by the Claus process. *CAMURE Conference*, September 2002, Lausanne, Switzerland, (Preprint).

Elsner, M.P., Menge, M., Müller, C. and Agar, D.W. (2003) The Claus process: teaching an old dog new tricks. *Catal. Today*, **79–80**, 487–494.

Falle, S.A.E.G., Kallrath, J., Brockmüller, B., Schreieck, A., Giddings, J.R., Agar, D.W. and Watzenberger, O. (1995) The dynamics of reverse flow chromatographic reactors with side stream feed. *Chem. Eng. Commun.*, **135**, 185–211.

Falk, T. and Seidel-Morgenstern, A. (1999) Comparison between a fixed-bed reactor and a chromatographic reactor. *Chem. Eng. Sci.*, **54**, 1479–1485.

Falk, T. and Seidel-Morgenstern, A. (2002) Analysis of a discontinuously operated chromatographic reactor. *Chem. Eng. Sci.*, **57**, 1599–1606.

Fish, B.B. and R.W. Carr (1989) An experimental study of the countercurrent moving-bed chromatographic reactor. *Chem. Eng. Sci.*, **44**(9), 1773–1783.

Fish, B., Carr, R.W. and Aris, R. (1986) The continuous countercurrent moving bed chromatographic reactor. *Chem. Eng. Sci.*, **41**(4), 661–668.

Fish, B.B., Carr, R.W. and Aris, R. (1988) Computer aided experimentation in countercurrent reaction chromatography and simulated countercurrent chromatography. *Chem. Eng. Sci.*, **43**, 1867–1873.

Fogler, H.S. (1992) *Elements of Chemical Reaction Engineering*, 2nd edn. Prentice-Hall, New Jersey.

Fox, J.B., Jr., Calhoun, R.C. and Eglinton, W.J. (1969) *J. Chromatogr.*, **43**, 48.

Fredeslund, A., Gmehling, J. and Rasmussen, P. (1977) *Vapor–Liquid Equilibria using UNIFAC*. Elsevier Science Publishers, Amsterdam.

Freeman, A., Woodley, J.M. and Lilly, M.D. (1993) In situ product removal as a tool for bioprocessing. *Biotechnology*, **11**, 1007–1012.

Fricke, J. and Schmidt-Traub, H. (2003) A new method supporting the design of simulated moving bed chromatographic reactors. *Chem. Eng. Proc.*, **42**, 237–248.

Fricke, J., Meurer, M., Dreisörner, J. and Schmidt-Traub, H. (1999) Effect of process parameters on the performance of a simulated moving bed chromatographic reactor. *Chem. Eng. Sci.*, **54**, 1487–1492.

Froment, G.F. and Bischoff, K.B. (1990) *Chemical Reactor Analysis and Design*, 2nd edn. John Wiley & Sons, New York.

Garcia-Bordeje, E., Kapteijn, F. and Moulijn, J.A. (2002) Preparation and characterization of carbon coated monoliths for catalyst support. *Carbon*, **40**(7), 1079–1088.

Gaziev, G.A., Roginskii, S.Z. and Yanovskii, M.I. (1963) USSR Patent 149, 398.

Geelen, H. and Wijffels, J.B. (1964) The use of a distillation column as a chemical reactor. In *Proceedings of the Third Eur. Symp. Chem. Reac. Eng.*, Pergamon Press, Oxford, UK, pp. 125–131.

Gelosa, D., Ramaioli, M., Valente, G. and Morbidelli, M. (2003) Chromatographic reactors: esterification of glycerol with acetic acid using acidic polymeric resins. *Ind. Eng. Chem. Res.*, **42**, 6536–6544.

Gianetto, A. and Silveston, P.L. (1986) *Multiphase Chemical Reactors Theory, Design and Scale Up*. Hemisphere Press, Washington, DC.

Glueckauf, E. (1955) Theory of chromatography. Part 10 Formulae for diffusion into spheres and their application to chromatography. *Trans. Faraday Soc.*, **51**, 1540–1551.

Glueckauf, E. and Kitt, G.P. (1957) Gas chromatographic separation of hydrogen isotopes. In D.H. Desty (ed.), *Vapor Phase Chromatography*. Butterworths, London, pp. 422–427.

Gomes, V.G. and Fuller, O.M. (1996) Dynamics of propene metathesis: physisorption and diffusion in heterogeneous catalysis. *AIChE J.*, **42**(1), 204–213.

Gomes, V.G. and Yee, K.W.K. (2002) A periodic separating reactor for propene metathesis. *Chem. Eng. Sci.*, **57**, 3839–3850.

Gore, F.E. (1967) Performance of chromatographic reactors in cyclic operation. *Ind. Eng. Chem. Proc. Des. Dev.*, **6**(1), 10–22.

Gorissen, H.J. (2003) A general approach for the conceptual design of counter-current reactive separations. *Chem. Eng. Sci.*, **58**, 809–814.

Goto, S. and Matsubara, M. (1977) Extraction parametric pumping with reversible reaction. *Ind. Eng. Chem. Fundam.*, **16**, 193–200.

Goto, S., Tagawa, T. and Oomiya, T. (1993) Dehydrogenation of cyclohexane in a PSA reactor using hydrogen storage alloy. *Kagaku Kogaku Ronbunshu*, **19**(6), 978–983 (in Japanese).

Goto, S., Tagawa, T. and Ohkawara, H. (1995) Promotion effect of hydrogen storage alloys on the catalytic dehydroaromatization of *n*-hexane. *Sekiyu Gakkaishi*, **38**(3), 167–172 (in Japanese).

Goto, S., Tagawa, T. and Ohkawara, H. (1996) Promotive effect of hydrogen storage metals on catalytic dehydroaromatization of propane in a PSA type reactor. *Sekiyu Gakkaishi*, **39**(6), 389–394 (in Japanese).

Graham, W.R.C. and Lynch, D.T. (1990) CO oxidation on Pt: variable phasing of inputs during forced composition cycling. *AIChE J.*, **36**, 1796–1806.

Han, C. and Harrison, D.P. (1994) Simultaneous shift reaction and carbon dioxide separation for the direct production of hydrogen. *Chem. Eng. Sci.*, **49**, 5875–5882.

Han, C. and Harrison, D.P. (1997) Multicycle performance of a single step process for H_2 production. *Sep. Sci. Technol.*, **32**(1–4), 681–697.

Hanika, J., Kolena, J. and Smejkal, Q. (1999) Butylacetate via reactive distillation modeling and experiment. *Chem. Eng. Sci.*, **54**, 5205–5210.

Harmsen, G.J. and Chewter, L.A. (1999) Industrial applications of multi-functional, multiphase reactors. *Chem. Eng. Sci.*, **54**, 1541–1545.

Harrison, D.P. and Peng, Z.-Y. (2003) Low-carbon monoxide hydrogen by sorption-enhanced reaction. *Intern. J. Chem. React. Eng.*, **1**, A37.

Hart, J. and Thomas, W.J. (1991) Gas separation by pulsed pressure swing adsorption. *Gas. Sep. Purif.*, **5**, 125–133.

Hartman, M. and Coughlin, R. (1972) SO_2 oxidation in a trickle bed reactor packed with carbon. *Chem. Eng. Sci.*, **27**, 867–880.

Hashimoto, K., Adachi, S., Nojima, H. and Ueda, Y. (1983) A new process combining adsorption and enzyme reaction for producing higher-fructose syrup. *Biotechnol. Bioeng.*, **25**, 2371–2393.

Hattori, T. and Murakami, Y. (1968) Study on the pulse reaction technique I. Theoretical study. *J. Catal.*, **10**, 114–122.

Haure, P.M., Hudgins, R.R. and Silveston, P.L. (1989) Periodic operation of a trickle-bed reactor. *AIChE J.*, **35**, 1437–1444.

Herbsthofer, R. and Bart, H.-J. (2003) Influence of reaction kinetics on the performance of a chromatographic reactor. *Chem. Eng. Technol.*, **26**, 874–879.

Higler, A.P., Taylor, R. and Krishna, R. (1999) The influence of mass transfer and mixing on the performance of a tray column for reactive distillation. *Chem. Eng. Sci.*, **54**(13), 2873–2881.

Ho, C., Ching, C.B. and Ruthven, D.M. (1987) A comparative study of zeolite and resin adsorbents for the separation of fructose–glucose syrup. *Ind. Eng. Chem. Res.*, **26**, 1407–1412.

Hoffmann, U. and Sundmacher, K. (1997) Multifunktionale Reaktoren. *Chem.-Ing.-Tech.*, **69**, 613–622.

Hufton, J.R., Mayorga, S. and Sircar, S. (1999) Sorption-enhanced reaction process for hydrogen production. *AIChE J.*, **45**(2), 248–256.

Hugo, A.J., Jakelski, D.M., Stanitsas, G., Sullivan, G.R., Hudgins, R.R. and Silveston, P.L. (1986) Long-term transients in a reactor under forced concentration cycling. *Can. J. Chem. Eng.*, **64**, 349–351.

Imai, H., Tagawa, T. and Kuraishi, M. (1985) Acceleration effect of hydrogen storage alloys for the catalytic dehydrogenation of cyclohexane. *Mater. Res. Bull.*, **20**, 511–516.

Itoh, N. (1987) A membrane reactor using palladium. *AIChE J.*, **33**, 1576–1578.

Jain, E., Hudgins, R.R. and Silveston, P.L. (1985) Effectiveness factor under cyclic operation of reactor. *Can. J. Chem. Eng.*, **63**, 165–169.

Jaree, A., Budman, H., Hudgins, R.R., Silveston, P.L., Yakhnin, V. and Menzinger, M. (2001) Temperature excursions in packed bed reactors with an axial variation of catalyst activity. *Catal. Today*, **69**, 137–146.

Jensen, K.F. (2001) Microreaction engineering Is small better? *Chem. Eng. Sci.*, **56**, 293–303.

Jensen, T.B., Reijns, T.G.P., Billet, H.A.H. and van der Wielen, L.A.M. (2000) Novel simulated moving-bed method for reduced solvent consumption. *J. Chromatogr.* A, **873**, 149–163.

Jeong, Y.O. and Luss, D. (2003) Pollutant destruction in a reverse-flow chromatographic reactor. *Chem. Eng. Sci.*, **58**, 1095–1102.

Jones, R.L. and Keller, G.E. (1981) Pressure swing parametric pumping a new adsorption process. *J. Sep. Proc. Technol.*, **2**, 17–23.

Julbe, A., Farrusseng, D. and Guizzard, C. (2001) Porous ceramic membranes for catalytic reactors overview and new ideas. *J. Membr. Sci.*, **181**, 3–20.

Kallrath, J., Schreieck, A., Brockmüller, B., Agar, D.W., Watzenberger, O., Falle, S.A.E.G. and Giddings, J.R. (1994) Simulation of chromatographic reactors. *Comput. Chem. Eng.*, **18**, 331–335.

Kato, Y., Ando, K. and Yoshizawa, Y. (2003) Study on a regenerative fuel reformer for a zero-emission vehicle system. *J. Chem. Eng. Jpn.*, **36**(7), 860–866.

Kawase, M., Suzuki, T.B., Inoue, K., Yoshimoto, K. and Hashimoto, K. (1996) Increased esterification conversion by application of the simulated moving-bed reactor. *Chem. Eng. Sci.*, **51**, 2971–2976.

Kawase, M., Inoue, Y., Araki, T. and Hashimoto, K. (1999) The simulated moving-bed reactor for production of bisphenol A. *Catal. Today*, **48**, 199–209.

Kawase, M., Pilgrim, A., Araki, T. and Hashimoto, K. (2001) Lactosucrose production using a simulated moving bed reactor. *Chem. Eng. Sci.*, **56**, 453–458.

Kim, D.H. (1989) Linear driving force formulas for diffusion and reaction in porous catalysts. *AIChE J.*, **35**(2), 343–346.

Kim, S.-I., Aida, T. and Niiyama, H. (2005) Binary adsorption of very low concentration ethylene and water vapor on mordenites and desorption by microwave heating. *Sep. Purif. Technol.*, in press.

Kirkby, N.F. and Morgan, J.E.P. (1994) A theoretical investigation of pressure swing reaction. *Trans. I. Chem. E*, **72**(A), 541–550.

Klinkenberg, A. (1961) Chromatography of substances undergoing slow reversible chemical reaction. *Chem. Eng. Sci.*, **15**, 255–259.

Kloppenburg, E. and E.D. Gilles (1999) A new concept for operating simulated moving bed processes. *Chem. Eng. Technol.*, **22**(10), 813–817.

Kocirik, M. (1967) Linear non-equilibrium chromatography with a chemical reaction of the first order. *J. Chromatogr.*, **30**, 459–468.

Kodde, A.J. and Bliek, A. (1997) Selectivity enhancements in consecutive reactions using the pressure swing reactor. *Stud. Surf. Sci. Catal.*, **109**, 419–428.

Kodde, A.J., Fokma, Y.S. and Bliek, A. (2000) Selectivity effects on series reactions by reactant storage and PSA operation. *AIChE J.*, **46**(11), 2295–2304.

Kowler, D.E. and Kadlec, R.H. (1972) The optimal control of a periodic adsorber. I. Experiment. *AIChE J.*, **18**(6), 1207–1219.

Krishna, R. (2002) Reactive separations: more ways to skin a cat. *Chem. Eng. Sci.*, **57**, 1491–1504.

Kruglov, A.V. (1994) Methanol synthesis in a simulated countercurrent moving-bed adsorptive catalytic reactor. *Chem. Eng. Sci.*, **49**(24A), 4699–4716.

Kruglov, A.V., Bjorklund, M.C. and Carr, R.W. (1996) Optimization of the simulated countercurrent moving-bed chromatographic reactor for the oxidative coupling of methane. *Chem. Eng. Sci.*, **51**, 2945–2950.

Kuczynski, M., van Doteghem, A. and Westerterp, K.R. (1986) Methanol adsorption by amorphous silica alumina in the critical temperature range. *Colloid Polym. Sci.*, **264**, 362–367.

Kuczynski, M., Browne, W.I., Fontein, H.J. and Westerterp, K.R. (1987a) Reaction kinetics for the synthesis of methanol from CO and H_2 on a copper catalyst. *Chem. Eng. Proc.*, **21**(4), 179–191.

Kuczynski, M., Oyeavaar, M.H., Pieters, R.T. and Westerterp, K.R. (1987b) Methanol synthesis in a countercurrent gas–solid–solid trickle flow reactor. An experimental study. *Chem. Eng. Sci.*, **42**(8), 1887–1896.

Kulprathipanja, S. (2001) *Reactive Separation Processes.* Taylor & Francis, New York.

Langer, S.H. and Patton, P.D. (1974) Chemical reactor application of the gas chromatographic column. In H. Purnell (ed.), *New Developments in Gas Chromatography.* John Wiley & Sons, New York, pp. 294–373.

Lax, P.D. (1957) Hyperboric systems of conservation laws II. *Commun. Pure Appl. Math.*, **10**, 537–566.

Lee, C.J., Yeh, H.J., Wang, W.J. and Kan, C.R. (1993) Extraction separation of penicillin G by facilitated transport via carrier supported liquid membranes. *Biotechnol. Bioeng.*, **42**, 527–534.

Lee, I.D. and Kadlec, R.H. (1989) Effects of adsorbent and catalyst distributions in pressure swing reactors. *AIChE Symp. Ser.*, **84**(264) 167–176.

Lee, J.-K., Hudgins, R.R. and Silveston, P.L. (1995) A cycled trickle bed reactor for SO_2 oxidation. *Chem. Eng. Sci.*, **50**(16), 2523–2530.

Lee, J.-K., Ferrero, S., Hudgins, R.R. and Silveston, P.L. (1996) Catalytic SO_2 oxidation in a periodically operated trickle-bed reactor. *Can. J. Chem. Eng.*, **74**, 706–712.

Lee, Y.Y., Fratzke, A.R., Wun, K. and Tsao, G.T. (1976) Glucose isomerase immobilized on porous glass. *Biotechnol. Bioeng.*, **18**, 389–413.

Liapis, A.I. and Rippin, D.W.T. (1979) Simulation of binary adsorption in continuous counter-current operation in comparison with other operating modes. *AIChE J.*, **25**, 455–460.

Liaw, J.-S. and Bailey, J.E. (1974) Regulation of lumped periodic processes. *AIChE J.*, **20**, 924–932.

Liden, G. and Vamling, L. (1989) Periodic operation of a tubular reactor: a simulation study of consecutive reactions in a chromatographic reactor. *Chem. Eng. J.*, **40**, 31–37.

Lode, F., Houmard, M., Migliorini, C., Mazzotti, M. and Morbidelli, M. (2001) Continuous reactive chromatography. *Chem. Eng. Sci.*, **56**(2), 269–291.

Lode, F., Mazzotti, M. and Morbidelli, M. (2003a) Comparing true countercurrent and simulated moving-bed chromatographic reactors. *AIChE J.*, **49**(4), 977–990.

Lode, F., Francesconi, G., Mazzotti, M. and Morbidelli, M. (2003b) Synthesis of methylacetate in a simulated moving-bed reactor: experiment and modeling. *AIChE J.*, **49**(6), 1516–1524.

Lopez Ortiz, A. and Harrison, D.P. (2001) Hydrogen production using sorption-enhanced reaction. *Ind. Eng. Chem. Res.*, **40**(23), 5102–5109.

Lu, Z.P. and Rodrigues, A.E. (1994) Pressure swing adsorption reactors: simulation of three-step, one-bed process. *AIChE J.*, **40**(7), 1118–1137.

Machocki, A. (1996) Methane oxidative coupling in an undiluted reaction mixture in a reactor–adsorber system with gas recirculation. *Appl. Catal.* A, **146**, 391–400.

Magee, E.M. (1961) Can. Pat. 631 882.

Magee, E.M. (1963) The course of a reaction in a chromatographic column. *Ind. Eng. Chem. Fundam.*, **2**(1), 32–36.

Mata, A.R. and Smith, J.M. (1981) Oxidation of sulfur dioxide in a trickle bed reactor. *Chem. Eng. J.*, **22**, 229–235.

Matros, Yu.Sh. (1985) *Unsteady Processes in Catalytic Reactors*, Elsevier, Amsterdam, The Netherlands.

Matsen, J.M., Harding, J.W. and Magee, E.M. (1965) Chemical reactions in chromatographic columns. *J. Phys. Chem.*, **69**(2), 522–527.

Matsuda, T., Koike, I., Kubo, N. and Kikuchi, E. (1993) Dehydrogenation of isobutene in a palladium membrane reactor. *Appl. Catal.*, **96**, 3–13.

Mazzotti, M., Kruglov, A., Neri, B., Gelosa, D. and Morbidelli, M. (1996) A continuous chromatographic reactor, SMBR. *Chem. Eng. Sci.*, **51**, 1827–1836.

Mazzotti, M., Neri, B., Gelosa, D. and Morbidelli, M. (1997a) Dynamics of a chromatographic reactor: esterification catalyzed by acidic resins. *Ind. Eng. Chem. Res.*, **36**, 3163–3172.

Mazzotti, M., Storti, G. and Morbidelli, M. (1997b) Supercritical fluid simulated moving bed chromatography. *J. Chromatogr.* A, **786**, 309–320.

Mazzotti, M., Storti, G. and Morbidelli, M. (1997c) Optimal operation of simulated moving bed units for nonlinear chromatographic separations. *J. Chromatogr.* A, **769**, 3–24.

Menge, M., Elsner, M.P. and Agar, D.W. (2003) Adsorption enhanced Claus reaction: an old process in a new guise. *4th International Conference on Unsteady-State Processes in Catalysis*, October 26–29, 2003, Montreal, Canada.

Mensah, P. and Carta, G. (1999) Adsorptive control of water in esterification with immobilized enzymes. Continuous operation in a periodic counter-current reactor. *Biotechnol. Bioeng.*, **66**(3), 137–146.

Mensah, P., Gainer, J.L. and Carta, G. (1998a) Adsorptive control of water in esterification with immobilized enzymes. I. Batch reactor behavior. *Biotechnol. Bioeng.*, **60**, 434–444.

Mensah, P., Gainer, J.L. and Carta, G. (1998b) Adsorptive control of water in esterification with immobilized enzymes. II. Fixed bed reactor behavior. *Biotechnol. Bioeng.*, **60**, 445–453.

Metzinger, J., Kühter, A., Silveston, P.L. and Gangwal, S.K. (1994) A novel periodic reactor for scrubbing SO_2 from industrial stack gases. *Chem. Eng. Sci.*, **49**(24A), 4533–4546.

Meurer, M., Altenhöner, U., Strube, J., Untiedt, A. and Schmidt-Traub, H. (1996) Dynamic simulation of a simulated moving bed chromatographic reactor for the inversion of sucrose. *Starch/Stärke*, **48**, 452–457.

Migliorini, C., Fillinger, M., Mazzotti, M. and Morbidelli, M. (1999a) Analysis of simulated moving-bed reactors. *Chem. Eng. Sci.*, **54**, 2475–2480.

Migliorini, C., Mazzotti, M. and Morbidelli, M. (1999b) Simulated moving-bed units with extra-column dead volume. *AIChE J.*, **45**(7), 1411–1421.

Migliorini, C., Meissner, J.P., Mazzotti, M. and Carta, G. (2000) Regioselective enzymatic diol esterification in batch and fixed bed adsorptive reactors. Experiments and modeling. *Biotechnol. Progr.*, **16**, 600–609.

Minotti, M., Doherty, M.F. and Malone, M.F. (1998) Design for simultaneous reaction and liquid–liquid extraction. *Ind. Eng. Chem. Res.*, **37**, 4748–4755.

Murakami, Y., Hattori, T. and Hattori, T. (1968) Study on the pulse reaction technique. II. Dealkylation and disproportionation of cumene. *J. Catal.*, **10**, 123–127.

Na-Ranong, D., Aida, T. and Niiyama, H. (2002) Kinetic study on NO–CO reaction over Pt/Al_2O_3 under unsteady state. *Appl. Catal. A: Gen.*, **234**, 103–110.

Nicoud, R.-M., Perrut, M. and Hotier, G. (1995) US Pat. 5 422 007.

Nieken, U., Kolios, G. and Eigenberger, G. (1994) Control of the ignited steady state in autothermal fixed-bed reactors for catalytic combustion. *Chem. Eng. Sci.*, **49**(24B), 5507–5518.

Nijhuis, T.A., Beers, A.W., Vergunst, T., Hoek, I., Kapteijn, F. and Moulijn, J.A. (2001) Preparation of monolithic catalysts. *Catal. Rev. Sci. Eng.*, **43**(4), 345–380.

Obenaus, F. and Droste, W. (1980) Hüls process: methyl tertiary butyl ether. *Erdöl Kohle. Erdgas. Petrochem. Brennstoff Chem.*, **33**, 271–275.

Özgülşen, F., Kendra, S.J. and Çinar, A. (1993) Nonlinear predictive control of periodically forced chemical reactors. *AIChE J.*, **39**, 589–598.

Pai, R.A., Doherty, M.F. and Malone, M.F. (2002) Design of reactive extraction systems for bioproduct recovery. *AIChE J.*, **48**(3), 514–526.

Petroulas, T., Aris, R. and Carr, R.W. (1985a) Analysis of the counter-current moving-bed chromatographic reactor. *Comput. Maths. Appl.*, **11**(1–3), 5–34.

Petroulas, T., Aris, R. and Carr, Jr., R.W. (1985b) Analysis and performance of a countercurrent moving-bed chromatographic reactor. *Chem. Eng. Sci.*, **40**(12), 2233–2240.

Piret, E.L., Penney, W.H. and Trambouze, P.J. (1960) Extractive reaction: batch and continuous-flow chemical reaction systems, dilute case. *AIChE J.*, **6**(3), 394–400.

Pritchard, C.L. and Simpson, G.K. (1986) Design of an oxygen concentrator using pressure-swing adsorption principle. *Chem. Eng. Chem. Res. Des.*, **64**, 467–471.

Rambeau, G. and Amariglio, H. (1981) Improvement in the catalytic performance of ruthenium powder in ammonia synthesis with the use of cyclic procedure. *Appl. Catal.*, **1**, 291–302.

Ray, A.K. and Carr, R.W. (1995a) Experimental study of a laboratory-scale simulated countercurrent moving-bed chromatographic reactor. *Chem. Eng. Sci.*, **50**(14), 2195–2202.

Ray, A.K. and Carr, R.W. (1995b) Numerical simulation of a simulated countercurrent moving-bed chromatographic reactor. *Chem. Eng. Sci.*, **50**(19), 3033–3041.

Ray, A.K., Tonkovich, A.L., Aris, R. and Carr, R.W. (1990) The simulated countercurrent moving bed chromatographic reactor. *Chem. Eng. Sci.*, **45**(8), 2431–2437.

Ray, A.K., Carr, R.W. and Aris, R. (1994) The simulated countercurrent moving bed chromatographic reactor: a novel reactor–separator. *Chem. Eng. Sci.*, **49**(4), 469–480.

Řeháček, J., Kubíček, M. and Marek, M. (1998) Periodic, quasiperiodic and chaotic spatiotemporal patterns in a tubular catalytic reactor with periodic flow reversal. *Comput. Chem. Eng.*, **22**(1–2), 283–297.

Reid, R.C., Prausnitz, J.M. and Sherwood, T.K. (1986) *The Properties of Gases and Liquids*, 4th edn. McGraw-Hill, New York.

Reschke, M. and Schügerl, K. (1984a) Reactive extraction of penicillin 1. Stability of penicillin G in the presence of carriers and relationships for the distribution coefficient and degrees of extraction. *Chem. Eng. J.*, **28**, B1–B10.

Reschke, M. and Schügerl, K. (1984b) Reactive extraction of penicillin 2. Distribution coefficient and degrees of extraction. *Chem. Eng. J.*, **28**, B11–B24.

Reschke, M. and Schügerl, K. (1984c) Reactive extraction of penicillin 3. Kinetics. *Chem. Eng. J. Biochem. Eng. J.*, **29**, B25–B29.

Richardson, J.F. and Zaki, W.N. (1954) Sedimentation and fluidisation: I. *Trans. Inst. Chem. Engrs.*, **32**, 35.

Roginskii, S.Z. and Rozental, A.L. (1962) Chemical reactions in a chromatographic regime. *Dokl. Akad. Nauk SSSR*, **146**, 152–155 (in Russian).

Roginskii, S.Z. and Rozental, A.L. (1964) Kinetics for reactions under chromatographic conditions on heterogeneous surfaces. *Kinetika i Kataliz*, **5**(1), 104–112 (in Russian).

Roginskii, S.Z., Yanovskii, M.I. and Gaziev, G.A. (1961) Chemical reactions under chromatographic conditions. *Dokl. Akad. Nauk SSSR*, **140**, 1125–1127 (in Russian).

Roginskii, S.Z., Yanovskii, M.I. and Gaziev, G.A. (1962) Catalytic reactions and catalysts under chromatographic conditions. *Kinetika i Kataliz*, **3**, 529–540 (English translation (1961), 041 109 500, I40, 771–785).

Ruthven, D.M. (1984) *Principles of Adsorption and Adsorption Processes.* John Wiley & Sons, New York.

Ruthven, D.M. and Ching, C.B. (1989) Counter-current and simulated counter-current adsorption separation processes. *Chem. Eng. Sci.*, **44**(5), 1011–1038.

Sad, M.R., Querini, C.A., Comelli, R.A., Figoli, N.S. and Parera, J.M. (1996) Dehydroisomerization of *n*-butane in a chromatographic reactor. *Appl. Catal.* A, **146**, 131–143.

Sadhankar, R.R. and Lynch, D.T. (1996) Transient and resonant behavior for NO reduction by CO over a Pt/Al_2O_3 catalyst during forced composition cycling. *Can. J. Chem. Eng.*, **74**, 674–682.

Saito, H., Murakami, Y. and Hattori, T. (1965) Conversion and product distribution in pulse reactor. *Kagaku Kogaku* (*Chem. Eng. Jpn.*), **29**, 585–590.

Salden, A. and Eigenberger, G. (2001) Multifunctional adsorber/reactor concept for waste-air purification. *Chem. Eng. Sci.*, **56**, 1605–1611.

Sanfillippo, D., Lupieri, M. and Ancillotti, F. (1992) European Patent EP 0 470 6651.

Saracco, G., Neomagos, H.W.J.P., Versteeg, G.F. and van Swaaij, W.P.M. (1999) High temperature membrane reactors: potential and problems. *Chem. Eng. Sci.*, **54**, 1997–2017.

Sardin M. and Villermaux, J. (1979) Estérification catalysée par une résine échangeuse de cation dans un réacteur chromatographique. *Nouv. J. Chim.*, **3**(4), 255–261 (in French).

Sarmidi, M.R. and Barker, P. (1993a) Saccharification of modified starch to maltose in a continuous rotating annular chromatograph (CRAC). *J. Chem. Tech. Biotechnol.*, **57**(3), 229–235.

Sarmidi, M.R. and Barker, P. (1993b) Simultaneous biochemical reaction and separation in a rotating annular chromatograph. *Chem. Eng. Sci.*, **48**(14), 2615–2623.

Satrio, J.A., Shanks, B.H. and Wheelock, T.D. (2005) Development of a novel combined catalyst and sorbent for hydrocarbon reforming. *Ind. Eng. Chem. Res.*, **44**, 3901–3911.

Schramm, H., Kaspereit, M., Kienle, A. and Seidel-Morgenstern, A. (2002) Improving simulated moving bed processes by cyclic modulation of the feed concentration. *Chem. Eng. Technol.*, **25**(12), 1151–1155.

Schramm, H., Grüner, S., Kienle, A. (2003a) Optimal operation of a simulated moving bed chromatographic process by means of simple feedback control. *J. Chromatogr.* A, **1006**, 3–13.

Schramm, H., Kienle, A., Kaspereit, M. and Seidel-Morgenstern, A. (2003b) Simulated moving bed processes with cyclic modulation of the feed concentration. *J. Chromatogr.* A, **1006**, 77–86.

Schramm, H., Kaspereit, M., Kienle, A. and Seidel-Morgenstern, A. (2003c) Improved operation of simulated moving bed processes through cyclic modulation of feed flow and feed concentration. *Chem. Eng. Sci.*, **58**, 5217–5227.

Schügerl, K., Hänsel, R., Schlichting, E. and Halwachs, W. (1988) Reactive extractions. *Int. Chem. Eng.*, **28**(3), 393–405.

Schweich, D. and Villermaux, J. (1978) The chromatographic reactor. A new theoretical approach. *Ind. Eng. Chem. Fundam.*, **17**(1), 1–7.

Schweich, D. and Villermaux, J. (1982a) The preparative chromatographic reactor revisited. *Chem. Eng. J.*, **24**, 99–109.

Schweich, D. and Villermaux, J. (1982b) Model for catalytic dehydrogenation of cyclohexane in a chromatographic reactor: comparison of theory and experience. *Ind. Eng. Chem. Fundam.*, **21**, 47–51.

Schweich, D. and Villermaux, J. (1982c) Evidence for a transient promoting catalytic effect in a chromatographic reactor. *Ind. Eng. Chem. Fundam.*, **21**(1), 51–55.

Schweich, D., Villermaux, J. and Sardin, M. (1980) An introduction to the nonlinear theory of adsorptive reactors. *AIChE J.*, **26**(3), 477–486.

Scott, C.D., Spence, R.D. and Sisson, W.G. (1976) Pressurised annular chromatograph for continuous separations. *J. Chromatogr.*, **126**, 381–400.

Seidlitz, F., Mathieu, C., Breysse, J. and Houzelot, J.L. (2001) Simultaneous synthesis and separation of a product by cooling crystallization in a multifunctional reactor. *In Proceedings of the Second International Conference on Multifunctional Reactors*, June 26–28, 2001, Nuremberg.

Semenenko, E.I., Roginskii, S.Z. and Yanovskii, M.I. (1964) Catalytic pulsed dehydrogenation of *n*-butene. *Kinetika i Kataliz*, **5**, 490–495 (in Russian).

Sheikh, J., Kershenbaum, L.S. and Alpay, E. (1998) Analytical basis for separation enhanced reaction in continuous flow processes. *Chem. Eng. Sci.*, **53**(16), 2933–2939.

Sheikh, J., Kershenbaum, L.S. and Alpay, E. (2001) 1-Butene dehydrogenation in rapid pressure swing reaction processes. *Chem. Eng. Sci.*, **56**, 1511–1516.

Sheldon, R. and Stringaro, J.-P. (1992) Eur. Pat. EP 0 396 650.

Shi, C., Sun, Q., Hu, H., Herman, R.G., Klier, K. and Wachs, I.E. (1996) Direct conversion of methane to methanol and formaldehyde over a double-layered catalyst bed in the presence of steam. *Chem. Commun.*, (5), 663–664.

Shieh, M.T. and Barker, P.E. (1995) Saccharification of modified starch to maltose in a semi-continuous counter-current chromatographic reactor–separator (SCCR-S). *J. Chem. Tech. Biotechnol.*, **63**, 125–134.

Shieh, M.T. and Barker, P.E. (1996) Combined bioreaction and separation in a simulated countercurrent chromatographic bioreactor–separator for the hydrolysis of lactose. *J. Chem. Tech. Biotechnol.*, **66**, 265–278.

Shinji, O., Misono, M. and Yoneda, Y. (1982) The dehydrogenation of cyclohexane by the use of a porous glass reactor. *Bull. Chem. Soc. Jpn.*, **55**, 2760–2764.

Siirola, J.J. (1996) Industrial application of chemical process synthesis, *Adv. Chem. Eng.*, **23**, 1–62.

Silaban, A. and Harrison, D.P. (1995) High temperature capture of carbon dioxide: characteristics of the reversible reaction between CaO(s) and CO_2(g). *Chem. Eng. Commun.*, **137**, 177–190.

Silaban, A., Narcida, M. and Harrison, D.P. (1996) Characteristics of the reversible reaction between CO_2(g) and calcined dolomite. *Chem. Eng. Commun.*, **146**, 149–162.

Silva, V.M.T.M. and Rodrigues, A.E. (2002) Dynamics of a fixed-bed adsorptive reactor for synthesis of diethylacetal. *AIChE J.*, **48**(3), 625–634.

Silveston, P.L. (1998) *Composition Modulation of Catalytic Reactors.* Gordon and Breach Scientific Publishers, Lausanne, Switzerland.

Silveston, P.L. and Hanika, J. (2004) Periodic operation of three-phase catalytic reactions a review. *Can. J. Chem. Eng.*, **82**, 1105–1142.

Sneesby, M.G., Tadé, M.D., Datta, R. and Smith, T.N. (1999) ETBE synthesis via reactive distillation I. Steady state simulation and design aspects. *Ind. Eng. Chem. Res.*, **36**, 1855–1869.

Snyder, J.D. and Subramanian, B. (1998) Numerical simulation of a reverse-flow NO_x–SCR reactor with side-stream ammonia addition. *Chem. Eng. Sci.*, **53**, 727–734.

Štěpánek, F., Kubíček, M., Marek, M. and Adler, P.M. (1999) Optimal design and operation of a separating microreactor. *Chem. Eng. Sci.*, **54**, 1493–1498.

Storti, G., Masi, M., Paludetto, R., Morbidelli, M. and Carrá, S. (1988) Adsorption separation processes: countercurrent and simulated countercurrent operations. *Comp. Chem. Eng.*, **12**, 475–482.

Storti, G., Mazzotti, M., Morbidelli, M. and Carrá, S. (1993) Robust design of binary countercurrent adsorption separation processes. *AIChE J.*, **39**(3), 471–492.

Struijk, J. (1992) Catalytic routes to cyclohexane. *PhD Thesis*, Dept. of Chem. Eng., Techn. University of Delft, The Netherlands.

Sun, Y.-M. and Khang, S.-J. (1988) Catalytic membrane for simultaneous chemical reaction and separation applied to a dehydrogenation reaction. *Ind. Eng. Chem. Res.*, **27**, 1136–1142.

Sundmacher, K. and Hoffmann, U. (1996) Development of a new catalytic distillation process for fuel ethers via a detailed non-equilibrium model. *Chem. Eng. Sci.*, **51**(10), 2359–2368.

Sundmacher, K., Uhde, G. and Hoffmann, U. (1999) Multiple reactions in catalytic distillation processes for the production of the fueloxygenate MTBE and TAME Analysis by rigorous model and experimental verification. *Chem. Eng. Sci.*, **54**(13), 2839–2848.

Suzuki, M. (1990) *Adsorption Engineering*. Kodansha Ltd., Tokyo.

Takeuchi, K. and Uraguchi, Y. (1976a) Separation conditions of the reactant and the product with a chromatographic moving bed reactor. *J. Chem. Eng. Jpn.*, **9**(2), 164–166.

Takeuchi, K. and Uraguchi, Y. (1976b) Basic design of chromatographic moving-bed reactors for product refining. *J. Chem. Eng. Jpn.*, **9**(3), 246–248.

Takeuchi, K. and Uraguchi, Y. (1977a) The effect of the exhausting section on the performance of a chromatographic moving-bed reactor. *J. Chem. Eng. Jpn.*, **10**(1), 72–74.

Takeuchi, K. and Uraguchi, Y. (1977b) Experimental studies of a chromatographic moving-bed reactor catalytic oxidation of carbon monoxide on activated alumina as a model reaction. *J. Chem. Eng. Jpn.*, **10**(6), 455–460.

Takeuchi, K., Miyauchi, T. and Uraguchi, Y. (1978) Computational studies of a chromatographic moving-bed reactor for consecutive and reversible reactions. *J. Chem. Eng. Jpn.*, **11**(3), 216–220.

Tanaka, K., Yoshikawa, R., Ying, C., Kita, H. and Okamoto, K. (2002) Application of zeolite T membrane to vapor-permeation-aided esterification of lactic acid with ethanol. *Chem. Eng. Sci.*, **57**, 1577–1584.

Thomas, C.O. and Smith, H.A. (1959) Gas chromatography with hydrogen and deuterium. *J. Phys. Chem.*, **63**, 427–432.

Tonkovich, A.L.Y. and Carr, R.W. (1994a) A simulated countercurrent moving-bed chromatographic reactor for the oxidative coupling of methane: experimental results. *Chem. Eng. Sci.*, **49**(24A), 4647–4656.

Tonkovich, A.L.Y. and Carr, R.W. (1994b) Modeling of the simulated countercurrent moving-bed chromatographic reactor used for the oxidative coupling of methane. *Chem. Eng. Sci.*, **49**(24A), 4657–4665.

Tonkovich, A.L., Carr, R.W. and Aris, R. (1993) Enhanced C_2 yields from methane oxidative coupling by means of a separative chemical reactor. *Science*, **262**, 221–223.

Trambouze, P.J. and Piret, E.L. (1960) Extractive reaction: Batch- and continuous-flow reaction systems, concentrated case. *AIChE J.*, **6**(4), 574–578.

Tuchlenski, A., Beckmann, A., Reusch, D., Düssel, R., Weidlich, U. and Janowsky, R. (2001) Reactive distillation Industrial applications, process design & scale-up. *Chem. Eng. Sci.*, **56**, 387–394.

Turnock, P.H. and Kadlec, R.H. (1971) Separation of nitrogen and methane via periodic adsorption. *AIChE J.*, **17**, 335–342.

Unger, B.D. and Rinker, R.G. (1976) Ammonia synthesis reaction in the chromatographic column. *Ind. Eng. Chem. Fundam.*, **15**, 225–227.

van den Bussche, K.M. and Froment, G.F. (1996) The STAR configuration for methanol synthesis in reversed flow reactors. *Can. J. Chem. Eng.*, **74**(5), 729–734.

van der Wielen, L.A.M., Potters, J.J.M., Straathof, A.J.J. and Luyben, K.Ch.A.M (1990) Integration of bioconversion and continuous product separation by means of countercurrent adsorption. *Chem. Eng. Sci.*, **45**(8), 2397–2404.

van der Wielen, L.A.M., Diepen, P.J., Houwers, J. and Luyben, K.Ch.A.M. (1996) A countercurrent adsorptive reactor for acidifying bioconversions. *Chem. Eng. Sci.*, **51**, 2315–2325.

van der Wielen, L.A.M., Sjauw-Koen-Fa, A.W.K.G., Potters, J.J.M. and Luyben, K.Ch.A.M. (1997a) Transient behavior of pulsed particulate fluidized beds. *AIChE J.*, **43**(3), 625–630.

van der Wielen, L.A.M., van Dam, M.H.H. and Luyben, K.Ch.A.M. (1997b) Trickle flow of dense particles through a fluidized bed of others. *Chem. Eng. Sci.*, **52**(4), 553–565.

van der Wielen, L.A.M., van Nunen, C.A.P.M. and Luyben, K.Ch.A.M. (1998) Pulsed adsorbent transport in a countercurrent adsorptive reactor. *AIChE J.*, **44**, 83–92.

van Noorden, T.L., Verduyn Lunel, S.M. and Bliek, A. (2002) Acceleration of the determination of periodic states of cyclically operated reactors and separators. *Chem. Eng. Sci.*, **57**, 1041–1055.

van Noorden, T.L., Verduyn Lunel, S.M. and Bliek, A. (2003a) Optimization of cyclically operated reactors and separators. *Chem. Eng. Sci.*, **58**, 4115–4127.

van Noorden, T.L., Verduyn Lunel, S.M. and Bliek, A. (2003b) The efficient computation of periodic states of cyclically operated chemical processes. *IMA J. Appl. Math.*, **68**, 149–166.

Vaporciyan, G.G. and Kadlec, R.H. (1987) Equilibrium-limited periodic separating reactors. *AIChE J.*, **33**(8), 1334–1343.

Vaporciyan, G.G. and Kadlec, R.H. (1989) Periodic separating reactors: experiments and theory. *AIChE J.*, **35**(5), 831–844.

Velardi, S.A. and Barresi, A.A. (2002) Methanol synthesis in a forced unsteady-state reactor network. *Chem. Eng. Sci.*, **57**, 2995–3004.

Viswanathan, S. and Aris, R. (1974) Countercurrent moving bed chromatographic reactors. *ACS Symp. Ser.*, **133**, 191–204.

Vos, H.J. (1990) Design of a continuous reactor for immobilized biocatalysis. *PhD Thesis*, Delft University of Technology, The Netherlands.

Vos, H.J., Groen, D.J., Potters, J.J.M. and Luyben, K.Ch.A.M. (1990a) Countercurrent multistage fluidized bed reactor for immobilized biocatalysis: I. Modeling and simulation. *Biotechnol. Bioeng.*, **36**, 367–378.

Vos, H.J., Zomerdijk, M., Groen, D.J. and Luyben, K.Ch.A.M. (1990b) Countercurrent multistage fluidized bed reactor for immobilized biocatalysis: II. Operation of a laboratory-scale reactor. *Biotechnol. Bioeng.*, **36**, 377–386.

Vos, H.J., van Houwelingen, C., Zomerdijk, M. and Luyben, K.Ch.A.M. (1990c) Countercurrent multistage fluidized bed reactor for immobilized biocatalysis: III. Hydrodynamic aspects. *Biotechnol. Bioeng.*, **36**, 387–396.

Waldron, W.E., Hufton, J.R. and Sircar, S. (2001) Production of hydrogen by cyclic sorption enhanced reaction process. *AIChE J.*, **47**, 1477–1479.

Wankat, P.C. (1974) Cyclic separation processes. *Sep. Sci.*, **9**(2), 85–116.

Wankat, P.C. (1986) *Large Scale Adsorption and Chromatography.* CRC Press, Boca Raton, FL.

Wardwell, A.W., Carr, Jr., R.W. and Aris, R. (1982) Continuous reaction gas chromatography: the dehydrogenation of cyclohexane over $Pt/\gamma\text{-}Al_2O_3$. In J. Wei and C. Georgakis (eds.), Proceedings *Chemical Reaction Engineering*, Boston. *ACS Symp. Ser.*, **196**, 297–306.

Wasewar, K.L., Pangarkar, V.G., Heesink, A.B.M. and Versteeg, G.F. (2003) Intensification of enzymatic conversion of glucose to lactic acid by reactive extraction. *Chem. Eng. Sci.*, **58**, 3385–3393.

Westerterp, K.R. (1992) Multifunctional reactors. *Chem. Eng. Sci.* **47**(9–11), 2195–2206.

Westerterp, K.R. and Kuczynski, M. (1987a) A model for a counter-current gas–solid–solid trickle flow reactor for equilibrium reactions. The methanol synthesis. *Chem. Eng. Sci.*, **42**, 1871–1887.

Westerterp, K.R. and Kuczynski, M. (1987b) Gas–solid trickle flow hydrodynamics in a packed column. *Chem. Eng. Sci.*, **42**(7), 1539–1551.

Westerterp, K.R, Kuczynski, M. and Kamphuis, C. (1989) Synthesis of methanol in a reactor system with interstage product removal. *Ind. Eng. Chem. Res.*, **28**, 763–771.

Westerterp, K.R., van Swaaij, W.P.M. and Beenackers, A.A.C.M. (1993) *Chemical Reactor Design and Operation.* John Wiley & Sons, New York.

Wetherold, R.G., Wissler, E.H. and Bischoff, K.B. (1974) An experimental and computational study of the hydrolysis of methyl formate in a chromatographic reactor. *Adv. Chem. Ser.*, **133**, 181–190.

Wu, J.-Y. (1998) Quick analysis of enantioselectivity by a chromatographic reactor. *AIChE J.*, **44**, 474–476.

Wu, J.-Y. and Liu, S.-W. (1999) Quick analysis of enantioselectivity by a chromatographic reactor: II. Experimental verification. *J. Chem. Technol. Biotechnol.*, **74**, 974–978.

Xiu, G.-H., Li, P. and Rodrigues, A.E. (2002a) Sorption-enhanced reaction process with reactive regeneration. *Chem. Eng. Sci.*, **57**, 3893–3908.

Xiu, G.-H., Soares, J.L., Li, P. and Rodrigues, A.E. (2002b) Simulation of five-step, one-bed sorption-enhanced reaction process. *AIChE J.*, **48**(12), 2817–2832.

Xiu, G., Li, P. and Rodrigues, A.E. (2003a) Adsorption-enhanced steam methane reforming with intraparticle-diffusion limitations. *Chem. Eng. J.*, **95**, 83–93.

Xiu, G.-H., Li, P. and Rodrigues, A.E. (2003b) New generalized strategy for improving sorption-enhanced reaction process. *Chem. Eng. Sci.*, **58**, 3425–3437.

Xu, J. and Froment, G.F. (1989) Methane steam reforming, methanation and water-gas shift, I. Intrinsic kinetics. *AIChE J.*, **35**, 89–96.

Yabannavar, V.M. and Wang, D.I.C. (1991) Extractive fermentation for lactic acid production. *Biotechnol. Bioeng.*, **37**, 1095–1103.

Yagi, S. and Kunii, D. (1964) Studies on heat transfer near wall surfaces in packed beds. *AIChE J.*, **6**, 97–103.

Yamada, H. and Goto, S. (1997) Periodic operation of trickle bed reactor for hydrogenolysis in gas–liquid–liquid–solid four phases. *J. Chem. Eng. Jpn.*, **30**(3), 478–483.

Yang, C. and Cussler, E.L. (2000) Reactive extraction of penicillin G in hollow-fiber and hollow-fiber fabric modules. *Biotechnol. Bioeng.*, **69**(1), 66–73.

Yang, R.T. (1987) *Gas Separation by Adsorption Processes.* Butterworths Scientific Publishers, Boston, MA.

Yongsunthon, I. and Alpay, E. (1998a) Conversion–temperature trajectories for well-mixed adsorptive reactors. *Chem. Eng. Sci.*, **53**(4), 691–696.

Yongsunthon, I. and Alpay, E. (1998b) Design and optimization of temperature cycled adsorptive reactors. *Comput. Chem. Eng.*, **22**, S733–S736.

Yongsunthon, I. and Alpay, E. (1999) Design of periodic adsorptive reactors for the optimal integration of reaction, separation and heat exchange. *Chem. Eng. Sci.*, **54**, 2647–2657.

Yongsunthon, I. and Alpay, E. (2000) Total connectivity models for adsorptive reactor design. *Chem. Eng. Sci.*, **55**, 5643–5656.

Yu, S., Zhou, A. and Tan, Q. (1997) Simulation of multistage catalytic stripping with a non-equilibrium stage model. *Comp. Chem. Eng.*, **21**, 409–415.

Zabransky, R. and Anderson, R. (1977) Simulated moving bed alkylation process. US Pat. 4 008 291.

Zafar, I. and Barker, P.E. (1988) An experimental and computational study of a biochemical polymerization reaction in a chromatographic reactor separator. *Chem. Eng. Sci.*, **43**(9), 2369–2375.

Zang, Y. and Wankat, P. (2002a) SMB operation strategy partial feed. *Ind. Eng. Chem. Res.*, **41**(10), 2504–2511.

Zang, Y. and Wankat, P. (2002b) Three zone simulated moving bed with partial feed and selective withdrawal. *Ind. Eng. Chem. Res.*, **41**(21), 5283–5289.

Zhang, Z., Hidajat, K. and Ray, A.K. (2001) Application of simulated countercurrent moving bed chromatographic reactor for MTBE synthesis. *Ind. Eng. Chem. Res.*, **40**, 5305–5316.

Zhang, Z., Hidajat, K. and Ray, A.K. (2002) Multiobjective optimization of simulated countercurrent moving bed chromatographic reactor (SCMCR) for MTBE synthesis. *Ind. Eng. Chem. Res.*, **41**, 3212–3232.

Zhang, Z., Mazzotti, M. and Morbidelli, M. (2003) PowerFeed operation of simulated moving bed units: the flow rates during the switching interval. *J. Chromatogr.* A, **1006**, 87–99.

Zhu, B., Wang, Y. and Wang, H. (1999) Modelling catalytic reaction–absorption coupling process of ethylene oxide synthesis in slurry reactor. *Chem. Eng. Sci.*, **54**, 1531–1534.

Zhu, Y., Minet, R.G. and Tsotsis, T.T. (1996) A continuous pervaporation membrane reactor for the study of esterification reactions using a composite polymer/ceramic membrane. *Chem. Eng. Sci.*, **51**(17), 4103–4113.

Ziaka, Z.D., Minet, R.G. and Tsotsis, T.T. (1993) Propane dehydrogenation in a packed bed membrane reactor. *AIChE J.*, **39**, 526–529.

Author Index

Numbers in italic type indicate the pages appearing in the reference section.

Subject Index

Numbers in bold type indicate the first page of a chapter section or sub-section devoted to the subject.